T0189914

Communications
in Computer and Information Science   1036

Commenced Publication in 2007
Founding and Former Series Editors:
Phoebe Chen, Alfredo Cuzzocrea, Xiaoyong Du, Orhun Kara, Ting Liu,
Krishna M. Sivalingam, Dominik Ślęzak, Takashi Washio, and Xiaokang Yang

More information about this series at http://www.springer.com/series/7899

K. C. Santosh · Ravindra S. Hegadi (Eds.)

Recent Trends in Image Processing and Pattern Recognition

Second International Conference, RTIP2R 2018
Solapur, India, December 21–22, 2018
Revised Selected Papers, Part II

 Springer

Editors
K. C. Santosh
Department of Computer Science
University of South Dakota
Vermillion, SD, USA

Ravindra S. Hegadi
Solapur University
Solapur, India

ISSN 1865-0929 ISSN 1865-0937 (electronic)
Communications in Computer and Information Science
ISBN 978-981-13-9183-5 ISBN 978-981-13-9184-2 (eBook)
https://doi.org/10.1007/978-981-13-9184-2

This Springer imprint is published by the registered company Springer Nature Singapore Pte Ltd.
The registered company address is: 152 Beach Road, #21-01/04 Gateway East, Singapore 189721, Singapore

Preface

It is our great pleasure to introduce the collection of research papers in the *Communication in Computer and Information Science* (CCIS) Springer series from the second Biennial International Conference on Recent Trends in Image Processing and Pattern Recognition (RTIP2R). The RTIP2R conference event took place at the Solapur University, Maharastra, India, during December 21–22, 2018, in collaboration with the Department of Computer Science, University of South Dakota (USA) and Universidade de Evora (Portugal). Further, the conference had a very successful workshop titled Pattern Analysis and Machine Intelligence (PAMI): Document Engineering to Healthcare, with more than 70 participants.

As announced in the Call For Paper, RTIP2R attracted current and/or recent research on image processing, pattern recognition, and computer vision with several different applications, such as document understanding, biometrics, medical imaging, and image analysis in agriculture. Altogether, we received 371 submissions and accepted 173 papers based on our thorough review reports. We followed a double-blind submission policy and therefore the review process was extremely solid. On average, there were at least three reviews per paper except the few that had desk rejections, and therefore we had 859 review reports. We also made the authors aware of plagiarism, and rejected a few of them even after making review reports.

During the event, we hosted more than 200 participants from more than 29 different countries, such as USA, Vietnam, Australia, Russia and Sri Lanka (not just limited to India). In brief, the event was found to be a great platform bringing together research scientists, academics, and industry practitioners throughout the world. We categorized the papers into five different tracks: (a) computer vision and applications; (b) machine learning and applications; (c) document analysis; (d) healthcare and medical imaging; (e) biometrics and applications; (f) data mining, information retrieval and applications; (g) image processing; and (h) image analysis in agriculture.

We also selected the best papers based on the review reports, review scores, and presentations at the conference, and provided authors an opportunity to publish their extended works in the following journals: (a) *Multimedia Tools and Applications* (Springer); (b) *International Journal of Ambient Computing and Intelligence* (IGI Global); and (c) *Journal of Healthcare Informatics Research* (Springer).

The conference event was full of new ideas, including keynote speeches that were from (a) Sameer Antani, National Institutes of Health; (b) Mohan Gundeti, University of Chicago Medicine; and (c) Ernest Cachia, University of Malta.

April 2019

K. C. Santosh
Ravindra S. Hegadi

Organization

Patron

M. M. Fadnavis

Organizing Chairs

V. B. Ghute
V. B. Patil
B. C. Shewale

Honorary Chairs

P. Nagabhushan   IIIT, Allahabad, India
P. S. Hiremath   KLE University, Hubballi, India
B. V. Dhandra   Symbiosis University, India

General Chairs

Jean-Marc Ogier   University of la Rochelle, France
Laurent Wendling   University of Paris Descartes, France
Sameer Antani   US National Library of Medicine, USA
D. S. Guru   University of Mysore, India

Conference Chairs

Ravindra Hegadi   Solapur University, India
Teresa Goncalves   Universidade de Evora, Portugal
K. C. Santosh   University of South Dakota, USA

Area Chairs

Szilard Vajda   Central Washington University, USA
Mickael Coustaty   University of La Rochelle, France
Nibaran Das   Jadavpur University, India
Nilanjan Dey   Techno India College of Technology, India
Jude Hemanth   Karunya University, India

Publicity Chairs

Hubert Cecotti	California State University, USA
Odemir Martinez Bruno	University of Sao Paulo, Brazil
Alba Garcia Seco de Herrera	University of Essex, UK
Sheng-Lung Peng	National Dong Hwa University, Taiwan
Do T. Ha	VNU University of Science, Vietnam
B. Uyyanonvara	Thammasat University, Thailand
Sk Md. Obaidullah	University of Evora, Portugal
V. Bevilacqua	Polytechnic of Bari, Italy
R. S. Mente	Solapur University, India
Pratim P. Roy	Indian Institute of Technology (IIT), India
Manjunath T. N.	BMSIT, Bangalore, India
Nadra Ben Romdhane	University of Sfax, Tunisia
M. A. Jabbar	Vardhaman College of Engineering, India

Finance Chairs

A. R. Shinde	Solapur University, Solapur, India
S. D. Raut	Solapur University, Solapur, India

Advisory Committee

Daniel P. Lopresti	Lehigh University, USA
Rangachar Kasturi	University of South Florida, USA
Sargur N. Srihari	CEDAR, USA
K. R. Rao	University of Texas at Arlington, USA
Ishwar K. Sethi	Oakland University, USA
G. K. Ravikumar	CVS Health/Wipro, Texas, USA
Jose Flores	University of South Dakota, USA
Rajkumar Buyya	University of Melbourne, Australia
Arcot Sowmya	UNSW, Sydney, Australia
Antanas Verikas	Halmstad University, Sweden
Diego Liberati	Politecnico di Milano, Italy
B. B. Chaudhuri	Indian Statistical Institute, Kolkata, India
Atul Negi	University of Hyderabad, India
Arun Agarwal	University of Hyderabad, India
Hemanth Kumar	University of Mysore, India
K. V. Kale	Dr. BAMU, Aurangabad, India
B. V. Pawar	NMU, Jalgaon, India
R. R. Deshmukh	Dr. BAMU, Aurangabad, India
Karunakar A. K.	MIT, Manipal, India
Suryakanth Gangashetty	IIIT Hyderabad, India
Kaushik Roy	West Bengal University, India
Mallikajrun Hangarge	KASCC, Bidar, India
T. Devi	Bharathiar University, Coimbatore, India

G. R. Sinha	IIIT, Bangalore, India
U. P. Kulkarni	SDMCE, Dharwad, India
Rajendra Hegadi	IIIT, Dharwad, India
S. Basavarajappa	IIIT, Dharwad, India
B. P. Ronge	SVERI'S College of Engineering, India

Technical Program Committee (Country-Wise)

Randy C. Hoover	South Dakota School of Mines and Technology, USA
Sivarama Krishnan Rajaraman	US National Library of Medicine, NIH, USA
Yao-Yi Chiang	University of Southern California - LA, USA
Ullas Bagci	University of Central Florida, USA
Yuhlong Lio	University of South Dakota, USA
Eugene Borovikov	Intelligent Automation Inc., USA
Szilard Vajda	Central Washington University, USA
Hubert Cecotti	California State University, USA
Sema Candemir	US National Library of Medicine, NIH, USA
Md Mahmudur Rahman	Morgan State University, USA
Gabriel Picioroaga	University of South Dakota, USA
Peter Dolan	University of Minnesota Morris, USA
Michael Clement	York University, Canada
Alba Garca Seco de Herrera	University of Essex, UK
Nico Hochgeschwender	University of Luxembourg, Luxembourg
Benoit Naegel	University of Strasbourg, France
Vincent Bombardier	CRAN, University of Lorraine, France
Isabelle Debled-Rennesson	LORIA, University of Lorraine, France
Camille Krutz	University Institutes of Technology (IUT de Paris), France
Jean Cousty	University Paris-Est, France
Jonathan Weber	University of Haute-Alsace, France
Sabine Barrat	University of Tours, France
Muhammad Muzzamil Luqman	University of La Rochelle, France
Mickael Coustaty	University of La Rochelle, France
Jean-Pierre Salmon	University of Bordeaux Montaigne, France
Victor Codocedo	University de Lyon, CNRS, INSA-Lyyon, France
Diego Liberati	Politecnico di Milano, Italy
Vitoantonio Bevilacqua	Polytechnic of Bari, Italy
Salim Jouili	Euro Nova, Belgium
Paulo Quaresma	University of Evora, Portugal
Luis Rato	University of Evora, Portugal
Joao Barroso	University of Tras-os-Montes e Alto Douro, Portugal
Vitor M Filipe	University of Tras-os-Montes e Alto Douro, Portugal
Mohamed-Rafik Bouguelia	Halmstad University, Sweden
Marcal Rusinol	Universitat Autonoma de Barcelona, Spain

Margit Antal	Sapientia University, Romania
Laszlo Szilagyi	Sapientia University, Romania
Srikanta Pal	Griffith University, Australia
Alireza Alaei	Griffith University, Australia
M. Cerda Villablanca	University of Chile, Chile
B. Uyyanonvara	SIIT, Thammasat University, Thailand
V. Sornlertlamvanich	Thammasat University, Thailand
S. Marukatat	Thammasat University, Thailand
I. Methasate	NECTEC, Thailand
C. Pisarn	Rangsit University, Thailand
Makoto Hasegawa	Tokyo Denki University, Japan
P. Shivakumara	University of Malaya, Malaysia
Sophea Prum	National R&D Center in ICT, Malaysia
Lalit Garg	University of Malta, Malta
Nadra Ben Romdhane	University of Sfax, Tunisia
Nafaa Nacereddine	Centre de Recherche en Techno. Industrielles (CRTI), Algeria
Aicha Baya Goumeidane	Centre de Recherche en Techno. Industrielles (CRTI), Algeria
Ameni Boumaiza	Qatar foundation, Qatar
Nguyen Thi Oanh	Hanoi University of Science Technology, Vietnam
Do Thanh Ha	VNU University of Science, Vietnam
Tien-Dat Nguyen	FPT Corp., Vietnam
T. Kartheeswaran	University of Jaffna, Sri Lanka
Shaikh A. Fattah	Bangladesh University of Engineering and Technology, Bangladesh
Pratim P. Roy	Indian Institute of Techno (IIT), India
Surekha Borra	KS Institute of Technology, (KSIT), India
Ajit Danti	JNN College of Engineering, Shimoga, India
Lalita Rangarajan	University of Mysore, Mysore, India
Manjaiah D. H.	Mangalore University, Mangalore, India
V. S. Malemath	KLE Engineering College, Belagavi, India
B. H. Shekar	Mangalore University, Mangalore, India
G. Tippeswamy	BMSIT, Bangalore, India
Aziz Makandar	Akkamahadevi Women's University Karnataka, Vijayapura, India
Mallikarjun Holi	BDT College of Engineering, Davangere, India
S. S. Patil	Agriculture University, Bangalore, India
H. S. Nagendraswamy	University of Mysore, Mysore, India
Shivanand Gornale	Ranichannamma University, Belagavi, India
S. Shivashankar	Karnatak University, Dharwad, India
Ramesh K.	Akkamahadevi Women's University Karnataka, Vijayapura, India
H. L. Shashirekha	Mangalore University, Mangalore, India
Dayanand Savakar	Ranichannamma University, Belagavi, India
S. B. Kulkarni	SDM College of Engineering, Dharwad, India

M. T. Somashekhar	Bangalore University, Bangalore, India
Manjunath Hiremath	Christ University, Bangalore, India
Sridevi Soma	PDA College of Engineering, Gulbarga, India
V. M. Thakare	SGB Amravati University, Amaravati, India
G. V. Chaudhari	SRTM University, Nanded, India
R. K. Kamat	Shivaji University, Kolhapur, India
Ambuja Salgaonkar	University of Mumbai, India
Praveen Yannavar	Dr. BAM University, India
R. R. Manza	Dr. BAM University, Aurangabad, India
A. S. Abhyankar	SP Pune University, India
V. T. Humbe	SRTMU Sub-Centre, Latur, India
P. B. Khanale	SRTMU, Nanded, India
M. B. Kokre	GGSIET, Nanded, India
Gururaj Mukrambi	Symbiosis International University, Pune, India
S. R. Kolhe	North Maharashtra University, Jalgaon, India
M. Sundaresan	Bharathiar University, Coimbatore, India
C. P. Sumathi	SDNBV College for Women, Chennai, India
J. Satheeshkumar	Bharathiar University, Coimbatore, India
Britto Ramesh Kumar	St. Joseph's College, Tiruchirappalli, India
Neeta Nain	Malaviya National Institute of Technology (MNIT), Jaipur, India
A. A. Desai	Veer Narmad South Gujarat University, Gujarat, India
Chandra Mouli P. V. S. S. R.	VIT University, Vellore, India
Nagartna Hegde	Vasavi Eng. College, Hyderabad, India
B. Gawali	Dr. BAM University, Aurangabad, India
K. T. Deepak	IIIT, Dharwad, India
P. M. Pawar	SVERI'S College of Eng., India
S. R. Gengaje	Walchand Inst. of Technology, Solapur, India
B. Ramadoss	National Inst. of Technology, Tamil Nadu, India

Local Organizers

P. Prabhakar	C. G. Gardi
S. S. Suryavanshi	P. M. Kamble
V. B. Ghute	D. D. Sawat
R. B. Bhosale	A. B. Jagtap
B. J. Lokhande	D. D. Ruikar
G. S. Kamble	P. P. Gaikwad
J. D. Mashale	

Additional Reviewers

Abdullah Mohammed Kaleem
Abhinav Muley
Addepalli Krishna
Adithya Pediredla
Aditya Patil
Ajay Nagne
Ajeet A. Chikkamannur
Ajit Danti
Ajju Gadicha
Akbaruddin Shaikh
Alba García Seco De Herrera
Alessia Saggese
Alexandr Ezhov
Almas Siddiqui
Ambika Annavarapu
Amol Vibhute
Amruta Jagtap
Anagha Markandey
Anderson Santos
Andrés Rosso-Mateus
Aniket Muley
Anita Dixit
Anita Khandizod
Anitha H.
Anitha J.
Anitha N.
Ankita Dhar
Anupriya Kamble
Archana Nandibewoor
Arjun Mane
Arunkumar K. L.
Ashish Mourya
Atish Patel
Aznul Qalid Md Sabri
Balachandran K.
Balaji Sontakke
Balamurugan Karnan
Basavaprasad B.
Basavaraj Dhandra
Bb Patil
Benoit Naegel
Bharath Bhushan
Bharathi Pilar

Bharatratna Gaikwad
Bhausaheb Pawar
Bindu V. R.
Brian Keith
C. Namrata Mahender
C. P. Sumathi
Camille Kurtz
Chandrashekhara K. T.
Chetan Pattebahadur
Daneshwari Mulimani
Daniel Caballero
Darshan Ruikar
Dattatray Sawat
Dericks Shukla
Diego Bertolini
Diego Liberati
Dnyaneshwari Patil
E. Naganathan
Ebenezer Jangam
Evgeny Kostyuchenko
G. P. Hegde
G. R. Sinha
G. S. Mamatha
Ganesh Janvale
Ganesh Magar
Ganga Holi
Gireesh Babu
Girish Chowdhary
Gururaj Mukarambi
H. L. Shashirekha
Hajar As-Suhbani
Hanumant Gite
Haripriya V.
Harshavardhana Doddamani
Hayath Tm
Hemavathy R.
Himadri Mukherjee
Hubert Cecotti
Ignazio Gallo
Jayendra Kumar
João Cardia
Jonathan Weber
Joseph Abraham Sundar K.

Jude Hemanth
Jyoti Patil
K. K. Chaturvedi
K. C. Santosh
Kalman Palagyi
Kalpana Thakare
Kapil Mehrotra
Kartheeswaran Thangathurai
Kasturi Dewi Varathan
Kaushik Roy
Kavita S. Oza
Kiran Phalke
Kwankamon Dittakan
Laszlo Szilagyi
Latchoumi Thamarai
Lingdong Kong
Lorenzo Putzu
Lp Deshmukh
Lucas Alexandre Ramos
Luis Rato
M. T. Somashekhar
Madhu B.
Mahesh Solankar
Mahmudur Rahman
Mainak Sen
Maizatul Akmar Ismail
Mallikarjun Hangarge
Mallikarjun Holi
Manasi Baheti
Manisha Saini
Manjunath Hiremath
Manjunath T. N.
Manohar Madgi
Manoj Patil
Mansi Subhedar
Manza Ramesh
Marçal Rusiñol
Margit Antal
Masud Rana Rashel
Md Obaiduallh Sk
Md. Ferdouse Ahmed Foysal
Md. Rafiqul Islam
Michael Clement
Midhula Vijayan
Miguel Alberto Becerra Botero
Mikhail Tarkov

Minakshi Vharkate
Minal Moharir
Mohammad Idrees Bhat Bhat
Mohammad Shakirul Islam
Mohan Vasudevan
Mohd. Saifuzzaman
Monali Khachane
Muhammad Muzzamil Luqman
Mukti Jadhav
Nadra Ben Romdhane
Nafis Neehal
Nagaraj Cholli
Nagaratna Hegde
Nagsen Bansod
Nalini Iyer
Nico Hochgeschwender
Nita Patil
Nitin Darkunde
Nitta Gnaneswara Rao
P. P. Patavardhan
Pankaj Agrawal
Parag Bhalchndra
Parag Kaveri
Parag Tamhankar
Parashuram Bannigidad
Parashuram Kamble
Parminder Kaur
Paulo Quaresma
Peter Dolan
Pooja Janse
Poonam Ghuli
Poornima Patil
Prabhakar C. J.
Pradeep Udupa
Prajakta Dhamdhere
Prakash Hiremath
Prakash Khanale
Prakash Unki
Praneet Saurabh
Prasanna Vajaya
Prasanth Vaidya
Pratima Manhas
Praveen K.
Pravin Metkewar
Pravin Yannawar
Prema T. Akkasaligar

Priti Singh
Pushpa Patil
Pushpa S. K.
Qazi Fasihuddin
Rafaela Alcântara
Rajendra Hegadi
Rajesh Dhumal
Rajivkumar Mente
Rajkumar Soundrapandiyan
Rajkumar Yesuraj
Rakesh K.
Ramya D.
Rashmi Somshekhar
Ratnadeep Deshmukh
Ratnakar Ghorpade
Ravi Hosur
Ravi M.
Ravindra Babu Tallamraju
Ravindra Hegadi
Rim Somai
Ritu Prasad
Rodrigo Nava
Rohini Bhusnurmath
Rosana Matuk Herrera
Rupali Surase
S. Basavarajappa
S. Ramegowda
S. B. Kulkarni
Sachin Naik
Sahana Das
Sameer Antani
Sanasam Inunganbi
Sangeeta Kakarwal
Sanjay Jain
Santosh S. Chowhan
Sarika Sharma
Satish Kolhe
Sema Candemir
Shajee Mohan
Shankru Guggari
Shanmugapriya Padmanabhan
Shanthi D. L.
Sharath Kumar
Shaveta Thakral
Sheikh Abujar
Shilpa Bhalerao
Shiva Murthy Govindaswamy

Shivani Saluja
Shivashankar S.
Shridevi Soma
Shrikant Mapari
Siddanagouda Patil
Siddharth Dabhade
Sivarama Krishnan Rajaraman
Slimane Larabi
Smriti Bhandari
Srikanta Pal
Sudha Arvind
Suhas Sapate
Sunanda Biradar
Suneeta Budihal
Sunil Nimbhore
Swapnil Waghmare
Szilard Vajda
Tejaswi Potluri
Thanh Ha Do
Ujwala Suryawanshi
Ulavappa B. Angadi
Umakant Kulkarni
Urmila Pol
Usha B. A.
Vaibhav Kamble
Veerappa Pagi
Víctor Codocedo
Vidyagouri Hemadri
Vijay Bhaskar Semwal
Vijaya Arumugam
Vikas Humbe
Vilas Naik
Vilas Thakare
Vinay T. R.
Vincent Bombardier
Virendra Malemath
Vishal Waghmare
Vishweshwarayya Hallur
Yao-Yi Chiang
Yaru Niu
Yoanna Martínez-Díaz
Yogesh Gajmal
Yogesh Rajput
Yogish H. K.
Yuhlong Lio
Zati Hakim Azizul Hasan

Contents – Part II

Biometrics and Applications

Healthcare and Medical Imaging

Contrast Stretching-Based Unwanted Artifacts Removal from CT Images

Darshan D. Ruikar[1]($\boxtimes$), K. C. Santosh[2], and Ravindra S. Hegadi[1]

[1] Department of Computer Science, Solapur University, Solapur, 413255,
Maharastra, India
{ddruikar,rshegadi}@sus.ac.in
[2] Department of Computer Science, University of South Dakota, 414 E Clark St,
Vermillion, SD 57069, USA
santosh.kc@ieee.org

Abstract. This paper presents a contrast stretching-based image enhancement technique to remove unwanted artifacts, such as flesh and to enhance bony regions from Computed Tomographic (CT) images. Our technique is based on enhancing the dynamic range of the image by linear contrast stretching through histogram modeling and intensity transformation function. The intensity range: low and high-intensity values are heuristically computed, and squared shape mask is moved to clean the image further. Experiments are carried out on several patient-specific CT images (source: Prism Medical Diagnostics lab, Chhatrapati Shivaji Maharaj Sarvopachar Ruganalay and Ashwini Hospital, India). Our results show that the technique provides the reliable promising results. Besides, the tool is simple, faster and easy to implement.

Keywords: Image enhancement · Intensity transformation function ·
Contrast stretching · Histogram modeling · Flesh removal · CT images

1 Introduction

Real-time, patient-specific data collection in the form of medical images (CT, Magnetic Resonance (MR), X-ray images for instance), i.e., data acquisition is the initial step in the development of expert systems for healthcare solutions like anomaly detection, disease growth (severity) estimation, and surgical training simulator development [15, 21]. More generally medical images are collected from radiology centers or publicly available databases. The images collected from such sources cannot be used directly for analysis and application development. Because these images are very raw; it may contain noise and other artifacts which must be removed before further processing or else it may reduce overall accuracy of the application [10, 24].

Nowadays, along with the orthopedic surgeons, the researchers developing expert systems in orthopedic filed; mostly prefer CT images for trauma analysis and prognosis. CT images provide a better cross-sectional view of the scanned

© Springer Nature Singapore Pte Ltd. 2019
K. C. Santosh and R. S. Hegadi (Eds.): RTIP2R 2018, CCIS 1036, pp. 3–14, 2019.
https://doi.org/10.1007/978-981-13-9184-2_1

organ by making use of X-ray and computer. CT image is constructed by scanning the desired area from different angles which interns provide detailed information about the object without cutting, and it preserves the actual anatomic structure [14,22,23]. These features of CT imaging help surgeons and researchers to visualize the geometric structure of involved bone accurately moreover to gain a precise understanding of severe bone trauma like communicated fractures[1]. Intensity inhomogeneities [29], low resolutions [5], and presence of other things like flesh (i.e., soft tissues), CT bed and wires [28] are the significant hurdles to develop an automated trauma analysis and prognosis system. Presences of such things reduce the subjective image quality and suppress the expected features. Hence the efficient pre-processing technique is required to eliminate the unwanted artifacts and noise moreover to enhance application specific features (bony regions, for instance) present in that image [10].

Image pre-processing techniques can be used to perform gray level and contrast manipulation, noise reduction, edge crisping and sharpening, filtering, interpolation, and magnification [7]. Efficient image pre-processing technique that not only removes unwanted artifacts from the medical image but also enhances the required portion. These two are the prime need of several medical image analysis based applications [2,19]. This paper presents an effective CT image preprocessing technique, which removes unwanted artifacts by considering image contents moreover highlights the bone tissue along with complete fracture lines and dislocated pieces precisely. The presented method uses histogram modeling techniques and intensity transformation functions efficiently.

This paper is organized as follows: Sect. 2 provides detailed information about CT imaging. Several exiting medical image pre-processing techniques are enlisted in Sect. 3. The detailed description of the proposed technique is discussed in Sect. 4. Section 5 contains the comparison of the proposed method with some other commonly used medical image pre-processing techniques. In addition to this, the results of proposed pre-processing technique on real patient-specific images are also presented in this section. Section 6 gives conclusions and directions for future work.

2   CT Imaging

Computerized axial transverse (CAT) or Computerized tomography (CT) scanning is a novel medical imaging technique developed by Godfrey Hounsfield in 1972 [11]. The X-ray generators emitting the X-ray beam are rotated around the scanning material. The X-ray detectors fitted exactly opposite position in a circle computes the attenuation coefficient based on the density of scanning material. This process is the basis for CT image, i.e., slice formation. Hounsfield unit, the linear transformation of the tissue attenuation coefficient is the measurement unit to measure the material density. The Hounsfield scale varies according to

---

[1] Communicated fracture: Type of fracture in which bone is broken into multiple pieces possibly with dislocation.

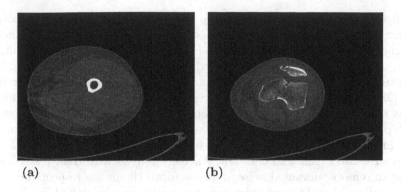

(a)   (b)

Fig. 1. CT images of the same patient at (a) diaphysis (b) epiphysis

the density of the material. For air, water and dense bone it is −1000, 0 and 1000 respectively [27].

In the formed CT image, each pixel is represented by the gray intensity value proportional to the density of the tissue. Any bone is made-up of two of type tissues namely, cortical tissue and cancellous tissue and can be divided into distinct regions epiphysis (end part) and diaphysis (middle part). In eight-bit gray-scale image intensity values of cortical tissues are ranging from 220 to 255, i.e., they are brighter, and they are present at outer part of bone whereas cancellous tissues are ranging from 90 to 210 and they are present at inner part of the bone. Soft tissues (flesh) also fall under similar intensity range [5]. Over the CT stack, both cortical and cancellous tissues show high variation in their intensity values. Moreover, the intensity values of both the tissue may not be the same in all the slices [29]. Figure 1 shows the intensity variation of cortical tissues in the same patients CT stack. They are brighter and dense at diaphysis (Fig. 1(a)) and appear fuzzy and thinner near epiphysis, i.e., at the joint area (Fig. 1(b)). In addition to this, CT bed is present at the lower part of both the images. To remove these intrinsic difficulties and other artifacts present in the CT images, there is a need of efficient pre-processing technique, which will enhance the bone region by removing the surrounding fleshy part and other artifacts.

3   Literature Survey

There are several papers focused on automatic segmentation to extract a region of interest like a healthy or fractured bone from CT images. In those authors used different pre-processing techniques to remove unwanted artifacts and to enhance the region of interest.

Abdelsamea et al. [1] applied a median filter to remove noise and to enhance the image. Hemanth et al. [10] used morphological erosion and filling operation to remove skull tissue which is surrounded by brain tissue. 5 × 5 ball-shaped structuring element is used for erosion. Later to connect broken boundaries filling logic is implemented by interpolating neighborhood pixels. Lin et al. [16] adopted an anisotropic diffusion filter for image smoothening. Along with anisotropic diffusion filter, curvature diffusion filter is used by Ruggieri et al. [20] to increase the signal to noise ratio and to increase the contrast of the region of interest. On the same note, Felix et al. [18] applied an anisotropic diffusion method based curvature flow filter on CT images for image smoothening and edge preservation.

In [28] authors developed a pre-processing technique to separate abdominal regions from nearby unnecessary artifacts like CT table and cables. Firstly an accepted image is converted to a binary version and various morphological operations are performed. Then blob analysis is done to remove objects having a smaller size. Later Gaussian filter and Wavelet analysis with Haar transform are applied to remove unwanted noise and to smooth the image respectively.

A novel multi-scale sheet enhancement measurement method is adapted to segment sinus bone from CT images. Local shape information extracted from an Eigen-value decomposition of the Hessian is used to enhance the image [3]. Harders et al. [8] and Fornaro et al. [6] also used a sheetness measure based on a modified Hessian matrix method to extract humerus and pelvis bone from CT images. The region of interest is extracted from CT images by plotting and analyzing the histogram [13,17]. They extracted two threshold values to label each pixel in the image. Pixels with intensity value less than the low threshold are marked as soft tissue and pixels having intensity value more than high threshold are labeled as bone tissue.

In summary, there are several pre-processing methods available in the literature. Various medical image pre-processing techniques applied to remove noise and to enhance the image quality are enlisted in [2]. Most of these techniques are applied independently of image content. Adopting smoothening filters (like Gaussian filters to remove noise) without image understanding may lose valuable information. After smoothening, it may be possible that fracture pieces get wrongly connected, and the fracture lines near joint area appear fuzzier. Hence there is a need for image content based pre-processing and enhancement technique which will remove noise and unwanted artifacts without affecting the region of interest. In addition to this, such techniques will help to enhance bone tissue and in crisping fuzzy fracture lines.

4   Methodology

A detailed explanation of a systematic approach for CT image pre-processing for bone tissue enhancement by removing surrounding soft tissue is discussed in the following paragraphs.

4.1   Data Acquisition

In the presented work, twenty-eight patient-specific CT stacks are collected from several radiology centers and hospitals having CT scan machines with different specifications. To test the correctness and robustness of the proposed technique, intentionally images are collected from different sources. Each CT stack is a comprehensive collection of hundreds of CT slices (ranging from one hundred to five hundred). The count of slices per stack depends upon the area under supervision, the severity of the injury, the thickness of each slice and distance between two slices. In total eight thousand images are collected in the DICOM[2] format. Each slice is 8 bit, 512×512 pixels sized and it contains a scanned portion of long bone like femur with or without fractures. However, these files contain patient-specific information like a patient number, name, and age. Disclosing this information may violate the ethical laws like HIPPA[3] laws or IRB[4] protocol. So to erase the patient-specific data, the MATLAB script discussed in [12] is adapted to remove patient-specific information from CT image. The script uses pre-defined function dicomanon (part of the MATLAB Image Processing Toolbox) to remove confidential medical data from a dicom file.

4.2   Contract Stretching

A histogram is plotted (Fig. 2(b)) for sample CT image (Fig. 2(a)) to obtain global information about the image. From observing the plotted histogram, it is clear that CT images are low contrast images; as maximum pixels cover a small range of gray level. That is, ample amount of intensity values are concentrated mostly towards middle-intensity values (i.e., from 95 to 110). The reason behind this is the flesh occupies a maximum region of the image.   To enhance the image

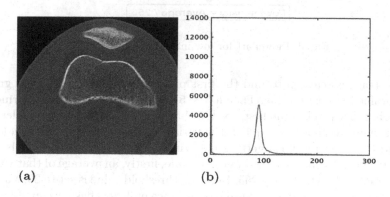

(a)   (b)

Fig. 2. (a) CT image, (b) Its histogram

quality moreover to highlight the bone tissue application of contrast enhancement is essential. The contrast enhancement can be achieved by applying one of the intensity transformation techniques, known as contrast stretching. This technique is responsible for enhancing the contrast by distributing intensity values that cover a wide range [7]. For that purpose, it is necessary to predict image specific low and high-intensity values.

---

[2] DICOM: Digital Imaging and Communications in Medicine.
[3] HIPPA: Health Insurance Portability and Accountability Act.
[4] IRB: Institutional Review Board.

Low-Intensity Value Computation. To find the low-intensity value the histogram plots of the input image are observed. By inspecting histogram beans, the first bean with the highest value is skipped as it represented dark background pixels and considered the second bean with the highest number of pixels for predicting low-intensity value. The intensity value associated with the second highest bean is the basis for predicting low-intensity value. The detailed procedure to compute the low-intensity value is depicted in Fig. 3.

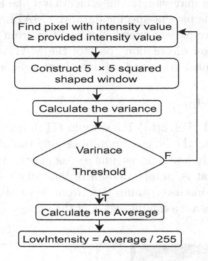

Fig. 3. Flowchart for low intensity computation

The image is scanned to find the first pixel whose intensity value is greater than obtained intensity value. Then a 5 × 5 square shaped window is formed by considering that pixel as the first pixel in that window and variance is calculated by the formula as shown in Eq. 1. If the variance is more than the threshold, the same process repeated until getting the window with variance is less than the threshold. Then to obtain a low-intensity value, firstly, an average of that window is calculated and divided it by 255. Here the threshold value is set to 20. Variance should be as less as possible because less variance indicates flash is equally spread in that region. Such a region is more suitable to identify low-intensity value which will be used to remove flesh from the image.

$$Variance = HighIntensity - lowIntensity \qquad (1)$$

High-Intensity Value Computation. Same like low-intensity value computation, the image is scanned to find the first pixel whose intensity value is greater than 220 (this value is the initial intensity value of cortical tissue). Then a 5 × 5 square shaped window is formed if the first four pixels also follow the same criteria, i.e., intensity value more than 220. After getting such a window, an average of that window is calculated and divided it by 255 to obtain high-intensity value. The detailed procedure to compute high-intensity value is depicted in Fig. 4.

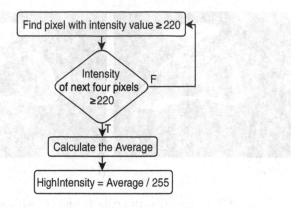

Fig. 4. Flowchart for high intensity computation

Intensity Transformation Function. After obtaining the low and high-intensity values, an intensity transformation is applied to distribute the pixels over a wide range (i.e. 0 to 1). The formula for intensity transformation function is as shown in Eq. 2 [7].

$$I_{out} = ((I_{in} - i_{il})./OriginalRange. \times DesiredRange) + i_{ol} \qquad (2)$$

$$OriginalRange = i_{ih} - i_{il} \qquad (3)$$

$$DesiredRange = i_{oh} - i_{ol} \qquad (4)$$

where

$i_{il} = Computed\ Low\ Intensity;\ i_{ih} = Computed\ High\ Intensity$
$i_{ol} = Desired\ Low\ Intensity\ (i.e.\ 0;)\ i_{oh} = Desired\ High\ Intensity\ (i.e.\ 1).$

The transformation function shifts and stretches the gray level range of the input image to occupy the entire dynamic range (0, 1).

4.3 Mask Formation

After application of intensity transformation function, most of the artifacts get erased, and the image is enhanced nicely. However, in a few images, wherein flesh part is a little bit brighter or dense; some artifacts may remain as it. Usually, these are the borderlines between the flesh part and background shown in Fig. 5(a). The entire image is scanned once by creating a 3 × 3 squared shaped window To remove remaining artifacts. If the window has less number of non-zero pixels (proposed technique considers value 4), then all pixels in that window is reset to 0. The Fig. 5(b) shows the cleaned image.

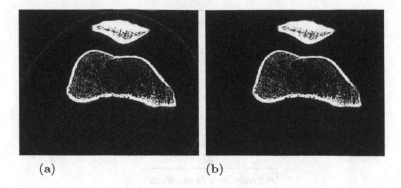

(a)   (b)

Fig. 5. CT image (a) with artifacts and (b) without artifacts

5   Experimental Results

5.1   Comparison with Other Methods

In order to test the performance of the presented approach, it is compared with some other methods used in literature. Gaussian filter, multi-scale Wiener filter [9,14] and adaptive Wiener filter with Wavelet packet thresholding methods [4] are adapted to remove noise and to improve visual image quality. In addition to this, median filter and anisotropic diffusion filters are used to smooth the image. However, in most of the bone trauma analysis applications, the smoothening operation is not always preferable. The reason behind this is smoothening operation blurs the image little bit. This may cause joining of fractured pieces which are much closer to each other and fuzzy boundaries near joint area become fuzzier. The resultant images are shown in Fig. 6(b) and (c), after applying the Gaussian flier and Wiener filter respectively. Salt-paper noise or Gaussian noise present in the image is removed successfully. Figure 7 shows the resultant images obtained by applying a median filter and the anisotropic diffusion filter. The label I in the Fig. 7(b) shows the blurring of boundary lines near the joint area and label I in the Fig. 7(c) shows that the fractured pieces appear connected due

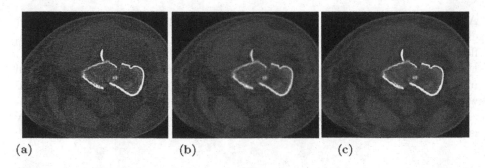

(a)   (b)   (c)

Fig. 6. (a) Original image. Image after applying (b) Gaussian filter (c) Wiener filter

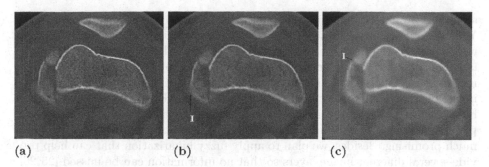

Fig. 7. (a) Original image. Image after applying (b) Median filter (c) Anisotropic diffusion filter

to smoothing. Confirming to literature survey, no previous attempts are made to clean the image by removing other artifacts, majorly the flesh surrounded by bone tissue. So the researchers can focus on the development of trauma or anomaly analysis applications. The proposed method will help to remove all such artifacts by inspecting the image.

5.2 Application to Real Patient-Specific Images

In addition to comparisons with some other methods, it is tested on real-time, patient-specific CT images those are collected from several radiology centers and hospitals from India. Figure 8 shows the output of proposed method on real-time, patient-specific images. Images shown in Fig. 8(a) and (b) are original CT images having cortical and cancellous tissue respectively, whereas in same figure (c) and (d) are the images obtained after applying proposed method. The resultant images are cleaned nicely and no artifact is present in it.

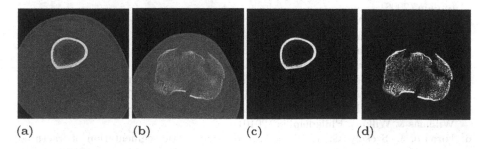

Fig. 8. (a) and (b) Original images. (c) and (d) Resultant images.

6   Conclusion and Future Work

Image pre-processing or image enhancement is an initial and important step in almost all image processing based applications. The accuracy of subsequent steps and entire application majorly depends on the result of the pre-processing step. Yet the significance of these techniques remains unnoticed in medical imaging-related applications. In the presented work, the novel pre-processing technique is implemented to remove unwanted artifacts from the CT image. The presented method is based on histogram modeling and stretching the dynamic range of the image by applying intensity transformation function. The image specific low and high-intensity values are computed heuristically by making use of the histogram plots. Confirming to experimental results, the result of the proposed method is much promising. Besides, we plan to apply fuzzy binarization that can help provide several different image layers so that no information can be missed [25,26]. In the near future, more successful research attempts must be made to develop pre-processing techniques for other sources of medical images like MR and X-ray images. In addition to this, future studies would be useful focusing on the edge enhancement techniques especially bone regions near the joint area, automatic segmentation and labeling of fractured CT image and automatic fracture classification based on fracture line.

Acknowledgments. Authors thank the Ministry of Electronics and Information Technology (MeitY), New Delhi for granting Visvesvaraya Ph.D. fellowship through file no. PhD-MLA\4(34)\201-1. Dated: 05/11/2015.

The first author would like to thank Dr. Jamma and Dr. Jagtap for providing expert guidance on bone anatomy. Along with this, he also would like to thank, Prism Medical Diagnostics lab, Chhatrapati Shivaji Maharaj Sarvopachar Ruganalay and Ashwini Hospital for providing patient-specific CT images.

References

1. Abdelsamea, M.M.: An automatic seeded region growing for 2D biomedical image segmentation (2014). arXiv preprint, arXiv:1412.3958
2. Bankman, I.: Handbook of Medical Image Processing and Analysis. Elsevier, Amsterdam (2008)
3. Descoteaux, M., Audette, M., Chinzei, K., Siddiqi, K.: Bone enhancement filtering: application to sinus bone segmentation and simulation of pituitary surgery. Comput. Aided Surg. **11**(5), 247–255 (2006)
4. Diwakar, M., Kumar, M.: CT image noise reduction based on adaptive Wiener filtering with wavelet packet thresholding. In: 2014 International Conference on Parallel, Distributed and Grid Computing (PDGC), pp. 94–98. IEEE (2014)
5. Egol, K.A., Koval, K.J., Zuckerman, J.D.: Handbook of Fractures. Lippincott Williams & Wilkins, Philadelphia (2010)
6. Fornaro, J., Székely, G., Harders, M.: Semi-automatic segmentation of fractured pelvic bones for surgical planning. In: Bello, F., Cotin, S. (eds.) ISBMS 2010. LNCS, vol. 5958, pp. 82–89. Springer, Heidelberg (2010). https://doi.org/10.1007/978-3-642-11615-5_9
7. Gonzalez, R.C., Woods, R.E.: Digital Image Processing. Prentice Hall, Upper Saddle River (2012)
8. Harders, M., Barlit, A., Gerber, C., Hodler, J., Székely, G.: An optimized surgical planning environment for complex proximal humerus fractures. In: MICCAI Workshop on Interaction in Medical Image Analysis and Visualization, vol. 30 (2007)

9. Hegadi, R.S., Navale, D.I., Pawar, T.D., Ruikar, D.D.: Multi feature-based classification of osteoarthritis in knee joint X-ray images. In: Medical Imaging: Artificial Intelligence, Image Recognition, and Machine Learning Techniques, chap. 5. CRC Press (2019). ISBN 9780367139612
10. Hemanth, D.J., Anitha, J.: Image pre-processing and feature extraction techniques for magnetic resonance brain image analysis. In: Kim, T., Ko, D., Vasilakos, T., Stoica, A., Abawajy, J. (eds.) FGCN 2012. CCIS, vol. 350, pp. 349–356. Springer, Heidelberg (2012). https://doi.org/10.1007/978-3-642-35594-3_47
11. Hounsfield, G.N.: Computed medical imaging. Med. Phys. **7**(4), 283–290 (1980)
12. Hunter, E.J., Palaparthi, A.K.R.: Removing patient information from MRI and CT images using MATLAB. National Repository for Laryngeal Data Technical Memo No. 3 (version 2.0), pp. 1–4 (2015)
13. Kang, Y., Engelke, K., Kalender, W.A.: A new accurate and precise 3D segmentation method for skeletal structures in volumetric CT data. IEEE Trans. Med. Imaging **22**(5), 586–598 (2003)
14. Ke, L., Zhang, R.: Multiscale Wiener filtering method for low-dose CT images. In: 2010 3rd International Conference on Biomedical Engineering and Informatics (BMEI), vol. 1, pp. 428–431. IEEE (2010)
15. Lai, J.Y., Essomba, T., Lee, P.Y.: Algorithm for segmentation and reduction of fractured bones in computer-aided preoperative surgery. In: Proceedings of the 3rd International Conference on Biomedical and Bioinformatics Engineering, pp. 12–18. ACM (2016)
16. Lin, Z., Jin, J., Talbot, H.: Unseeded region growing for 3D image segmentation. In: Selected Papers from the Pan-Sydney Workshop on Visualisation, vol. 2, pp. 31–37. Australian Computer Society, Inc. (2000)
17. Mancas, M., Gosselin, B., Macq, B.: Segmentation using a region-growing thresholding. In: Image Processing: Algorithms and Systems IV, vol. 5672, pp. 388–399. International Society for Optics and Photonics (2005)
18. Paulano, F., Jiménez, J.J., Pulido, R.: 3D segmentation and labeling of fractured bone from CT images. Vis. Comput. **30**(6–8), 939–948 (2014)
19. Ritter, F., et al.: Medical image analysis. IEEE Pulse **2**(6), 60–70 (2011)
20. Ruggieri, V.G., et al.: CT-scan images preprocessing and segmentation to improve bioprosthesis leaflets morphological analysis. Med. Hypotheses **81**(1), 86–93 (2013)
21. Ruikar, D.D., Hegadi, R.S., Santosh, K.C.: A systematic review on orthopedic simulators for psycho-motor skill and surgical procedure training. J. Med. Syst. **42**(9), 168 (2018)
22. Ruikar, D.D., Santosh, K.C., Hegadi, R.S.: Automated fractured bone segmentation and labeling from CT images. J. Med. Syst. **43**(3), 60 (2019). https://doi.org/10.1007/s10916-019-1176-x
23. Ruikar, D.D., Santosh, K.C., Hegadi, R.S.: Segmentation and analysis of CT images for bone fracture detection and labeling. In: Medical Imaging: Artificial Intelligence, Image Recognition, and Machine Learning Techniques, chap. 7. CRC Press (2019). ISBN 9780367139612
24. Ruikar, D.D., Sawat, D.D., Santosh, K.C., Hegadi, R.S.: 3D imaging in biomedical applications: a systematic review. In: Medical Imaging: Artificial Intelligence, Image Recognition, and Machine Learning Techniques, chap. 8. CRC Press (2019). ISBN 9780367139612
25. Santosh, K.C., Roy, P.P.: Arrow detection in biomedical images using sequential classifier. Int. J. Mach. Learn. Cybern. **9**(6), 993–1006 (2018)

26. Santosh, K.C., Wendling, L., Antani, S., Thoma, G.R.: Overlaid arrow detection for labeling regions of interest in biomedical images. IEEE Intell. Syst. **31**(3), 66–75 (2016)
27. Shapurian, T., Damoulis, P.D., Reiser, G.M., Griffin, T.J., Rand, W.M.: Quantitative evaluation of bone density using the hounsfield index. Int. J. Oral Maxillofac. Implants **21**(2) (2006)
28. Vasilache, S., Najarian, K.: Automated bone segmentation from pelvic CT images. In: 2008 IEEE International Conference on Bioinformatics and Biomeidcine Workshops, pp. 41–47. IEEE (2008)
29. Willis, A., Anderson, D., Thomas, T., Brown, T., Marsh, J.L.: 3D reconstruction of highly fragmented bone fractures. In: Medical Imaging 2007: Image Processing, vol. 6512, p. 65121P. International Society for Optics and Photonics (2007)

Comparison with Evaluation of Intra Ocular Pressure Using Different Segmentation Techniques for Glaucoma Diagnosis

Dnyaneshwari D. Patil[1]([✉]), Ramesh R. Manza[2], Rakesh J. Ramteke[3], Yogesh Rajput[4], and Sanjay Harke[5]

[1] MGM's Institute of Biosciences and Biotechnology, Aurangabad, India
dnyaneshwari03patil@gmail.com
[2] Department of CS and IT, Dr. Babasaheb Ambedkar Marathwada University, Aurangabad, India
[3] Department of Information Technology, North Maharashtra University, Jalgaon, India
[4] MGM's G. Y. Pathrikar College of CS and IT, Aurangabad, India
[5] MGM's Institute of Biosciences and Biotechnology, Aurangabad, India

Abstract. In the process of automatic Glaucoma diagnosis, we have used freely available database for research work, like DRIONS-DB, RIM-ONE, MESSIDORE (Base 1 to Base 12), DRISHTY, HRF (High Resolution Fundus Images) total 2866 retinal fundus images we have used for evaluation of Intra ocular pressure using different image segmentation techniques, like HAAR wavelet, median filter, morphological opening, and top-hat filters, from these techniques first we have extracted features important to diagnose Glaucoma like retinal blood vessels with the fine features of arteries, capillaries and veins. After extracting features we have calculated statistical features important to diagnose glaucoma like area, diameter, length, thickness and tortuosity to measure the intra ocular pressure generated in retinal blood vessels, all these procedures are divided in to two separate experiments. Performed statistical calculations and feature extraction separately then in advance procedure of diagnosing we have applied K-Means calcification and clustering methods separately on both the experiments to measure the intensity of disease. Then on the basis of comparison, we have concluded that Top-hat filter method or experiment number two gives better result than another one, overall we got highest 85.

Keywords: CDR - Cup to Disc Ratio · RDR - Rim to Disc Ratio · IOP - Intra Ocular Pressure · ROI - Region of Interest

1 Introduction

Glaucoma is a cluster of syndrome that injures the eye's optic nerve and can effect in visualization loss and sightlessness. However, with preliminary detection

© Springer Nature Singapore Pte Ltd. 2019
K. C. Santosh and R. S. Hegadi (Eds.): RTIP2R 2018, CCIS 1036, pp. 15–32, 2019.
https://doi.org/10.1007/978-981-13-9184-2_2

and healing, people can often defend their eyes in opposition to serious visualization loss. Glaucoma is the analysis given to a bunch of optical conditions that contribute to the defeat of retinal nerve fibers with a corresponding defeat of visualization. Glaucoma is a syndrome of the optic nerve, the nerve bundle which transmit images from the eye retina to the brain.

Glaucoma is an eye syndrome in which the optic nerve reimbursements by the height in the intra ocular pressure in the interior portion of eye caused by a build-up of overload fluid. This pressure can harm vision by causing irreversible hurt to the optic nerve and to the retina. It can escort to the blindness if it is not detected and treated in appropriate time. Glaucoma result in insignificant vision crush, and is an especially dangerous eye condition because it frequently advancements without noticeable indications. This often referred to as "The Silent Thief of Sight" (Fig. 1).

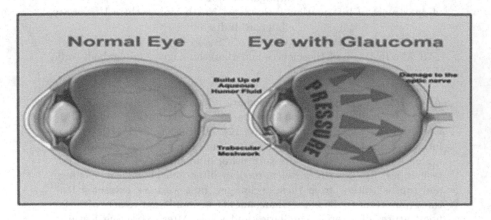

Fig. 1. Normal and Glaucoma Retina

2   Methodology

For this research paper we have present here our two experiments with different methodologies used for feature extraction and glaucoma diagnosis, some of steps are same in both experiments like preprocessing with enhancement and filtering.

2.1   ROI (Region of Interest)

Glaucoma is a bunch of eye syndrome traditionally differentiated by central intra ocular pressure (IOP). However, glaucoma is supplementary exactly defined as an optic neuropathy than a disease of high pressure. The aim of the study is to asses' retinal blood vessels breadth in glaucomatous eyes using retinal fundus images and to study their correlations with glaucomatous harm. The principle of preliminary glaucoma suspect, with Optic Cup to Disc ratio of more than 0.6

at least in one eye, intra ocular pressure (IOP) elevated than 20 mm Hg (mm Hg. Is a unit micro millimeter in mercury), and age more than 20 years.

Withdrawal of retinal blood vessels is very essential for recognition of Glaucoma, Diabetic Retinopathy (DR), and Hypertension etc. [1]. It is look like hierarchy. For withdrawal of retinal blood vessels we have used 2866 images of dissimilar database. For withdrawal of retinal blood vessels we have recommend two algorithms that determined on Haar Wavelet and TOP-HAT filter. In these experiments first RGB image is transformed in to the green channel image. Then apply Contrast-limited adaptive histogram equalization on the green channel, and then we have applied morphological opening operation, after that used the TOP-HAT filter and extracted retinal blood vessels [2] and in another we have used Haar wavelet to extract retinal blood vessels. We have also applied the boundary setting on the extracted output. After extractions of retinal blood vessels we have calculate the statistical value as area, diameter, thickness, length and tourtosity of final RNFL (retinal nerve fiber layer) extracted images healthy images and Glaucomatous images.

3   Workflow Diagrams for both Experiments

From the above Figs. 2 and 3 there are main five steps from preprocessing to statistical calculations used for both the experiments, some of these common implementation also like preprocessing with image resizing, green channel extraction,

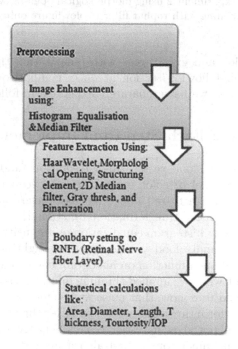

Fig. 2. Work flow for experiment 1. (Color figure online)

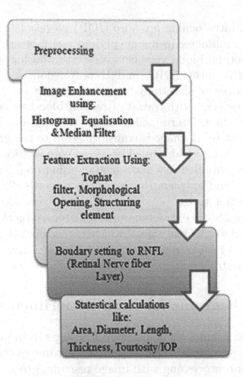

Fig. 3. Work flow for experiment 2 using morphological opening with 2D median filter using morphological opening with tophat filter (Color figure online)

histogram equalization and gray thresh with binarization these common steps we have explained in following section at once, comparison starts in feature extraction methods that we have separately explained in following section.

4 Explanation of Each Step from Implementation

For extraction of retinal blood vessels, high resolution fundus images is taken then extract green channel of color image because green channels shows the intensity image as compare to red and blue respectively, after green channel extraction perform intensity transformation function for enhancement of the fundus image, then apply histogram equalization on intensity transformed image for highlighting the retinal blood vessels, after highlighting the blood vessels perform morphological open function on histogram equalized image for thinning the blood vessels. But when we applied thinning operation some salt and pepper kind of noise get added to remove that noise we use median filter, for extraction of blood vessels we perform threshold operation. After threshold we get extracted blood vessels but it is not showing exact vessels network that's why we apply Haar wavelet with four dimensions in first algorithm and reconfirm it with top-hat filter in another experiment.

4.1   Original Image and Resize Image

We have applied our proposed algorithms on different datasets like DRIONS-DB, DRISHTI, MESSIDORE, RIM-ONE and HRF (High Resolution Fundus images). We have used different data sets with different image sizes, for making equality concept we have to resize all the images in equal size i.e. 400 × 600. Before any operation performs on retinal fundus image we have observed original image. The sampled original image is shown in following figure (Fig. 4).

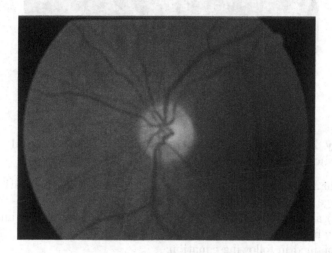

Fig. 4. Sampled original image from DRISHTI dataset

4.2   Green Channel Extraction

As we have seen previously our region of interest is retinal blood vessels and for extraction of that, we have need of greatest intensity to work on segmentation of edges or boundary present in image. For higher intensity we have to choose green channel from RGB image because intensity of green channel is greater as compare to red and blue. Following figure shows an example of green channel image (Fig. 5).

4.3   Complement of Green Channel Image

In the second step of preprocessing after extraction of green channel we have complemented that green channel for visualizing minute characteristics present in image.

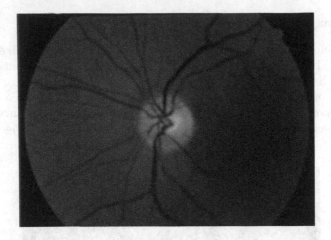

Fig. 5. Output of green channel image

4.4   Apply Histogram Equalization on Green Image and Morphological Opening

In image enhancement first step we have applied histogram equalization on green channel image by using following formulas.

Let $ps(s)$ and $pd(d)$ represent the standard image and desired image probability density functions, respectively. The histogram equalization of the standard image is explained in following equation:

$$u = T(s) = s0ps(x)dx \tag{1}$$

The histogram equalization of standard image is obtained by a similar transformation function as follows:

$$= Q(d) = d0pd(x)dx \tag{2}$$

The values of d for the standard image are obtained are explain as follows:

$$d = Q\,1\,[u] = Q1\,[T(s)] \tag{3}$$

A standard retinal image is used as a indication for histogram specification method in concurrence with the professional ophthalmologist [3]. In the morphological Opening execute an erosion procedure followed by a dilation procedure using a predefined neighborhood or structuring element. Following image symbolize an illustration of illustration of histogram equalized image or improved image (Fig. 6).

4.5   Remove Optic Disc

As per our region of interest we have concentrated on retinal blood vessels other than any features present on retinal fundus image. For moving one step ahead

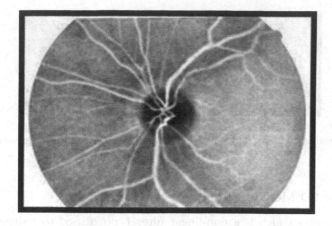

Fig. 6. Sampled histogram equalized image.

at feature extraction i.e. our retinal blood vessels we have removed optic disc from the fundus image. And then apply histogram on that image for visualizing the changes after optic disc removed. Following figures shows both the images of optic disc removed binary fundus image and histogram of optic disc removed image (Fig. 7).

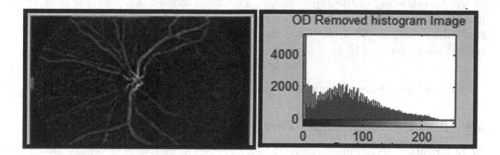

Fig. 7. OD removed output image and OD removed histogram image.

4.6   Apply Histogram Equalization on OD Removed Image:

After removing optic disc from the retinal fundus image we have enhanced our OD removed image for better visualization of retinal blood vessels. After this step it is going to be easier to extort retinal blood vessels from the retinal fundus image. Following figure shows sampled OD removed histogram equalized image and also background removed image (Fig. 8).

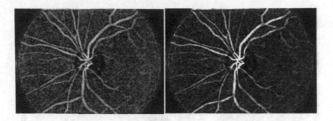

Fig. 8. OD removed histogram equalized image. And background removed sampled output image.

4.7   Apply 2D Median Filter

The median filter which is a non-linear filter type is used to reduce the effect of noise without blurring the sharp edge. The operation of the median filter is first position the pixel values in moreover the ascending or descending order and then computes the median value of the neighborhood pixels.

4.8   Gray Thresh on Adjusted Image

The gray thresh method disregard any nonzero imaginary fraction. Gray thresh technique returns the effectiveness metric, as the second output argument. The efficiency metric is a value in the range [0 1] that indicates the effectiveness of the thresholding of the input image. The lower bound is achievable only by images having a single gray level, and the upper bound is achievable merely by two-valued images. Then we have use the Threshold method for pull out the retinal blood vessels [4].

$$T = M(m1 + m2) \tag{4}$$

Here m1 & m2 are the Intensity principles [5].

4.9   Binaries Adjusted and Thresh Image

In binarization technique it converts the grayscale image to a binary image. In this procedure the output image restores all pixels in the input image with luminance less than level with the value 0 (black) and replaces all other pixels with the value 1 (white And the fundamental idea at the back of wavelets is to examine the signal at dissimilar scales or resolutions, which is called multi-resolution.). Specify level in the range [0, 1]. This assortment is qualified to the signal levels probable for the image's class. Therefore, a level value of 0.5 is central way between black and white, apart from the class. Then in next step we have applying morphological opening on binaries image.

4.10   Median Filter on Binaries Image

After performing morphological opening operation on binariese image, then we have uninvolved the optic disc from the image. After uninvolving the optic disc, we have applied 2D Median filter procedure on image for eliminate the noise [6]. From the following formula gives us some idea about 2D median filtering.

$$y = [m, n] = medianx[i, j], (I, j)\epsilon\omega \tag{5}$$

The above formula represents a neighborhood centered on location (m, n) in the image [4]. Following figure shows sampled 2D median filter image which is obtained from after applying median filter method (Fig. 9).

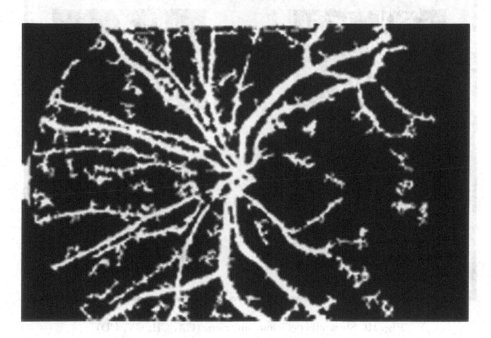

Fig. 9. Sampled median filter on binaries image

4.11 Haar Wavelet

Multi-resolution is the basic idea in the wake of wavelets that are working on data by cut it into dissimilar frequency mechanism. And the essential idea in the wake of wavelets is to scrutinize the signal at diverse measures or resolutions. The most noteworthy characteristic of wavelet transform is it tolerates multi-resolution decomposition [7].

After the extraction of grey threshold of image we have exploit HAAR wavelets Mothers wavelet function [8].

$$0 \le t \le 1/2\psi(t) = 1011/2 \le t \le 1 \ otherwise \tag{6}$$

Its Scaling function ϕ (t) is given as,

$$0 \le t \le 1/2\psi(t) = 10 \ otherwise \tag{7}$$

After applying the HAAR wavelet's mother function it provide four types of output and for the supplementary processing we have used approximate image. There are four types of outputs are show in following figure (approximate, horizontal, vertical, diagonal) (Fig. 10).

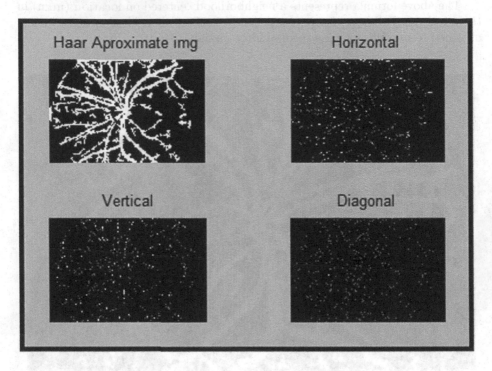

Fig. 10. Show all coefficient matrices (CA, CH, CV, CD)

4.12   Apply dwt2 (Discrete Wavelet Transform)

The dwt2 function execute a single-level two-dimensional wavelet decomposition with respect to either a exacting wavelet or exacting wavelet decomposition filters (Lo_D i.e. Low pass filter and Hi_D i.e. High pass filter) specified by us.

$[cA, cH, cV, cD] = dwt2$ calculate the rough calculations coefficients matrix cA and details coefficients matrices cH, cV, and cD (horizontal, vertical, and diagonal, respectively), obtain by wavelet decomposition of the input matrix. The 'haar' string contains the wavelet name [9].

On the experimental images, there exists an algorithm similar to the one-dimensional case for two-dimensional wavelets and scaling functions obtained from one-dimensional ones by tonsorial product [10]. The fundamental decomposition stepladder for images is described in following chart:

The basic decomposition steps for images are describe in following chart (Fig. 11):

Two-Dimensional DWT

Decomposition step

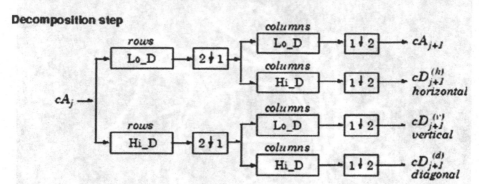

Fig. 11. Decomposition steps of two dimensional discrete wavelet transform

Where $2\downarrow1$ Down sample columns: Keep the even indexed columns

$1\uparrow2$ Down samples rows: Keep the even indexed rows

rows
X Convolve with filter X rows of entry

columns
X Convolve with filter X columns of entry

Initialization CAo= s for the decomposition initialization

This type of two-dimensional DWT escorts to a decomposition of approximation coefficients at stage j in four components: the approximation at rank j + 1, and the details in three orientations (horizontal, vertical, and diagonal).

4.13   Apply Top-Hat Filter

After getting the output of morphological open operation we have applied TOP-hat filter on the image for filtering purpose and get more clear extracted retinal blood vessels. Top-Hat transform is operations that extracts small elements and details from given image [6]. TOP-HAT transform mathematically define as follows:

Let from the following formula f: E→R be a grayscale image, recording points from a Euclidean space or distinct grid E (such as R2 or Z2) into the real line. Let b(X) be a grayscale structuring element [6]. Then TOP-HAT transform is defining as

$$Tw\ (f) = f - f \circ b \qquad (8)$$

Where denotes the opening operation (Fig. 12).

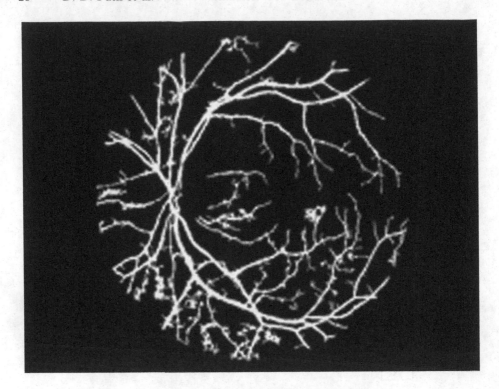

Fig. 12. Output image of top-hat filter

4.14 Boundary Setting on Retinal Blood Vessels

After extracting the four dimensions' images of retinal blood vessels we get our
sufficient output from approximation coefficients matrix CA so we have ignore
other coefficients like horizontal vertical and diagonal. And for reconfirmation we
have applied top-hat filter from another experiment, then for proper calculation
of physical appearance we have set the boundary to vessels by using canny edge
detection method. And the following image represents the sampled images of
boundary setting to retinal blood vessels output from approximate view from
haar wavelet and another one from top-hat filter algorithm (Fig. 13).

5 Results

5.1 Calculate Area, Diameter, Length, Thickness, and Tortuosity (IOP) of Retinal Blood Vessels

After extracting the ROI i.e. our region of interest from the retinal fundus image
after processing on each image, we have to perform some mathematical calculations for exact diagnosis of disease. So in mathematical calculations first we
have calculated area of retinal blood vessels, then length, diameter and thickness

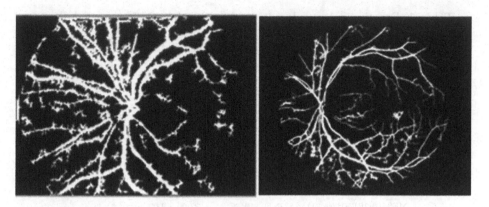

Fig. 13. Boundary set from Haar wavelet and boundary set from top-hat algorithm output

and on the basis of all these features we have calculated tortuosity i.e. our IOP (Intra Ocular Pressure) by using following formulas.

Area

$$Area = \pi \times r2 \tag{9}$$

Diameter

$$Diameter = \sqrt{Area/\pi} \tag{10}$$

Length

$$Lenght = Area^2 \tag{11}$$

Thickness

$$Thickness = Area \times Length \tag{12}$$

Tortuosity

$$Tortuosity = Length \times Distance \tag{13}$$

After mathematical calculations applied on extracted features, we get dense statistical values for every parameter like area, diameter, length, thickness, and tortuosity; with these values we can predict the level of glaucoma. Following Tables 1 and 2 shows sample statistical values for each parameter from all datasets. Table 1 shows resultant parameters for first experiment, and Table 2 shows resultant parameters for second experiment.

Table 1. Experiment 1 sample statistical results obtained from each dataset

Sr. no.	Database name	Area	Diameter	Length	Thickness	Tortuosity/IOP
1	DRIONS-DB	14652.25	68.29	68.00	2.00	56.47
2	HRF Normal Images	12460.50	62.98	106.00	2.00	20.69
2.1	HRF Glaucoma	10046.00	56.55	63.00	2.00	48
3	RIM-ONE Normal	9564.25	55.18	16.00	2.00	21.98
3.1	RIM-ONE Glaucoma	9453.00	54.85	45.00	2.00	13.68
4	DRISHTI Training	7704.38	49.52	34.00	2.00	19.76
4	DRISHTI Testing	4514.63	37.91	5.00	2.00	22.36
5.1	MESSIDORE (Base 11)	11187.88	59.68	64.00	2.00	64.98
5.2	MESSIDORE (Base 12)	9466.50	54.89	206.00	2.00	39.61
5.3	MESSIDORE (Base 13)	7655.38	49.36	60.00	2.00	47.43
5.4	MESSIDORE (Base 14)	13586.50	65.76	35.00	2.00	44.81
5.5	MESSIDORE (Base 21)	8275.00	51.32	27.00	2.00	31.39
5.6	MESSIDORE (Base 22)	5797.75	42.96	6.00	2.00	26.83
5.7	MESSIDORE (Base 23)	11393.50	60.22	217.00	2.00	30.80
5.8	MESSIDORE (Base 24)	11401.75	60.24	6.00	2.00	21.21
5.9	MESSIDORE (Base 31)	9946.38	56.27	12.00	2.00	11.94
5.10	MESSIDORE (Base 32)	6137.88	44.20	41.00	2.00	22.24
5.11	MESSIDORE (Base 33)	10744.75	58.48	45.00	2.00	28.75
5.12	MESSIDORE (Base 34)	7942.38	50.28	96.00	2.00	24.65

Table 2. Experiment 2 sample statistical results obtained from each dataset

Sr. no.	Database name	Area	Diameter	Length	Thickness	Tortuosity/IOP
1	DRIONS-DB	20.67	2.56	68.00	2.00	56.47
2	HRF Normal	34760.00	105.19	106.00	2.00	20.69
2.1	HRF Glaucoma	24580.13	88.45	63.00	2.00	47.90
3	RIM-ONE Normal	21.33	2.61	16.00	2.00	21.98
3.1	RIM-ONE Glaucoma	11807.00	61.30	45.00	2.00	13.68
4	DRISHTI Training	5141.50	40.45	34.00	2.00	19.76
4.1	DRISHTI Testing	9625.88	55.35	5.00	2.00	22.36
5.1	MESSIDORE (Base 11)	24051.25	87.50	64.00	2.00	64.98
5.2	MESSIDORE (Base 12)	20906.38	81.58	206.00	2.00	39.61
5.3	MESSIDORE (Base 13)	19808.00	79.40	60.00	2.00	47.43
5.4	MESSIDORE (Base 14)	25784.88	90.60	35.00	2.00	44.81
5.5	MESSIDORE (Base 21)	23139.50	85.82	27.00	2.00	31.39
5.6	MESSIDORE (Base 22)	14790.13	68.61	6.00	2.00	26.83
5.7	MESSIDORE (Base 23)	24814.63	88.87	217.00	2.00	30.80
5.8	MESSIDORE (Base 24)	27960.00	94.34	6.00	2.00	21.21
5.9	MESSIDORE (Base 31)	19771.88	79.33	12.00	2.00	11.94
5.10	MESSIDORE (Base 32)	11907.38	61.56	41.00	2.00	22.24
5.11	MESSIDORE (Base 33)	22428.00	84.49	45.00	2.00	28.75
5.12	MESSIDORE (Base 34)	20455.50	80.69	12.00	2.00	24.00

Above Table 3 shows results in system classified form for Experiment 1
Table 4 shows results in system classified form for Experiment 2 with the help
of these tables results shows in six ways with the number of images for normal,
glaucoma with mild, mode, and moderate stages again with percentages also for
whole datasets used for this research work.

Table 3. Results in system classified form for Exp 1

Sr. no.	Database name	Normal	Glaucoma			Glaucoma %	Normal %
			Mild	Mode	Mode rate		
	DRIONS-DB	24	26	11	49	78.18	21.82
2.1	HRF (Normal)	09	00	01	05	40.00	60.00
2.2	HRF (Glaucoma)	03	03	03	06	80.00	20.00
3.1	RIM-ONE (Normal)	134	26	19	46	40.44	59.56
3.2	RIM-ONE Glaucoma)	89	28	23	60	55.50	44.50
4.1	DRISHTI (Training)	22	10	05	14	56.86	43.14
4.2	DRISHTI (Testing)	27	12	00	12	47.06	52.94
5.1	MESSIDORE (Base11)	33	27	14	26	60.00	33.00
26	MESSIDORE (Base12)	30	26	11	33	70.00	30.00
5.3	MESSIDORE (Base13)	25	27	13	35	75.00	25.00
5.4	MESSIDORE (Base14)	40	22	05	33	60.00	40.00
5.5	MESSIDORE (Base21)	38	25	14	23	62.00	38.00
5.6	MESSIDORE (Base22)	42	30	11	17	58.00	42.00
5.7	MESSIDORE (Base23)	31	26	06	27	69.00	31.00
5.8	MESSIDORE (Base24)	42	23	08	27	58.00	42.00
5.9	MESSIDORE (Base31)	41	14	20	25	59.00	41.00
5.10	MESSIDORE (Base32)	33	22	14	31	67.00	33.00
5.11	MESSIDORE (Base33)	29	22	25	24	71.00	29.00
5.12	MESSIDORE (Base34)	38	24	09	29	62.00	38.00

Table 4. Results in system classified form for Exp 2

Normal	Glaucoma			Glaucoma %	Normal %
05	48	09	48	95.4	4.55
12	01	00	02	20.0	80.0
02	03	03	07	86.6	13.3
193	11	05	16	14.2	85.7
89	29	23	59	55.5	44.5
21	06	12	12	58.8	41.1
17	18	00	16	66.6	33.3
16	15	33	36	84.0	16.0
12	19	26	43	88.0	12.0
15	12	27	46	85.0	15.0
27	11	20	42	73.0	27.
22	12	31	35	78.0	22.0
26	12	35	27	74.0	26.0
31	34	06	29	69.0	31.0
40	23	09	28	60.0	40.0
15	16	45	24	85.0	15.0
05	19	49	27	95.0	5.00
27	33	19	21	73.0	27.0
36	29	09	26	64.0	36.0

5.2   Literature Survey on K-Means Classification

K-means uses a two-phase iterative algorithm to minimize the sum of point-to-centroid distances, summed over all k clusters: Authors Hoboken, N. J and John Wiley explain the concept of K-Means, "The primary period uses batch apprises, where each recurrence consists of reassigning points to their neighboring cluster centroid, all at once, followed by recalculation of cluster centroids. This stage infrequently does not congregate to solution that is a local minimum, that is, a separation of the data where affecting any particular point to a dissimilar cluster increases the total sum of distances. This is more likely for minuscule data sets. The batch stage is fast, but potentially only approximates explanation as a starting point for the second stage" [11].

Author J. Goldschmidt explain this in his published work "The second segment uses online updates, where points are independently transported if doing so will reduce the amount of distances, and cluster centroids are recomputed behind every reassignment. Each iteration throughout the second segment consists of one overtake though all the points. The second segment will converge to a confined minimum, although there may be other local minima with minor total amount of distances. The problem of finding the global minimum can only be solved in common by an comprehensive (or clever, or lucky) choice of initial points, but using numerous replicates with random starting points typically results in a solution that is a inclusive minimum" [6].

5.3   K-Means Clustering and Classification Applies on Obtained Results:

K-Mean algorithm is an unsupervised clustering algorithm that classifies the enter data points into several classes based on their intrinsic reserve from each other. The algorithm assumes that the data features from a vector space and attempts to discover clustering in them.

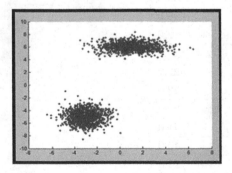   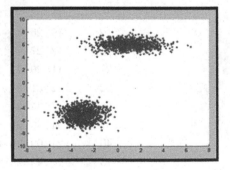

Fig. 14. K-Means classification   Fig. 15. K-Means clustering

Above Fig. 14 shows graphical representation of K-Means classification in two separate groups, upper group of blue points represents abnormal class of dataset images and lower represents normal group of images. And Fig. 15 shows graphical representation of K-means clustering, upper first cluster shows normal group while lower three clusters shows the intensity of disease like mild mode and moderate, and last opposite directional cluster shows unclassified data.

6   Discussions and Conclusion

This is our scientific contribution towards society we presented here robust and efficient system. The objective of retinal blood vessels segmentation for glaucoma detection is to find out early detection of glaucoma because as the symptoms occurs when the dieses is quite advance, it leads to loss of vision, so early detection of glaucoma is very essential. This approach presents efficient and objective method for automatically classifying digital fundus images into either normal or glaucomatous types in order to facilitate ophthalmologists, for that we have used fuzzy classification based methods for retinal blood vessels segmentations for glaucoma screening.

For evaluation we have used retinal images from clinical database. The system has been implemented using Haar wavelet for experiment one and for another experiment Top-hat filter, on the basis of feature extraction we got more clear results from experiment two as compare to one and other supportive feature extraction techniques with two different fuzzy classifiers are used to classify image as glaucoma suspicious or normal. Comparison being made between two experiments and K-Means classifier among those K-Means gives best result which has better result rate than other. The system mainly provides an efficient method for glaucoma detection and is aimed to be highly beneficial for any person or ophthalmologist can help to find the severance of the system. Experimental result shows the proposed method is more effective and helpful to user and ophthalmologist as it reduces false rate.

Acknowledgement. We would like to acknowledge all the medical dataset provider agencies for research purpose, like DRIONS-DB, DRISHTI, MESSIDORE, RIN-ONE and HRF. If the learner wanted to implement and develop their own graphical user interface (GUI), then please kindly refer, "Understanding Programming Aspects of Pattern Recognition Using MATLAB", and the same kind of experiments also available in "Glaucoma Diagnosis in the Vision of Biomedical Image Analysis" and "Projects in digital image processing", By the same authors Dr. Ramesh R. Manza & Dr. Dnyaneshwari D. Patil, Shroff Publisher & Distributer Pvt. Ltd.

32   D. D. Patil et al.

References

1. Zhang, B., Zhang, L., Zhang, L., Karray, F.: Retinal vessel extraction by matched filter with first-order derivative of Gaussian. Comput. Biol. Med. **40**, 438–445 (2010)
2. Zana, F., Klein, J.-C.: Segmentation of vessel-like patterns using mathematical morphology and curvature evaluation. IEEE Trans. Image Process. **10**(7), 1010–1019 (2001)
3. Emedicinemedscape.com. Retinal Anatomy. http://emedicine.medscape.com/article/2019624-overview. Accessed 3 Nov 2016
4. K-NN Classification Algorithm. http://en.wikipedia.org/wiki/K-nearest_neighbors_algorithm
5. Rajput, Y.M., Manza, R.R., Patwari, M.B., Deshpande, N.: Retinal blood vessels extraction using 2D median filter. In: National Conference in Advances in computing (NCAC 2013), 05–06 March 2013
6. Spath, H.: Cluster Dissection and Analysis: Theory, FORTRAN Programs, Examples (Translated by J. Goldschmidt). Halsted Press, New York (1985). https://in.mathworks.com/help/stats/kmeans.html. Accessed Aug 2016
7. Chrastek, R., et al.: Automated segmentation of the optic nerve head for diagnosis of glaucoma. Med. Image Anal. **9**(4), 297–314 (2014)
8. Rathod, D.D., Manza, R.R., Rajput, Y.M., Patwari, M.B., Saswade, M., Deshpande, N.: Localization of optic disc and macula using multilevel 2-D wavelet decomposition based on Haar wavelet transform. Int. J. Eng. Res. Technol. (IJERT) **3**(7) (2014). ISSN 2278-0181
9. Daubechies, I.: Ten lectures on wavelets. CBMS-NSF Conference Series in Applied Mathematics, SIAM Ed (1992). http://epubs.siam.org/doi/book/10.1137/1.9781611970104
10. Mallat, S.: A theory for multiresolution signal decomposition: the wavelet representation. IEEE Pattern Anal. Mach. Intell. **11**(7), 674–693 (1989)
11. Seber, G.A.F.: Multivariate Observations. Wiley, Hoboken (1984). https://doi.org/10.1002/9780470316641.ch2/summaryMay2016. Accessed 27 May 2008

Image Enhancement Using Filters on Alzheimer's Disease

Aziz Makandar and Rashmi Somshekhar$^{(\boxtimes)}$

Akkamahadevi Women's University, Vijayapura, India
azizkswuv@gmail.com, rashmiraj.rathod@gmail.com

Abstract. The Alzheimer's disease (AD) is a neurological disorder. It is a slow ongoing brain disease that starts well before clinical symptoms emerge. The subsequent decline in cognitive functions worsens with time and eventually can lead to death. In this study, we worked on T2 sequence Magnetic Resonance Images (MRI) of the Brain AD affected images. The brain may be affected by the AD in different regions. Proposed work concentrates on the hippocampus region affected by the AD using MRI imaging modality. In order to analysis the medical images with no noise the images are enhanced by using different filtering techniques. The work carried out in this paper discuss about different filtering techniques like Gaussian, Median, Wiener and order statistical filter. The performance of these algorithm were examined on PSNR and RMSE values. After the computation the median filter is consider to be the good filter as per the results for the MRI images of AD.

Keywords: Alzheimer's disease ·
Gaussian, Median, Wiener and order statistical filter

1 Introduction

Alzheimer's disease (AD) is an irreversible that slowly destroys memory and thinking skills. In many people the Alzheimers symptoms rst appear in their mid-60s. Recent estimates indicate that the disorders may rank third just after the heart disease and cancer as a cause for the death of people. The regions affected by the AD are Hippocampus [1]: This brain region which is responsible for forming new memories. Amygdale is the region which is responsible for the experience and expressions of emotion. Cerebellums are responsible for coordination of movements posture, speech and motor functions. The frontal lobe is responsible for executive and management function. Parietal lobe is responsible for handling spatial relationship and magnitude [1] and also responsible for the sense of ones body. Occipital lobe this helps to interpret what the eyes are seeing. Corpus callous this region helps in transformation of information between the brains left and right hemisphere. Thalamus is responsible for relaying motor and sensory signals to the cerebral cortex and regulates sleep and alertness (Fig. 1).

Different diagnoses methods exist for the detection of Alzheimer's. Here in this paper we are working on hippocampus region where the person affected

K. C. Santosh and R. S. Hegadi (Eds.): RTIP2R 2018, CCIS 1036, pp. 33–41, 2019.
https://doi.org/10.1007/978-981-13-9184-2_3

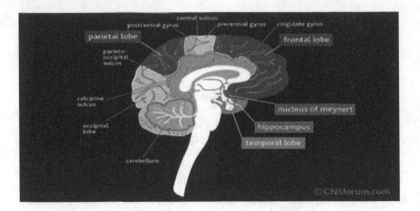

Fig. 1. Shows the different regions of the brain.

by the AD will have loss of memory. Human brain have two side hippocampus one in each side of the brain it is located under cerebral cortex (all cortical) and in the medial temporal lobe [2]. In Alzheimer disease the rest region to suffer from damage is hippocampus the early symptoms includes short-term memory loss and disorientation. Injury to the hippocampus may result from oxygen starvation, encephalitis (in ammation of the brain) or epilepsy. Person with more bilateral hippocampus damage may suffer from anterograde amnesia (the inability to form and retain new memories) (Fig. 2).

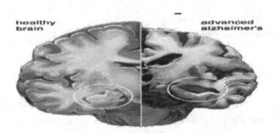

Fig. 2. Shrinking of hippocampus.

In hippocampus region shrinkage is severe, an area of the cortex that plays a key role in formation of new memories. The Data used for the work is obtained from (http://www.loni.ucla.edu/ADNI) Alzheimer's disease neuroimaging Initiative (ADNI) database [3]. The images used are MRI in the axial plain with T2 sequence concentrating on hippocampus regions with 20 images. In order to enhance the image must be noise free so this can be accomplished by applying different filtering techniques. The different filtering techniques generally used to analysis medical images are Gaussian, Median, Wiener [25], Adaptive, contrast stretching [23,24] and order statistic filters. Our work concentrates on Gaussian, Median, Wiener and order statistic filters and the result is computed for the

different techniques using quantitative analysis methods like PSNR and RMSE
values and the best technique is considered to be median filter. Once the images
are pre processed we will be applying arrow detection method that works on
signature-based technique. The signature will be compared with arrow(region)
signature and then the similarity score indicates the presence of the arrow that
points to connected components [4] which will be extracted by binarization of
images using a fuzzy binarization tool that take the connected components at
every level of binarization and compute the overheads by using an arrow to select
proper region, [4,5] this would help in pruning of noise in connected components
[4] for the MRI images for finding out the region of interest ROI, [4–6] and
then differentiating the foreground and background regions for the segmentation process.

2   Filtering Techniques

The aim of filtering techniques is used to reduce noise and improve the visual
quality of the image. Below are some of the filtering techniques used in medical
imaging applied on MRI images.

2.1   Gaussian Filter

The working of Gaussian filter is considered to be non uniform low pass filter
here lower frequency is preserved and sharp edges are removed [7]. Its a filter
that filters blur edges and reduce contrast. When working with the images we
need to use the two dimensional Gaussian function. This is simply the product
of two 1D Gaussian functions and is given by the Eq. (1).

$$G(x + y) = \frac{1}{2\pi\sigma^2}e - \frac{(x^2 + y^2)}{2\sigma^2} \tag{1}$$

Where G(x, y) is an image, sigma () is standard deviation of the distribution
function, larger the value of wider the peak of gaussian and larger blurring. The
noise is been added to known its mean and variance, such that we will come to
know its statistics, i.e. its behavior. Then after filtering we can estimate how the
mean, variance and the rest of the statistics are changing. So we have added the
Gaussian noise of 0.09 and then applied the Gaussian filter. Figure 3 shows the
Gaussian filter for the AD MRI image.

2.2   Median Filter

Median filter is a nonlinear method. It is used to remove noise from images and
for preserving edges. It is primarily used to eliminate impulse type of noise. The
impulse term is also used for salt and pepper noise. The working of median filter
done by moving the each pixel in the image and by replacing each value with
the median value of the neighboring pixel. This pattern of neighbors is called as

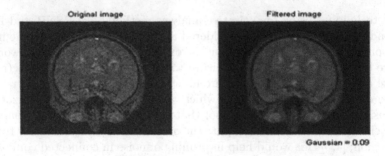

Fig. 3. Gaussian filter applied on MRI image.

the "window", which slides, pixel by pixel over the entire image and then the median value is calculated by sorting all the pixels values from the window into numerical order and then replacing the pixel that is being considered with the middle (median) pixel value [7,8]. Figure 4 shows the images with applied salt and pepper noise to the original image and then the filtered image.

Fig. 4. Original image, noisy image and median filter applied on MRI image.

2.3 Wiener Filter

The significance of wiener filter is to decrease the amount of noise present in an image by comparison with an estimation of the desired noiseless signal. It is based on a statistical approach to achieve the noise free image. Before implementing wiener filter it is assumed that user knows the spectral properties of the original image and noise. The aim of the process is to have minimum square error. That is, the difference the original signal and the new signal should be less as possible. The equation of the Wiener filter in frequency domain as W(u, v). The restored image is given by Eq. (2).

$$Xn(u, v) = W(u, v).Y(u, v) \qquad (2)$$

Where Y(u, v) is the received signal and Xn(u, v) is the restored image. The perspective of minimizing the degradation at a time which induces to develop a restoration algorithm. The Wiener filtering executes an optimal trade of between inverse filtering and noise smoothing [9–11]. It eliminates the additive noise and reverses the blurring simultaneously. The below MRI images shows the noisy image with salt and pepper noise and the wiener filtered image (Fig. 5).

Fig. 5. Wiener filter applied to the MRI image.

2.4   Order Statistic Filters

Order statistic filters are nonlinear spatial filter it is based on the ordering of the pixels in the image area. By replacing the value in the central pixel with the determined result. The equation of order filter is as given in the Eq. 3.

$$B = \text{ordfilt2} (A, \text{order}, \text{domain}) \tag{3}$$

By replacing each element in A by the order the element in the sorted set of neighbors specified by the nonzero elements in domain. It can be used to reduce the additive white noise, signal dependent noise and impulse noise. Below figure shows the order statistic filter applied on the AD MRI image (Fig. 6).

Fig. 6. Order statistic filter applied to the MRI image.

3   Experimental Results

The work is done with the comparison of different filtering techniques and the results are computed with the different values of PSNR and RMSE as shown in the given table. Also the graphical representation shows the variations of the values. In the (PSNR) peak signal to noise ratio the peak value possible for any one pixel or element is taken and then the values are computed by using the 4th equation [12].

$$\text{snr} = 10 * \log 10((\text{imaximum} - \text{iminimum})./\text{mse}) \tag{4}$$

The (RMSE) root mean square error is the mean square difference between the original and recreated one by using the following Eq. 5 [11].

$$\text{mse} = \text{sqrt}(\text{mean}(\text{mean}(I - I2))) \tag{5}$$

Table 1. PSNR values for the different filtering techniques with image resolution 2048 x 2048

Sl. No.	Images	Gaussian filter	Median filter	Wiener filter	Order statistic filters
1	ADT2	23.0927	22.663	22.87785	18.7632
2	AD1T2	22.247	22.552	22.0423	19.7491
3	AD2T2	23.3193	22.9591	23.3903	18.9913
4	AD3	22.1007	22.5606	22.3304	20.147
5	AD3T2	22.4096	22.9315	22.4343	20.0361
6	AD4T2	22.0893	22.563	22.0496	19.8586
7	AD5T2	22.3726	22.8962	22.8196	19.7632
8	AD6T2	22.9486	22.7917	23.1041	19.2574
9	AD7T2	22.3511	22.8848	22.6082	19.7747
10	ADT3	22.2168	22.3381	21.9646	19.4233
11	6AD6T2	22.9486	22.7756	23.1041	19.1759
12	6AD7T2	22.3511	22.6963	22.6082	19.7597
13	6AD8T2	22.7707	23.2671	23.1573	19.7348
14	AD	23.504	23.9954	24.3376	21.041
15	13AD	23.504	24.0022	24.3376	21.0393
16	39AD	23.3226	23.6419	24.2132	20.9719
17	303AD	23.3235	24.2158	24.1109	20.8534
18	179AD	23.6042	24.3334	24.5513	20.6529
19	MCI3	22.7633	23.0414	23.3733	20.0739
20	MCI4T2	22.7117	23.0445	22.9571	19.9925

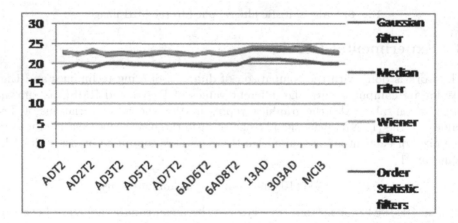

Fig. 7. PSNR values represented graphically for the different filtering technique.

The results show the higher PSNR values and the lower RMSE values for the good quality of the images. Different filtering techniques when compared the median filter technique is considered to be the good filtering technique for the AD MRI images. Tables 1 and 2 gives the computational results of PSNR and RMSE and the Figs. 7 and 8 shows the graphical representation of PSNR and RMSE values for different filters. Below Table 1 and Fig. 7 gives the computational result of peak signal to noise ratio (PSNR) with the comparison for the different filters. The statistic behavior shows that median filter and wiener filter has slight difference both the filters show the higher PSNR values but which shows less RMSE value that filter is considers to be the good filter for these MRI images. In Table 2 and Fig. 8 clearly shows that the statistical behavior of the Median filter is less as compared to other filters.

The graph shows the statistical behavior obtained by PSNR values for the different filter.

Below the Table 2 shows the root mean square error calculated for the different filtering techniques with the image resolution 2048 × 2048.

Table 2. RMSE values for the different filtering techniques

Sl. No.	Images	Gaussian filter	Median filter	Wiener filter	Order statistic filters
1	ADT2	1.1922	1.0927	1.1532	1.6393
2	AD1T2	1.4783	1.3027	1.4371	2.2258
3	AD2T2	1.1083	1.0248	1.0766	1.5663
4	AD3	1.3625	1.1159	1.2513	1.8209
5	AD3T2	1.424	1.2296	1.3531	2.1016
6	AD4T2	1.5329	1.334	1.4472	2.3385
7	AD5T2	1.2913	1.1293	1.1441	2.0866
8	AD6T2	1.1918	1.029	1.1205	1.8465
9	AD7T2	1.3094	1.1076	1.1903	2.0757
10	ADT3	1.5126	1.3379	1.4186	2.3617
11	6AD6T2	1.1918	1.0312	1.1205	1.8326
12	6AD7T2	1.3094	1.1281	1.1903	2.0749
13	6AD8T2	1.3209	1.1371	1.1746	2.0693
14	AD	1.1246	0.928	0.9135	1.6004
15	13AD	1.1246	0.914	0.9135	1.5919
16	39AD	1.1214	0.9132	0.8983	1.6076
17	303AD	1.1863	0.9422	0.9779	1.635
18	179AD	1.0902	0.8476	0.8661	1.487
19	MCI3	1.2543	1.1558	1.0578	2.1098
20	MCI4T2	1.3122	1.1569	1.1895	2.1201

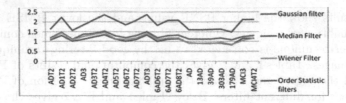

Fig. 8. RMSE values represented graphically for the different filtering technique.

The root mean square error value is shown in the Fig. 8 by the graphical representation for the different filters. Here we can find that the median filters RMSE value is less as compared to other filters.

4   Conclusion

Alzheimer disease symptoms are diagnosed most probably at the mid 60s. To diagnosis the disease at the earlier would be difficult for the physician using imaging modalities since it is usually diagnosed by the clinical traits. However, enhancing the images with noise free and applying the different filtering technique would be the one to get information about the regions affected by the AD. Our work compared different filtering techniques used for the medical imaging and computed the results using PSNR and RMSE values the median filter is considered to produced the noise free image since its mean square error is less as compared to the other filters. Which helps to get good quality images. Since we have compared result with the existing filters and computed the results our next step could be working on designing a filter for better results.

References

1. Alzheimer's disease facts, figures. Alzheimer's and dementia. J. Alzheimer's Assoc. **13**(4), 325–373 (2017). https://doi.org/10.1016/j.jalz.2017.02.001
2. Amaral, D., Lavenex, P.: Hippocampus neuroanatomy. In: Anderson, P., Morris, R., Amaral, D., Bliss, T., O'Keefe, J. (eds.) The Hippocampus Book, 1st edn, p. 37. Oxford University Press, New York (2007). ISBN 978-0-19-510027-3
3. Alzheimer's Disease Neuroimaging Initiative. ADNI for Alzheimer's disease database. http://adni.loni.usc.edu
4. Santosh, K.C., Wendling, L., Antani, S., Thoma, G.R.: Overlaid arrow detection for labeling regions of interest in biomedical images. IEEE Intell. Syst. **31**(3), 66–75 (2016). Browse Journal and Magazines
5. Santosh, K.C., Alam, N., Roy, P.P., Wendling, L., Antani, S., Thoma, G.R.: A simple and efficient arrowhead detection technique in biomedical images. Int. J. Pattern Recogn. Artif. Intell. **30**(05), 1657002 (2016)
6. Santosh, K.C., Roy, P.P.: Arrow detection in biomedical images using sequential classifier. Int. J. Mach. Learn. Cybern. **9**(6), 993–1006 (2018)
7. Kumar, N., Nachamai, M.: Noise removal and filtering techniques used in medical images. Orient. J. Comput. Sci. Technol. **10**(1), 103–113 (2017). ISSN 0974-6471

8. Ali, H.M.: MRI Medical Image Denoising by Fundamental
9. Temizel, A., Vlachos, T.: Wavelet domain image resolution enhancement using cycle-spinning. Electron. Lett. **41**(3), 119–121 (2005)
10. Temizel, A., Vlachos, T.: Image resolution upscaling in the wavelet domain using directional cycle spinning. J. Electron. Imaging **14**(4), 040501 (2005)
11. Anbarjafari, G., Demirel, H.: Image super resolution based on interpolation of wavelet domain high frequency subbands and the spatial domain input image. ETRI J **32**(3), 390–394 (2010)
12. Chaurasia, K., Sharma, N.: Performance evaluation and comparison of different noise, apply on PNG image format used in deconvolution Wiener filter (FFT) algorithm. Evol. Trends Eng. Technol. **4**, 8–14 (2015). https://doi.org/10.18052/www.scipress.com/ETET.4.8. ISSN 2349-915X
13. Gonzalez, R.C., Woods, R.E.: Digital Image Processing Using MATLAB, 3rd edn. Pearson Education, London (2003)
14. Sonka, M., Hlavac, V., Boyle, R.: Image Processing, Analysis and Machine Vision, 2nd edn. Thomson Learning, Boston (2001)
15. Makandar, A., Somshekhar, R.: Approaches of pattern recognition in analysing the neurological disorders: a review. Int. J. Trend Sci. Res. Dev. (IJTSRD) Int. Open Access J. **2**(3), 1729–1733 (2018). http://www.ijtsrd.com. ISSN 2456-6470
16. Yalabık, N.: Medical applications of pattern recognition. In: HIBIT 2010, Antalya, April 2010
17. Makandar, A., Halalli, B.: Threshold based segmentation technique for mass detection in mammography. J. Comput. **11**(6), 472–478 (2016)
18. Makandar, A., Patrot, A.: Malware image analysis and classification using support vector machine. Int. J. Trends Comput. Sci. Eng. **4**(5), 01–03 (2015)
19. Makandar, A.U.R., Karibasappa, K.: Wavelet based medical image compression using SPHIT. J. Comput. Sci. Math. Sci. **1**, 769–775 (2010)
20. Alzhimer's Association. https://www.alz.org/alzheimers_diseaseAlzheimersAssociation
21. Hippocampus Wikipedia. https://en.wikipedia.org/wiki/Hippocampus
22. Anderson, P., Morris, R., Amaral, D., Bliss, T., O'Keefe, J. (eds.): The Hippocampus Book. Oxford University Press, Oxford (2007). ISBN 978-0-19-510027-3
23. Ruikar, D.D., Santosh, K.C., Hegadi, R.S.: Automated fractured bone segmentation and labeling from CT images. J. Med. Syst. (2019). https://doi.org/10.1007/s10916-019-1176-x
24. Ruikar, D.D., Santosh, K.C., Hegadi, R.S.: Segmentation and analysis of CT images for bone fracture detection and labeling. In: Medical Imaging: Artificial Intelligence, Image Recognition, and Machine Learning Techniques, chap. 7. CRC Press, Boca Raton (2019). ISBN 9780367139612
25. Hegadi, R.S., Navale, D.I., Pawar, T.D., Ruikar, D.D.: Multi feature-based classification of osteoarthritis in knee joint X-ray images. In: Medical imaging: Artificial Intelligence, Image Recognition, and Machine Learning Techniques chap. 5. CRC Press, Boca Raton (2019). ISBN 9780367139612

Visualizing Salient Network Activations in Convolutional Neural Networks for Medical Image Modality Classification

Sivaramakrishnan Rajaraman[(✉)] and Sameer Antani

Lister Hill National Center for Biomedical Communications,
National Library of Medicine, Bethesda, MD 20894, USA
sivaramakrishnan.rajaraman@nih.gov

Abstract. Convolutional neural network (CNN) has become the architecture of choice for visual recognition tasks. However, these models are perceived as black boxes since there is a lack of understanding of their learned behavior from the underlying task of interest. This lack of transparency is a drawback since poorly understood model behavior could adversely impact subsequent decision-making. Researchers use novel machine learning (ML) tools to classify the medical imaging modalities. However, it is poorly understood how these algorithms discriminate the modalities and if there are implicit opportunities for improving visual information access applications in computational biomedicine. In this study, we visualize the learned weights and salient network activations in a CNN based Deep Learning (DL) model to determine the image characteristics that lend themselves for improved classification with a goal of developing informed clinical question-answering systems. To support our analysis we cross-validate model performance to reduce bias and generalization errors and perform statistical analyses to assess performance differences.

Keywords: Image modality · Classification · Visualization · Saliency ·
Deep learning · Machine learning

1 Introduction

Medical images serve as a vital source of information for disease screening/diagnosis and an indispensable source of knowledge in clinical decision making/research [1]. The size of medical image repositories has been growing exponentially with the widespread use of digital imaging in clinics and hospitals. These voluminous repositories provide opportunities for researchers to extract meaningful information and develop computerized tools for screening/diagnosis. Medical images have also become an imperative part of the content in several biomedical publications [2–6]. This makes automated medical image classification and retrieval significant in a clinical decision support system, as well as for educational purposes. The images are from diverse medical imaging modalities such as common radiological imagery (e.g., X-rays, Computed Tomography (CT), Magnetic Resonance Imaging (MRI), ultrasound, among others), other medical images that often captured in the visual image spectrum as "photographs" but have distinct clinical classes (e.g., retinal funduscopy, endoscopy,

© Springer Nature Singapore Pte Ltd. 2019
K. C. Santosh and R. S. Hegadi (Eds.): RTIP2R 2018, CCIS 1036, pp. 42–57, 2019.
https://doi.org/10.1007/978-981-13-9184-2_4

different kinds of microscopic images, surgical photography, photographs of medical devices and systems, among others), and also other graphical drawings and statistical charts of clinical importance found in biomedical journals.

Traditional image classification/retrieval methods tend to be text-based that index images based using descriptive metadata and manual annotations. While these are useful, the textual information may not be consistent with the visual content and modality classification becomes a concern. Also, text-based image classification doesn't consider the representation of image content. Images from different modalities have unique visual patterns, not contained in text labels. Thus, classification tools that arrive at decisions based on visual similarity are a meaningful complement and key for hybrid (text + image) information access systems. Conventionally, rule-based, hand-crafted features are extracted from the images toward visual recognition tasks [7]. However, the process is laborious, computationally expensive, demands iterative labeling and calibration. Hand-engineered features are specific to the modality and are often not transferable to other tasks resulting in suboptimal classification. Under these circumstances, data-driven machine learning (ML) approaches like Deep Learning (DL) becomes a handy tool [8–13].

To overcome challenges of devising hand-engineered features that capture variations in the underlying data, convolutional neural networks (CNN), a class of DL models are used in visual recognition tasks, to discover/learn hierarchical feature representations from raw input pixels [14]. CNNs learn these representations through their inherent mechanisms of receptive fields, shared weights and pooling [15]. These models learn to abstract lower-level features to construct higher-level features and learn complex, non-linear decision-making functions toward performing simultaneous feature extraction and classification. In 2012, the AlexNet model proposed by Krizhevsky [15] won the ImageNet Large Scale Visual Recognition Challenge (ILSVRC) [16] and brought the initial breakthrough in visual recognition tasks. This led to the use of several CNNs with varying architecture and depths including VGGNet [17], Inception [18], ResNet [19], Xception [20], and DenseNet [21]. At present, CNNs are delivering promising results in visual recognition tasks at levels exceeding human performance [22]. The encouraging performance of CNNs is accompanied by the availability of a large amount of annotated data. When annotated data are sparse, like in the case of medical images, transfer learning methods are preferred. In this approach, the CNNs are trained on a large selection of stock photographic images like ImageNet that contain more than 1.2 million annotated stock photographic images across 1000 categories [16]. The rich feature representations learned in the form of generic features from these large-scale datasets are transferred to the current task. The pre-trained model weights are fine-tuned in the deeper layers to learn the representations from the new task. The number of layers to fine-tune depends on the availability of annotated data. Literature studies have demonstrated that fine-tuning is a robust and effective strategy to adopt for a variety of classification/recognition tasks [23]. Fine-tuning would be a promising option for medical modality classification, the pre-trained weights could serve as a promising initialization for the new task [24]. The pre-trained CNNs are also used as feature extractors to extract the learned features from the current task. The extracted features are used to train a classifier to make predictions [25].

While state-of-the-art studies elucidate diverse methods for classification/recognition, none of them provide insights into the learned behavior of CNNs or rationalize their performance. It is equitable to declare that the current studies leave room for progress in visualizing and understanding CNN models. The unsettled issue of interpreting the learned behavior of CNNs toward classifying image modalities has become significant and is the focus of this work. In this study, we visualized the learned weights and salient network activations in an optimally trained CNN model applied to the task of medical modality classification. We evaluated the performance of state-of-the-art CNNs including VGG-16, ResNet-50, Xception, Inception-V3, and DenseNet-121, customized for the current task. We cross-validated the performance of the models to reduce bias/generalization errors and statistically validated for the presence/absence of significant differences in their performance. The following paper is organized as follows: Sect. 2 discusses the related work, Sect. 3 particularizes on the materials and methods, Sect. 4 discusses the results, and Sect. 5 concludes the work.

2 Related Work

Modality classification has become an imperative research task in recent years. Evaluation campaigns like Image Cross Language Image Retrieval (ImageCLEF) has been providing collections of annotated medical images for a variety of challenges including modality classification, compound figure separation, image captioning, and visual question and answering [26–30]. However, the collections are sparse in comparison to ImageNet and other large-scale data collections. Conventional methods for modality classification use rule-based, handcrafted feature descriptors toward representing the image characteristics. In [31], the authors used the ImageCLEF2011 modality classification dataset and a combination of SIFT and Bag-of-Colors based feature descriptors toward classifying the modalities. The authors demonstrated a classification accuracy of 72.5% and highlighted the importance of using color descriptors. In another study [32], the authors used the ImageCLEF2015 modality classification dataset and extracted SIFT and Bag-of-Words features to train a multi-class support vector machine (SVM) and obtained a classification accuracy of 60.91%. The authors in [33] used the same dataset, extracted 1st and 2nd order color features, and a manifold-based sparse classification method toward obtaining an accuracy of 73.24%. In another study [34], the authors participated in the ImageCLEF2016 sub-figure classification task, extracted features including contrast, gradient orientations, color, and local pattern distributions and trained a multi-class neural network classifier to obtain a classification accuracy of 72.46%. Literature reveals that the studies used rule-based features that suffer from limitations owing to lack of sufficient human experts to perform manual annotations, inter-/intra-observer variability, inter-class similarity and intra-class variances [35].

At present, DL is delivering promising results as compared to hand-engineered feature extraction in visual recognition tasks. The authors in [36] used multiple, customized CNNs to classify the ImageCLEF2013 medical modality collection with an accuracy of 74.90%. The authors in [37] used an ensemble of fine-tuned CNNs including AlexNet and GoogLeNet and obtained an accuracy of 82.48% in classifying the modality data collection from the ImageCLEF2016 evaluation challenge. The authors in [38] used a combination of pre-trained and customized CNNs toward classifying the sub-figure classification dataset from ImageCLEF2015 and Image-CLEF2016 evaluation challenges and obtained an accuracy of 76.87% and 87.37% respectively. In another study [39], the authors extracted features using a pre-trained ResNet and trained an SVM classifier to obtain an accuracy of 85.38% toward ImageCLEF2016 modality classification task. The authors in [40] participated in the ImageCLEF2016 subfigure classification challenge, employed a synergic signal method to merge the features of multiple fine-tuned ResNet models, and obtained an accuracy of 86.58%. While current literature explains promising methods for modality classification using CNNs, none of them interprets their learned behavior. The indispensable need for visualizing and interpreting the learned behavior of CNNs toward classifying image modalities is the motivation behind this study.

3 Materials and Methods

3.1 Data Collection and Preprocessing

In this study, an 11-class modality classification was performed, the modalities include Computed Tomography (CT), Magnetic Resonance Imaging (MRI), Positron Emission Tomography/CT fusion (PET/CT), ultrasound, 2-D radiography (X-ray), Scanning Electron Microscopy (SEM), Fluorescence Microscopy (FM), Light Microscopy (LM), retinal funduscopy, colonoscopy and statistical graphs. We pooled data from various resources including ImageCLEF2013 modality classification challenge [31], Open Access Biomedical Image Search Engine (OpenI®) and the Internet. For PET/CT, we collected additional images belonging to soft-tissue sarcoma [41], breast [42], head and neck cancer [43], non-small cell lung cancer [44] and prostate cancer [45], from the Cancer Imaging Archive [46]. For colonoscopy, we collected colonoscopy procedure videos from the Web and applied a frame grabbing algorithm to generate images. Additional retinal funduscopy images were pooled from the Messidor [47] dataset that includes 1200 images, acquired using color CCD cameras on a non-mydriatic retinograph. Sample images from different image modalities (Fig. 1). The distribution of data across the modalities is tabulated in Table 1. We evaluated the performance of state-of-the-art CNNs including VGG-16, ResNet-50, Xception, Inception-V3, and DenseNet-121 through five-fold cross-validation toward optimal model selection. As observed in Table 1, the distribution of data was imbalanced across the classes. To mitigate this issue, data was augmented by introducing class-specific perturbations in the training samples [15]. Data augmentation has been shown to improve the robustness, generalization ability of the predictive models, reduce bias and overfitting. The perturbations were reproducible that did not alter the image semantics but helped in generating new samples during model training. The data was augmented with rotations in the range (−3, 3), width/height shifts (−2, 2), horizontal and vertical flips. The categorical cross-entropic loss was modified by observing the distribution of class labels and producing weights to equally penalize the under/over-represented classes to assist the models to learn equally well across the classes [48]. Images were resampled to 300×300 pixel resolutions and mean-normalized to assist the models in faster convergence. The models were trained and tested on an NVIDIA DGX-1 system having Tesla V-100 GPUs with computational tools including Python® 3.6.3, Keras® 2.1.2 with Tensorflow® 1.4.0 backend, and CUDA 8.0/cuDNN 5.1 dependencies for GPU acceleration.

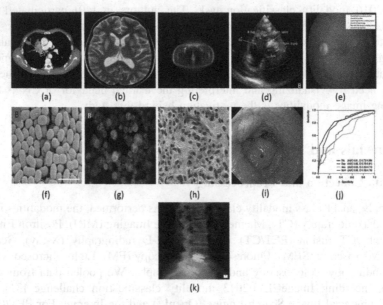

Fig. 1. Sample images from different modalities: (a) CT, (b) MRI, (c) PET/CT, (d) ultrasound, (e) retinal funduscopy, (f) SEM, (g) FM, (h) LM, (i) colonoscopy, (j) statistical graphs, (k) X-ray.

Table 1. Distribution of data across the image modalities.

Modality	#Images
CT	17,055
MRI	12,720
PET/CT	5,510
X-ray	20,030
Ultrasound	6,230
Colonoscopy	14,000
SEM	2,200
FM	5,000
LM	3,900
Retinal funduscopy	2,650
Statistical graphs	2,190

3.2 Model Configuration

The performance of CNNs including VGG-16, ResNet-50, Xception, Inception-V3, and DenseNet-121 was evaluated in this study. The models were initialized with pre-trained ImageNet weights and truncated at their deepest convolutional layer. A convolutional layer with 1024 filters of 3×3 dimensions, followed by a global average pooling (GAP) and Softmax layer was added to the truncated models as shown in Fig. 2. The pre-trained weights were fine-tuned using small weight updates to learn the representations of the image modalities, as established in the following steps: (a) instantiating the convolutional base of the pre-trained models and loading their weights; (b) truncating the models at the deepest convolutional layer; (c) adding the convolutional, GAP layer and top-level classifier; and (d) fine-tuning the models alongside the newly added layers to learn the representations for the current task. The models were optimized for hyper-parameters by a randomized grid search method [49]. Search ranges including [1e−3 10e−2], [0.8 0.95] and [1e−10 10e−2] were used for the learning rate, stochastic gradient descent (SGD) momentum and L2-regularization respectively. A mini-batch size of 10 was used, 9148 iterations were performed per epoch. The performance of the predictive models were evaluated through five-fold cross-validation with the following performance metrics: (a) accuracy, (b) area under receiver operating characteristic (ROC) curve, (c) sensitivity, (d) specificity, (e) F1-score, and (f) Matthews Correlation Coefficient (MCC) [50].

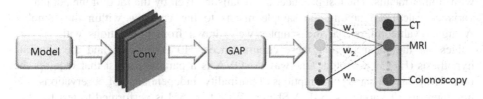

Fig. 2. Model architecture.

3.3 Visualization Studies

DL models are perceived as black boxes since the learned representations are difficult to extract and present in a human-readable form. However, the representation learned by CNNs are highly amenable to visualization because they represent visual concepts. The convolutional layers learn template matching filters whose output gets maximized on observing similar patterns in the input samples [51]. The layers at the beginning of the model are simple to interpret by visualizing the weights as an image. Filters in the deeper layers abstract the outputs from the earlier layers, making interpretation hard. Several methods have been developed for visualizing and interpreting the learned representations of the CNN models including visualizing the learned weights and intermediate layer activations. Visualizing the weights helps in interpreting the visual patterns/concepts learned by the filters in a given layer. The process helps to identify the robustness and generalization ability of the trained model by generating input images that maximize the output of the given layer corresponding to the expected class.

This is accomplished by performing gradient ascent in the input space by applying gradient descent to the value of the input image pixels in order to maximize the response of a specific filter. A loss function is constructed to maximize the value of the filter in a given convolution layer and is minimized during iterations of gradient descent. Visualizing intermediate layer activations helps in understanding successive data transformations and gaining an intuition of the learned patterns. Given an input, activation visualization helps to display the feature maps that are output by the convolutional and pooling layers toward interpreting the relatively independent features learned by the filters. The learned feature maps are visualized by independently plotting the filter contents as a 2D image.

3.4 Statistical Analyses

Statistical analyses help to validate the performance of the predictive models by describing the nature of the data, identifying the trends, and summarizing their relationships. Skewness and kurtosis give a measure of data variability [52]. Skewness is a measure of symmetry. Kurtosis is a measure of whether the samples are heavy-tailed/light-tailed relative to the normal distribution. Skewness and kurtosis measures for a normal distribution should be as close to zero as possible. Statistical tests like one-way analysis of variance (ANOVA) help in identifying the statistically significant differences between the means of two or more unrelated groups [53]. The null hypothesis (H0) infers the samples in the unrelated groups are drawn from populations with similar means. The test produces an F-statistic given by the ratio of the population variance calculated among the sample means to the variance within the samples. A higher value infers that the samples were drawn from populations with varying values for the mean. Under these circumstances, H0 is rejected and the alternate hypothesis (H1) is accepted. One-way ANOVA is a parametric test that requires the underlying data to satisfy assumptions of normality, independence of observations, and homogeneity of variances [54]. A Shapiro-Wilk test [55] is performed to test for data normality and Levene's test [56] to observe the homogeneity of variances. One-way ANOVA is an omnibus test that doesn't reveal where exactly the statistically significant difference exists. A post-hoc analysis like Tukey post-hoc is performed to identify the models that demonstrate statistically significant differences in the mean values for the different performance metrics under study [57].

4   Results and Discussion

4.1 Performance Metrics Evaluation

Training was stopped after 30 epochs ($\approx$274K iterations) when the validation accuracy ceased to improve. From the randomized grid search, the optimal values for the learning rate, SGD momentum, and L2 regularization were found to be 1e−4, 0.9 and 1e−6 respectively. The CNNs converged to an optimal solution due to hyper-parameter optimization and regularization. It was observed from the cross-validation studies that

VGG-16 gave promising results for accuracy, sensitivity, precision, and F1-score compared to other models, as shown in Table 2.

4.2   Statistical Analyses Interpretation

While performing statistical analyses, it was observed that the skewness and kurtosis measures were close to 0 to signify that the values of the performance metrics were a little skewed and kurtotic but did not significantly differ from normality. It was also observed that the results of Levene's test were not statistically significant ($p > 0.05$) to signify that the homogeneity of variances was not violated. Shapiro-Wilk test ($p > 0.05$) and a visual inspection of the histograms and box plots showed that the values for the different performance metrics were approximately normally distributed. Hence, the parametric one-way ANOVA analysis was performed to observe for the significant differences in the performance metrics for the models under study. The values for the statistical measures and the consolidated results of one-way ANOVA are shown in Table 3. It was observed that, in terms of accuracy, no statistically significant difference in performance existed between the different models ($F_{(4, 20)} = 0.788$, $p = .547$). Similar results were observed for AUC ($F_{(4, 20)} = 2.221$, $p = .103$), sensitivity ($F_{(4, 20)} = 0.814$, $p = .531$), precision ($F_{(4, 20)} = 0.777$, $p = .553$), F1-score ($F_{(4, 20)} = 0.805$, $p = .537$) and MCC ($F_{(4, 20)} = 0.692$, $p = .606$). However, in terms of the minimum and maximum values, VGG-16 outperformed the other CNNs except for AUC and MCC where ResNet-50 and DenseNet-121 demonstrated higher values respectively. Considering the balance between sensitivity and precision as demonstrated by F1-score, VGG-16 delivered promising results than the other models under study.

Table 2. Performance metrics computed for the predictive CNN models.

Models	Accuracy	AUC	Sensitivity	Precision	F1-score	MCC
VGG-16	0.990 ± 0.003	0.998 ± 0.001	0.990 ± 0.01	0.990 ± 0.007	0.990 ± 0.008	0.985 ± 0.008
ResNet-50	0.986 ± 0.009	1.0 ± 0.001	0.986 ± 0.009	0.987 ± 0.009	0.986 ± 0.009	0.995 ± 0.009
Xception	0.984 ± 0.01	1.0 ± 0.001	0.984 ± 0.01	0.985 ± 0.01	0.984 ± 0.01	0.988 ± 0.011
Inception-V3	0.988 ± 0.008	0.999 ± 0.002	0.989 ± 0.008	0.988 ± 0.008	0.988 ± 0.008	0.990 ± 0.007
DenseNet-121	0.980 ± 0.013	0.994 ± 0.009	0.980 ± 0.013	0.981 ± 0.012	0.980 ± 0.013	0.986 ± 0.015

4.3   Visualizing Convolutional and Dense Layers

Visualization studies were performed with the optimally trained VGG-16 model for the current task. Inputs that maximized the filter activations in different layers of the model were visualized. It was observed that each layer learned a collection of filters that got increasingly refined and complex in the deeper layers. As observed in Fig. 3, the filters in the first convolutional layer (block1-conv1) encoded simple directional edges and colors/colored edges. The filters in the deeper layers (Fig. 4) abstracted the features from the earlier layers to form complex patterns.

Table 3. Summary of statistical measures.

Performance metrics	Models	Skewness		Kurtosis		Shapiro-Wilk	One-way ANOVA
Accuracy	VGG-16	−0.398	0.913	−1.052	2.0	0.478	$F(4,20) = 0.788$, $p = .547$
	ResNet-50	−0.782	0.913	0.094	2.0	0.814	
	Xception	−0.734	0.913	−0.378	2.0	0.772	
	Inception-V3	−0.439	0.913	−0.743	2.0	0.735	
	DenseNet-121	−0.778	0.913	−1.271	2.0	0.403	
AUC	VGG-16	−1.540	0.913	2.581	2.0	0.113	$F(4,20) = 2.221$, $p = .103$
	ResNet-50	−0.512	0.913	−2.963	2.0	0.167	
	Xception	−0.588	0.913	−2.898	2.0	0.111	
	Inception-V3	−0.932	0.913	−1.290	2.0	0.148	
	DenseNet-121	−1.586	0.913	2.520	2.0	0.145	
Sensitivity	VGG-16	−0.398	0.913	−1.052	2.0	0.478	$F(4,20) = 0.814$, $p = .531$
	ResNet-50	−0.818	0.913	0.268	2.0	0.806	
	Xception	−0.725	0.913	−0.382	2.0	0.782	
	Inception-V3	−0.313	0.913	−1.077	2.0	0.747	
	DenseNet-121	−0.775	0.913	−1.264	2.0	0.413	
Precision	VGG-16	−0.418	0.913	−1.077	2.0	0.522	$F(4,20) = 0.777$, $p = .553$
	ResNet-50	−0.698	0.913	−0.204	2.0	0.844	
	Xception	−0.654	0.913	−0.562	2.0	0.815	
	Inception-V3	−0.506	0.913	−0.559	2.0	0.730	
	DenseNet-121	−0.673	0.913	−1.741	2.0	0.397	
F1-Score	VGG-16	−0.398	0.913	−1.066	2.0	0.483	$F(4,20) = 0.805$, $p = .537$
	ResNet-50	−0.762	0.913	0.064	2.0	0.829	
	Xception	−0.689	0.913	−0.468	2.0	0.799	
	Inception-V3	−0.410	0.913	−0.845	2.0	0.723	
	DenseNet-121	−0.765	0.913	−1.357	2.0	0.401	
MCC	VGG-16	−1.538	0.913	3.068	2.0	0.209	$F(4,20) = 0.692$, $p = .606$
	ResNet-50	1.214	0.913	2.302	2.0	0.457	
	Xception	−1.285	0.913	1.747	2.0	0.450	
	Inception-V3	0.440	0.913	1.422	2.0	0.747	
	DenseNet-121	−1.697	0.913	3.152	2.0	0.154	

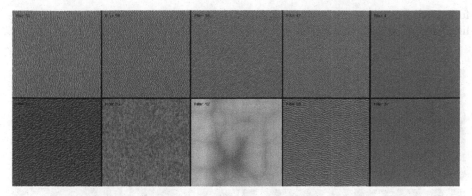

Fig. 3. Visualizing random filters in the first convolutional layer in the first convolutional block.

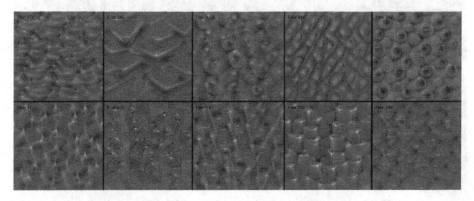

Fig. 4. Visualizing random filters in the third convolutional layer in the fifth convolutional block.

The filters functioned as a basis of vectors to compactly encode the inputs that became more intricate as they begun to incorporate information from an increasingly larger spatial extent, converging to specific patterns in the image modalities under study. The final dense layer was visualized as shown in Fig. 5. Filters 0 to 10 represent CT, SEM, colonoscopy, FM, retinal funduscopy, statistical graphs, LM, MRI, PET/CT, ultrasound, and X-ray respectively. It was observed that for CT, the filters maximally responded to the CT-like contour patterns in the input. For SEM and LM, higher activations were observed for blob-like cell patterns. For colonoscopy, the input patterns that mimicked tissue color and texture maximally activated the filters. For FM, the patterns exhibiting green fluorescence captured by the cells produced higher activations. Axes like patterns maximally activated the filters for the expected statistical graphs class. For MRI, the input patterns simulating the neo-cortical convolutions led to high values of activations. For PET/CT fusion, the input patterns mimicked scan contours. For ultrasound, the patterns mimicked the tissue texture and shape of image formation. For the X-ray class, the filters were maximally activated for rib and bone-like patterns in the input image.

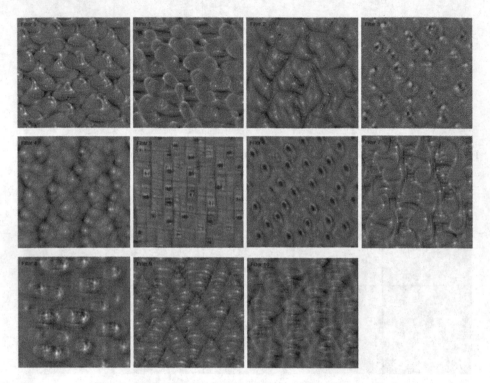

Fig. 5. Visualizing the final dense layer. (Color figure online)

The trained model's notion of the expected classes were found to be at the level of local textures to demonstrate that they do not understand the concept of the classes analogous to humans. The model learned the decomposition of the visual space as a hierarchical-modular network of filters, a probabilistic mapping between filter combinations and a set of class labels. However, the nature of human vision is not purely convolutional. The human visual cortex is complex, active, structured into cortical columns with multifaceted functionality, and involves motor control [58].

4.4 Visualizing Intermediate Activations

The activations of the intermediate layers of the trained VGG-16 model was visualized to gain further insight into its performance and learned behavior. Given an input image of CT and MRI for instance, random filters in the feature maps were extracted and plotted to visualize the activations as shown in Figs. 6 and 7. It was observed that the filters in the earlier layers acted as a collection of various edge detectors. The activations retained almost all of the information present in the original image. At the deeper layers, the activations became increasingly abstract and less visually interpretable. Higher representations carried increasingly less information about the visual contents, and increasingly more information related to the expected class.

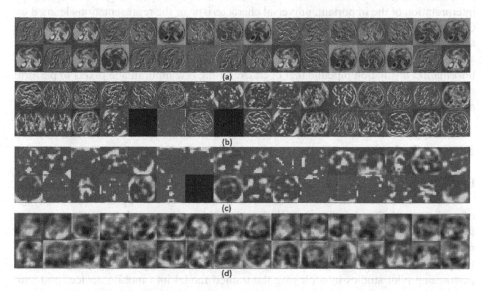

Fig. 6. Visualizing intermediate activations for an input CT image: (a) block1-conv1, (b) block3-conv1, (c) block5-conv3, and (d) deepest convolutional layer.

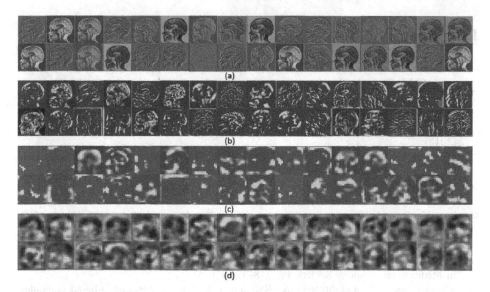

Fig. 7. Visualizing intermediate activations for an input MRI image: (a) block1-conv1, (b) block3-conv1, (c) block5-conv3, and (d) deepest convolutional layer.

The sparsity of activations increased with the depth of the layer: in the first convolutional layer of the first convolutional block, almost all channels were activated by the input image; in the deeper layers, many of the channels were blank that inferred that the pattern encoded by these filters was not found in the input image. This led to the interpretation of the important, universal characteristic of the representations learned by CNNs; the features became increasingly abstract with the depth of the layer. The activations of deeper layers carried subtle information about the specific input being seen, and more information about the imaging modality. The CNN model acted as an information distillation pipeline that operated on the raw input image pixels to perform repeated transformations so that irrelevant information was filtered and useful information pertaining to the modality of the image was magnified and retained.

5 Conclusion

In this study, we visualized the learned weights and salient network activations in a trained CNN model toward understanding its learned behavior, applied to the challenge of medical modality classification. We interpreted how the model distilled information through transformations and retained characteristic features pertaining to the imaging modality toward decision-making. We also statistically validated the performance of the predictive CNNs for optimal model selection and deployment. We are currently performing pilot studies in deploying the trained model into mobile devices and analyzing performance. We currently worked with images containing a single modality. Often, figures contain multiple panels describing different modalities. In the future, we plan to work with multi-panel images, perform panel segmentation so that each sub-panel contains a single modality and improve the classification performance.

References

1. Ben Abacha, A., Gayen, S., Lau, J.J., Rajaraman, S., Demner-Fushman, D.: NLM at ImageCLEF 2018 visual question answering in the medical domain. In: CEUR Workshop Proceedings, p. 2125 (2018)
2. Demner-Fushman, D., Antani, S., Thoma, G.R., Simpson, M.: Design and development of a multimodal biomedical information retrieval system. J. Comput. Sci. Eng. **6**, 168–177 (2012)
3. Rajaraman, S., Candemir, S., Kim, I., Thoma, G.R., Antani, S.: Visualization and interpretation of convolutional neural network predictions in detecting pneumonia in pediatric chest radiographs. MDPI Appl. Sci. **8**(10), 1715 (2018)
4. Rajaraman, S., et al.: Understanding the learned behavior of customized convolutional neural networks toward malaria parasite detection in thin blood smear images. J. Med. Imag. **5**(3), 034501 (2018)
5. Rajaraman, S., et al.: A novel stacked generalization of models for improved TB detection in chest radiographs. In: Proceedings of the International Conference of the IEEE Engineering in Medicine and Biology Society, pp. 718–721 (2018)
6. Thamizhvani, T.R., Lakshmanan, S., Rajaraman, S.: Mobile application-based computer-aided diagnosis of skin tumours from dermal images. Imaging Sci. J. **66**(6), 382–391 (2018)

7. Khan, S., Yong, S.P.: A comparison of deep learning and hand crafted features in medical image modality classification. In: Proceedings of the International Conference on Computer and Information Sciences, pp. 633–638 (2016)
8. LeCun, Y., Bengio, Y., Hinton, G.: Deep learning. Nature **521**, 436–444 (2015)
9. Rajaraman, S., et al.: Pre-trained convolutional neural networks as feature extractors toward improved malaria parasite detection in thin blood smear images. PeerJ **6**, e4568 (2018)
10. Rajaraman, S., et al.: Comparing deep learning models for population screening using chest radiography. In: Proceedings of the SPIE Medical Imaging: Computer-aided Diagnosis, p. 105751E (2018)
11. Rajaraman, S., Antani, S., Xue, Z., Candemir, S., Jaeger, S., Thoma, G.R.: Visualizing abnormalities in chest radiographs through salient network activations in deep learning. In: Proceedings of the IEEE Life Sciences Conference, pp. 71–74 (2017)
12. Rajaraman, S., Antani, S., Jaeger, S.: Visualizing deep learning activations for improved malaria cell classification. Proc. Mach. Learn. Res. **69**, 40–47 (2017)
13. Xue, Z., Rajaraman, S., Long, L.R., Antani, S., Thoma, G.R.: Gender detection from spine x-ray images using deep learning. In: Proceedings of the IEEE International Symposium on Computer-based Medical Systems, pp. 54–58 (2018)
14. Simard, P., Steinkraus, D., Platt, J.C.: Best practices for convolutional neural networks applied to visual document analysis. In: Proceedings of the 7th International Conference on Document Analysis and Recognition, pp. 958–963 (2003)
15. Krizhevsky, A., Sutskever, I., Hinton, G.E.: ImageNet classification with deep convolutional neural networks. In: Proceedings of the Advances in Neural Information Processing Systems, pp. 1–9 (2012)
16. Deng, J., Dong, W., Socher, R., Li, L.J., Li, K., Li, F.F.: ImageNet: a large-scale hierarchical image database. In: Proceedings of the IEEE Conference on Computer Vision and Pattern Recognition, pp. 248–255 (2009)
17. Simonyan, K., Zisserman, A.: Very deep convolutional networks for large-scale image recognition. arXiv preprint arXiv:1409.1556 (2015)
18. Szegedy, C., Vanhoucke, V., Ioffe, S., Shlens, J., Wojna, Z.: Rethinking the inception architecture for computer vision. In: Proceedings of the IEEE Conference on Computer Vision and Pattern Recognition, pp. 2818–2826 (2016)
19. He, K., Zhang, X., Ren, S., Sun, J.: Deep residual learning for image recognition. In: Proceedings of the IEEE Conference on Computer Vision and Pattern Recognition, pp. 770–778 (2016)
20. Chollet, F.: Xception: Deep Learning with Separable Convolutions. arXiv preprint arXiv: 1610.02357 (2016)
21. Huang, G., Liu, Z., Weinberger, K.Q., van der Maaten, L.: Densely Connected Convolutional Networks. arXiv preprint arXiv:1608.06993 (2017)
22. Mnih, V., et al.: Human-level control through deep reinforcement learning. Nature **518**, 529–533 (2015)
23. Margeta, J., Criminisi, A., Lozoya, R.C., Lee, D.C., Ayache, N.: Fine-tuned convolutional neural nets for cardiac MRI acquisition plane recognition. Comput. Methods Biomech. Biomed. Eng. Imaging Vis. **5**, 339–349 (2017)
24. Lynch, S., Ng, A.: Why AI is the new electricity. https://news.stanford.edu/thedish/2017/03/14/andrew-ng-why-ai-is-the-new-electricity/
25. Razavian, A.S., Azizpour, H., Sullivan, J., Carlsson, S.: CNN features off-the-shelf: an astounding baseline for recognition. In: IEEE Conference on Computer Vision and Pattern Recognition, pp. 512–519 (2014)
26. De Herrera, A., Schaer, R., Bromuri, S., Müller, H.: Overview of the ImageCLEF 2016 medical task. In: CEUR Workshop Proceedings, p. 1609 (2016)

27. Apostolova, E., You, D., Xue, Z., Antani, S., Demner-Fushman, D., Thoma, G.R.: Image retrieval from scientific publications: Text and image content processing to separate multipanel figures. J. Am. Soc. Inf. Sci. Tec. **64**, 893–908 (2013)
28. Santosh, K.C., Aafaque, A., Antani, S., Thoma, G.R.: Line segment-based stitched multipanel figure separation for effective biomedical CBIR. Int. J. Pattern Recogn. Artif. Intell. **31**(6), 1757003 (2017)
29. Santosh, K.C., Xue, Z., Antani, S., Thoma, G.R.: NLM at ImageCLEF 2015: biomedical multipanel figure separation. In: CEUR Workshop Proceedings, p. 1391 (2015)
30. Santosh, K.C., Antani, S., Thoma, G.R.: Stitched multipanel biomedical figure separation. In: IEEE International Symposium on Computer-based Medical Systems, pp. 54–59 (2009)
31. De Herrera, A., Markonis, D., Müller, H.: Bag–of–colors for biomedical document image classification. In: Greenspan, H., Müller, H., Syeda-Mahmood, T. (eds.) MCBR-CDS 2012. LNCS, vol. 7723, pp. 110–121. Springer, Heidelberg (2013). https://doi.org/10.1007/978-3-642-36678-9_11
32. Pelka, O., Friedrich, C.M.: FHDO biomedical computer science group at medical classification task of ImageCLEF 2015. In: CEUR Workshop Proceedings, p. 1391 (2015)
33. Cirujeda, P., Binefa, X.: Medical image classification via 2D color feature based covariance descriptors. In: CEUR Workshop Proceedings, p. 1391 (2015)
34. Li, P., et al.: UDEL CIS at ImageCLEF medical task 2016. In: CEUR Workshop Proceedings, p. 1609 (2016)
35. De Herrera, A., Kalpathy-Cramer, J., Fushman, D.D., Antani, S., Müller, H.: Overview of the imageCLEF 2013 medical tasks. In: CEUR Workshop Proceedings, p. 1179 (2013)
36. Yu, Y., et al.: Modality classification for medical images using multiple deep convolutional neural networks. J. Comput. Inf. Syst. **11**(15), 5403–5413 (2015)
37. Kumar, A., Kim, J., Lyndon, D., Fulham, M., Feng, D.: An ensemble of fine-tuned convolutional neural networks for medical image classification. IEEE J. Biomed. Heal. Inf. **21**, 31–40 (2017)
38. Yu, Y., Lin, H., Meng, J., Wei, X., Guo, H., Zhao, Z.: Deep transfer learning for modality classification of medical images. MDPI Inf. **8**(3), 91 (2017)
39. Koitka, S., Friedrich, C.M.: Traditional feature engineering and deep learning approaches at medical classification task of ImageCLEF 2016. In: CEUR Workshop Proceedings, p. 1609 (2016)
40. Zhang, J., Xia, Y., Wu, Q., Xie, Y.: Classification of Medical Images and Illustrations in the Biomedical Literature Using Synergic Deep Learning. arXiv preprint arXiv:1706.09092 (2017)
41. Vallières, M., Freeman, C.R., Skamene, S.R., El Naqa, I.: A radiomics model from joint FDG-PET and MRI texture features for the prediction of lung metastases in soft-tissue sarcomas of the extremities. Phys. Med. Biol. **60**, 5471–5496 (2015)
42. Bloch, B., Jain, A., Jaffe, C.: Data From BREAST-DIAGNOSIS. https://wiki.cancerimaging archive.net/display/Public/BREAST-DIAGNOSIS#9e4592af79b249bfaff992eceebbf842
43. Vallières, M., et al.: Radiomics strategies for risk assessment of tumour failure in head-and-neck cancer. Sci. Rep. **7**(1), 10117 (2017)
44. Gevaert, O., et al.: Non-small cell lung cancer: identifying prognostic imaging biomarkers by leveraging public gene expression microarray data-methods and preliminary results. Radiology **264**, 387–396 (2012)
45. Kurdziel, K.A., et al.: The kinetics and reproducibility of 18F-sodium fluoride for oncology using current pet camera technology. J. Nucl. Med. **53**, 1175–1184 (2012)
46. Clark, K., et al.: The cancer imaging archive (TCIA): maintaining and operating a public information repository. J. Digit. Imaging **26**, 1045–1057 (2013)

47. Decencière, E., et al.: Feedback on a publicly distributed image database: the messidor database. Image Anal. Stereol. **33**, 231–234 (2014)
48. Wang, X., Peng, Y., Lu, L., Lu, Z., Bagheri, M., Summers, R.: ChestX-ray8: hospital-scale chest X-ray database and benchmarks on weakly-supervised classification and localization of common thorax diseases. In: Proceedings of the IEEE Conference on Computer Vision and Pattern Recognition, pp. 1–19 (2017)
49. Bergstra, J., Bengio, Y.: Random search for hyper-parameter optimization. J. Mach. Learn. Res. **13**, 281–305 (2012)
50. Matthews, B.W.: Comparison of the predicted and observed secondary structure of T4 phage lysozyme. BBA - Protein Struct. **405**, 442–451 (1975)
51. Zeiler, M.D., Fergus, R.: Visualizing and understanding convolutional networks. In: Fleet, D., Pajdla, T., Schiele, B., Tuytelaars, T. (eds.) Computer Vision – ECCV 2014, LNCS, vol. 8689, pp. 818–833. Springer, Cham (2014). https://doi.org/10.1007/978-3-319-10590-1_53
52. Groeneveld, R.A., Meeden, G.: Measuring skewness and kurtosis. Statistician **33**, 391–399 (1984)
53. Rossi, J.S.: One-way ANOVA from summary statistics. Educ. Psychol. Meas. **47**, 37–38 (1987)
54. Daya, S.: One-way analysis of variance. Evid. Based Obstet. Gynecol. **5**, 153–155 (2003)
55. Shapiro, S.S., Wilk, M.B.: An analysis of variance test for normality (complete samples). Biometrika **52**, 591 (1965)
56. Gastwirth, J.L., Gel, Y.R., Miao, W.: The Impact of levene's test of equality of variances on statistical theory and practice. Stat. Sci. **24**, 343–360 (2009)
57. Kucuk, U., Eyuboglu, M., Kucuk, H.O., Degirmencioglu, G.: Importance of using proper post hoc test with ANOVA. Int. J. Cardiol. **209**, 346 (2016)
58. Bressler, S.L.: Large-scale cortical networks and cognition. Brain Res. Rev. **20**(3), 288–304 (1995)

Comparison of Deep Feature Classification and Fine Tuning for Breast Cancer Histopathology Image Classification

D. Sabari Nathan, R. Saravanan$^{(\boxtimes)}$, J. Anbazhagan,
and Praveen Koduganty

Cognizant Technology Solutions India Private Ltd., Chennai, India
Saravanan.Radhakrishnan@cognizant.com

Abstract. Convolutional Neural Networks (ConvNets) are increasingly being used for medical image diagnostic applications. In this paper, we compare two transfer learning approaches - **Deep Feature classification** and **Fine-tuning** ConvNets for Diagnosing Breast Cancer malignancy. BreaKHis dataset is used to benchmark our results with ResNet-50, InceptionV2 and DenseNet-169 pre-trained models. Deep feature classification accuracy ranges from 81% to 95% using Logistic Regression, LightGBM and Random Forest classifiers. Fine-tuned DenseNet-169 model accuracy outperformed all other classification models with 99.25 ± 0.4%.

Keywords: Breast cancer · Medical image diagnosis · Histopathology ·
Convolutional Neural Network · Transfer learning ·
Deep feature classification · Fine-tuning

1 Introduction

In 2015, Out of 2.4 million cases of Breast Cancer in the US, 523,000 deaths were reported. In the US, it is estimated that approximately 260,000 new cases of invasive breast cancer will be diagnosed in 2018 [1, 2], with about 40,920 women mortalities. Worldwide, Breast cancer claims the maximum mortality rates among all cancer diseases afflicting women.

Early screening and diagnosis can improve treatment and survival rates [47]. Initial screening is generally done by breast palpation and regular check-ups using mammography or ultrasound imaging, followed by detailed diagnosis with breast tissue biopsy and histopathology analysis and clinical screening. Hematoxylin and eosin (H&E) stained biopsy tissues are analyzed under the microscope for various parameters like nuclear atypia, tubules, and mitotic counts. Visual identification using H&E stained biopsies is non-trivial, tedious and can be exceedingly subjective, with average diagnostic concordance between pathologists approximately 75% [3]. Whole slide imaging (WSI) scanners are increasingly being used for digitizing histopathology slides enabling automated image processing and machine-learned methods for image

© Springer Nature Singapore Pte Ltd. 2019
K. C. Santosh and R. S. Hegadi (Eds.): RTIP2R 2018, CCIS 1036, pp. 58–68, 2019.
https://doi.org/10.1007/978-981-13-9184-2_5

enhancement, normalization, localization of the tissue, segmentation, quantitative analysis, detection, and diagnosis.

Convolution neural networks [4–8, 49] are the de-facto choice for researchers in this field and have outperformed conventional machine learning algorithms in many other medical image applications [9–12] including diabetic retinopathy, bone diseases detection [44], bone fracture detection [45, 46], pneumonia detection, etc. Deep networks require large training data to generalize though and publicly available annotated breast cancer datasets are small, thereby needing special methods to be viable. Data augmentation techniques like flipping, rotation, patching etc. and transfer learning approaches are promising. Conventional machine learning with handcrafted [13–17] features for Medical Imaging diagnosis doesn't generalize in the real world due to variations in tissue preparation, staining and slide digitization which has a significant impact on the tissue/image appearance. Pre-trained deep networks [18] have been used as a feature extractor in many real world applications for Diabetic Retinopathy [19], Handwritten digits recognition [20, 21], image retrieval [22, 23], Remote sensing [24, 42], Mammography breast cancer image classification [25, 26].

In ICIAR 2018 [27] Grand Challenge, 400 microscopy and whole-slide images from the BreAst Cancer Histology (BACH) extended dataset were classified into normal, benign, in-situ carcinoma and invasive carcinoma. Rakhlin et al. [28] report deep feature classification with multiple pre-trained deep networks, with the best accuracy of 93.8% on this dataset. Also, Rakhlin et al. [28] report that deep feature classification outperforms fine-tuning approach on ICIAR 2018 Grand Challenge dataset.

Habibzadeh et al. [29] use fine-tuning on pre-trained Inception (V1, V2, V3, and V4) and ResNet (V1 50, V1 101, and V1 152) to classify H&E stained microscopy images from BreaKHis dataset as benign or malignant. Their best-reported result for classifying into benign and malignant is from ResNet V1 101 with fine-tuning all layers with 98.4% confidence. Despite a lot of studies available on transfer learning and fine-tuning ConvNets [30], and to the best of our knowledge, we find no literature evaluating or comparing these two approaches, pre-trained deep feature classification and fine-tuning ConvNets on the same Breast Cancer dataset. In this paper, we evaluate these two approaches using BreaKHis dataset [31].

2 Dataset

The dataset we have used is the Breast Cancer Histopathological Database (BreaKHis) [31] which has 7,909 microscopic images of breast biopsy images collected from 82 patients across multiple magnifying factors (40x, 100x, 200x, and 400x). This dataset has two classes, 2480 benign and 5429 malignant images. Height and width of each image are 700×460 pixels, 3-channel RGB, 8-bit depth in each channel, PNG format. This dataset was provided to us by Fabio et al. [31] from the P&D Laboratory, Parana, Brazil (Table 1).

Table 1. Image distribution by magnification factor and class [31]

Magnification	Benign	Malignant	Total
40x	652	1,370	1,995
100x	644	1,437	2,081
200x	623	1,390	2,013
400x	588	1,232	1,820
Total	2,480	5,429	7,909

3   Data Augmentation and Pre-preprocessing

Data augmentation is an important step to create diverse, supplemented training dataset from small datasets to train deep networks. The training images are augmented by flipping the images along their horizontal and vertical axes and also rotating them by 90, 180, 270°. In the pre-processing step, the Mean image is calculated by averaging the images and the mean image is subtracted from all train and test images for brightness normalization. After mean subtraction, all the images are resized to (224 × 224 × 3), recommended image size for InceptionV2, ResNet-50, and DensNet-169 architectures.

4   Methods

4.1   Deep Feature Extraction and Classification

We used Pre-trained deep networks trained on ImageNet [32] – a dataset for object recognition for 1000 object classes and trained on 1.2 Million images. These pre-trained ConvNet models are used as generalized feature extractors since the top layers extract discriminant features like edges, textures etc. By removing the last fully connected output layer from the pre-trained deep network and extract feature vectors called Deep Features from the truncated network. The similar approach was used in [48]. These Deep Features are then used as input to standard classifiers like Random Forest, Logistic Regression etc., this is known as Deep Feature Extraction and Classification.

We use standard pre-trained DenseNet-169 [33], ResNet-50 [34] and InceptionV2 [35] networks from Keras distribution [36] trained on ImageNet. These pre-trained networks are used as fixed deep feature extractors for the breast cancer dataset by removing the last fully-connected (bottleneck features) and softmax classifier layers.

The extracted deep feature vectors (CNN codes) - InceptionV2 (1 × 38400), ResNet-50 (1 × 2048), DenseNet-169(1 × 94080) are then classified by traditional machine learning classifiers. We split the dataset 70% for training, 30% for testing. We build three different machine learning model to classify the deep features using Logistic Regression [37], LightGBM [38] and Random [39] Forest. The models were trained on NVIDIA Quadro K630 GPU [43] (Fig. 1 and Table 2).

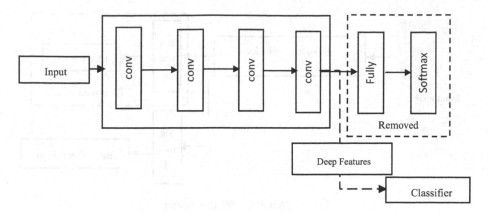

Fig. 1. Deep feature classification flow chart

Table 2. Classifier with model parameters

Classifier	Model parameters
Logistic regression	Multi class = one-vs-rest, L2 normalization and Lib-Linear optimizer, Regularization Parameters = 0.95
LightGBM	Boosting = Gradient Boosting Decision Tree, Learning Rate = 0.07, bagging fraction = 0.95, feature fraction = 0.7, Metric = Binary Log Loss, number of boosting iterations = 2000
Random forest	Criterion = Gini impurity measure, maximum depth = 2, minimum samples split = 2, minimum samples leaf = 1, maximum features = sqrt (no.features)

4.2 DenseNet-169 Fine Tuning

Fine-tuning is another promising transfer learning technique for medical image classification, Habibzadeh et al. [29] report fine-tuned ResNet classification accuracy of 98.7% and Spanhol et al. report 90.0% accuracy using AlexNet fine-tuning. A continuation of these techniques, we select DenseNet-169 [33] to fine-tune, since they are easier and faster to train with no loss of accuracy due to improved gradient flow as compared to other networks [40, 41]. We took DenseNet-169 pre-trained on ImageNet, freeze the top layers because they capture universal features, remove the last softmax layer and replace it with an output sigmoid layer (binary classification). We fine-tune the last layer with small learning rate on cancer images as shown in Fig. 2. The dataset is divided into three parts, training (60%), validation (20%) and testing (20%). In the training phase, the data augmentation is applied to increase the training images. We use Stochastic Gradient Descent (SGD) optimizer with - learning rate = 0.0005, decay = $1e^{-6}$ and Momentum = 0.9. Each epoch operates on a batch of 16 images that are randomly sampled from the training set and the network is trained for 12 epochs. The models are trained on NVIDIA Quadro K630 GPU [43].

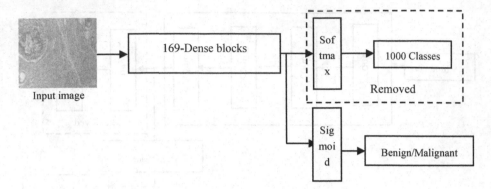

Fig. 2. DenseNet-169 fine-tuning

5 Results

We report standard classification metrics including classification accuracy, F1 score, Sensitivity (SN) & Specificity (SP). Sensitivity (SN) also called True Positive Rate, measures the proportion of actual positives (malignant) that are correctly identified as such, and represents the model's ability to not overlook actual positives (malignant) (Tables 3 and 4).

Table 3. Accuracy of deep features classification in percentage

Model	Magnification	Logistic regression	Light GBM	Random forest
ResNet50	40x	94.01	93.24	88.10
	100x	93.34	92.85	73.17
	200x	95.04	94.34	87.64
	400x	94.96	93.24	88.46
DenseNet-169	40x	94.12	92.54	86.36
	100x	92.43	92.04	88.23
	200x	92.12	93.09	90.90
	400x	91.97	91.58	89.10
Inception V2	40x	91.05	90.71	82.19
	100x	88.93	87.51	81.43
	200x	88.76	88.91	82.23
	400x	87.42	86.01	81.11

Table 4. Accuracy of DenseNet-169 fine-tuned model

Magnification	Accuracy in %
40x	99.60
100x	99.81
200x	99.40
400x	98.68

Specificity (SP) also called the True Negative Rate, on the other hand, measures the proportion of actual negatives (benign) that are correctly identified as such, and represents the model's ability to not overlook actual negatives (benign). ResNet-50 with Logistic Regression classifier consistently outperforms other deep feature classification models across all magnification factors. Higher magnification factors perform poorly for deep feature classification method. Fine-tuned DenseNet-169 with last layer tuning demonstrated the best accuracy among all models with **99.25 ± 0.4%** (Figs. 3, 4 and Tables 5, 6).

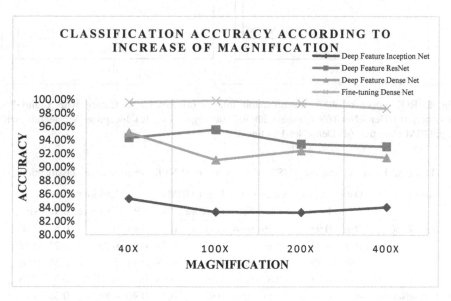

Fig. 3. Overall classification accuracies for deep feature classification and fine-tuned DenseNet-169. In Figure, we select logistic regression accuracies of all deep features to compare with the DenseNet-169 fine-tuned model result, as it outperforms other classifiers. We visualize fine-tuning approach performs better compare to Deep feature classifiers for all the magnification.

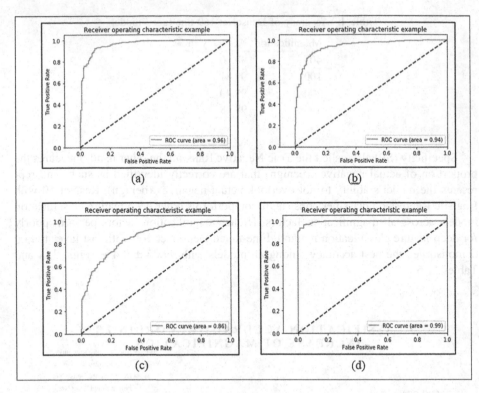

Fig. 4. ROC curve of 400x magnification image, (a) ResNet-50 feature with LightGBM classifier, (b) DenseNet-169 feature with logistic regression, (c) Inception V2 feature with LightGBM classifier, (d) DenseNet-169 fine-tuning

Table 5. F1 score, specificity (SP) and sensitivity (SN) for deep feature classification

Model	Data	Logistic regression			Light GBM			Random forest		
		F1 score	SN	SP	f1 score	SN	SP	F1 score	SN	SP
ResNet-50	40x	0.94	0.89	0.94	0.93	0.85	0.93	0.88	0.5	0.96
	100x	0.93	0.89	0.96	0.92	0.78	0.95	0.73	0.21	0.97
	200x	0.95	0.91	0.95	0.94	0.88	0.94	0.87	0.38	0.98
	400x	0.94	0.89	0.94	0.93	0.86	0.93	0.88	0.5	0.96
DenseNet 169	40x	0.94	0.81	0.96	0.92	0.72	0.96	0.86	0.36	0.97
	100x	0.92	0.78	0.94	0.92	0.76	0.95	0.88	0.50	0.97
	200x	0.92	0.94	0.77	0.93	0.96	0.75	0.90	0.67	0.93
	400x	0.91	0.83	0.90	0.91	0.77	0.93	0.89	0.68	0.93
Inceptionv2	40x	0.91	0.76	0.92	0.90	0.64	0.94	0.82	0.06	0.98
	100x	0.88	0.67	0.90	0.87	0.64	0.91	0.81	0.02	0.99
	200x	0.88	0.67	0.90	0.88	0.62	0.92	0.82	0.03	0.99
	400x	0.87	0.62	0.90	0.86	0.58	0.90	0.81	0.01	0.99

Table 6. F1 Score, specificity and sensitivity for DenseNet-169 fine-tuning

Magnification	F1 score	Sensitivity	Specificity
40x	0.9985358	0.99354	0.999987129
100x	0.992987377	0.99375	0.988826816
200x	0.913767	0.922580645	0.870317003
400x	0.961602671	0.97260274	0.938110749

6   Conclusion

In this paper, we benchmark two transfer learning approaches using popular pre-trained networks namely ResNet, Inception and DenseNet for Breast Cancer Benign/Malignant classification. Deep Features extracted from pre-trained ResNet-50 and logistic regression classifier performs better among all the deep network feature classification and the accuracy is 94 ± 1%. In another experiment, a continuation of the literature [29], fine-tuned the DenseNet-169 with strong data augmentation. The average accuracy of the DenseNet-169 fine-tuned model is 99.3% and it is an improvement of 3% to 5% higher than the deep network feature classification and shows better performance compared to other proposals in literature.

As per the study [28], Deep feature classification performs better when the dataset is small. Our experiment presents that Fine-tuning approach with strong augmentation techniques outperforms deep feature classification when the dataset size is moderate or large. The outcomes are expected to be more comprehensively evaluated in the future considering DenseNet-169 fine-tuned model will be used for semantic segmentation on whole-slide histopathology images.

References

1. Veta, M., Pluim, J., Van Deist, P.J., Viergever, M.A.: Breast cancer histopathology image analysis: a review. IEEE Trans. Biomed. Eng. **61**(5), 1400–1411 (2014). https://doi.org/10.1109/TBME.2014.2303852
2. U.S. Breast Cancer statistics. http://www.breastcancer.org/symptoms/understand_bc/statistics
3. Elmore, J.G., Longton, G.M., Carney, P.A., Geller, B.M., Onega, T., Tosteson, A.N.A., et al.: Diagnostic concordance among pathologists interpreting breast biopsy specimens. J. Am. Med. Assoc. **313**, 1122–1132 (2015). https://doi.org/10.1001/jama.2015.1405
4. Krizhevsky, A., Sutskever, I., Hinton, G.E.: ImageNet classification with deep convolutional neural networks. In: Proceedings of 26th Annual Conference on Neural Information Processing Systems (NIPS), December 2012, pp. 1106–1114 (2012)
5. Spanhol, F.A., Oliveira, L.S., et al.: Breast cancer histopathological image classification using convolutional neural networks. In: Proceedings of 2016 IEEE International Joint Conference on Neural Network, July 2016, pp. 2560–2567. IEEE (2016). https://doi.org/10.1109/ijcnn.2016.7727519

6. Abu Samah, A., Fauzi, M.F.A., Mansor, S.: Classification of benign and malignant tumors in histopathology images. In: Proceedings of 2017 IEEE International Conference on Signal and Image Processing Application, September 2017, pp. 43–48 (2017). https://doi.org/10.1109/icsipa.2017.8120587

7. Sun, J., Binder, A.: Comparison of deep learning architectures for H&E histopathology images. In: Proceedings of 2017 IEEE Conference on Big Data and Analytics, November 2017. https://doi.org/10.1109/icbdaa.2017.8284105

8. Bayramoglu, N., Kannala, J., Heikkila, J.: Deep learning for magnification independent breast cancer histopathology image classification. In: 23rd International Conference on Pattern Recognition, pp. 2440–2445 (2016). https://doi.org/10.1109/icpr.2016.7900002

9. Santosh, K.C., Vajda, S., Antani, S.: Edge map analysis in chest X-rays for automatic pulomary abnormality screening. JCARS **11**, 1637 (2016). https://doi.org/10.1007/s11548-016-1359-6

10. Dhandra, B.V., Hegadi, R.: Classification of abnormal endoscopic images using morphological watershed segmentation. In: International Conference on Cognition and Recognition, (ICCR-2005), Mysore, India, 22–23 December 2005, pp. 695–700 (2005). ISBN 81-7764-952-3

11. Dhandra, B.V., Hegadi, R.: Active contours without edges and curvature analysis for endoscopic image classification. Int. J. Comput. Sci. Secur. **1**(1), 19–32 (2007)

12. Hiremath, P.S., Dhandra, B.V., Humnabad, I., Hegadi, R., Rajput, G.G.: Detection of esophageal Cancer (Necrosis) in the Endoscopic images using color image segmentation. In: Second National Conference on Document Analysis and Recognition (NCDAR-2003), Mandya, India, pp. 417–422 (2003)

13. Nahid, A., Kong, Y.: Local and global feature utilization for breast image classification by convolutional neural network. In: 2017 International Conference on Digital Image Computing: Techniques and Applications, NSW, pp. 1–6 (2017). https://doi.org/10.1109/dicta.2017.8227460

14. Wan, S., Huang, X., Lee, H.C., Fujimoto, J.G., Zhou, C.: Spoke-LBP and ring-LBP: new texture features for tissue classification. In: 2015 IEEE 12th International Symposium on Biomedical Imaging (ISBI), pp. 195–199. IEEE (2015). https://doi.org/10.1109/isbi.2015.7163848

15. Vajda, S., Karargyris, A., Jaeger, S., et al.: Feature selection for automatic tuberculosis screening in frontal chest radiograph. J. Med. Syst. **42**, 146 (2018). https://doi.org/10.1007/s10916-018-0991-9

16. Santosh, K.C., Antani, S.: Automated chest x-ray screening: can lung region symmetry help detect pulmonary abnormalities. IEEE Trans. Med. Imaging **37**(5), 1168–1177 (2018). https://doi.org/10.1109/TMI.2017.2775636

17. Karargyris, A., Siegelman, J., Tzortzis, D., Jaeger, S., et al.: Combination of texture and shape features to detect pulmonary abnormalities in digital chest X-rays. Int. J. Comput. Assist. Radiol. Surg. **11**, 99 (2016). https://doi.org/10.1007/s11548-015-1242-x

18. Chatfield, K., Simonyan, K., Vedaldi, A., Zisserman, A.: Return of the devil in the details: delving deep into convolutional nets. arXiv preprint arXiv:1405.3531 (2014)

19. Li, X., Pang, T., Xiong, B., Liu, W., Liang, P., Wang, T.: Convolutional neural networks based transfer learning for diabetic retinopathy fundus image classification. In: 2017 10th International Congress on Image and Signal Processing, BioMedical Engineering and Informatics (CISP-BMEI), Shanghai, pp. 1–11 (2017). https://doi.org/10.1109/cisp-bmei.2017.8301998

20. Niu, X.-X., Suen, C.Y.: A novel hybrid CNN–SVM classifier for recognizing handwritten digits. Pattern Recogn. **45**(4), 1318–1325 (2012). https://doi.org/10.1016/j.patcog.2011.09.021

21. Ukil, S., Ghosh, S., Obaidullah, S.M., Santosh, K.C., et al.: Deep learning for word-level handwritten Indic script identification. arXiv preprint arXiv:1801.01627 (2018)
22. Babenko, A., Lempitsky, V.: Aggregating local deep features for image retrieval. In: Proceedings of the IEEE International Conference on Computer Vision, pp. 1269–1277 (2015). https://doi.org/10.1109/iccv.2015.150
23. Alzu'bi, A., Amira, A., Ramzan, N.: Content-based image retrieval with compact deep convolutional features. Neurocomputing **249**(2), 95–105 (2017). https://doi.org/10.1016/j.neucom.2017.03.072
24. Penatti, O.A, Nogueira, K., dos Santos, J.A.: Do deep features generalize from everyday objects to remote sensing and aerial scenes domains. In: Computer Vision and Pattern Recognition Workshop. IEEE (2015). https://doi.org/10.1109/cvprw.2015.7301382
25. Charan, S., Khan, M.J., Khurshid, K.: Breast cancer detection in mammograms using convolutional neural network. In: 2018 International Conference on Computing, Mathematics and Engineering Technologies (iCoMET), Sukkur, pp. 1–5 (2018). https://doi.org/10.1109/icomet.2018.8346384
26. Jiao, Z., Gao, X., Wang, Y., Li, J.: A deep feature based framework for breast masses classification. Neurocomputing **12**, 221–231 (2016). https://doi.org/10.1016/j.neucom.2016.02.060
27. ICIAR 2018 Grand Challenge on Breast Cancer Histology (BACH) images. https://iciar2018-challenge.grand-challenge.org/
28. Rakhlin, A., Shvets, A., Iglovikov, V., Kalinin, A.A.: Deep convolutional neural networks for breast cancer histology image analysis. In: Campilho, A., Karray, F., ter Haar Romeny, B. (eds.) ICIAR 2018. LNCS, vol. 10882, pp. 737–744. Springer, Cham (2018). https://doi.org/10.1007/978-3-319-93000-8_83
29. Habibzadeh, M., Motlagh, N.H., Jannesary, M., Aboulkheyr, H., Khosravi, P.: Breast cancer histopathological image classification: a deep learning approach. bioRxiv 242818 (2018). https://doi.org/10.1101/242818
30. Han, Z., Wei, B., Zheng, Y., Yin, Y., Li, K., Li, S.: Breast cancer multi-classification from histopathological images with structured deep learning model. Sci. Rep. **7**, 4172 (2017). https://doi.org/10.1038/s41598-017-04075-z
31. Spanhol, F., Oliveira, L.S., Petitjean, C., Heutte, L.: A dataset for breast cancer histopathological image classification. IEEE Trans. Biomed. Eng. **63**(7), 1455–1462 (2016). https://doi.org/10.1109/tbme.2015.2496264
32. Guo, Y., Liu, Y., Oerlemans, A., Lao, S., Wu, S., Lew, M.S.: Deep learning for visual understanding: a review. Neurocomputing **187**, 27–48 (2016). https://doi.org/10.1016/j.neucom.2015.09.116
33. Huang,G., Liu, Z., van der Maaten, L.: Densely connected convolutional network. In: IEEE Conference on Computer Vision and Pattern Recognition, pp. 2261–2269 (2017). https://doi.org/10.1109/cvpr.2017.243
34. He, K., Zhang, X., Ren, S., Sun, J.: Deep residual learning for image recognition. In: Proceedings of the IEEE Conference on Computer Vision and Pattern Recognition, pp. 770–778 (2016). https://doi.org/10.1109/cvpr.2016.90
35. Szegedy, C., Vanhoucke, V., Ioffe, S., Shlens, J., Wojna, Z.: Rethinking the inception architecture for computer vision. In: Proceedings of the IEEE Conference on Computer Vision and Pattern Recognition, pp. 2818–2826 (2016)
36. Keras: The Python Deep Learning Library. https://keras.io/
37. Haifley, T.: Linear logistic regression: an introduction. In: IEEE International Integrated Reliability Workshop Final Report, pp. 184–187 (2002). https://doi.org/10.1109/irws.2002.1194264

38. Ke, G., et al.: LightGBM: a highly efficient gradient boosting decision tree. In: Advances in Neural Information Processing Systems, pp. 3149–3157 (2017)
39. Breiman, L.: Random forests. J. Mach. Learn. **45**(1), 5–32 (2001). https://doi.org/10.1023/a:1010933404324
40. Yosinski, J., Clune, J., Bengio, Y., Lipson, H.: How transferable are features in deep neural networks. In: Advances in Neural Information Processing Systems, pp. 3320–3328 (2014)
41. Tajbakhsh, N., et al.: Convolutional neural networks for medical image analysis: full training or fine tuning. IEEE Trans. Med. Imaging **35**(5), 1299–1312 (2016). https://doi.org/10.1109/tmi.2016.2535302
42. Hu, F., Xia, G.-S., Hu, J., Zhang, L.: Transferring deep convolutional neural networks for the scene classification of high-resolution remote sensing imagery. MDPI Remote Sens. **7**(11), 14680–14707 (2015). https://doi.org/10.3390/rs71114680
43. NVIDIA Corporation: Nvidia tesla product literature (2018). https://www.nvidia.com/en-us/design-visualization/quadro-desktop-gpus/
44. Hegadi, R.S., Navale, D.I., Pawar, T.D., Ruikar, D.D.: Multi feature-based classification of osteoarthritis in knee joint X-ray images, Chap. 5. In: Medical Imaging: Artificial Intelligence, Image Recognition, and Machine Learning Techniques. CRC Press (2019). ISBN 9780367139612
45. Ruikar, D.D., Santosh, K.C., Hegadi, R.S.: Automated fractured bone segmentation and labeling from CT images. J. Med. Syst. **43**(3), 60 (2019). https://doi.org/10.1007/s10916-019-1176-x
46. Ruikar, D.D., Santosh, K.C., Hegadi, R.S.: Segmentation and analysis of CT images for bone fracture detection and labeling, Chap. 7. In: Medical Imaging: Artificial Intelligence, Image Recognition, and Machine Learning Techniques. CRC Press (2019). ISBN 9780367139612
47. Ruikar, D.D., Hegadi, R.S., Santosh, K.C.: A systematic review on orthopedic simulators for psycho-motor skill and surgical procedure training. J. Med. Syst. **42**(9), 168 (2018)
48. Sawat, D.D., Hegadi, R.S.: Unconstrained face detection: a Deep learning and Machine learning combined approach. CSI Trans. ICT **5**(2), 195–199 (2017)
49. Jagtap, A.B., Hegadi, R.S.: Feature learning for offline handwritten signature verification using Convolution Neural Network. Int. J. Technol. Hum. Interact. (IJTHI). ISSN 1548-3908

Gabor Filter Based Classification of Mammography Images Using LS-SVM and Random Forest Classifier

Mantragar Vijaya Madhavi[1](✉) and T. Christy Bobby[2]

[1] East Point College of Engineering and Technology, Bengaluru, Karnataka, India
vijayamadhavi79@gmail.com
[2] M. S. Ramaiah University of Applied Sciences, Bengaluru, Karnataka, India
christy.ec.et@msruas.ac.in

Abstract. Breast cancer is formed in breast cells and it occurs primarily in women. Mammogram is x-ray of the breast that provides structural details. In the proposed work, attempt has been made to discriminate normal and abnormal mammograms belonging to fatty, fatty-glandular or dense-glandular breast tissue density. Breast x-ray images acquired in Medio-Lateral-Oblique (MLO) view (n = 197) chosen from mini-MIAS database are preprocessed using median filter to remove noise and adaptive histogram equalization has been applied to improve image contrast. The preprocessed images are subjected to Pectoral Muscle Removal (PMR) algorithm to obtain Region of Interest (RoI) comprising of only breast tissue region. For the obtained RoI, Gabor filtering has been employed at 5 various scales and 8 orientations leading to (5×8) 40 filter responses. Magnitude, phase information and statistical features such as energy, entropy, variance, kurtosis and skewness are computed for the obtained filter responses. In the next step, reduced feature set is obtained by implementing two schemes. In scheme1, the dominant scale and orientations of Gabor filter responses are determined by employing Kernel Principal Component Analysis (KPCA) and features are considered for the obtained dominant scale and orientation coefficients. Further, absolute difference value of extracted features are computed and compared to choose the significant features. In scheme2, sequential forward selection algorithm is employed to obtain significant features that discriminates normal and abnormal subjects. The derived significant features belonging to each category of breast density are fed to Least Square-Support Vector Machine (LS-SVM) and Random Forest (RF) classifier to classify subjects as normal and abnormal. The results indicate that it is possible to discriminate normal and abnormal subjects using Gabor and statistical features. LS-SVM classifier performs better with an accuracy of 93.54%, sensitivity of 95.24% and specificity of 90% for fatty-glandular tissues.

Keywords: Mammogram · Gabor filter · Random forest classifier

© Springer Nature Singapore Pte Ltd. 2019
K. C. Santosh and R. S. Hegadi (Eds.): RTIP2R 2018, CCIS 1036, pp. 69–83, 2019.
https://doi.org/10.1007/978-981-13-9184-2_6

1   Introduction

Breast cancer accounts for 30% of cancers among female group and is a major disease affecting them [1]. As per World Health Organization (WHO) by 2020, global cancer rates would increase by 50% leading to 15 million new cases [2]. Breast cancer originates in breast cells, multiples, invades, progresses into neighboring tissues destroying them and may even spread to surrounding lymph node leading to metastasis. Early detection of breast cancer becomes essential as it improves survival rate and also increases treatment options. Mammography is an imaging modality considered as a gold standard and a good quality mammogram can identify 85–90% of breast cancers [3], however it may produce false-negative or false-positive results in younger women and women with dense breasts [4]. During primeval stages of breast cancer, the cancerous region is small, confined to the organ and does not involve lymph nodes hence screening mammogram would be suitable in identifying premature indications of cancer.

Mammography employs low energy X-rays (around 30 kVp) to examine breast or mammary gland for screening or diagnosis purpose [5]. Mammogram images are usually acquired in two standard views: Cranio-Caudal (CC) and Medio-Lateral-Oblique (MLO) view. In MLO view, the middle portion of the mammary gland is notable and allows imaging of majority of breast belonging to upper-outer quadrant where almost 43% of tumors appear [6]. It is very essential to perform pre-processing as poor x-ray quality may cause some of the lesions to go undetected. Preprocessing involves removal of undesired elements such as artifact and label performed using connected component labelling [7,42] or combination of thresholding and morphological operation ([8] and [9]), de-noising was performed using median filtering ([8] and [10]), adaptive wiener filtering [11,44], Gaussian filtering [12] or shock filtering [13] and contrast enhancement was performed using histogram equalization [14,43], morphological top hat transform [15] or Contrast Limited Adaptive Histogram Equalization (CLAHE) ([10,16] and [17]).

Pectoral muscle or chest muscle is visible in MLO view and is located in upper posterior margin of the image. In breast x-ray images, coexistence of Pectoral Muscle (PM) along with breast tissues may affect cancer findings as PM has intensity values same as mammary gland tissues. Hence identification and exclusion of PM in mammogram is an essential task. Various approaches used for PM removal are canny edge detection & straight line approximation [18], region growing procedure [9], marker-controlled watershed transform [13] and thresholding [19]. However, most of the approaches failed to identify the complete pectoral region as this region comprises of pixels with varying intensity. Also in most of the images, the frontier of PM cannot be estimated as a linear contour because the pectoral contour looks like a vaulted line. Hence some robust approach has to be adopted for accurate removal of PM. Other procedures such as adaptive fuzzy binarization [20] or robust edge map technique [21] can be utilized for improved PM boundary identification.

Gabor filters are orientation perceptive band-pass filters that provide strong response to sharp intensity transitions and texture changes. When oriented in a particular scale and orientation, they give strong response for targets or objects having structures in that direction. In breast cancer analysis, Gabor filtering has been employed for identifying architectural distortion ([16,22] and [23]), for extracting texture properties of masses ([24] and [25]) and for attribute extraction using edge histogram descriptors [26] or Local Binary Pattern (LBP) [17]. Various statistical features extracted from magnitude response of Gabor filterbank includes mean, Standard Deviation (SD) [27], skewness ([25] and [28]), kurtosis [29], entropy and energy [30].

In applications related to image processing, feature extraction techniques are applied to obtain features or attributes useful for classification and recognition of regions in an image. Various attribute extraction approaches employed in mammography images are Gray Level Co-occurrence Matrix (GLCM) ([10] and [31]), Local Binary Pattern (LBP) [32], Law's texture measures [10], Local Configuration Pattern (LCP) [32], statistical features ([24] and [31]), fractal dimension [10], Histogram of Gradient (HOG) [33] and Gray Level Run Length Matrix (GLRLM) [31].

Classifiers learn from input data called as training data, uses it to map testing data so as to produce an output depicting class label. The success of the classifier depends on the separable nature of the employed classes. Different classifiers employed for categorization of breast x-ray images as normal-abnormal, benign-malignant and normal-benign-malignant are Support Vector Machine (SVM) ([10] and [25]), Random Forest (RF) [12], Artificial Neural Network (ANN) [31] and K-Nearest Neighbor (KNN) [26].

The main objective of the proposed work is breast cancer finding using MLO-view mammograms containing fatty, fatty-glandular or dense-glandular breast tissues. It is a challenging job as the density distribution of these three types of breast tissues varies causing adversity in accurately classifying the subjects as normal or abnormal. In the projected work, breast x-ray images are pre-processed by means of median filter, connected component analysis and contrast enhancement is performed using CLAHE. A novel pectoral muscle removal algorithm is suggested and applied to choose the RoI. The extracted RoI is further subjected to Gabor filtering and attributes such as magnitude, phase and statistical features are extracted for all filter outputs. Feature reduction is performed by implementing two different schemes. In scheme1, dominant scale & orientation is determined by employing KPCA, features are considered in dominant scale and orientation, Absolute Difference (AD) value is calculated and Significant Feature Set1 (SFS1) is determined. In scheme2, sequential forward selection which is a type of wrapper based attribute selection method is employed to determine Significant Feature Set2 (SFS2). These two sets of significant features are fed to Least Square SVM (LS-SVM) and RF classifier to classify subjects as normal and abnormal.

2 Methodology

Breast x-ray images provide 2D visualization of mammary gland structure and MLO is a standard view acquired during screening phase. Hence MLO view mammograms are considered for the analysis. Breast images of size 1024×1024 pixels are chosen from mini-MIAS database [34]. The data set consists of 197 mammogram images (n = 197), out of which 130 are normal and 67 are abnormal (benign or malignant). Also the images are categorized into three types based on the tissue density, namely fatty-glandular (42 normal, 24 abnormal), fatty (47 normal, 24 abnormal) and dense-glandular (41 normal, 19 abnormal). Figure 1 represents framework for breast cancer screening using image analysis.

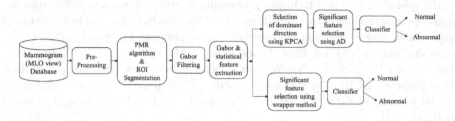

Fig. 1. Framework for breast cancer screening using image analysis

2.1 Image Pre-processing

During image acquisition, mammograms are corrupted by random variation of intensity values known as noise and the image also contains labels and artifacts. Noise removal is performed using nonlinear 2D median filtering that requires computing the median value in 3×3 neighborhood of pixel under consideration and replacing the center pixel with the computed value. This process is repeated for all the picture elements in the image. Thresholding is performed on the filtered image to obtain a 2-level image and the largest connected element is selected which automatically eliminates labels and artifacts that are smaller in dimension when compared to the breast region. Contrast enhancement is performed using CLAHE as it accentuates edges and other image features. It operates on sub-regions of the image by applying contrast limiting procedure in the confined region thereby improving local contrast. This process is repeated over different sub-regions to cover the complete image.

2.2 Pectoral Muscle Removal and Region of Interest (RoI) Segmentation

A novel Pectoral Muscle Removal (PMR) algorithm has been designed with an assumption that pectoral muscle is almost triangular in shape, it is located at top left corner and the pixel intensity values along the boundary of pectoral muscle is almost similar. Flip the mammogram of right breast prior to employment of PMR algorithm so that PM region is always located at top-left corner of the image. Steps involved in PMR algorithm are

1. Select seed point along the boundary of pectoral muscle region.
2. Construct binary image within gray-level tolerance of seed point.
3. Perform image reconstruction to reconstruct the connected section.
4. Dilate the image using suitable Structural Element (SE) to obtain the boundary of pectoral muscle.
5. Multiply the dilated image with pre-processed gray-scale image to obtain boundary superimposed image.
6. All the pixels to the left of the boundary is assigned zero to obtain pectoral muscle removed mammogram.

In the resultant pectoral muscle removed image, all the blank rows and columns are removed to obtain RoI comprising of only breast region.

2.3   Gabor Filtering

Gabor filters are 2D band pass filters that can apprehend structures in an image. They are chosen for breast x-ray image analysis as they are invariant to rotation, scale and translation. Gabor filter is represented in spatial domain as

$$g(x, y) = \frac{f_0^2}{\pi \gamma \eta} \exp[-(\alpha^2 x'^2 + \beta^2 y'^2)] \exp j 2\pi f_0 x' \tag{1}$$

where $x' = x\cos\theta + y\sin\theta$, $y' = -x\sin\theta + y\cos\theta$, $\alpha = \frac{f_0}{\gamma}$, $\beta = \frac{f_0}{\eta}$, f_0 is the central frequency of the filter, θ is the rotation angle of Gaussian major axis & plane wave and spatial aspect ratio of Gaussian is given by $(\frac{\eta}{\gamma})$ [35].

$$f_0 = \frac{f_{max}}{(\sqrt{2})^{i-1}} \tag{2}$$

$$\theta = \frac{\pi(j-1)}{\gamma} \tag{3}$$

Gabor filter bank is obtained by tuning filter parameters such as filter orientation (θ), central frequency or scale (f_0), maximum frequency (f_{max}), sharpness along minor axis (η) and sharpness along major axis (γ). Usually $\eta = \gamma = \sqrt{2}$, f_{max} is determined by computing Fast Fourier Transform (FFT) of an image and the average value obtained is 0.9. The filter-bank comprising of 40 filters are obtained by choosing number of scales as 5 and number of orientations as 8 to obtain (5×8) 40 filter responses. For five scales, values of f_0 are $\{0.9, 0.6364, 0.45, 0.3182, 0.225\}$ and for eight orientations, values of are $\{0°, 22.5°, 45°, 67.5°, 90°, 112.5°, 135°, 157.5°\}$. Dominant direction denotes the dominant scale and orientation where maximum filter response occurs. In the projected work, both magnitude and phase extracted from filter-bank response is used to determine dominant direction as they help in efficient detection of interest points.

2.4   Feature Extraction

Features such as magnitude & phase and statistical features such as energy, entropy, variance, kurtosis & skewness are extracted [30] from dominant filter responses obtained using KPCA. Filter response has the size similar to the input and each pixel in the response image is a complex quantity containing real and imaginary parts. Magnitude of a pixel with coordinates (x,y) in the response image is computed as

$$A(x,y) = \sqrt{[Re(g(x,y))]^2 + [Im(g(x,y))]^2} \tag{4}$$

Magnitude of an image is the mean of pixel magnitudes in the response image. Phase of a pixel with coordinates (x,y) in the response image is computed as

$$\phi(x,y) = \arctan[\frac{Re(g(x,y))}{Im(g(x,y))}] \tag{5}$$

Phase of an image is the mean of pixel phases in the response image.

2.5   Feature Reduction Using Kernel Principal Component Analysis (KPCA) and Wrapper Based Method

Gabor filter responses results in enormous amount of redundant data. In order to reduce redundancy, two different schemes of feature selection are used. Scheme1 employs kernel technique that projects the data onto feature space of larger dimension. Polynomial kernel has finite expansion leading to finite dimensional mapping and is suitable for normalized data. Hence in this work polynomial kernel is used to determine dominant scale and dominant orientation where filter response is maximum. Polynomial kernel of degree h is represented as

$$k(x_i, x_j) = \langle x_i \cdot x_j \rangle^h \tag{6}$$

Principal components of x are obtained by computing projections onto eigen vectors [36]. The magnitude and phase information of all 40 filter responses are fed to KPCA and three dominant directions are selected based on largest eigen value corresponding to principal components. Finally absolute difference of average feature values among normal ($avg_{fnormal}$) and abnormal ($avg_{fabnormal}$) categories is represented mathematically as

$$absolute\_difference = |avg_{fnormal} - avg_{fabnormal}| \tag{7}$$

Three features taking larger value of absolute difference in each category of breast tissues are considered as SFS1.

In scheme2, wrapper based feature selection is employed that considers subset of attributes for which classification algorithm performs superlative [37]. Sequential forward selection is a type of wrapper method that is implemented by means of KNN classifier employing 10-fold cross-validation. It is an iterative approach wherein process starts with an empty set and features are appended until certain objective function is achieved. In this work, subset of attributes are selected such that misclassification rate is minimized for each category of breast tissues to obtain SFS2.

2.6   Classification

Objective of classifier is to categorize the given data set into diverse groups represented by class labels. Two classifiers LS-SVM and RF based on diverse approaches are explored for analysis. LS-SVM belongs to a class of kernel based learning methods that implements curvilinear decision regions and converges to smallest mean squared error. It employs equality constraints and its solution requires solving set of linear equations making it easier to implement [38]. RBF kernel is chosen as it provides better separation between classes of data by establishing hyper-spheres or circular boundaries. Random forests (RF) is an ensemble based learning technique that operates by constructing a group of decision trees at the training phase and produces a class that is mode of the classes output produced by the individual trees in the testing phase. Since results obtained from different decision trees are combined, it overcomes the problem of over-fitting and is more accurate but is computationally complex and time consuming process due to construction of large number of decision trees [39]. The classifiers are trained and tested with data count mentioned in Table 1. Tuning parameters of LS-SVM classifier employing RBF kernel are regularization parameter, $\gamma = 1$, gaussian bandwidth, $\sigma^2 = 2.5$ and for RF classifier, number of trees, n = 10.

Table 1. Training and testing data for LS-SVM and RF classifier

Breast tissue	Class	Training	Testing
Fatty-glandular	Normal	21	21
	Abnormal	14	10
Dense-glandular	Normal	23	18
	Abnormal	10	9
Fatty	Normal	25	22
	Abnormal	12	12

3   Results and Discussion

Representative mammogram of right breast containing fatty-glandular, fatty and dense-glandular tissues respectively is shown in Fig. 2. We can observe that appearance of these types of breast densities is different due to variation in proportion of glandular, connective and fatty tissues. The mammograms are flipped so that pectoral muscle is on top left corner of the image and then preprocessed with nonlinear 2D median filtering & CLAHE is represented in Fig. 3. The resultant obtained has enhanced contrast with accentuated edges. The typical fatty-glandular, fatty and dense-glandular mammogram images, obtained after PMR algorithm is shown in Fig. 4. The resultant images obtained are the complete breast region in the nonappearance of pectoral muscle. It is the desired

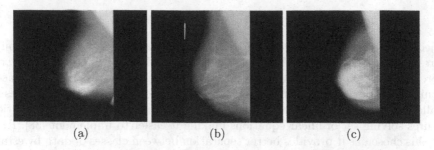

Fig. 2. Representative mammogram (normal case) (a) fatty-glandular (mdb071) (b) fatty (mdb251) (c) dense-glandular (mdb037)

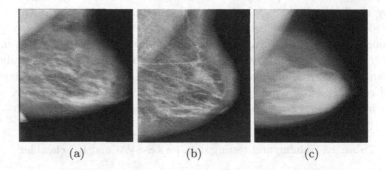

Fig. 3. Pre-processed mammogram (a) fatty-glandular (b) fatty (c) dense-glandular breast tissue

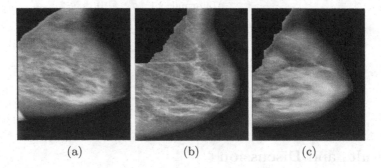

Fig. 4. Pectoral muscle removed mammogram (RoI) (a) fatty-glandular (b) fatty (c) dense-glandular breast tissue

RoI for the further analysis. A diagram of Gabor filter bank and its magnitude and phase characteristics of filter-bank comprising of 40 filters is shown in Fig. 5 with rows representing 5 different scales and columns representing 8 different orientations.

In feature extraction step, magnitude, phase and five statistical features are computed for obtained filtered responses leading to 280 features (7 features × 40

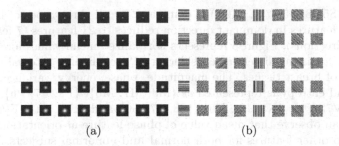

(a)   (b)

Fig. 5. Gabor filter-bank (a) magnitude characteristics (b) phase characteristics

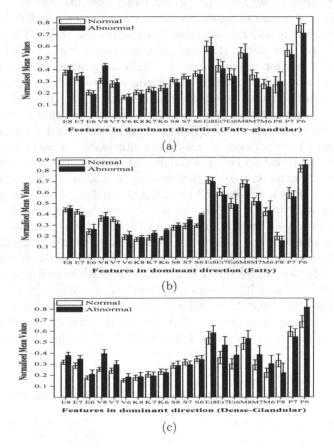

Fig. 6. Features in dominant directions for (a) fatty-glandular (b) fatty (c) dense-glandular breast tissue

responses). SFS1 is achieved by initially determining dominant scale and orientation by employing KPCA with the polynomial kernel of degree (h) = 0.0005 to the attained filter bank response. The results shows that scale 5 with orientation

8, 7 and 6 (S5O8, S5O7 & S5O6) is dominant for all three types of breast tissues. Hence features in dominant direction reduces to 21 features (7 features × 3 dominant directions). Figure 6 represents normalized mean values for attributes obtained in dominant directions considering 95% confidence interval for all the three types of breast tissues. The magnitude, phase, energy, variance, kurtosis, skewness and entropy is represented as {M8, M7, M6}, {P8, P7, P6}, {E8, E7, E6}, {V8, V7, V6}, {K8, K7, K6}, {S8, S7, S6} and {Et8, Et7, Et6} respectively. We can observe that mean value of phase feature at orientation 6 is high compared to other features for both normal and abnormal subjects. This may be due to sensitiveness of phase to variations in pixel values near edges.

Further feature reduction is performed using absolute difference value and three significant features obtained for fatty-glandular case is V8, P6 and P7, fatty case is S6, K6 and S7 and dense-glandular case is V8, P6 and Et7 to obtain SFS1. We can notice that variance, skewness, kurtosis and entropy is predominantly large for abnormal subjects indicating heterogeneous region representing structural variation due to the presence of lesion or tumor in an image. Phase is also an important descriptor representing variation in edge pixel values along with above statistical features to discriminate normal and abnormal subjects. SFS2 is obtained by employing sequential forward selection and the features obtained are variance at scale5, orientation8 (var_S5_O8), variance at scale5, orientation6 (var_S5_O6) and variance at scale3, orientation5 (var_S3_O5) for fatty-glandular breast tissues, skewness at scale5, orientation6 (ske_S5_O6) and variance at scale5, orientation3 (var_S5_O3) for fatty breast tissue, entropy at scale5, orientation8 (Et_S5_O8), entropy at scale5, orientation4 (Et_S5_O4) and variance at scale3, orientation4 (var_S3_O4) for dense-glandular breast tissues. We can observe that variance is an important attribute contributing to edges in abnormal images depicted by variation in pixel intensity values. SFS1 and SFS2 are fed as feature vector to LS-SVM and RF classifier. Classification results for LS-SVM and RF classifier in terms of sensitivity (sens), specificity (spec), accuracy (acc) and Area Under Curve (AUC) is shown in Table 2.

From Table 2, we can observe that for SFS1 average-accuracy, average-sensitivity and average-specificity is found to be 91.45%, 96.89% and 80.92% for

Table 2. Classification results for LS-SVM and RF classifier

Breast tissue	Significant feature set1				Significant feature set2			
	sens (%)	spec (%)	acc (%)	AUC	sens (%)	spec (%)	acc (%)	AUC
Fatty-glandular	95.24	90	93.54	0.926	90.47	40	74.19	0.652
	85.71	100	90.32	0.928	95.23	70	87	0.826
Dense-glandular	100	77.77	92.59	0.889	100	44.44	81.48	0.722
	88.88	88.88	88.88	0.889	88.88	44.44	74.07	0.666
Fatty	95.45	75	88.23	0.852	100	58.33	85.29	0.791
	95.45	75	88.23	0.852	100	75	91.11	0.875

LS-SVM, 89.14%, 90.01% and 87.96% for RF classifier respectively whereas for SFS2 average-accuracy, average-sensitivity and average-specificity is found to be 80.32%, 96.82% and 47.59% for LS-SVM, 84.06%, 94.7% and 63.15% for RF classifier respectively. Best performance results are obtained for fatty-glandular case with SFS1 employing LSSVM with an accuracy of 93.54%, sensitivity of 95.24%, specificity of 90% and AUC of 0.926. Usually sensitivity of dense-glandular mammogram is low due to larger breast density [40], but in the proposed methodology, the sensitivity of 100% is achieved for both the SFSs due to precise localization of edges in dominant direction. The specificity value obtained for fatty breast tissues is relatively less (75%). This may be due to dark appearance of fatty breast tissues in mammograms characterized by less fibro-glandular tissue compared to fat tissue producing false positive result. The performance measures obtained by proposed method are comparable with performance measures attained by other authors employing Gabor based filtering with different number of scales and orientations [17, 25–29]. Also the proposed system achieved performance measures comparable with the existing system and is depicted in Table 3.

Table 3. Comparison of proposed system with existing system

Method	acc (%)	sens (%)	spec (%)
Local & global features + SVM-RBF [10]	93.17	92.71	93.46
GLCM, GLRLM, histogram & shape features + FFANN [31]	96	83	98
LBP features + KNN [32]	95.3	93.8	97.8
GLCM + RF [12]	89.02	90.56	86.20
Proposed work: Gabor & statistical features + LSSVM-RBF	91.45	96.89	80.92

4 Conclusion

Breast cancer is the second principal reason for death amongst female group and mammography is an imaging procedures extensively employed for mass screening. In the proposed methodology, Gabor filtering has been applied to MLO view mammograms. Pre-processing has been performed to enhance the appearance of an image, pectoral boundary is identified, pectoral region is removed and ROI is extracted. Gabor filtering is applied on extracted RoI's and statistical features, magnitude and phase information is extracted from filter responses. SFS1 is obtained by determining dominant direction using KPCA, considering features in dominant directions and further feature reduction is performed using AD value. SFS2 is obtained by utilizing sequential forward selection and are fed to LSSVM and RF classifier to classify subjects as normal and abnormal. Average accuracy, sensitivity and specificity obtained for SFS1 are 91.45%, 96.89%,

80.92% for LSSVM and 89.14%, 90.01%, 87.96% for RF classifier, average accuracy, sensitivity and specificity obtained for SFS2 are 80.32%, 96.82% and 47.59% for LS-SVM, 84.06%, 94.7% and 63.15% for RF classifier. Also best performance results are obtained for fatty-glandular breast tissues with SFS1 employing LS-SVM with an accuracy of 93.54%, sensitivity of 95.24%, specificity of 90% and AUC of 0.926. Specificity obtained for fatty breast density is less compared to other types of breast densities. By combining mammography with other imaging modalities such as breast tomo-synthesis or ultrasound, the specificity of the system can be improved. The proposed work can be employed for mass screening purpose. This work can be extended to classify abnormal subjects as benign or malignant which can be further categorized as mass or calcification. The healthcare filed dramatically inculcating the technological advancements (computer aided diagnosis (CAD) system, pre-operative outcome prediction system, intra-operative assistant system and surgery simulators for instance) in daily clinical process [41]. So in future we aim to develop and deploy the CAD system for Breast cancer screening purpose.

References

1. Female Breast Cancer: Facts and figures. https://data.web.health.state.mn.us/cancer_breast
2. Hoffman and Hoffmann Worldwide. http://www.hoffmanpr.com/press-release/global-cancer-rates-could-increase-50-by-2020
3. Mina, L.M., Isa, N.A.M.: A review of computer-aided detection and diagnosis of breast cancer in digital mammography. J. Med. Sci. **15**(3), 110–121 (2015). https://doi.org/10.3923/jms.2015.110.121
4. The Susan G. Komen Breast Cancer Foundation, Inc. https://ww5.komen.org/BreastCancer/AccuracyofMammograms.html
5. Saha, D., Bhowmik, M.K., De, B.K., Bhattacharjee, D.: A survey on imaging-based breast cancer detection. In: Das, K.N., Deep, K., Pant, M., Bansal, J.C., Nagar, A. (eds.) Proceedings of Fourth International Conference on Soft Computing for Problem Solving. AISC, vol. 335, pp. 255–266. Springer, New Delhi (2015). https://doi.org/10.1007/978-81-322-2217-0_22
6. Lee, R.J., Vallow, L.A., McLaughlin, S.A., Tzou, K.S., Hines, S.L., Peterson, J.L.: Ductal carcinoma in situ of the breast. Int. J. Surg. Oncol. 1–12 (2012). https://doi.org/10.1155/2012/123549
7. Pragathi, J., Patil, H.T.: Segmentation method for ROI detection in mammographic images using Wiener filter and Kittler's method. In: IJCA Proceedings on International Conference on Recent Trends in Engineering and Technology 2013 ICRTET (4), pp. 27–33 (2013)
8. Sukassini, M.P., Velmurugan, T.: Noise removal using morphology and median filter methods in mammogram images. In: The 3rd International Conference on Small and Medium Business 2016, Hochiminh, Vietnam, pp. 413–419 (2016)
9. Alam, N., Islam, M.J.: Pectoral muscle elimination on mammogram using k-means clustering approach. Int. J. Comput. Vis. Signal Process. **4**(1), 11–21 (2014)
10. Phadke, A.K., Rege, P.P.: Fusion of local and global features for classification of abnormality in mammograms. Sadhana **41**(4), 385–395 (2016). https://doi.org/10.1007/s12046-016-0482-y

11. Vieira, M.A.C., Bakic, P.R., Maidment, A.D.A., Schiabel, H., Mascarenhas, N.D.A.: Filtering of poisson noise in digital mammography using local statistics and adaptive Wiener filter. In: Maidment, A.D.A., Bakic, P.R., Gavenonis, S. (eds.) IWDM 2012. LNCS, vol. 7361, pp. 268–275. Springer, Heidelberg (2012). https://doi.org/10.1007/978-3-642-31271-7_35
12. Singh, V.P., Srivastava, A., Kulshreshtha, D., Chaudhary, A., Srivastava, R.: Mammogram classification using selected GLCM features and random forest classifier. Int. J. Comput. Sci. Inf. Secur. 14(6), 82–87 (2016)
13. Babu, J.S., Sukumar, L.B., Anandan, K.: Quantitative analysis of digitized mammograms using nonsubsampled contourlets and evolutionary extreme learning machine. J. Med. Imaging Health Inform. 3(2), 206–213 (2013). https://doi.org/10.1166/jmihi.2013.1146
14. Langarizadeh, M., et al.: Improvement of digital mammogram images using histogram equalization, histogram stretching and median filter. J. Med. Eng. Technol. 35(2), 103–108 (2011). https://doi.org/10.3109/03091902.2010.542271
15. Rajkumar, K.K., Raju, G.: Enhancement of mammograms using top hat filtering and wavelet decomposition. J. Comp. Math. Sci. 2(6), 812–818 (2011)
16. Kamra, A., Jain, V.K., Singh, S.: Extraction of orientation field using Gabor filter and gradient based approach for the detection of subtle signs in mammogram. J. Med. Imaging Health Inform. 4(3), 374–381 (2014). https://doi.org/10.1166/jmihi.2014.1266
17. Alquodl, A., Jaffar, M.A.: Hybrid Gabor based local binary patterns texture features for classification of breast mammograms. Int. J. Comput. Sci. Netw. Secur. 16(2), 16–21 (2016)
18. Kinoshita, S.K., Azevedo-Marques, P.M., Pereira, R.R., Rodrigues, J.A.H., Rangayyan, R.M.: Radon-domain detection of nipple and pectoral muscle in mammograms. J. Digit. Imaging 21(1), 37–49 (2007). https://doi.org/10.1007/s10278-007-9035-6
19. Sreedevi, S., Sherly, E.: A novel approach for removal of pectoral muscles in digital mammograms. Procedia Comput. Sci. 46, 1724–1731 (2015). https://doi.org/10.1016/j.procs.2015.02.117
20. Santosh, K.C., Alam, N., Roy, P.P., Wendling, L., Antani, S., Thoma, G.R.: A simple and efficient arrowhead detection technique in biomedical images. Int. J. Pattern Recogn. Artif. Intell. 30(5), 1657002-1–1657002-16 (2016). https://doi.org/10.1142/S0218001416570020
21. Santosh, K.C., Vajda, S., Antani, S., Thoma, G.R.: Edge map analysis in chest x-rays for automatic pulmonary abnormality screening. Int. J. Comput. Assist. Radiol. Surg. 11(9), 1637–1646 (2016). https://doi.org/10.1007/s11548-016-1359-6
22. Bailur, A., Pandey, A.K., Sharma, A.K., Saseendran, S.: Modified Gabor filter with control chart and image plots for identifying architectural distortion of mammogram images. Int. J. Comput. Sci. Telecommun. 5(4), 12–19 (2014)
23. Rangayyan, R.M., Ayres, F.J.: Gabor filters and phase portraits for the detection of architectural distortion in mammograms. Med. Bio. Eng. Comput. 44, 883–894 (2006). https://doi.org/10.1007/s11517-006-0088-3
24. Hussain, M., Khan, S., Muhammad, G., Berbar, M., Bebis, G.: Mass detection in digital mammograms using Gabor filter bank. In: IET Conference on Image Processing, London, pp. 1–5 (2012). https://doi.org/10.1049/cp.2012.0465
25. Khan, S., Khan, A., Maqsood, M., Aadil, F., Ghazanfar, M.A.: Optimized Gabor feature extraction for mass classification using cuckoo search for big data e-health care. J. Grid Comput. (2018). https://doi.org/10.1007/s10723-018-9459-x

26. Zheng, Y.: Breast cancer detection with Gabor features from digital mammograms. Algorithms **3**, 44–62 (2010). https://doi.org/10.3390/a3010044
27. Abdel-Naseer, M., Moreno, A., Puig, D.: Towards cost reduction of breast cancer diagnosis using mammography texture analysis. J. Exp. Theor. Artif. Intell. (2015). https://doi.org/10.1080/0952813X.2015.1024496
28. Prasad, R.K., Basha, M.S.: Effective texture feature model for classification of mammogram images. ARPN J. Eng. Appl. Sci. **13**(3), 961–967 (2018)
29. Torrents-Barrena, J., Puig, D., Melendez, J., Vallas, A.: Computer-aided diagnosis of breast cancer via Gabor wavelet bank and binary class SVM in mammographic images. J. Exp. Theor. Artif. Intell. (2015). https://doi.org/10.1080/0952813X.2015.1024491
30. Jaffar, M.A.: Hybrid texture based classification of breast mammograms using AdaBoost classifier. Int. J. Adv. Comput. Sci. Appl. **8**(5), 321–327 (2017). https://doi.org/10.14569/IJACSA.2017.080540
31. Sheba, K.U., Raj, S.G.: An approach for automatic lesion detection in mammograms. Cogent Eng. **5**, 1–16 (2018). https://doi.org/10.1080/23311916.2018.1444320
32. Gardezi, S.J.S., Faye, I., Adjed, F., Kamel, N., Hussain, M.: Mammogram classification using chi-square distribution on local binary pattern features. J. Med. Imaging Health Inform. **7**(1), 1–5 (2017). https://doi.org/10.1166/jmihi.2017.1982
33. Ergin, S., Kilinc, O.: A new feature extraction framework based on wavelets for breast cancer diagnosis. Comput. Biol. Med. **51C**, 171–182 (2014). https://doi.org/10.1016/j.compbiomed.2014.05.008
34. Mini-MIAS database for mammograms. http://peipa.essex.ac.uk/info/mias.html
35. Kamarainen, J.K., Kyrki, V., Kalviainen, H.: Invariance properties of Gabor filter-based features - overview and applications. IEEE Trans. Image Proc. **15**(5), 1088–1099 (2006). https://doi.org/10.1109/TIP.2005.864174
36. Schölkopf, B., Smola, A., Müller, K.-R.: Kernel principal component analysis. In: Gerstner, W., Germond, A., Hasler, M., Nicoud, J.-D. (eds.) ICANN 1997. LNCS, vol. 1327, pp. 583–588. Springer, Heidelberg (1997). https://doi.org/10.1007/BFb0020217
37. Vajda, S., et al.: Feature selection for automatic tuberculosis screening in frontal chest radiographs. J. Med. Syst. **42**, 146 (2018)
38. Suykens, J.A.K., Vandewalle, J.: Least square support vector machine classifiers. Neural Process. Lett. **9**(3), 293–300 (1999). https://doi.org/10.1023/A:1018628609742
39. Breiman, L.: Random forests. Mach. Learn. **45**, 5–32 (2001). https://doi.org/10.1023/A:1010933404324
40. Institute of Medicine and National Research Council: Saving Women's Lives: Strategies for Improving Breast Cancer Detection and Diagnosis (Consensus Study Report). The National Academies Press, Washington, DC (2005). https://doi.org/10.17226/11016
41. Ruikar, D.D., Hegadi, R.S., Santosh, K.C.: A systematic review on orthopedic simulators for psycho-motor skill and surgical procedure training. J. Med. Syst. **42**(9), 168 (2018)
42. Ruikar, D.D., Santosh, K.C., Hegadi, R.S.: Automated fractured bone segmentation and labeling from CT images. J. Med. Syst. **43**(3), 60 (2019). https://doi.org/10.1007/s10916-019-1176-x

43. Ruikar, D.D., Santosh, K.C., Hegadi, R.S.: Segmentation and analysis of CT images for bone fracture detection and labeling. In: Medical Imaging: Artificial Intelligence, Image Recognition, and Machine Learning Techniques, chap. 7. CRC Press, Boca Raton (2019). ISBN 9780367139612
44. Hegadi, R.S., Navale, D.I., Pawar, T.D., Ruikar, D.D.: Multi feature-based classification of osteoarthritis in knee joint X-ray images. In: Medical Imaging: Artificial Intelligence, Image Recognition, and Machine Learning Techniques, chap. 5. CRC Press, Boca Raton (2019). ISBN 9780367139612

A Review of Contemporary Researches on Biomedical Image Analysis

Pravin R. Lokhande[1]($\boxtimes$), S. Balaguru[1], G. Deenadayalan[1],
and Ratnakar R. Ghorpade[2]

[1] Vel Tech Rangarajan Dr. Sagunthala R and D Institute of Science and Technology,
Chennai, India
pravinrlokhande@gmail.com
[2] MIT-World Peace University, Pune, India

Abstract. The objectives of this article was to review the literature on image analysis studies. The review article discussed various contemporary topics and studies performed by researchers in last five years. The various topics discussed are Advances in Biomedical imaging, Big data work flow for biomedical image analysis, Biomedical Image Analysis of Micro-bubbles in Dental Ultrasonic Scalars, Dynamic Programming Based Segmentation in Biomedical Imaging, Thermal Image Analysis using Serpentine method, A Review of Novel Approaches In Orthopaedic And Endoscopy and Biomedical Image Analysis of Obturated Root Canal: A Proposed Approach etc. The advances in biomedical image analysis are discussed based on the transform type used and fusion type used. The various transforms such as Laplace, Wavelet, Shearlet, Hilbert, Warbler, Tunable and Q Hadamard etc. The various types of fusions are used by authors to calculate the accuracy but there are certain limitations which we have discussed. The big data workflow process is discussed in detail. The biomedical image analysis for micro-bubble of dental ultrasonic scalar is reviewed. M-tracking is used for calculating the bubble radius and speed of bubble for analysis purpose. The M-tracking plugin helps to track the location of bubble. The cavitation is one of the most effective method to remove the bio-film of biomedical surfaces. The dynamic based programming helps to highlight the lines, contour and organ margin or location. The biomedical image analysis has its four quadrant viz. physics, medical imaging, machine learning and image processing and graphics. All the above discussed studies provides sound basis for future research. The biomedical image analysis of obturated root canal using pixel programme is proposed work by the authors of this review article, is also discussed. This review article is helpful and informative to Ph.D. scholars, researchers, decision makers and experts in the field of biomedical image analysis. The review is also useful in interdisciplinary fields which are concerned with biomedical image analysis. In future the biomedical image analysis can be effectively implemented for faster diagnosis, qualitative analysis of obturation.

Keywords: Biomedical image analysis of obturated root canal ·
Dental ultrasonic scalars · Big data work flow · Serpentine method

© Springer Nature Singapore Pte Ltd. 2019
K. C. Santosh and R. S. Hegadi (Eds.): RTIP2R 2018, CCIS 1036, pp. 84–96, 2019.
https://doi.org/10.1007/978-981-13-9184-2_7

1   Introduction

Biomedical image analysis has become the prime concerns in natural sciences and medical field. Biomedical signal analysis and processing have widened their interests in academic as well as research and development departments [1]. Processing the biomedical information is fast growing field. These signals are used by designers for designing the biomedical systems. The biomedical signals are of two important uses viz. diagnosis and interpretation of health information of patient. There are various advanced bio-image analysis techniques are available viz. electrocardiogram, electroneurogram, electromyogram, electrogastrogram, phonocardiogram, signal from catheter sensor, vibromyogram, vibroarthrogram, otoacoustic emission signal, ultra-sound imaging, micro-computed tomography, positron imaging and functional magnetic resonance imaging etc. [2]. The big data process is classified into two broad types viz. data management and analytics and storage. There are various steps for processing big data viz. clinical image acquisition, extraction, annotation, cleaning, integration, representation, modeling, classification, prediction, decision making, validation, transformation and share/storage etc. [3]. The cavitations is one of the most effective method to remove the bio-film of biomedical surfaces [4]. Biomedical image analysis for micro-bubbles in dental ultrasonic scalars is used to perform this task. To avoid the noisy data and fuzzy edges the dynamic programming based segmentation in biomedical imaging is done. This helps automatically detection of organ boundaries, contour and lines etc. Thermal imaging is method of improving the visibility of object in dark room by detecting object infrared radiations and it creates image based on that information [5].

Under the title of this review article, we are going to discuss Advances in Biomedical imaging, Big data work flow for biomedical image analysis, Biomedical Image Analysis of Micro-bubbles in Dental Ultrasonic Scalars, Dynamic Programming Based Segmentation in Biomedical Imaging, Thermal Image Analysis using Serpentine method, A novel approaches in orthopaedic and endoscopy and the Proposed Biomedical Image Analysis of Obturated Root Canal etc. (Figs. 1 and 2).

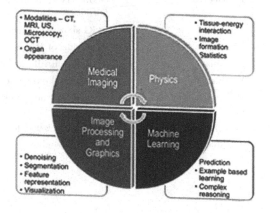

Fig. 1. Overview of biomedical image analysis.

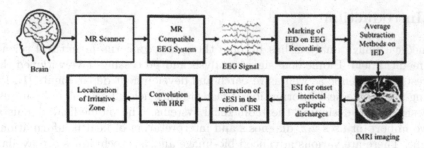

Fig. 2. Modified fusion process of fMRI and EEG signal imaging - a salient approach [1].

2   A Review of Advances in Biomedical Image Analysis

Processing the biomedical information is fast growing filed of natural sciences. These signals are used by designers for designing the biomedical systems. The biomedical signals are of two important uses viz. diagnosis and interpretation of health information of patient. The biomedical imaging signals are categorized into two broad classes viz. event potential and action potential. The Electrocardiogram, Electro-neurogram, Electro-myogram are considered as action potential. The Electrogastrogram, Phonocardiogram, signal from catheter sensor, vibromyogram, vibroarthrogram, Oto acoustic emission signal are event potential. Apart from these there are various advanced imaging techniques used are Ultrasound imaging, Micro-computed tomography, Positron imaging, Functional magnetic resonance imaging etc. (Tables 1 and 2).

Table 1. Review of biomedical image analysis studies.

Author	Type of transform	Aim of study
Gutierrej et al. [6]	Wavelet	ECG signal categorization
He et al. [7]	Laplace	Geometry estimation from body potential
Jero et al. [8]	Curvelet	ECG signal information security
Sahoo et al. [9]	Hilbert	Detection of R peak of ECG signal
Annavarapu et al. [10]	Hadamard	Artificial fibrillation detection
Kazemi et al. [11]	Warbler	Vital signal extraction from ECG
Bian et al. [12]	Wavelet packet decomposition	Evoked and non-evoked potential from ECG
Amorim et al. [13]	Shearlet and contourlet	Epileptic seizure detection
Patidar et al. [14]	Tunable Q	Epileptic seizures

Table 2. Review of fusion studies in biomedical image analysis.

Author	Type of fusion	Limitation
Mjahad et al. [15]	Time frequency and ECG	Computationally more complex
Arenja et al. [16]	fMRI and ECG	Unavailability of modality comparison
Vulliemoz et al. [17]	fMRI and EEG	Limitation not available
Hinterberger et al. [18]	fMRI and CBI	Limitation not available
Zhang et al. [19]	fMRI and EEG	Unilateral condition applicability
Darbari et al. [20]	fMRI and EEG	Sample size issue
Cagnie et al. [21]	fMRI and EMG	Testing under controlled condition is required
Hassanien et al. [22]	fMRI and EMG	Abduction problem

3 Big Data Workflow for Biomedical Image Analysis

This section discussed the review of method to process big data in biomedical image analysis. The big data process is classified into two broad types viz. data management and analytics and storage. However, there is further explanation of these broad classes is given below:

3.1 Clinical Image Acquisition

Computed tomography, Magnetic resonance imaging, X-ray, Photo-acoustic imaging, fluoroscopy have got much better importance in clinical settings [33, 34]. The images taken by these techniques provides higher resolution and large image size. These techniques gives the internal body parts information as output. The smart phones and cameras can be used as acquisition devices along with these techniques. Such image captured are then transferred to big data processing system.

3.2 Extraction, Annotation and Cleaning

In extraction, only useful biological data is taken out from entire raw data and can be used for further processing. In cleaning, the unwanted noise is eliminated with the help of filter. Annotation refers to the addition of the information externally which is related to concern technique.

3.3 Integration and Representation

In this stage the auto-clustering of the images in big database is done. Image preview can be done and after this analysis can be performed.

3.4 Modeling

Modelling provides accessory steps and can be applied when required. The modelling has to rely upon the computational algorithms and mathematical models. Modelling makes it convenient to convert 3D models into 2D.

3.5 Classification

The classification refers to divide the data under different heading as per the concerns of data. In case of big data work flow the classification is done with help of support vector machine based on supervised learning algorithm.

3.6 Decision and Prediction

The prediction and decision making is done at this stage. The convolutional neural network algorithm can be used to perform this task.

3.7 Validation

The validation is done with help of specificity and sensitivity factor.

$$Sensitivity = TruePositive(TP)/Positive(P) \tag{1}$$

$$Sensitivity = TrueNegative(TN)/Negative(N) \tag{2}$$

where, TP = No. of symptoms correctly predicted for image; P = Total symptoms; TN = No. of symptoms correctly predicted for benign; N = Total no. of benign symptoms.

3.8 Transformation

Transformation refers to transformation of images. This is also called as Compression.

3.9 Share/Storage

This is big data application used for both SQL and NoSQL technologies (Fig. 3).

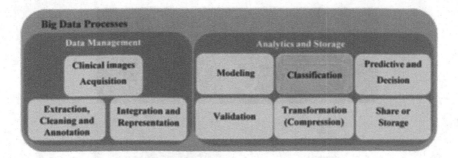

Fig. 3. Big data workflow for biomedical image analysis [23].

4   Biomedical Image Analysis of Micro-bubbles in Dental Ultrasonic Scalars

The mechanism of individual bubble cavitation is of prime importance in hydraulics, ultrasound cleaning and biomechanics. The cavitation is one of the most effective method to remove the bio-film of biomedical surfaces. Vyas et al. conducted study with aim of characterization of the individual micro-bubbles using imaging and analysis. The author used two important calculation viz. bubble radius calculation and bubble wall speed. They calculated the bubble radius at each and every point and plotted verses the time as function. However the bubble collapse and growth calculated using M track manual software plug-in in Fiji. The formula for calculating the bubble radius is given as,

$$r = \sqrt{a/\pi} \tag{3}$$

where, a = Bubble area under consideration.

With the help of high speed image analysis performed in this article, ultra-sonic scaler shown the cavitation in form of clouds. The individual micro-bubbles

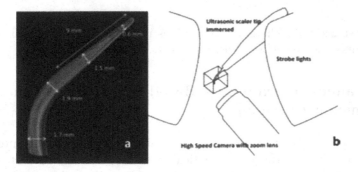

Fig. 4. (a) Example of m-track plug in showing collapse and growth of 8 point bubble (b) Mean speed measurement (c) M-tracking bubble used for tracking (d) Original image sequence of bubble [24].

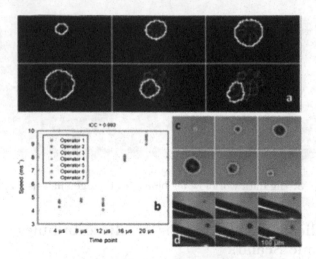

Fig. 5. (a) Micro-computed tomography 3 dimensional construction of tip (b) Experimental setup schematic with high speed camera [24].

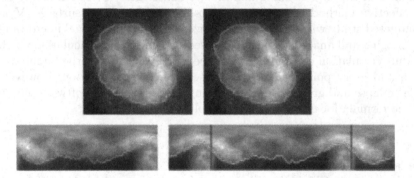

Fig. 6. (a) Image analysis using shortest path (b) Image analysis using circular shortest path (c) Shortest path using polar space (d) Circular shortest path in polar space [25].

also got visualization. The image analysis is really helpful way to calculate the bubble radius and speed over entire life cycle (Fig. 4).

5   Dynamic Programming Based Segmentation in Biomedical Imaging

There are wide range of biomedical application which demands automation. It is expected by experts to automatically detect the lines, contour or organ boundaries etc. But there is always problem arises viz. noisy data and fuzzy edges. To solve this issue dynamic programming based segmentation is performed by Ungru et al. (Figs. 5 and 6).

5.1 Shortest Path Problem

In graph notes are connected with edges. The path is connection of these several nodes with edges. Each edges is related with some weight called as cost. A shortest path algorithm is applied to solve the minimization problems. The energy level is sum of energy of data and energy of prior.

5.2 Circular Shortest Path Problem

The circular shortest path algorithm is special category which having condition that start and end nodes must be connected. The circular structures are converted into polar representation for analysis (Fig. 7).

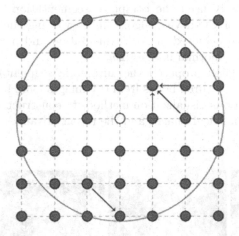

Fig. 7. The nodes and neighbour nodes ordered by polar angle [25].

6   A Review of Novel Approaches in Orthopedic and Endoscopy Using Image Analysis

Hegadi et al. in 2017, proposed the automatic segmentation algorithm for synovial activity using image processing technique and used this algorithm arthritis called osteoarthritis. The osteoarthritis is primary cause of the protective cartilage bone loss because of wear and tear. The steps of the automatic segmentation algorithm are Image acquisition, Image preprocessing, Thresholding, Cropping and detection of ROI and Segmentation of synovial cavity region etc. The study used X-ray images to perform the experimentation. The automatic segmentation techniques is compared with the manual segmentation technique to get the segmented part of synovial cavity region [26, 35].

Hegadi et al. in 2010, worked on Discrete Topological Derivative approach for segmentation of tumours using Endoscopic images. The topological derivatives devised for topology optimization problems. The algorithm when applied to image processing it has proved to be effective tool for segmentation process. Thus for the segmentation process the information is gathered from Discrete Topological Derivative function [27].

Navale et al. in 2015, used the block based texture analysis approach for knee osteoarthritis using support vector machine. The X-ray images of 20 patients were collected and were segmented into blocks. The each block then applied with texture analysis algorithm. The result of the investigation indicated that the texture analysis algorithm is useful to detect the disease and helpful to describe the image information of osteoarthritis [28].

Ravi et al. in 2015, have proposed the glomerulosclerosis in diabetic nephropathy. The study used the computer segmentation. The study consist of the modifications of Chan-Vese algorithm by varying the parameter values. This modified method achieved better segmentation and provided the solution to glomerulosclerosis abnormal images [29].

Ruikar et al. in 2019 proposed the automatic segmentation technique to extract and label the fractured bones from CT images [33, 34]. Further in [36, 37] they surveyed the various visualization methods to construct anatomically accurate 3D model from medical images and the recent advances in orthopedic simulator development.

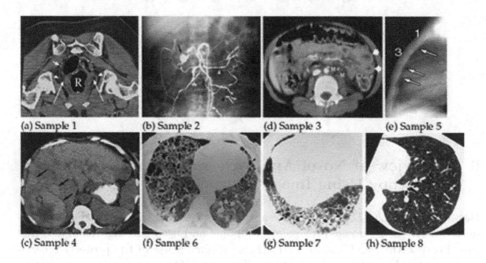

(a) Sample 1   (b) Sample 2   (d) Sample 3   (e) Sample 5

(c) Sample 4   (f) Sample 6   (g) Sample 7   (h) Sample 8

Fig. 8. Arrows in biomedical image [31].

Santosh et al. in 2014, presented the novel method for detection of lung rotation using chest radiograph. The method used generalized line histogram technique to achieve the quality control which also helped to detect the abnormality. The proposed technique can also be used for computer aided diagnosis research. The study achieved the accuracy of 90% [30]. In 2015, he used the overlaid arrow detection technique to label the region of interest for captured biomedical images. The study used the geometric based arrow annotations for generating the biomedical image samples. The sample image of the arrows in biomedical image is shown in Fig. 8.

7 The Biomedical Image Analysis of Obturated Root Canal Using Pixel Programs - A Proposed Approach

The image analysis using the radio graphs and X ray is commonly being used by medical practitioners for easy and speedy diagnosis of patient health. The proposed work by the authors of this review article consist of estimation of the percentage filling of the obturated root canal using pixel program. The estimation of percentage filling of the canal has got the prime importance since the success of every root canal treatment depends upon the micro leakage of oral fluids. The incomplete filling of the canal leads to micro leakage. It will be helpful for the Dental practitioners if they get the percentage filling of the canal in few minutes. The image analysis can help if used to estimate percentage filling of the canal using the MATLAB pixel programs. The proposed procedure is to divide the image into the number of the small squares. While dividing the image into squares we have to divide the image into as small as dimension square since the accuracy of estimation of percentage filling is depend upon the size of square. The white colour indicates the presence of the filling materials, if we count the squares which are black and divide with total number of squares we can get the percentage filling. For obtaining greater accuracy the image should be divided into the small squares but it is difficult to count the number of squares manually. The MATLAB pixel programming can be used to deal with this difficulty. The number of white squares and total number of squares thus can be counted with MATLAB pixel program. We can write the program to get percentage filling in MATLAB. However, this approach is suitable only for images where dual colour pattern of image taken is available.

8 Discussion

This review article is helpful and informative to Ph.D. scholars, researchers, decision makers and experts in the field of biomedical image analysis. The review is also useful in interdisciplinary fields which are concerned with biomedical image analysis.

The review article discussed various contemporary topics and studies performed by researchers in last five years. The various topics discussed are Advances in Biomedical imaging, Big data work flow for biomedical image analysis, Biomedical Image Analysis of Micro-bubbles in Dental Ultrasonic Scalars, Dynamic Programming Based Segmentation in Biomedical Imaging and Thermal Image Analysis using Serpentine method etc. A novel approaches in orthopaedic and endoscopy and the Proposed Biomedical Image Analysis of Obturated Root Canal are discussed in detail.

The advances in biomedical image analysis are discussed based on the transform type used and fusion type used. The various transforms such as Laplace, Wavelet, Shearlet, Hilbert, Warbler, Tunable and Q Hadamard etc. The various types of fusions are used by authors to calculate the accuracy but there are certain limitations which we have discussed. The big data work flow process is discussed in detail. The biomedical image analysis for micro-bubble of dental ultrasonic scalar is reviewed. M-tracking is used for calculating the bubble radius and speed of bubble for analysis purpose. The M-tracking plugin helps to track the location of bubble. The cavitation is one of the most effective method to remove the bio-film of biomedical surfaces. The dynamic based programming helps to highlight the lines, contour and organ margin or location. The biomedical image analysis has its four quadrant viz. physics, medical imaging, machine learning and image processing and graphics. All the above discussed studies provides sound basis for future research.

Acknowledgements. Author acknowledges the support and guidance received from Dr. Vivek Hagade of M. A. Rangoonwala College of Dental Sciences and Research Centre Pune, India, Dr. Srinidhi S. R. of Sinhgad Dental College and Hospital Pune, India, Shivam Dental Laboratory Pune, India, Sculpt dent Dental Laboratory Ghorpadi, Pune, India for this PhD research work.

Author thanks to Dr. S. Koteeswaran (Dean-research studies), Dr. A. T. Ravichandran (Head of mechanical engineering department) of Veltech University, Chennai, India for approval of topic and for their insightful comments, encouragement and love. Author sincerely acknowledges the support received from Dr. S. D. Lokhande, Principal, Sinhgad College of Engineering Pune, India.

References

1. Rajeswari, J., Jagannath, M.: Advances in biomedical signal and image processing-a systematic review. Inform. Med. Unlocked **8**, 13–19 (2017)
2. Ciaccio, E.J.: Biomedical Signal and Image Processing, Review of Biomedical Signal and Image Processing. CRC Press, Taylor and Francis Group, Boca Raton (2013). Review by Edward J. Ciaccio
3. Schindelin, J., Rueden, C.T., Hiner, M.C., Eliceiri, K.W.: The ImageJ ecosystem: an open platform for biomedical image analysis. Mol. Reprod. Dev. **82**(7–8), 518–529 (2015)
4. Timp, S., Karssemeijer, N.: A new 2D segmentation method based on dynamic programming applied to computer aided detection in mammography. Med. Phys. **31**(5), 958–971 (2004)
5. Ring, E.F.J., Ammer, K.: Infrared thermal imaging in medicine. Physiol. Meas. **33**(3), R33 (2012)
6. Gutiérrez-Gnecchi, J.A., et al.: DSP-based arrhythmia classification using wavelet transform and probabilistic neural network. Biomed. Signal Process. Control **32**, 44–56 (2017)
7. He, B., Li, G., Lian, J.: A spline Laplacian ECG estimator in a realistic geometry volume conductor. IEEE Trans. Biomed. Eng. **49**(2), 110–117 (2002)

8. He, B.: Brain electric source imaging: scalp Laplacian mapping and cortical imaging. Crit. Rev. Biomed. Eng. **27**(3–5), 149–188 (1999)
9. Sahoo, S., Biswal, P., Das, T., Sabut, S.: De-noising of ECG signal and QRS detection using Hilbert transform and adaptive thresholding. Procedia Technol. **25**, 68–75 (2016)
10. Annavarapu, A., Kora, P.: ECG-based atrial fibrillation detection using different orderings of Conjugate Symmetric-Complex Hadamard Transform. Int. J. Cardiovasc. Acad. **2**(3), 151–154 (2016)
11. Kazemi, S., Ghorbani, A., Amindavar, H., Morgan, D.R.: Vital-sign extraction using bootstrap-based generalized warblet transform in heart and respiration monitoring radar system. IEEE Trans. Instrum. Meas. **65**(2), 255–263 (2016)
12. Bian, Y., Li, H., Zhao, L., Yang, G., Geng, L.: Research on steady state visual evoked potentials based on wavelet packet technology for brain-computer interface. Procedia Eng. **15**, 2629–2633 (2011)
13. Amorim, P., Moraes, T., Fazanaro, D., Silva, J., Pedrini, H.: Electroencephalogram signal classification based on shearlet and contourlet transforms. Expert Syst. Appl. **67**, 140–147 (2017)
14. Patidar, S., Panigrahi, T.: Detection of epileptic seizure using Kraskov entropy applied on tunable-Q wavelet transform of EEG signals. Biomed. Signal Process. Control **34**, 74–80 (2017)
15. Mjahad, A., Rosado-Muñoz, A., Bataller-Mompeán, M., Francés-Víllora, J.V., Guerre-ro-Martínez, J.F.: Ventricular Fibrillation and Tachycardia detection from surface ECG using time-frequency representation images as input dataset for machine learning. Comput. Methods Programs Biomed. **141**, 119–127 (2017)
16. Arenja, N., et al.: Right ventricular long axis strain-validation of a novel parameter in non-ischemic dilated cardiomyopathy using standard cardiac magnetic resonance imaging. Eur. J. Radiol. **85**(7), 1322–1328 (2016)
17. Vuilleumier, P., Pourtois, G.: Distributed and interactive brain mechanisms during emotion face perception: evidence from functional neuroimaging. Neuropsychologia **45**(1), 174–194 (2007)
18. Hinterberger, T., Weiskopf, N., Veit, R., Wilhelm, B., Betta, E., Birbaumer, N.: An EEG-driven brain-computer interface combined with functional magnetic resonance imaging (fMRI). IEEE Trans. Biomed. Eng. **51**(6), 971–974 (2004)
19. Zhang, C.H., Lu, Y., Brinkmann, B., Welker, K., Worrell, G., He, B.: Lateralization and localization of epilepsy related hemodynamic foci using presurgical fMRI. Clin. Neurophysiol. **126**(1), 27–38 (2015)
20. Darbari, D.S., et al.: Frequency of hospitalizations for pain and association with altered brain network connectivity in sickle cell disease. J. Pain **16**(11), 1077–1086 (2015)
21. Cagnie, B., Dirks, R., Schouten, M., Parlevliet, T., Cambier, D., Danneels, L.: Functional reorganization of cervical flexor activity because of induced muscle pain evaluated by muscle functional magnetic resonance imaging. Manual Ther. **16**(5), 470–475 (2011)
22. Hassanien, O.A., Younes, R.L., Dawoud, R.M., Younis, L.M., Hamoda, I.M.: Reliable MRI and MRN signs of nerve and muscle injury following trauma to the shoulder with EMG and Clinical correlation. Egypt. J. Radiol. Nucl. Med. **47**(3), 929–936 (2016)
23. Kouanou, A.T., Tchiotsop, D., Kengne, R., Tansaa, Z.D., Adele, N.M., Tchinda, R.: An optimal big data workflow for biomedical image analysis. Inform. Med. Unlocked **11**, 68–74 (2018)

24. Vyas, N., Dehghani, H., Sammons, R.L., Wang, Q.X., Leppinen, D.M., Walmsley, A.D.: Imaging and analysis of individual cavitation microbubbles around dental ultrasonic scalers. Ultrasonics **81**, 66–72 (2017)
25. Ungru, K., Jiang, X.: Dynamic programming based segmentation in biomedical imaging. Comput. Struct. Biotechnol. J. **15**, 255–264 (2017)
26. Hegadi, R.S., Navale, D.I.: Quantification of synovial cavity from knee X-ray images. In: 2017 International Conference on Energy, Communication, Data Analytics and Soft Computing (ICECDS), pp. 1688–1691. IEEE, August 2017
27. Hegadi, R.S.: Segmentation of tumors from endoscopic images using topological derivatives based on discrete approach. In: 2010 International Conference on Signal and Image Processing (ICSIP), pp. 54–58. IEEE, December 2010
28. Navale, D.I., Hegadi, R.S., Mendgudli, N.: Block based texture analysis approach for knee osteoarthritis identification using SVM. In: 2015 IEEE International WIE Conference on Electrical and Computer Engineering (WIECON-ECE), pp. 338–341. IEEE, December 2015
29. Ravi, M., Hegadi, R.S.: Detection of glomerulosclerosis in diabetic nephropathy using contour-based segmentation. Procedia Comput. Sci. **45**, 244–249 (2015)
30. Santosh, K.C., et al.: Automatically detecting rotation in chest radiographs using principal rib-orientation measure for quality control. Int. J. Pattern Recogn. Artif. Intell. **29**(02), 1557001 (2015)
31. Santosh, K.C., Wendling, L., Antani, S., Thoma, G.R.: Overlaid arrow detection for labeling regions of interest in biomedical images. IEEE Intell. Syst. **31**(3), 66–75 (2016)
32. Santosh, K.C., Wendling, L.: Angular relational signature-based chest radiograph image view classification. Med. Biol. Eng. Comput. 1–12 (2018). https://doi.org/10.1007/s11517-018-1786-3
33. Ruikar, D.D., Santosh, K.C., Hegadi, R.S.: Automated fractured bone segmentation and labeling from CT images. J. Med. Syst. **43**(3), 60 (2019). https://doi.org/10.1007/s10916-019-1176-x
34. Ruikar, D.D., Santosh, K.C., Hegadi, R.S.: Segmentation and analysis of CT images for bone fracture detection and labeling. In: Medical Imaging: Artificial Intelligence, Image Recognition, and Machine Learning Techniques, chap. 7. CRC Press, Boca Raton (2019). ISBN 9780367139612
35. Hegadi, R.S., Navale, D.I., Pawar, T.D., Ruikar, D.D.: Multi feature-based classification of osteoarthritis in knee joint X-ray images. In: Medical Imaging: Artificial Intelligence, Image Recognition, and Machine Learning Techniques, chap. 5. CRC Press, Boca Raton (2019). ISBN 9780367139612
36. Ruikar, D.D., Sawat, D.D., Santosh, K.C., Hegadi, R.S.: 3D imaging in biomedical applications: a systematic review. In: Medical Imaging: Artificial Intelligence, Image Recognition, and Machine Learning Techniques, chap. 8. CRC Press, Boca Raton (2019). ISBN 9780367139612
37. Ruikar, D.D., Hegadi, R.S., Santosh, K.C.: A systematic review on orthopedic simulators for psycho-motor skill and surgical procedure training. J. Med. Syst. **42**(9), 168 (2018)

Osteoarthritis Detection and Classification from Knee X-Ray Images Based on Artificial Neural Network

Ravindra S. Hegadi, Dattatray I. Navale, Trupti D. Pawar, and Darshan D. Ruikar[✉]

Department of Computer Science, Solapur University,
Solapur 413255, Maharashtra, India
{rshegadi,dinavale,tdpawar,ddruikar}@sus.ac.in

Abstract. Osteoarthritis is a chronic and degenerative disorder of multifactorial etiology. Knee, hip and spine joints are majorly affected by the Osteoarthritis due to loss of articular cartilage, bone remodeling (due to accidents, for instance) and heavy weight-bearing on joints. In orthopedics, it is one of the most occurring disorder nowadays. Hence an effective computer-aided diagnosis (CAD) system is required to diagnose the Osteoarthritis. This paper presents a simple artificial neural network (ANN) based classification system to differentiate between healthy and affected knee X-ray images. In the proposed system, guided filter and adaptive histogram equalization techniques are respectively used for noise removal and image enhancement. Global thresholding-based segmentation technique is adapted to extract synovial cavity region from an image and curvature values (like mean, standard deviation, range, and skewness) are computed. To draw a fair conclusion, the experiments are conducted on real patient-specific images collected from local hospitals in India. By confirming the results, the proposed method accurately classifies the inputted image to their respective classes.

Keywords: Artificial neural network · Osteoarthritis ·
Curvature analysis · Learning

1 Introduction

Osteoarthritis (OA) frequently occurs in the weight-bearing joints like the hips, knees, and spine [18]. Along with this, some other organs such as large toe, neck, fingers, and thumb may also affect. Reduction of articular cartilage in joint is the major reason behind OA. Joint aching, soreness and pain, especially during joint movement or after long periods of inactivity stiffness are the some common symptoms of OA [10]. The osteoarthritis may occur at any stage of life; but, more likely chances are more old age. According to [18] some degree of OA is detected in most of people with age more than forty, but the severity may with individual. Even people in their early age may also suffer from osteoarthritis,

© Springer Nature Singapore Pte Ltd. 2019
K. C. Santosh and R. S. Hegadi (Eds.): RTIP2R 2018, CCIS 1036, pp. 97–105, 2019.
https://doi.org/10.1007/978-981-13-9184-2_8

due to joint injury or repetitive joint stress from overuse. Commonly after fifty, the probability of osteoarthritis affection is relatively more in women than men [8]. The Fig. 1 shows healthy and affected knee joint. Depending on severity, the Osteoarthritis can be categorized into five grades, from zero to four [16].

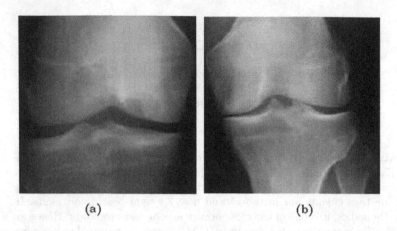

(a)   (b)

Fig. 1. Knee X-ray images of (a) healthy and (b) affected by osteoarthritis

Commonly experts uses X-ray images to diagnose the osteoarthritis [10]. Because changes in inflammation at the joints affected by osteoarthritis is better visible in X-ray images. The correct diagnosis and finding optimal recovery plan for osteoarthritis in the higher stage is a very cumbersome task for experts, inaccuracy may lead to permanent damage. So nowadays the field orthopedics is adapting the technological advancement to analyze and diagnose the injury moreover to decide suitable recovery plan [12,15]. In this paper, we proposed a simple, ANN-based knee osteoarthritis classification system to classify input X-ray images to appropriate class (healthy or affected). To achieve accurate results guided filter and an effective contrast limiting adaptive histogram equalization (CLAHE) based [1,6] noise removal and image enhancement techniques is developed to remove unwanted artifacts from the image. A global thresholding-based segmentation technique is adapted to segment synovial cavity, and derivative-based curvature computation is performed to classify the image to their respective classes (health and affected).

The paper is organized as follows. Section 2 explores the related work. Section 4 describes the proposed methodology in detail. Experimental results are demonstrated in Sect. 4. Conclusion is presented in Sect. 5.

2   Related Work

From last few decades, many researchers are actively working to devise a precise classification system to classify given X-ray image to appropriate class according to the severity of knee osteoarthritis.

Hegadi et al. [10] used block-based texture analysis approach to identify the synovial cavity region and support vector machine (SVM) classifier is adapted to classify the images. Nine equally sized blocks of single inputted images are considered to extract four different texture features (skewness, kurtosis, standard deviation and energy). In total 36 features are extracted and stored in feature vector to achieve desired accuracy. The proposed method demonstrated 80% and 86.7% accuracy for the healthy and affected images respectively.

Antony et al. [3] proposed a deep convolution neural networks (CNN) based classification technique to automatically quantify the severity of knee osteoarthritis (OA) from radio-graphs. The authors demonstrated 95.2% accuracy on Osteoarthritis Initiative (OAI) dataset obtained from 4,476 participants using linear SVM classifier with 5-fold cross validation.

Anifah et al. [2] proposed classification technique based on the self-organizing map. On the proposed work contrast limited adaptive histogram equalization (CLAHE) is used to increase the contrast of the image. The decision of whether an image is of left or right knee is based on template matching technique. To identify the synovial cavity region, they segmented knee image using Gabor kernel, template matching, row sum graph and gray level center of the mass method. Gray level co-occurrence matrix (GLCM) features are used for further classification of data in five grades of OA. They concluded that the accuracy of the classification of images having grade 0 and 4 type of osteoarthritis is high than the images having grade 1 to grade 3. They used three features like accuracy; specificity and sensitivity for the evaluation of research.

Knee X-ray image-based OA severity staging system is presented in [17]. The region-proposal and object classification neural network is respectively used to identify exact knee-joint region and to classify the image. The proposed method demonstartes 99.9% accuracy.

A volumetric CNN-based knee cartilage segmentation and evaluation technique is developed in [11]. The proposed method shows better accuracy on the MICCAI SKI10 public challenge using 5-fold cross-validation.

In the literature several pre-processing techniques are used to increase the contrast and to improve the dynamic range of inputted image [13,14]. Different segmentation methods were adapted to extract synovial cavity region from the input image. At last some classifier is applied to segregate the input to correct class (normal and abnormal) moreover to assign grade to these images.

3   Methodology

A detailed explanation of the proposed classification technique is discussed in this section.

3.1   Preprocessing

In the presented work, patient-specific X-ray images are collected in DICOM[1] format and converted those into a grayscale format for further processing. During X-ray image formation, the image is affected by noise. The guided filter [4] is used to reduce the noise. There are three main reasons for adapting the guided filter for noise removal. These are (a) time efficiency of the guided filter is more than the linear filtering, (b) guided filter does selective filtering (i.e., edges and high varying frequencies components are preserved during filtering process) and (c) the amount of smoothing is based on amount on local noise i.e. smoothing will be lesser for higher local noise and it is more for lower local noise.

The next step is contrast enhancement. Contrast-limited adaptive histogram equalization (CLAHE) [1] is used to increase the dynamic range of image through histogram equalization. In CLAHE, image is divided into smaller regions to enhance the intensity values of particular pixels. The bilinear interpolation technique is used to join the enhanced region.

The 3×3 block size opts for fine level local processing whereas the parameter clip limit is set to 0.01 to balance the contrast.

After noise removal and contrast enhancement, the images are suitable for extracting the synovial cavity region by using the appropriate segmentation method. In the proposed work, a global thresholding-based segmentation method is adapted to extract the synovial cavity region from an enhanced image. Figure 2 shows the resultant images obtained after applying various preprocessing stages. Figure 2(a) is original image whereas Fig. 2(b) and (c) shows the resultant images obtained after preprocessing step (i.e., noise removal and adaptive contrast enhancement process) and segmentation step respectively.

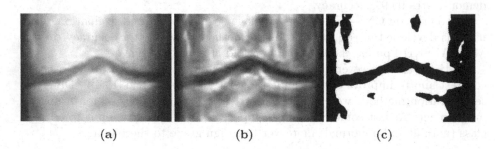

(a)   (b)   (c)

Fig. 2. Stages of image pre-processing (a) original image, image after (b) noise removal and (c) segmentation

3.2   Boundary Extraction

The basic purpose of extracting boundaries from the segmented images is to locate the boundary of the synovial cavity region. The synovial cavity is present

---

[1] DICOM: Digital Imaging and Communications in Medicine.

in the gap of two bones: femur and tibia. However, there are possibilities for generating the boundaries of false objects which are outside to region of the synovial cavity. In general the edges are smaller in size. Such false contours are removed, whereas the remaining (i.e. longer) are considered for further processing. To remove some unwanted smaller objects area open morphological operation is performed. Further to trace exact boundary locations simple morphological boundary extraction tools are used with 3×3 structuring element. The Eq. 1 is used to extract the boundaries from the segmented image.

$$I_{Boundary} = I - (I \ominus B) \tag{1}$$

where I, B and $\ominus$ is inputted image, 3×3 structuring element and the morphological erosion operation respectively. The result of extracted boundaries is shown in Fig. 3.

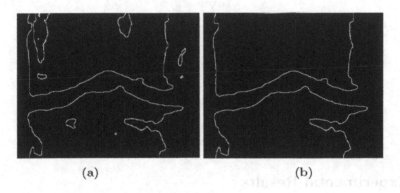

(a)   (b)

Fig. 3. Stages of boundary extraction process (a) with and (b) without the presence of false object boundaries

3.3   Edge Curvature Computation

The classification of input image to correct class (i.e. normal or affected) is based the features of the synovial cavity region. More generally curvature values are shows less variation in healthy image and the variation in curvature values are more in the affected image. In the proposed work, curvature computation is based on derivatives [7]. The Eq. 2 represents the curvature of a continuous curve C. The curve is conveyed as $\{x(s), y(s)\}$, where the length of edge points is denoted by s.

$$k(s) = \frac{\dot{x}\ddot{y} - \ddot{x}\dot{y}}{(\dot{x}^2 + \dot{y}^2)^{3/2}} \tag{2}$$

where $\dot{x} = dx/ds$, $\ddot{x} = d^2x/ds^2$, $\dot{y} = dy/ds$ and $\ddot{y} = d^2y/ds^2$. The above expression of curvature is in continuous form. The continuous curvature values are considered as set of equally spaced grid samples in Cartesian coordinate system for digital implementation. The Eq. 3 represents the derivatives in Eq. 2 in first and second order derivatives form.

$$\dot{x} = x_i - x_{i-1}, \dot{y} = y_i - y_{i-1}, \ddot{x} = x_{i-1} - 2x_i + x_{i+1} \ddot{y} = y_{i-1} - 2y_i + y_{i+1}. \quad (3)$$

3.4 Classification

The features like, mean, slandered deviation, range and skewness features are calculated for each image in the dataset using the Eqs. 4, 5, 6 and 7. These features are passed to two layer feed-forward neural network system [5,9] to classify inputted image to appropriate class.

$$\bar{x} = \frac{\sum_{i=1}^{n} x_i}{n} \quad (4)$$

$$\sigma = \sqrt{\frac{\sum_{i=1}^{n} (x_i - \bar{x})^2}{n}} \quad (5)$$

$$Range = MaximumValue - MinimumValue \quad (6)$$

$$skewness = (\bar{x} - mode)/\sigma. \quad (7)$$

4 Experimental Results

For the experimentation purpose 42 images are collected from Chidgupkar hospital, India. In those 50% i.e. 21 images are with a different level of osteoarthritis and rest 50% are normal. In addition to image collection, expert guidance of radiologist is also taken to make fare decision. Personal Computer having I7 processor and 8 GB RAM with Matlab R2016a is used for experimentation.

To train neural network (NN) classifier 80% i.e. 34 input samples are used and 10% i.e. 4 samples each are taken for testing and validation. The classifier gives its best performance at 14 epoch yielding 100% training and testing accuracy. The cross entropy is around and error (miss-classification) rate is 0%. The receiver operating characteristic (ROC) curve for trained classifier is shown in Fig. 4.

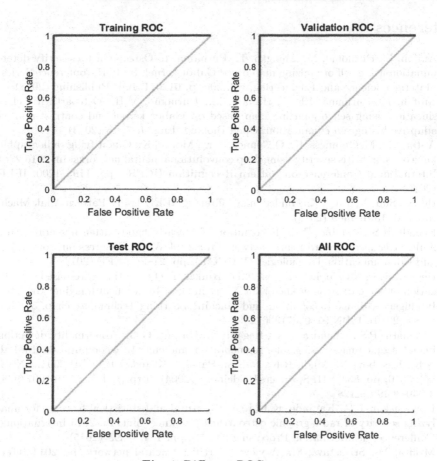

Fig. 4. Different ROC curves

5   Conclusion

Development of expert systems for osteoarthritis detection and classification are in demand. The proposed method precisely extracts the synovial cavity region. Based on segmentation results it rightly classify the images into two classes: normal and affected. Confirming to results, proposed method yield 100% accuracy. In the future we aim to develop a robust expert system to classify images into different grades of OA.

Acknowledgment. We would like to thank Chidgupkar Hospital Pvt. Ltd., India, for providing us the x-ray images and valuable guidance to conduct our experimentation.

References

1. Anifah, L., Purnomo, M., Mengko, T., Purnama, I.: Osteoarthritis severity determination using self organizing map based Gabor kernel. In: IOP Conference Series: Materials Science and Engineering, vol. 306, p. 012071. IOP Publishing (2018)
2. Anifah, L., Purnama, I.K.E., Hariadi, M., Purnomo, M.H.: Osteoarthritis classification using self organizing map based on Gabor kernel and contrast-limited adaptive histogram equalization. Open Biomed. Eng. J. **7**, 18 (2013)
3. Antony, J., McGuinness, K., O'Connor, N.E., Moran, K.: Quantifying radiographic knee osteoarthritis severity using deep convolutional neural networks. In: 2016 23rd International Conference on Pattern Recognition (ICPR), pp. 1195–1200. IEEE (2016)
4. He, K., Sun, J., Tang, X.: Guided image filtering. IEEE Trans. Pattern Anal. Mach. Intell. **6**, 1397–1409 (2013)
5. Hegadi, R.S., Kamble, P.M.: Recognition of Marathi handwritten numerals using multi-layer feed-forward neural network. In: 2014 World Congress on Computing and Communication Technologies (WCCCT), pp. 21–24. IEEE (2014)
6. Hegadi, R.S., Navale, D.I., Pawar, T.D., Ruikar, D.D.: Multi feature-based classification of osteoarthritis in knee joint X-ray images. In: Medical Imaging: Artificial Intelligence, Image Recognition, and Machine Learning Techniques, chap. 5. CRC Press (2019). ISBN 9780367139612
7. Hiremath, P.S., Dhandra, B.V., Hegadi, R., Rajput, G.G.: Abnormality detection in endoscopic images using color segmentation and curvature computation. In: Pal, N.R., Kasabov, N., Mudi, R.K., Pal, S., Parui, S.K. (eds.) ICONIP 2004. LNCS, vol. 3316, pp. 834–841. Springer, Heidelberg (2004). https://doi.org/10.1007/978-3-540-30499-9_128
8. Kawathekar, P.P., Karande, K.J.: Use of textural and statistical features for analyzing severity of radio-graphic osteoarthritis of knee joint. In: 2015 International Conference on Information Processing (ICIP), pp. 1–4. IEEE (2015)
9. Mishra, M., Srivastava, M.: A view of artificial neural network. In: 2014 International Conference on Advances in Engineering and Technology Research (ICAETR), pp. 1–3. IEEE (2014)
10. Navale, D.I., Hegadi, R.S., Mendgudli, N.: Block based texture analysis approach for knee osteoarthritis identification using SVM. In: 2015 IEEE International WIE Conference on Electrical and Computer Engineering (WIECON-ECE), pp. 338–341. IEEE (2015)
11. Raj, A., Vishwanathan, S., Ajani, B., Krishnan, K., Agarwal, H.: Automatic knee cartilage segmentation using fully volumetric convolutional neural networks for evaluation of osteoarthritis. In: 2018 IEEE 15th International Symposium on Biomedical Imaging (ISBI 2018), pp. 851–854. IEEE (2018)
12. Ruikar, D.D., Hegadi, R.S., Santosh, K.: A systematic review on orthopedic simulators for psycho-motor skill and surgical procedure training. J. Med. Syst. **42**(9), 168 (2018)
13. Ruikar, D.D., Santosh, K.C., Hegadi, R.S.: Automated fractured bone segmentation and labeling from CT images. J. Med. Syst. **43**(3), 60 (2019). https://doi.org/10.1007/s10916-019-1176-x
14. Ruikar, D.D., Santosh, K.C., Hegadi, R.S.: Segmentation and analysis of CT images for bone fracture detection and labeling. In: Medical Imaging: Artificial Intelligence, Image Recognition, and Machine Learning Techniques, chap. 7. CRC Press (2019). ISBN 9780367139612

15. Ruikar, D.D., Sawat, D.D., Santosh, K.C., Hegadi, R.S.: 3D imaging in biomedical applications: a systematic review. In: Medical Imaging: Artificial Intelligence, Image Recognition, and Machine Learning Techniques, chap. 8. CRC Press (2019). ISBN 9780367139612

16. Shamir, L., et al.: Knee X-ray image analysis method for automated detection of osteoarthritis. IEEE Trans. Biomed. Eng. **56**(2), 407–415 (2009)

17. Suresha, S., Kidziński, L., Halilaj, E., Gold, G., Delp, S.: Automated staging of knee osteoarthritis severity using deep neural networks. Osteoarthritis Cartilage **26**, S441 (2018)

18. Arthritis Research UK: Osteoarthritis in general practice: data and perspectives. Arthritis Research UK, Chesterfield (2013)

Classification of Pathology Images
of Breast Cancer

Bhagirathi Halalli[1][✉] and Aziz Makandar[2]

[1] Department of Computer Science, GFGC, Raibag, Karnataka, India
bhagyaigali@gmail.com
[2] Department of Computer Science, AWU, Vijayapura, Karnataka, India

Abstract. Diagnosis and grading of breast cancer are done through pathology examination which remains the traditional way in cancer diagnosis. Pathology test is the primary standard for finding severity of abnormality in many cancer diagnoses and also plays the key role in diagnostic assessments. Recently, computerized techniques have been evolving in diagnosing digital pathology for emerging applications related to nuclei detection, segmentation, and classification. Nuclei segmentation remains the challenging task because of cell morphology and architectural distribution. Pathological studies have been conducted for detection and grading many cancers like cervix, lung, brain, prostate and breast cancer and many more. In many of the cancer diagnosis, computer-based diagnostic approaches are playing an important role in reducing human intervention, providing an accurate clinical report. Hence, medical practitioner considers computer-based detection as second opinions. This paper focused on cluster analysis using K-means and classification of breast cancer in the pathological image. The proposed method helps to classify histopathology image in four class by using Support Vector Machine with the combined set of features. The pathology image classified as normal, benign, in situ malignant, invasive malignant by hybrid feature set. The experimental result conducted on publically available dataset INESC-TEC Breast Histology Dataset. The results have proved that the proposed algorithm is suitable for classification of Histopathology image with high accuracy.

Keywords: Breast cancer · Pathology image · Segmentation · K-means cluster · Classification

1 Introduction

Detection of cancer on the earliest has been usually the principle problem in the scientific area. Approximately 32% of the Indian population is tormented by most cancers of their lifetime. It became not unusual causes for death in the U.S. Because of their current way of life they consume more tobacco and nutritional habits. Now in India additionally the most cancers emerge as common purpose a boom in mortality fee of 0.3 million in one year [1]. First of all, in

© Springer Nature Singapore Pte Ltd. 2019
K. C. Santosh and R. S. Hegadi (Eds.): RTIP2R 2018, CCIS 1036, pp. 106–115, 2019.
https://doi.org/10.1007/978-981-13-9184-2_9

lots of instances, the cancer is deadly due to past due diagnosis and a shortage of signs. The early detection may additionally decrease the mortality charge [2]. Locating an abnormality in an image is subjective in nature and shortage of particular features. The automated diagnosis systems help in reading, interpreting and assuaging the functions to categorize cancerous and noncancerous pix [3–5]. An extensive range of imaging frameworks associated with breast malignancy evaluation, for instance, Mammography, pc tomography (CT), Ultrasound, Magnetic resonance imaging (MRI), Positron Emission Tomography (PET) [6]. These strategies assist in screening degree and exam of disease. Mammography is the low-degree X-Ray imaging and its miles the simple screening take a look at and powerful approach simply noticeably distinctive strategies. In those imaging structures radiologist broadly speaking facilities on harm display inside the breast tissue and abnormality in breast tissue [7]. Scarcely, they've centered on mass type as benign or malignant. In each unmarried case threat observed after obsessive take a look at just, the percentages of getting passing are on account commonly in willpower.

The number one goal of medicinal analysis the usage of virtual imaging will increase the possibility of getting a cure [8]. The selection of remedy is for the maximum component is predicated upon the degree of hazard. For genuine locating the pathologist's makes use of a histopathology biopsy picture which analyzes infinitesimal tissue, cellular and cores shape. In this manner, the biopsy imaging is the crucial approach in the detection of tumor and its spread. Histology is taken a look at associated with cell structure, its indications for sickness. To visualize the one of a kind parts of the tissue under a microscope, the biopsy element on slide dyed with distinctive color additives. Hematoxylin-Eosin is staining component the principle aim is to visualize the evaluation level of cells. Hematoxylin is used to stain the nuclei which can be blue in shade whereas Eosin is used to stain the cytoplasm and connective tissues which might be in red in shade [4]. Some investigations for the type and grading of breast most cancers in histopathological tissue exams particular to exceptional organs were discovered within the present work. In an expansive feel, those works had been basically organized towards gland and/or nuclei segmentation [9], and ensuring characteristic extraction and class. In some works a sub-photo which changed into taken because the agent of the entire picture was subjected to characteristic extraction [10]. In a few different works a solitary image changed into part into diverse sub-images and the functions from these photos had been combined into a solitary set to characterize the complete photo. The function set comprised of both one of the following features, Fractal functions, depth primarily based capabilities, Texture, Morphological capabilities, Topological functions, or a combination of at the least two had been used for classification [11–16]. The critical boost in extracting the cells' morphological functions is its segmentation. The simple concept is to perceive the abnormality is by way of the cell and cellular cluster, medical doctors and radiologist specifically study a lot of characteristics like nuclei structure, shape, color, length and percentage to the cytoplasm in the micro-scopic image. The high-decision image affords the dependable records concerning the presence of most cancers. As a result, the proposed set of features facilitates to categorize pathology images into 4 classes.

2 Materials and Methodology

2.1 Dataset

The study considered tissue samples from breast cancer biopsies stained with Hematoxylin and Eosin. The dataset used for histopathology processing from publically available dataset Breast Histology Dataset [17] and hospital data. This dataset contains breast histology images from four classes: normal, benign, in situ carcinoma and invasive carcinoma.

2.2 Methodology

Detection and classification of nuclei in the Histopathology image of breast cancer. The process of finding cancer and classifying microscopic images to classify as normal, benign, in situ malignant, invasive malignant carried with following steps. In the first stages involves stages represented in Fig. 1 like sample collection, digitization, preprocessing, segmentation, feature extraction,and classification. These steps reviewed in detail in the following sections.

Preprocessing: Image preprocessing is a mandatory step in medical diagnostic imaging because it influences in detection and classification. In addition to this, problem specific preprocessing technique help to segment desired portion with grater accuracy. The main objective of preprocessing is boosting image quality. Biopsy images acquired from microscope hence it may consist of some deficient like uneven staining, low contrast, dust particles, air bubbles, and tissue folding. The basic task in preprocessing is noise reduction and contrast enhancement of the region of interest. Histopathology images have been enhanced by top-bottom hat transformation. Stained images in color hence enhancement is done on individual color spaces [18–22]. Breast Enhancement is a crucial part in classification because almost all medical images consist the label and patient details. The need to be labeled region of interest to further key analysis [23]. For this purpose, we have preprocessed original images with diamond shape structuring element of size 5. Tophat filtering is applied on the image then dilate the image followed by erosion. Then add this image with original this process enhances the masses in the breast. Apply the bottomhat filter on Original Image then take the difference between Bottomhat filtered and Previous (Fig. 2).

K-Means Clustering: K-means is a cluster-based analysis of the image it is a more efficient technique adopted in biomedical imaging. In this paper, color pathology image converted to LAB color space to convert an RGB color profile might be non-standard into LAB color space because it is a device-independent color profile. Then the image converted to 2 channel called A and B and then K-means clustering applied repeatedly up to the clustering 3 times to avoid local minima value which helps to get desired feathers in color space.

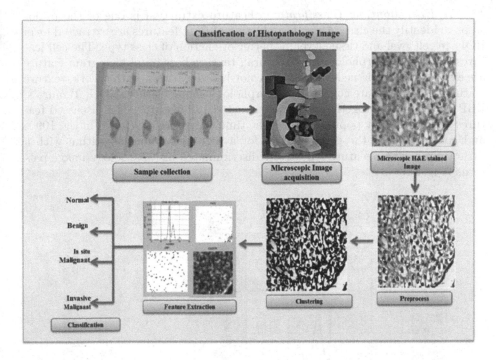

Fig. 1. Design of breast cancer classification using pathology image

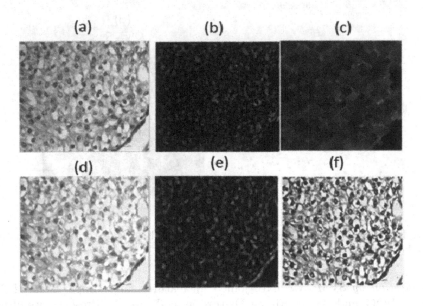

Fig. 2. Results of preprocessing (a) Original image (b) Top hat (c) Tophat+Dilated+ eroded (d) c+Original (e) Bottomhat (f) Bottomhat+d

Feature Extraction and Classification: Feature extraction is one of the major steps to identify the class. After the detection of ROI, features are extracted form ROI are cell level and tissue level for better prediction of class type. The cell features are shape, morphological, structural, texture based and histogram feature are extracted. The tissue level features color-based features. In this work, texture features considered are color based, morphology shape feature, HOG, Tamura's, LBP and Gabor. Altogether we have extracted 130 features. The discussed features were extremely popular around the time they were developed in the 1990s and continue to be the go-to method for a high-performing algorithm with a little tuning. These features effectively discriminated histopathology image patterns (Fig. 3).

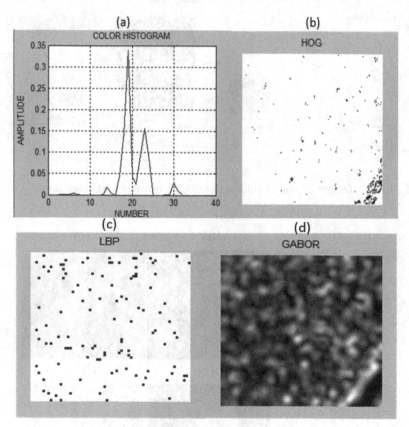

Fig. 3. Results of feature extraction (a) Color feature (b) Hog feature (c) LBP feature (d) Gabor

Classification by SVM: We have joined all features for example Color, LBP, Tamura, HOG, and Gabor. All out Feature set ends up 130. Presently we utilize direct SVM for characterization for variation from the norm finding on pathology images. The numeric information factors (x) in information (the segments) shape a n-dimensional space. The hyper-plane is gained from preparing information utilizing partner enhancement system that expands the edge [24]. The preparation of the hyper plane in direct SVM is done by redesigning the issue utilizing some variable based on features that is out of the extent of this prologue to SVM.

A ground-breaking knowledge is that the direct SVM will be reworded utilizing the speck result of any two given perceptions, rather than the perceptions themselves. The spot item between two vectors is that the aggregate of the duplication of each join of information esteems [25, 26]. The condition for making an expectation for information utilizing the scalar item between the information (x) and each help vector (xi) is determined as:

$$f(x) = B0 + sum(ai * (x, xi)) \tag{1}$$

This is a relating condition that includes ascertaining the internal results of another feature vector (x) with all support vectors in training data. The coefficients B0 and ai (for each information) ought to be measurable from the preparation data by the learning calculation. The dot item is known as the portion and can be re-composed as:

$$K(x, xi) = sum(x * xi)(2) \tag{2}$$

The portion characterizes the likeness or a separation measure between new data and feature vectors. The scalar item is that the closeness measure utilized for linear SVM or a direct bit because of the hole might be a dot product of the features and support vectors.

3   Results and Discussion

There are 285 samples selected from INSEC-TECH online histopathology dataset for analysis of the performance of the proposed methodology. The details of the samples used are as follows (Table 1).

Table 1. Data used for the analysis of proposed method

Total images	285
Normal patient's images	64
Benign patient's images	78
In situ carcinoma patient's images	72
Invasive carcinoma patient's images	71

The classification performance of histology images measured with the following equations. P-total images of positive class, N-indicates total images of negative class, TP-No. of images detected of the actual class, FN-No. of images

Table 2. Confusion matrix for the results of the proposed method

Actual or predicted	Normal	Benign	In situ	Invasive
Normal	63	1	0	0
Benign	1	76	1	0
In situ	1	0	71	0
Invasive	2	0	0	69

detected as a wrong class and TN-other class images detected as other only. TPR-True positive rate, TNR-true negative rate, FPR-False positive rate, and FNR-False negative rate (Table 2).

$$TPR = TP/P \tag{3}$$

$$TNR = TN/N \tag{4}$$

$$FPR = FP/N \tag{5}$$

$$FNR = FN/P \tag{6}$$

$$ACC(x) = (TP + TN)/(P + N) \tag{7}$$

where x = 1, 2, 3, 4: 1-Normal, 2-Benign, 3-in situ and 4-invasive

$$Accuracy = ((x = 1)^4 * Acc)/4 * 100 \tag{8}$$

Table 3. The accuracy of the proposed method

Actual/predicted	Accuracy
Normal	98.24%
Benign	98.94%
In situ	99.64%
Invasive	99.29%

By the above data accuracy achieved by the proposed method is 99.03% and the individual class accuracy is 98.24% of Normal, 98.94% of Benign, 99.64% In situ carcinoma and 99.29% of invasive carcinoma (Table 3).

Considering the above data in Table 4 and Fig. 4, it suggests that the combined feature set gives better accuracy instead of individual feature values. This algorithm is also able to classify histology image. The main obstacle to the development of new histopathology image analysis methods is the lack of large, publicly available, annotated datasets. Annotated database is also crucial to develop and validate machine learning systems.

Table 4. Performance of classifier on individual features and combined set of features

Feature set	Average accuracy
HOG	83%
LBP	82%
CH	81%
GF	87%
HOG,LBP	85%
HOG,GF	89%
HOG,LBP,CH	88%
Proposed	99%

Where HOG - Histogram of Gradient, LBP - Local Binary Pattern, CH - Color Histogram, GF - Gabor Filter.

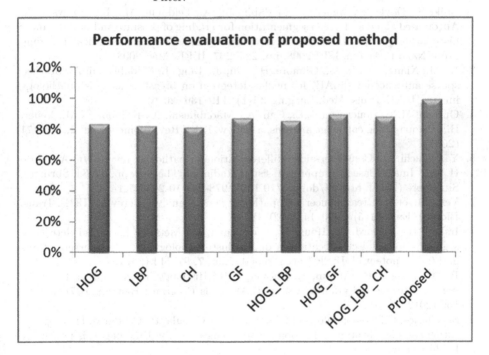

Fig. 4. Comparative analysis of proposed technique with different methods

4   Conclusion

Since from the last decade, many numbers of researches have been proposed in the field of pathology imaging, mainly focusing on cell segmentation and classification. Segmentation of nuclei/cell is the challenging one because of its irregular shape and structure. Moreover, it is not possible to numerically compare the experimental results because of the lack of a benchmark dataset; different studies have been reported using their own clinical datasets. In this paper, we have segmented nuclei by using the K-means clustering algorithm and classified as normal benign, in-situ malignant and invasive malignant by using SVM classifier. This paper suggests that computer image processing techniques show the significant changes in pathological image analysis. The performance of our system is similar or superior to the state-of-the-art methods, even though a smaller and more challenging dataset is used. In future segmentation can be improved by advanced segmentation techniques.

References

1. Mammography: Medline plus. https://medlineplus.gov/mammography.html. Accessed 09 June 2017
2. Naik, S., Doyle, S., Agner, S., Madabhushi, A., Feldman, M., Tomaszewski, J.: Automated gland and nuclei segmentation for grading of prostate and breast cancer histopathology. In: 5th IEEE International Symposium on Biomedical Imaging: From Nano to Macro. ISBI 2008, pp. 284–287. IEEE, May 2008
3. Xu, J., Xiang, L., Liu, Q., Gilmore, H., Wu, J., Tang, J., Madabhushi, A.: Stacked sparse autoencoder (SSAE) for nuclei detection on breast cancer histopathology images. IEEE Trans. Med. Imaging 35(1), 119–130 (2016)
4. Gurcan, M.N., Boucheron, L.E., Can, A., Madabhushi, A., Rajpoot, N.M., Yener, B.: Histopathological image analysis: a review. IEEE Rev. Biomed. Eng. 2, 147–171 (2009)
5. Yamaguchi, M.: Computer-aided differentiation for pathology images. In: Arimura, H. (ed.) Image-Based Computer-Assisted Radiation Therapy, pp. 67–84. Springer, Singapore (2017). https://doi.org/10.1007/978-981-10-2945-5_4
6. Veta, M., et al.: Breast cancer histopathology image analysis: a review. IEEE Trans. Biomed. Eng. 61(5), 1400–1411 (2014)
7. Irshad, H., Veillard, A., Roux, L., Racoceanu, D.: Methods for nuclei detection, segmentation, and classification in digital histopathology: a review-current status and future potential. IEEE Rev. Biomed. Eng. 7, 97–114 (2014)
8. He, L., Long, L.R., Antani, S., Thoma, G.R.: Histology image analysis for carcinoma detection and grading. Comput. Methods Programs Biomed. 107(3), 538–556 (2012)
9. Sakellarios, C., Kostopoulos, S., Glotsos, D., Ravazoula, P., Cavouras, D.: Segmentation and classification of histopathology images for abetting diagnosis in urinary bladder cancer
10. Diamond, J., Anderson, N.H., Bartels, P.H., Montironi, R., Hamilton, P.W.: The use of morphological characteristics and texture analysis in the identification of tissue composition in prostatic neoplasia. Hum. Pathol. 35(9), 1121–1131 (2004)
11. Esgiar, A.N., Naguib, R.N., Sharif, B.S., Bennett, M.K., Murray, A.: Microscopic image analysis for quantitative measurement and feature identification of normal and cancerous colonic mucosa. IEEE Trans. Inf. Technol. Biomed. 2(3), 197–203 (1998)
12. Huang, P.W., Lee, C.H.: Automatic classification for pathological prostate images based on fractal analysis. IEEE Trans. Med. Imaging 28(7), 1037–1050 (2009)

13. Gilmore, S., Hofmann-Wellenhof, R., Muir, J., Soyer, H.P.: Lacunarity analysis: a promising method for the automated assessment of melanocytic naevi and melanoma. PLoS One **4**(10), e7449 (2009)
14. Loukas, C.G., Wilson, G.D., Vojnovic, B., Linney, A.: An image analysis-based approach for automated counting of cancer cell nuclei in tissue sections. Cytometry Part A **55**(1), 30–42 (2003)
15. Janowczyk, A., Madabhushi, A.: Deep learning for digital pathology image analysis: a comprehensive tutorial with selected use cases. J. Pathol. Inf. **7** (2016)
16. Veta, M., Huisman, A., Viergever, M.A., van Diest, P.J., Pluim, J.P.: Marker-controlled watershed segmentation of nuclei in H&E stained breast cancer biopsy images. In: 2011 IEEE International Symposium on Biomedical Imaging: From Nano to Macro, pp. 618–621. IEEE, March 2011
17. Araújo, T., et al.: Classification of breast cancer histology images using convolutional neural networks. PLoS One (2017). https://doi.org/10.1371/journal.pone.0177544
18. Veta, M., van Diest, P.J., Kornegoor, R., Huisman, A., Viergever, M.A., Pluim, J.P.: Automatic nuclei segmentation in H&E stained breast cancer histopathology images. PloS One **8**(7), e70221 (2013)
19. Hartigan, J.A., Wong, M.A.: Algorithm AS 136: a k-means clustering algorithm. J. Roy. Stat. Soc. Series C (Appl. Stat.) **28**(1), 100–108 (1979)
20. Chitade, A.Z., Katiyar, S.K.: Color-based image segmentation using k-means clustering. Int. J. Eng. Sci. Technol. **2**(10), 5319–5325 (2010)
21. Recky, M., Leberl, F.: Windows detection using k-means in CIE-lab color space. In: 2010 20th International Conference on Pattern Recognition (ICPR), pp. 356–359. IEEE, August 2010
22. Heckbert, P.: Color image quantization for frame buffer display, vol. 16, no. 3, pp. 297–307. ACM (1982)
23. Santosh, K.C., Wendling, L., Antani, S., Thoma, G.R.: Overlaid arrow detection for labeling regions of interest in biomedical images. IEEE Intell. Syst. **31**(3), 66–75 (2016)
24. Support Vector Machines for Machine Learning. https://machinelearningmastery.com/support-vector-machines-for-machine-learning. Accessed 10 Oct 2018
25. A tour of the top 10 algorithms for machine learning newbies. https://towardsdatascience.com/a-tour-of-the-top-10-algorithms-for-machine-learning-newbies-dde4edffae11. Accessed 10 Oct 2018
26. SVM Theory. https://towardsdatascience.com/a-tour-of-the-top-10-algorithms-for-machine-learning-newbies-dde4edffae11. Accessed 10 Oct 2018

Nail Image Segmentation
for Disease Detection

Shweta Marulkar[1(✉)] and Rajivkumar Mente[2]

[1] Yashvantrao Chavan Institute of Science, Satara, India
khadilkarss96@gmail.com
[2] School of Computational Sciences, Solapur University, Solapur, India
rajivmente@rediffmail.com

Abstract. During modern era digital image processing in medical field is very popular concept. Number of diseases such as cancer, kidney disease, heart disease, brain haemorrhage etc. can be detected by capturing various types of images like MRI, CT Scan, PET, X-ray. Color feature of fingernail is used for disease detection. Various segmentation techniques like k-means segmentation, L*a*b color space, watershade segmentation are used to extract interested region. The segmented area used for the feature extraction and the disease can be detected.

Keywords: Fingernail disease · Digital image processing · Segmentation

1   Introduction

Disease detection in human body by capturing digital images is one of the advanced and popular technologies in medical image processing. Generally diseases are detected by techniques such as pathological test, MRI, CT Scan, Ultrasound, Positron Emission Tomography (PET) etc. but these techniques are quite painful and costly. The farthest ends of the body are the Nails and since they are farthest from the heart, receive the oxygen last. Due to this they are often the first to show signs of disease. Color, Texture and Shape of nails and Pliability of nails can be used to identify some of the diseases. Color analysis is mostly useful and highly preferred in diagnosis of diseases.

A method for the identification of plant diseases using SVM and k-means clustering is proposed by Sujatha *et al*. The experiment converts HIS image used in k-means clustering algorithm and for accurate detection of problem Support vector machine is used [1]. For the detection of Citrus Cancer Three diverse approaches like histogram comparison, Color Co-occurrence Matrix and k-means clustering methods presented by Sunny *et al*. To produce more accurate result Color co-occurrence Matrix and k-means clustering used [2]. Extraction of diseased portion made by conducting k-means and fuzzy-c means clustering algorithms proposed by Baghel *et al*. On the basis of output produces k number of colored images of infected region k-means Segmentation used [3]. Disease

© Springer Nature Singapore Pte Ltd. 2019
K. C. Santosh and R. S. Hegadi (Eds.): RTIP2R 2018, CCIS 1036, pp. 116–126, 2019.
https://doi.org/10.1007/978-981-13-9184-2_10

detection carried out based on color and texture, the said features were extracted and analyzed to state whether the body is healthy or not concluded after segmentation introduced by Sharma *et al.* [4]. Mechanical leaf disease detection conducted by clustering and region based segmentation reminiscent of region growing, region merging, region splitting and clustering based such as Meanshift, k-means, Fuzzy c-means correspondingly. During the test region growing is finest region based segmentation method [5]. Habitual method for disease recognition medical practitioners observes the tint of palm and nail by naked eyes. This shade can be determined by computer vision with-out any subjectivity [6]. Diverse image investigation and digital image processing methodology such as Image gaining, Image pre-processing, Segmentation, extraction of features etc are used to recognize features of nail. The nails outer shell of healthy human appears smooth, pink and shiny [7]. In conventional system medical practitioner predicts the disease by observing color of nails. Diverse colors of nails indicate symptoms of diseases as shown in the following Table 1.

Table 1. Nail color and diseases

Sr. No.	Color of nail	Symptoms of disease
1	Pale or white	Hemorrhage, cardiovascular or lung complications, anemic condition, helminth infection infestation, kidney disease, hepatic impairment
2	Red	Brain hemorrhage, heart diseases, high blood pressure, lung disease
3	Green	Bacillus infection, dermatitis
4	Copper color	Arsenic poisoning, infections by fungi
5	Black	Cancer, impairment of kidney function, vitamin B12 deficiency
6	Yellow	Diabetes, digestive problems, liver disease
7	Gray	Arthritis, enema, malnutrition, postoperative effects, glaucoma, lung problem
8	Purple	Oxygen deprivation, circulatory problems, congenital problems
9	Blue	Hepatitis, atherosclerosis, thick blood, liver disease, chronic obstructive pulmonary disease, cardiovascular disorders

2   Steps Involved in Disease Detection Using Nail Images

Generally, the disease detection using nail images consists of five steps (See Fig. 1). In the first step of Image Acquisition the nail images can be taken as input through scanner or digital camera. For accurate image analysis images with high resolution are required. Figure 2 shows a sample nail image. The image data is improved in the second step of pre-processing and by transforming input color image to color space which have intensity component the contrast enhancement is done.

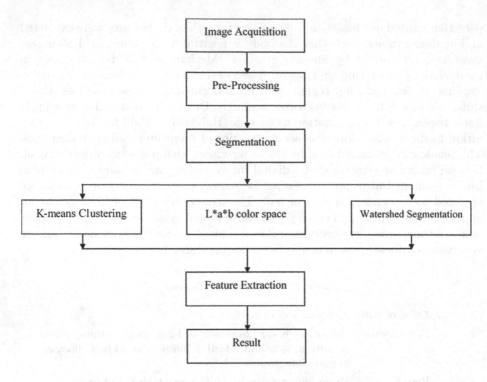

Fig. 1. Steps in disease detection using nail images

The input RGB image transformed to L*a*b color space using color transform functions. Further processing on luminosity layer 'L*' of the image without affecting color of original image, gives insight in intensity of the pixels.

In the Segmentation step image is divided into regions. Segmented image is used to extract features. These extracted features are used to detect

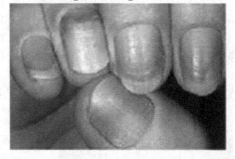

Fig. 2. Sample nail image.

the diseases. This paper presents three methods of segmentation of nail images viz. K-Means Clustering, L*a*b color space and Watershed segmentation.

3   Segmentation

Segmentation of image specifies the tool of converting images with low quality to high quality with respect to scene, appearance and matter. Segmentation methods applied in present work for extraction of nail sections from finger image are k-means clustering, L*a*b colorspace and Marker-Controlled Watershed Segmentation for better accuracy and precision.

3.1   Algorithm for K-Means Clustering

Step 1: Start.
Step 2: Input Color Image of nail.
Step 3: Produce L*a*b* color space by transforming the input image.
Step 4: k-means clustering used for color categorization of 'a*b*' color space.
Step 5: Mark each pixel to the output image from step 4.
Step 6: Stop.

The output of these steps are shown in Figs. 3 and 4.

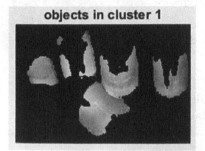
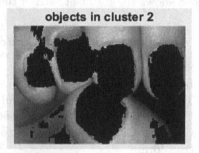
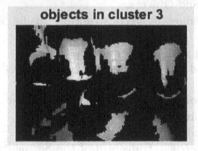

Fig. 3. Segmented input image by color (Color figure online)

Fig. 4. Results from K-means

3.2   Algorithm for L*a*b Colorspace

Step 1: Start.
Step 2: Input Color Image.
Step 3: Region wise calculate sample color in L*a*b* Color Space.
Step 4: Use Nearest Neighbor Rule for classification of pixel.
Step 5: Demonstrate arrangement results of Nearest Neighbor from Step 4.
Step 6: Shows Labeled Colors values for 'a*' and 'b*'.
Step 7: Stop.

Figure 5 shows the result of each of these steps.

3.3   Algorithm for Marker-Controlled Watershed Segmentation

1. Start.
2. Input Color Image.
3. Translate the image to Black and white image.
4. Use Sobel edge mask for Gradient Magnitude Segmentation.
5. Consider Gradient magnitude to Divide the image using the watershed transform.
6. Mark the Foreground Objects.
7. Calculate the opening by reconstruction (using imerode and imreconstruct).
8. Apply the opening with a closing for smoothness.
9. For obtaining good foreground markers calculate the regional maxima of step 8.
10. Overlay the foreground marker image on the input image.
11. Compute Background Markers.
12. Compute watershed ridge lines.
13. Calculate the Watershed Transform of the Segmentation Function.
14. Display Result.
15. Stop.

The result of each step is shown in Table 2

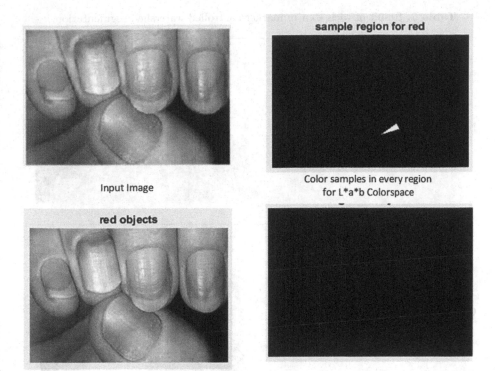

Input Image

sample region for red

Color samples in every region
for L*a*b Colorspace

red objects

Green, Purple, Magenta, Yellow objects

Results of Nearest Neighbour Classification

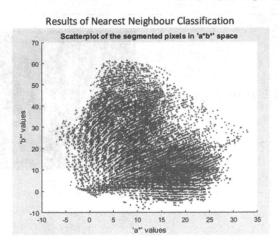

Fig. 5. Segmentation using L*a*b colorspace

Table 2. Result of each step in marker-controlled watershed segmentation.

Input Image	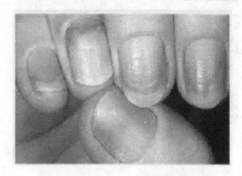
Gray scale Image	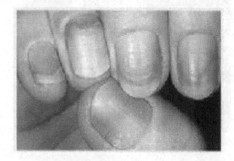
Gradient Magnitude as the Segmentation Function	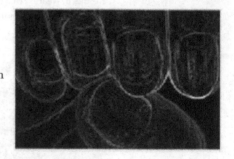
Watershed transform of gradient magnitude	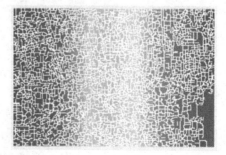

(*continued*)

Table 2. (*continued*)

Foreground Objects	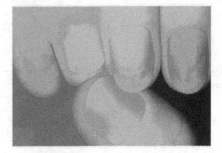
Result of opening-by-reconstruction	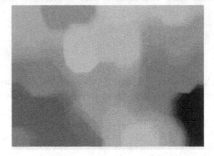
Remove dark spots and stem marks perform Opening with a closing	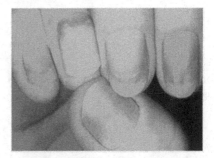
Complement the image input	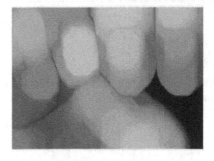

(*continued*)

Table 2. (*continued*)

To gain superior foreground markers apply regional maxima.	
Consign foreground marker image on the original image.	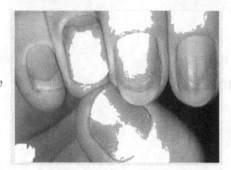
Result after cleaning the edges of the marker blobs and then shrink them a bit.	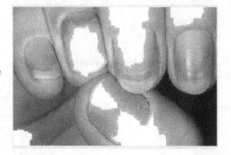
Result after Computing the Background Markers	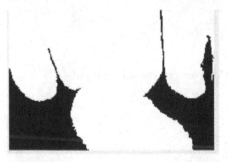

(*continued*)

Table 2. (*continued*)

Watershed ridge lines.

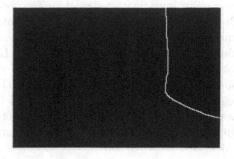

Result after dilation.

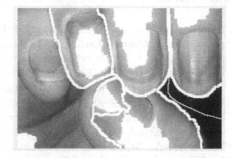

Result of visualization technique is to display the label matrix as a color image.

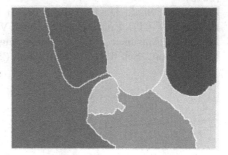

Result of transparency to super-impose this pseudo-color label matrix on top of the original intensity image.

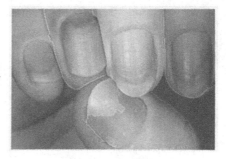

4   Conclusion

In present investigational work image pre-processing is carried out to study color characteristics of nails, by applying color transform functions to convert the image from RGB to L*a*b* color space. Commonly used segmentation techniques such as k-means segmentation, L*a*b color space, Watershed Segmentation are applied on enhanced image. From the literature survey it has been observed that k-means clustering algorithms can effectively extract the diseased region. Present work may be extended for calculation of shape attributes like area, diameter and perimeter of extracted nail segment images, which would further help in detection of chronic disease which proved fatal.

References

1. Sujatha, R., Sravan Kumar, Y., Akhil, G.U.: Leaf disease detection using image processing. J. Chem. Pharm. Sci. **10**(1), 670–672 (2017)
2. Sunny, S., Peter, R.: Detection of cancer disease on citrus leaves using image processing. Int. J. Comput. Eng. Appl. 129–134 (2016)
3. Baghel, J., Jain, P.: K-means segmentation method for automatic leaf disease detection. Int. J. Eng. Res. Appl. **6**(3), 83–86 (2016)
4. Sharma, V., Shrivastava, A.: System for disease detection by analyzing fingernails color and texture. Int. J. Adv. Eng. Res. Sci. (IJAERS) **2**(10), 1–6 (2015)
5. Darshana, A., Majumdar, J., Ankalaki, S.: Segmentation method for automatic leaf disease detection. Int. J. Innov. Res. Comput. Commun. Eng. **3**(7), 7271–7282 (2015)
6. Pandit, H., Shah, D.: Decision Support System for Healthcare Based on Medical Palmistry. GCET Engineering College, Vallabh Vidyanagar (2011)
7. Mente, R., Marulkar, S.V.: A review: fingernail images for disease detection. Int. J. Eng. Comput. Sci. **6**(11), 22830–22835 (2017)

Pathological Brain Tumour Detection Using Ridgelet Transform and SVM

Patil Ankita(✉) and Mansi Subhedar

Department of EXTC Engineering, Pillai HOC College of Engineering and
Technology, Rasayani, Dist. Raigad, Maharashtra, India
ankitapatil.3165@gmail.com, mansi_subhedar@rediffmail.com

Abstract. The identification, detection and classification of brain MRI
images into abnormal and healthful is a main pre-clinical step for
patients. Standard classification is tedious, valuable, inimitable, and time
consuming. Using simple imaging techniques, it is very difficult to have
vision about the normal and tumour cell due to the similarities between
them. The proposed brain tumour detection method employs ridgelet
transform and SVM to identify malignant and benign tumour. In this
work, gray level co-occurrence matrix (GLCM) based texture analysis
of discrete ridgelet transform coefficients is carried out. SVM classifier
is trained using textural features and intensity based features. Principal
component analysis (PCA) method is used to lessen the number of features used. SVM outputs the classified image and helps for automated
detection. Experimental results demonstrated the efficacy with respect
to precision, sensitivity, specificity and accuracy for tumour detection.

Keywords: MRI · Ridgelet transform · GLCM · SVM

1 Introduction

Brain tumour is common major factor for the increase in mortality among children and adults in the world. A brain tumour is abnormal and uncontrolled
growth of cell in brain region leads or around the brain itself, or spread from
cancers primarily located in other organs (metastatic tumours). Various types
of brain tumours exist. Many brain tumours are low grade tumours (benign),
and many are high grade tumours (malignant). Benign brain tumours have a
well define shape with regular smooth margin. Most benign tumours have no
encroachment in surrounding tissues. It does not comprise of cancer cells and
may be either radiologically regulated or completely evacuated surgically and
may not pursue again. The structure of malignant brain tumours has irregular margin and it contains cancer cells which can be healed with radiotherapy,
targeted therapy or a combination of both, steroids and anti-seizure medication
and are harmful to life. Craniopharyngiomas, astrocytoma and gliomas are the
examples of low grade tumours. Most common example for high grade tumour
is glioblastoma.

© Springer Nature Singapore Pte Ltd. 2019
K. C. Santosh and R. S. Hegadi (Eds.): RTIP2R 2018, CCIS 1036, pp. 127–133, 2019.
https://doi.org/10.1007/978-981-13-9184-2_11

The National Brain Tumour Foundation (NBTF) for research in United States computes that, in children, one quarter of all cancer deaths are caused due to brain tumour and the majority of the patients affected by brain tumours die within 9–12 months and less than 3% survive more than 3 years [1]. Explicit diagnosis and earlier prevision and detection of the brain tumour in an expedient time is vital and failing of which ends to death. Various studies have been carried out in literature using different image transforms and classifiers. Some of the significant transform based classification studies are reviewed and discussed here. Brain tumour segmentation technique was presented with the modified multitexton histogram features (MHF) and SVM classifier with hybrid kernel was employed to improve the classification accuracy. The authors achieved 86% of average classification accuracy using modified MHF [2]. Anantha et al. discussed another scheme that extracted features with Discrete Wavelet Transform (DWT), GLCM and law's texture features using Adaptive Neuro Fuzzy Inference System (ANFIS) classifier [3]. In another work presented in [4,15], textural features were extracted from curvelet transform and were reduced by PCA. Least square SVM was employed for classification which improved classification accuracy as compared to standard SVM. Fast Discrete Curvelet Transform (FDCT) based brain tumour detection was presented in a combination of a genetic algorithm and contourlet transform was used to extract texture and shape features with deep neural network and extreme learning employed for classification [5]. In order to improve detection accuracy further and obtain better classification, proposed work employed ridgelet transform based feature extraction. SVM is employed for classification. Figure 1 shows the different brain MR images.

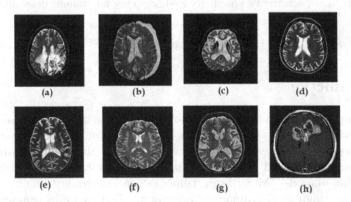

Fig. 1. Brain MR images: (a) Normal brain; (b) AIDS dementia; (c) Alzheimer's disease plus visual agnosia; (d) Alzheimer's disease; (e) Cerebral calcinosis; (f) Glioma; (g) Meningioma; (h) Glioblastoma

This article is arranged as follows. Section 2 represents details of proposed method. Section 3 deals with the simulation results and comparative analysis. In Sect. 4, a summary of the proposed work is presented.

2   Proposed Method

Use of computer aided technology for brain tumour detection is most popular domain nowadays due to necessity of the correct diagnosis. MRI images of patients are used here for analysis and study. Proposed method involves several steps like preprocessing, ridgelet decomposition, feature extraction and classification.

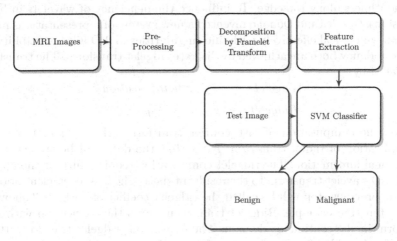

Fig. 2. Proposed method

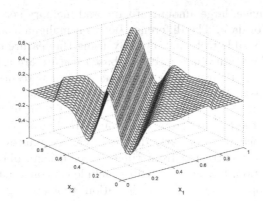

Fig. 3. An example of ridgelet function

Figure 2 shows the schematic for proposed method. In preprocessing, initially the input image is registered to a reference image. Multiplicative noise is present in the brain MRI image and its reduction is crucial to obtain better results. Noise removal is accomplished by median filter which enhances characteristic of brain

MRI images. Discrete wavelet transform is the well known tool for the analysis of images. The success of wavelets transform is due to the great achievement for piecewise smooth functions in 1-D. Two dimensional piecewise continuous signals show images have 1-D singularities. Smooth regions are separated by edges, and while edges are discontinuous across, they are ordinarily smooth curves. One dimensional wavelets tensor-product gives 2-D wavelets and they are thus excellent at differentiating the discontinuity across an edge, but will not see the smoothness along the edge. To influence the deficiency of wavelets in higher dimensions, Candes and Donoho invented a new system of representations named ridgelets which deal effectively with line singularities in 2-D [6]. Invertibility and non-redundancy both are achieved by discrete ridgelet transform. The transforms are related by,

$$Wavelets \rightarrow \psi_{scale};\ point\text{-}position \tag{1}$$

$$Ridgelets \rightarrow \psi_{scale};\ line\text{-}position \tag{2}$$

It needs the computation of fast Fourier transform (2D-FFT) first and then the application of radon transform. After that the data will be treated as one dimensional information. The ridgelet computation requires further the application of 1-D wavelet transform to the resultant data. The final coefficient obtained from the previous step will be called the ridgelet coefficients. Figure 3 shows the ridgelet function example. Ridgelet transform shows the connection with other transform in the continuous domain. The continuous ridgelet transform (CRT) in R^2 of integrable bivariate function $f(x)$ is given by,

$$CRT_f(a, b, \theta) = \int_{R^2} \psi_{a,b,\theta}(x) f(x) dx \tag{3}$$

To represent an image, large amount of time and memory space is required for the large amount of data. The difference between malignant and benign tissue may not be easily visible to human eye. By using the feature extraction in frequency domain, the texture and intensity based feature are extracted. Texture analysis differentiates tissues easily for human visual perception and machine learning. In texture method, Gray level co-occurrence matrix (GLCM) in four possible directions 0°, 45°, 90°, 135° & gray level different matrix (GLDM) are applied to extract the features from the brain MRI image [7]. By choosing the effective features, accuracy for early diagnosis can be improved. Textural and intensity features are extracted from ridgelet coefficients obtained by level 1 ridgelet decomposition. Extracted features include contrast, correlation, energy, homogeneity, entropy, mean, standard deviation, root mean square value, variance, smoothness, kurtosis, skewness. PCA is applied to the extracted features and principal components are used further for classification purpose.

The SVM algorithm is based on the study of a supervised learning technique and is applied to N number of class classification problem from one class classification problem. The error matrix establishing the terms TP, TN, FP, and FN from the anticipated outcome and ground truthing outputs for the assessment of accuracy, sensitivity, and specificity. A non-linear SVM is employed for classification.

3   Experimental Results

In this study, 80 T2 weighted MRI brain images were retrieved along the transaxial plane. Each images is of size 256×256. The dataset covers seven different illnesses of brain images, each disease from different subject. The sample test images were obtained from medical school of Harvard University (www.med.harvard.edu/aanlib/home). Performance of proposed method was evaluated using sensitivity, specificity, precision and accuracy. The metrics can be obtained as follow.

$$Sensitivity = \frac{TP}{TP + FN} \tag{4}$$

$$Specificity = \frac{TN}{TN + FP} \tag{5}$$

$$Precision = \frac{TP}{TP + FP} \tag{6}$$

$$Accuracy = \frac{TP + TN}{TP + FN + TN + FP} \tag{7}$$

where,

True Positive (TP): Non-cancerous brain accurately recognized as non-cancerous

True Negative (TN): Cancerous brain accurately recognized as cancerous

False Positive (FP): Cancerous brain inaccurately recognized as non-cancerous

False Negative (FN): Non-cancerous brain inaccurately recognized as cancerous

The proposed method achieved the accuracy of 97%, which is comparable with the recent studies and is better than some of the DWT based existing works as shown in Table 1.

Table 1. Comparison with existing work

Existing methods	Sensitivity (%)	Specificity (%)	Precision (%)	Accuracy (%)
CT + ZM + DNN [5]	51.48	40.19	-	88.8
GLRLM + Linear [8]	83.33	100	-	91.66
GLRLM + Poly [8]	66.66	100	-	83.33
GLCM + Bayesian [9]	91	83	91	88.2
GLCM + Tree [9]	93	100	100	96
DWT + PCA + A-NN [10]	98.3	81.8	-	95.7
FBB + GLCM + RBF [11]	68	40	-	61.8
FBB + GLCM + MLP [11]	93.1	81.33	-	96.63
DWT + PCA + RBF-NN [12]	92.47	72.00	98.25	91.3
WE + KSVM [13]	93.18	68.00	98.02	91.78
Proposed method	98.76	98.67	99.50	97

4   Conclusion

Computer aided detection of brain tumour has become most important research field in medical imaging. This proposed method extracts GLCM based texture features and intensity features from ridgelet coefficients. To classify brain image into malignant or benign, the extracted features are fed to the SVM classifier. The proposed methodology improves performance accuracy as compared to some of the existing methods. This work can be further extended for images with non-identical pathological condition, types and diseases with the help of overlaid arrow detection method using fuzzy binarization so that to produced several different image layers. This helps to avoid loss of data and make segmentation easier [14].

References

1. El-Sayed, A.E., Heba, M.M., Kenneth, R., Abdel-Badeeh, M.S.: Computer aided diagnosis of human brain tumour through MRI: a survey and a new algorithm. Expert Syst. Appl. **41**(11), 5526–5545 (2014)
2. Jayachandran, A., Dhanasekaran, R.: Brain tumour severity analysis using modified multi-texton histogram and hybrid kernel SVM. Int. J. Imaging Syst. Technol. **24**, 72–82 (2014)
3. Rufus, N.H.A., Selvathi, D.: Performance analysis of computer aided brain tumour detection system using ANFIS classifier. Int. J. Imaging Syst. Technol. **27**(3), 273–280 (2017). https://doi.org/10.1002/ima.22232
4. Deepak, R.N., Ratnakar, D., Banshidhar, M.: Pathological brain detection using curvelet features and least squares SVM. Multimed. Tools Appl. **77**(3), 3833–3856 (2016). https://doi.org/10.1007/s11042-016-4171-y
5. Anbarasa Pandian, A., Balasubramanian, R.: Fusion of contourlet transform and zernike moments using content based image retrieval for MRI brain tumour images. Indian J. Sci. Technol. **9**(29), 0974–5645 (2016). https://doi.org/10.17485/ijst/2016/v9i29/93837
6. Do, M.N., Vetterli, M.: The finite ridgelet transform for image representation. IEEE Trans. Image Process. **12**(1), 16–28 (2003)
7. Haralick, R.M., Shanmugam, K., Dinstein, I.H.: Textural features for image classification. IEEE Trans. Syst. Man Cybern. **6**, 610–621 (1973)
8. Parveen, A.S.: Detection of brain tumour in MRI images, using combination of fuzzy c-means and SVM. In: 2nd International Conference on Signal Processing and Integrated Networks (SPIN). IEEE Press (2015). 978-1-4799-5991-4/15/$31.00
9. El-Sayed, A.E., Abdel-Badeeh, M.S.: A hybrid technique for automatic MRI brain images classification. Studia Univ. Babes-Bolyal Inform. **54**(1), 55–67 (2009)
10. Anitha, V., Murugavalli, S.: Brain tumour classification using two-tier classifier with adaptive segmentation technique. IET Comput. Vis. 1–9 (2015). https://doi.org/10.1049/iet-cvi.2014.0193
11. Naik, J., Patel, S.: Tumour detection and classification using decision tree in brain MRI. IJCSNS Int. J. Comput. Sci. Netw. Sec. **14**(6), 87 (2014)
12. Joe, B.N., et al.: Brain tumour volume measurement: comparison of manual and semi automated methods. Radiology **212**, 811–816 (1999)

13. Praveen, G.B., Agrawal, A.: Hybrid approach for brain tumour detection and classification in magnetic resonance images. In: 2015 International Conference on Communication, Control and Intelligent Systems (CCIS). IEEE Press (2015). 978-1-4673-7541-2/15/$31.00
14. Santosh, K.C., Antani, S., et al.: Overlaid arrow detection for labelling regions of interest in biomedical images. IEEE Intell. Syst. **31**(3), 66–75 (2016). https://doi.org/10.1109/MIS.2016.24
15. Sawat, D.D., Hegadi, R.S.: Unconstrained face detection: a deep learning and machine learning combined approach. CSI Trans. ICT **5**(2), 195–199 (2017)

Color Transfer Method for Efficient Enhancement of Color Images and Its Application to Peripheral Blood Smear Analysis

M. Nandan Prasad, Keerthana Prasad(✉), and K. T. Navya

Manipal Academy of Higher Education, Manipal, Karnataka, India
nandanprasad22@gmail.com, {keerthana.prasad,navya.kt}@manipal.edu

Abstract. In this paper, we propose an efficient color transfer approach as a pre-processing step, and demonstrate use of the method for enhancement of peripheral blood smear images. Peripheral blood smear analysis is subjective, laborious and is error prone. Automation is highly desirable to obtain objective and accurate results. It also would help reduce the burden of pathologists. However, automation of peripheral blood smear analysis is challenging due to the fact that the microscopic images suffer from illumination variations. Also, variations in the process of manual staining leads to variations in colors expressed by the various components of the peripheral blood smear namely the red blood cells, white blood cells and platelets. Many research groups have reported various approaches to color transfer to enhance the quality of peripheral blood smear images to facilitate automation. Color transfer approach uses a template image with the desirable qualities and transfers the color characteristics of this image to the images under study. We propose an efficient color transfer approach and demonstrate its use for enhancement of peripheral blood smear images.

1 Introduction

In the case of peripheral blood smears, manually acquired microscopic images extensively suffer from color and illumination variations. However, the sophisticated image analysis tools developed for automation of the analysis of peripheral blood smear images require a completely automated laboratory workflow. These tools assume that the image is acquired under uniform illumination conditions and would fail to provide accurate results in the presence of color and illumination variations [1–9]. Many research groups have attempted to develop algorithms to tackle color and illumination variations in images using an approach called as color transfer or color normalization. Color transfer is a technique of transferring the color characteristics of a template image to the images under study to achieve desired characteristics in the image under study such as uniformity in color and illumination. It plays an important role in digital pathology since most of the microscopic images suffer from illumination variations. Specifically,

© Springer Nature Singapore Pte Ltd. 2019
K. C. Santosh and R. S. Hegadi (Eds.): RTIP2R 2018, CCIS 1036, pp. 134–142, 2019.
https://doi.org/10.1007/978-981-13-9184-2_12

many research groups have applied color normalization to tackle the variations in images of stained histological images [10–13]. The commonly used color normalization methods for images of histological specimen use the concept of deconvolution [15,16], decorrelated color channel statistics [17] etc. Extensive studies have been reported for application to histological images which work on each stain separately. For example, Vahadane et al. [14] reported a method based on stain density maps. Khan et al. [16] proposed a method of color transfer by using color deconvolution and color basis estimation. However, only one study, Zhang et al. [18] reports color transfer for peripheral blood smear image enhancement, using Reinhard method of color normalization for detection of white blood cells (WBCs) in peripheral blood smear images. This method is based on CIE-LAB color space which is computationally more complex than other color space transformations. We examine if a simpler color space transformation can be used which offers the same quality of color transfer results. There is a lack of a color normalization method which is designed to specifically address the variations of color and illumination in microscopic images of peripheral blood smear images.

2 Materials and Methods

2.1 Data Collection

We obtained the images of peripheral blood smears across various websites and internet resources [20–22]. We also used images from the LISC dataset [23] as described in [19]. A total of 104 images were collected. The dataset that we have used consists of images which widely vary in color and illumination levels.

2.2 Color Transfer Algorithm

The algorithm was implemented in MATLAB version 15. Reinhard et al. [17] used CIE-LAB color space representation for color correction of photography images. We adapted this color transfer method to suit the need of peripheral blood smear image analysis. We identified an appropriate template image with color characteristics for efficient segmentation of the red blood cells (RBCs) and the WBCs. The characteristics of this template image were represented in terms of the mean and standard deviations of the individual color components. All the images were subsequently mapped to the color characteristics of the template image. We performed the experiments using various color space representations for the application of enhancement of peripheral blood smear images. CIE-LAB color space has the advantage of separating the chrominance and luminance components to provide appropriate color correction [17]. However, it is computationally more complex in comparison with other color spaces such as RGB, CMYK, YIQ, YCbCr, HSV etc. Hence we considered YCbCr color space representation which is computationally much simpler than the CIE-LAB representation at the same time provides chrominance-luminance separation. Hence, we propose a similar color correction algorithm based on the YCbCr color space, the steps involved in which are as follows:

1. RGB to YCbCr color space conversion is applied to both the template image and the images under study.
2. Mean and standard deviation of each of the color components are obtained corresponding to the template image, and are denoted M_y, S_y, M_{Cb}, S_{Cb}, M_{Cr}, S_{Cr}.
3. Mean and standard deviation of each of the color components are obtained corresponding to the image under study, and are denoted M_{y1}, S_{y1}, M_{Cb1}, S_{Cb1}, M_{Cr1}, S_{Cr1}.
4. Color transfer is performed by modifying each of the components of the individual pixel with coordinates (i, j) of the image under study according to the following equations:

$$\text{Ratio-Y} = S_{y1}/S_y \tag{1}$$

$$Y_1(i,j) = (Y(i,j) - M_y) * \text{Ratio-Y} + M_{y1} \tag{2}$$

$$\text{Ratio-Cb} = S_{Cb1}/S_{Cb} \tag{3}$$

$$Cb_2(i,j) = (Cb(i,j) - M_{Cb}) * \text{Ratio-Cb} + M_{Cb1} \tag{4}$$

$$\text{Ratio-Cr} = S_{Cr1}/S_{Cr} \tag{5}$$

$$Cr_2(i,j) = (Cr(i,j) - M_{Cr}) * \text{Ratio-Cr} + M_{Cr1} \tag{6}$$

5. Convert Y_2, Cb_2 and Cr_2 to R_2 G_2 B_2
6. Color mapped image is represented by R_2 G_2 B_2.

2.3   Comparison with Different Color Spaces

We performed experiments on use of other color spaces such as YIQ and CIE-LAB color spaces for which we replaced the YCbCr components in the above steps by the color components of the other color spaces. We observed the time taken for processing all the images in the case of these different methods using different color spaces. We also observed the quality of the output images considering different template images with different characteristics. Based on these experiments we studied the performance of the color transfer approach for enhancement of the quality of the input image for segmentation of RBCs and WBCs in peripheral blood smear images.

2.4   Selection of Template Images for RBCs and WBCs

Considering that RBCs and WBCs have bright and saturated colors, we hypothesized that a template image with pale RBCs would be good for maintaining a distinct saturated color for the nucleus of the WBCs. Also, using template image with good contrast would result in a final output image with good contrast. Such an approach would be useful in the case of WBC segmentation or nucleus detection. However, in the case of RBC detection, an image with good contrast, and reasonably saturated and bright color for RBC would be good. To evaluate this assumption and to identify the appropriate characteristics of template images for RBC and WBC detection, we used a few images with different color characteristics as shown in Fig. 1, images labelled 1 to 4.

2.5   Quantitative Analysis of Performance of Color Transfer Method

We performed color transfer using the template image and modified the characteristics of the original images under study. Color renormalization also called 'Reverse color transfer' approach was used to analyse the performance of the color transfer algorithm. The steps involved in reverse color transfer is as follows:

1. Color transfer is applied on Image I using the template T to obtain the color transferred image I_T
2. Color transfer is again applied on I_T using the template as I to obtain I_{RT}.

The difference between I and I_{RT} was quantified by obtaining the total sum of pixel value differences between the two images. The difference value was divided by the number of pixels in the image to obtain the percentage error.

3   Results and Discussion

We considered 5 representative images as a template for color transfer. The experiments were repeated for all these templates using the YCbCr, CIE-LAB, YIQ and RGB representations. We observed the quality of output of the color transfer methods.

Figure 1 shows color transfer outputs corresponding to 4 input images labelled A-D for 4 template images, labelled 1–4 using YCbCr color space representation for the color transfer method. Image A is obtained by contrast and brightness enhancement of Image B. The template images vary in the colorfulness with Template 2 being highly colourful and Template 4 being very pale. It can be observed that the WBC cytoplasmic contents are more prominent when Template 1 and 3 are used. Template 3 could be the most appropriate among the four templates for nucleus or WBC segmentation. Template 1 would be a preferred choice for the purpose of RBC segmentation.

Figure 2 shows the results of applying color transfer method based on various color spaces on Image B and template images 1–4 as shown in Fig. 1. It can be observed that there is no visible difference in the quality of the output images.

The time taken for color transfer using the various color space representations for processing the dataset of 104 images in MATLAB was obtained and is listed in Table 1. It can be observed that CIE-LAB color space representation takes quite a large amount of time in comparison with the other color space representations.

Based on the time taken for processing, RGB color space based color transfer would be preferable. However, since the color components are not decorrelated, it results in color distortion in some cases, which may not be suitable in all scenarios. Figure 3 shows the color distortion caused by color transfer approach based on RGB color space. The top row represents the original image. The bottom row represents the color transferred image in which color distortion can be observed.

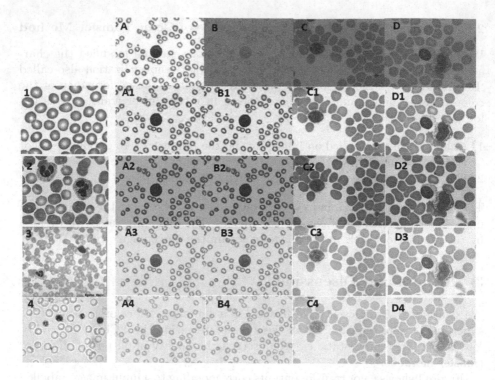

Fig. 1. Color transfer outputs corresponding to input images, labelled A-D, for template images, labelled 1–4 using YCbCr color space. (Color figure online)

YCbCr and YIQ color space representations are more suitable since their luminance and chrominance components are decorrelated and the possibility of color distortion is less. Also, the time taken for processing is much less in comparison with the color transfer method using the CIE-LAB color space representation which is also called the Reinhard color transfer method.

Quantitative analysis of performance of the color transfer method based on YCbCr and CIE-LAB color space was carried out using the reverse color transfer approach. Figure 4 shows the reverse color transfer being applied on to an image.

Original image and the reverse color transferred image were compared and the percentage error was calculated. The percentage errors corresponding to color transfers using 4 different template images were averaged and are tabulated in Table 2. It can be observed from Table 2 that the percentage errors in the case of both the color spaces are very close.

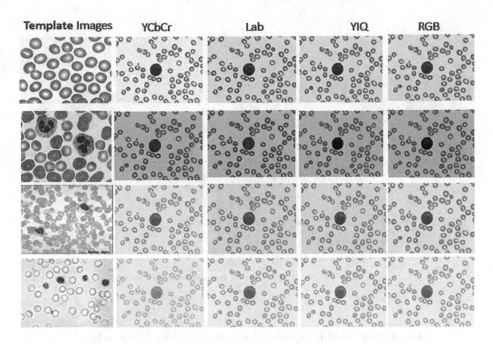

Fig. 2. Results of applying color transfer method based on various color spaces. (Color figure online)

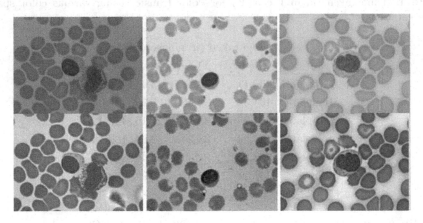

Fig. 3. Color distortion caused by color transfer approach based on RGB color space. (Color figure online)

Thus, we conclude that the color transfer method based on YCbCr color space is more computationally efficient and performs equally good in comparison with the Reinhard method and hence it would be a preferred choice for application to peripheral blood smear analysis as demonstrated in this paper.

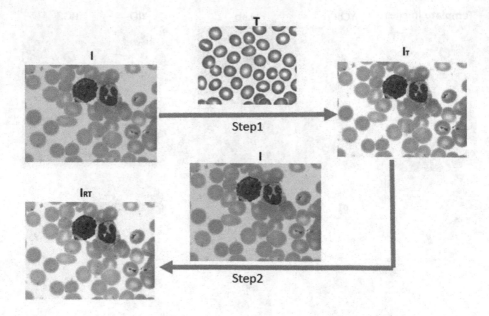

Fig. 4. Steps used in reverse color transfer. (Color figure online)

Table 1. Time taken (in milliseconds) for color transfer using various color space representations.

Images	Time taken (in milliseconds)			
	CIE-LAB	YCbCr	YIQ	RGB
Template1	139.11	34.21	35.48	18.85
Template2	134.34	34.95	25.79	19.30
Template3	135.39	33.11	27.16	18.84
Template4	138.23	33.24	32.08	18.91
Template5	133.96	32.33	34.46	18.88
Average	136.21	33.57	30.99	18.90

Table 2. Percentage errors corresponding to CIE-LAB and YCbCr color spaces.

Images	Average error (%)	
	CIE-LAB	YCbCr
1	0.178	0.177
2	0.0999	0.097
3	0.0677	0.0614
4	0.0677	0.063

4   Conclusion

An efficient color transfer method based on YCbCr color space for enhancement of peripheral blood smears was presented in this paper. We conclude that color transfer method based on CIE-LAB color space takes more time but provides results comparable to that of YCbCr color space based approach. Hence we conclude that the YCbCr color space based approach is more efficient and equally accurate and thus would be the preferred choice for enhancement of peripheral blood smear images.

References

1. Huang, D.C., Hung, K.D., Chan, Y.K.: A computer assisted method for leukocyte nucleus segmentation and recognition in blood smear images. J. Syst. Softw. **85**(9), 2104–2118 (2012)
2. Yang, L., Meer, P., Foran, D.J.: Unsupervised segmentation based on robust estimation and color active contour models. IEEE Trans. Inf. Technol. Biomed. **9**(3), 475–486 (2005)
3. Othman, M.Z., Ali, A.B.: Segmentation and feature extraction of lymphocytes WBC using microscopic images. Int. J. Eng. Res. Technol. **3**, 696 (2014)
4. Azam, B., Qureshi, R.J., Jan, Z., Khattak, T.A.: Color based segmentation of white blood cells in blood photomicrographs using image quantization. Res. J. Recent Sci. **3**(4), 34 (2014)
5. Amin, M.M., Kermani, S., Talebi, A., Oghli, M.G.: Recognition of acute lymphoblastic leukemia cells in microscopic images using k-means clustering and support vector machine classifier. J. Med. Signals Sens. **5**, 49 (2015)
6. Rezatofighi, S.H., Soltanian-Zadeh, H.: Automatic recognition of five types of white blood cells in peripheral blood. Comput. Med. Imaging Graph. **35**, 333–343 (2011)
7. Mathur, A., Tripathi, A.S., Kuse, M.: Scalable system for classification of white blood cells from Leishman stained blood stain images. J. Pathol. Inform. **4**(Suppl.), S15 (2013)
8. Marzuki, N.I.C., Mahmood, N.H., Razak, M.A.A.: Segmentation of white blood cell nucleus using active contour. J. Teknologi **74**(6), 115 (2015)
9. Chu, R., Zeng, X., Han, L., Wang, M.: Subimage cosegmentation in a single white blood cell image. In: Proceedings of IEEE 7th International Conference on Computational Intelligence, Communication Systems and Networks (2015). https://doi.org/10.1109/CICSyN.2015.36
10. Kayser, K., Gortler, J., Metze, K., Goldmann, T., Vollmer, E., Mireskandari, M.: How to measure image quality in tissue-based diagnosis (diagnostic surgical pathology). Diagn. Pathol. **3**(Suppl. 1), S11 (2008)
11. Tani, S., Fukunaga, Y., Shimizu, S., Fukunishi, M., Ishii, K., Tamiya, K.: Color standardization method and system for whole slide imaging based on spectral sensing. Anal. Cell. Pathol. (Amst.) **35**(2), 107–115 (2012)
12. Ruifrok, A.C., Johnston, D.A.: Quantification of histochemical staining by color deconvolution. Anal. Quant. Cytol. Histol. **23**(4), 291–299 (2001). The International Academy of Cytology and American Society of Cytology
13. Ruifrok, A.C., Katz, R.L., Johnston, D.A.: Comparison of quantification of histochemical staining by hue-saturation-intensity (HSI) transformation and color-deconvolution. Appl. Immunohistochem. Mol. Morphol. (AIMM) **11**(1), 85–91 (2003). Official publication of the Society for Applied Immunohistochemistry

14. Vahadane, A., Peng, T., Sethi, A., Albarqouni, S., Wang, L., Baust, M.: Structure-preserving color normalization and sparse stain separation for histological images. IEEE Trans. Med. Imaging **35**(8), 1962–1971 (2016)
15. Macenko, M., Niethammer, M., Marron, J.S., Borland, D., Woosley, J.T., Guan, X.: A method for normalizing histology slides for quantitative analysis. In: IEEE International Symposium on Biomedical Imaging: From Nano to Macro (2009)
16. Khan, A.M., Rajpoot, N., Treanor, D., Magee, D.: A nonlinear mapping approach to stain normalization in digital histopathology images using image-specific color deconvolution. IEEE Trans. Biomed. Eng. **61**(6), 1729–1738 (2014)
17. Reinhard, E., Ashikhmin, M., Gooch, B., Shirley, P.: Color transfer between images. IEEE Comput. Graph. Appl. **21**(5), 34–41 (2001)
18. Zhang, C., et al.: White blood cell segmentation by color-space-based k-means clustering. Sensors (Basel) **14**(9), 1612816147 (2014)
19. Rezatofighi, S.H., Soltanian-Zadeh, H.: Automatic recognition of five types of white blood cells in peripheral blood. Comput. Med. Imaging Graph. **35**(4), 333–343 (2011)
20. Peripheral Blood Smears. https://veteriankey.com/peripheral-blood-smears/. Accessed 10 Oct 2018
21. Heamatology Atlas. http://www3.med.unipmn.it/did/will/atlashem/node1.htm. Accessed 10 Oct 2018
22. American Journal of Laboratory Medicine. http://article.sciencepublishinggroup.com/html/10.11648.j.ajlm.20160103.12.html#paper-content-5. Accessed 10 Oct 2018
23. LISC dataset. http://users.cecs.anu.edu.au/~hrezatofighi/Data/Leukocyte%20Data.htm. Accessed 10 October 2018

Medical Image Encryption with Integrity Using DNA and Chaotic Map

Prema T. Akkasaligar[1] and Sumangala Biradar[2]([✉])

[1] Department of Computer Science and Engineering,
BLDEA'S V. P. Dr. P. G. Halakatti College of Engineering and Technology,
Vijayapur 586103, Karnataka, India
premasb@rediffmail.com

[2] Department of Information Science and Engineering,
BLDEA'S V. P. Dr. P. G. Halakatti College of Engineering and Technology,
Vijayapur 586103, Karnataka, India
biradarsumangala@gmail.com

Abstract. In current era the medical field is digitalizing. In digitalization, internet is playing the main role which is open source. Hence security is the major issue, particularly in medical field integrity, confidentiality and security are very significant issues. To provide this, a novel approach using SHA-256, Deoxyribonucleic Acid (DNA) cryptography and a chaotic map is proposed in this paper. The integrity for digital medical image is very important because slight modification also becomes very big problem. Diagnosing exact disease from slightly modified digital medical image is highly impossible. Hence in this paper we are mainly concentrating on integrity. For integrity purpose initially, the hash key is generated using SHA-256 and is hidden in the Least Significant Byte (LSB) of digitalized medical image. The image is encoded using DNA coding rules to provide security. The encoded DNA matrix pixels are shuffled and diffused using Chen's hyper chaotic map. The simulation results proved that the proposed method is resist against differential, exhaustive and statistical attacks.

Keywords: DNA · Digital medical image ·
Chen's hyper chaotic map · Encryption

1 Introduction

Due to advancement in the scientific world, the transformation of medical information and medical image over the internet has become explosive. The transmission of a digitalized medical image can be easily intercepted by hackers. That's why the encryption of digitalized medical image has a prime concern in the current period. The digital medical image encryption techniques are different from other type of image encryption techniques. The digitalized medical image contains very sensitive data and minor changes in an images affects the whole

K. C. Santosh and R. S. Hegadi (Eds.): RTIP2R 2018, CCIS 1036, pp. 143–153, 2019.
https://doi.org/10.1007/978-981-13-9184-2_13

image. Hence for digitalized medical image along with security, integrity also plays a very important role. The hash function is used along with the advanced encryption techniques like chaotic system and DNA sequence for medical image encryption.

The hash function generates hash key, which is used as a signature for the information [1]. With the help of the signature we can easily identify the integrity of the information. The hash key is mainly used to check the originality of the information. Hence hash function can fulfill the main requirement of digitalized medical image.

The chaotic system generates a nonlinear deterministic pseudorandom sequence, which is highly sensitive to initial conditions. It also has many characteristics, such as ergodicity, mixing property and structural complexity. The Chen's hyper chaotic map is an advanced technique in chaotic system. It has further positive Lyapunov exponent and has complicated dynamical uniqueness [2]. Due to these properties, "Chen's hyper chaotic system" is used in the field of cryptography for encrypting the image. But because of limited key space it can be easily attacked by the "key space" and "key sensitive" examination test. For this reason, a novel method named DNA technique is evolved in cryptography.

The DNA technique is fresh domain in cryptography. The DNA is used as an information carrier in image encryption. The major security depends on the structure of DNA. The structure of DNA is very complex hence computation of DNA on the image is very difficult. For image encryption, only sequences of DNA are used which is not sufficient. To defeat from these glitches, the image is encrypted using combination of "SHA-256", "Chen's hyper chaotic map" and "DNA sequence".

Several encryption algorithms for images have been put forward by using cryptographic and watermarking techniques. A slight work is carried out in DNA cryptography and chaotic system to transmit a digital medical image in a secure way.

In [3], a modular cosine number transform (CNT) arithmetic tool is used for the encryption of medical image. It is very sensitive to changes and also prevents the rounding-off errors. In [4], authors have proposed selective medical image encryption using 2D chaotic cat map. The digital image is separated into an area of interest and an area of the background. The area of interest blocks is identified and masked with a synthetic image to encrypt the sensitive part of the medical image. In [5], authors have proposed the watermarking technique for encryption. The cipher image is obtained by computing the mean and entropy of the digital medical image. The digital watermarking algorithm is used to create a water-marked image of the cipher image. In [6], authors have proposed the Rivest, Adi Shamir and Leonard Adleman (RSA) algorithm for encryption of magnetic resonance imaging (MRI) of medical images. Further, the K-means and watershed segmentation is used to extract the details of a tumor. In [7], the authors proposed a least significant bit (LSB) algorithm for embedding the patient confidential record in the high-frequency domain of a distorted image. Further, Linde, Buzo, and Gray (LBG) algorithm is used to the ensure security of data concealed image. In [8], the advanced encryption standard (AES) is used for the encryption of electronic patient records (EPR). The image is subdivided into an area of interest to protect from modification and an area of non-interest to prevent secret information. The EPR data is converted into cipher text and embedded in an area of non-interest. The discrete wavelet transforms (DWT) and inverse discrete wavelet transform (IDWT) is employed to a digital watermarked image. In [9], the logistic map and DNA sequence rules are used to create DNA mask. The best DNA mask is obtained by using a genetic algorithm. The digital watermarked image is obtained by embedding the electronic patient record into the DNA mask.

The above literature review highlights digital watermarking, steganography, and ancient cryptographic techniques. These techniques do not guarantee the robustness and integrity of digitalized medical images. Hence the novel method is proposed using SHA-256 for integrity, DNA for security due to its uniqueness and Chen's hyper chaotic map due its complex shuffling nature.

The rest of the paper is organized as follows: Sect. 2 mentions the encryption techniques suitable for medical digitalized image. Section 3 explains the proposed system and working principles in detail. Section 4 describes about performance analysis of the encryption method. Section 5 shows the experimental results we have performed on proposed system and Sect. 6 concludes this paper.

2   Encryption Techniques

The security, confidentiality, and integrity conditions are the main requirements for encrypting the digitalized medical images. The SHA-256 method is used to fulfil the integrity requirement. The Chen's hyper chaotic map and DNA methods are used to fulfil the security and confidentiality requirements.

2.1   SHA-256

The digitalized medical image contains very sensitive information and the small alteration in an image creates a very big cause. Hence integrity of the medical image is very important. The integrity is implemented using the SHA-256. It acts as a signature for medical image. It generates a 256-bit unique hash key for image. The DNA sequence is processed by 16 blocks with 32 bits in each block and each block requires 64 rounds of processing to generate a hash key of fixed length.

Let us consider DNA sequences M, which is subdivided into $(H_1, H_2, \ldots, H_{16})$ each of 32 bit blocks. The hash key is obtained by the XOR operation defined in Eq. 1.

$$\Delta(M) = H_1 \oplus H_2 \oplus \cdots \oplus H_{16} \tag{1}$$

where Δ(M) is 256-bit hash key. The hash key is used to check the integrity of the digital medical image in an encryption and decryption process.

2.2   Chen's Hyper Chaotic Map

The "Chen's hyper chaotic map" has spatiotemporal complexity and mixture property due to the positive Lyapunov exponent. The chaotic sequences obtained by hyper chaotic map are extremely complex and difficult to predict and explore. Hence the hyper chaotic system is used to provide security in image encryption techniques. The existing cryptography methods based on chaotic maps are categorized into two stages: permutation and diffusion. The position of pixels in the original image is altered by chaotic sequences in the permutation stage. The value of pixels in the original image is altered by chaotic sequences in the diffusion stage.

We have proposed digitalized medical image encryption using "Chen's hyper chaotic system". The "Chen's hyper chaotic system" is illustrated using Eqs. (2)–(5):

$$U_1 = e(v - u) \tag{2}$$

$$V_1 = uw + hu + gv - p \tag{3}$$

$$W_1 = uv - fw \tag{4}$$

$$P_1 = u + l \tag{5}$$

Where u, v, w and p, are state parameters and e, f, g and h are control factors. The value of l varies from -0.7 to 0.7. The U, V, W and P represents a chaotic sequence.

2.3   Biological Operation

The fundamental essentials of biological DNA is nucleotide. The nucleotides are separated into basic four alphabets due to a variant structure of chemical bonding. The alphabets are namely, 'Thymine (T)', 'Guanine (G)', 'Adenine (A)' and 'Cytosine (C)' [10]. Due to the key hydrogen, the chains are placed jointly to form a structure of twofold twist and one chain in the sequence-base is paired with another, like, T and A are a pair, C and G are a pair [11].

In binary form 1 is considered as opposite 0, similarly 11 is considered as opposite to 00 and 10 is considered as opposite to 01. The bases of nucleic acid T, G, A and C can be determined as 00, 10, 11 and 01 correspondingly. Using this concept, we can get $4! = 24$ varied encoding samples. However, due to the complement relation between DNA bases, only eight patterns of encoding satisfy the complementary base pairing as shown below:

R_1: G = 01 C = 10 T = 11 A = 00
R_2: G = 10 C = 01 T = 11 A = 00
R_3: G = 00 C = 11 T = 10 A = 01
R_4: G = 11 C = 00 T = 10 A = 01
R_5: G = 00 C = 11 T = 01 A = 10
R_6: G = 11 C = 00 T = 01 A = 10
R_7: G = 01 C = 10 T = 00 A = 11
R_8: G = 10 C = 01 T = 00 A = 11

3   Proposed Methodology

The medical image encryption is performed by combining the "SHA-256", "Chen's hyper chaotic system" and "DNA" operations.

To provide integrity and security, the combination of hash key, DNA and chaotic map is proposed. The DNA sequence of "Canis lupus" [12] of varying size is considered. The 256-bits fixed size unique hash key is generated for "Canis lupus" using SHA-256. The detailed hash key generation is shown in Fig. 1.

Fig. 1. Hash key generation

In the proposed model, the hash key is generated using SHA-256 and is embedded in the LSB bits of the digitalized original medical image as shown in Fig. 2. The medical image of size m × n is converted into an 8-bit binary image of size m rows and n × 8 columns. The sequence of DNA G = 11 C = 00 T = 10 and A = 01 is pertained on binary 8-bit medical image. The matrix of DNA encoded of size m rows and n × 4 columns is attained. The Chen's hyper chaotic map with state variables and control parameters are used to produce the sequences of hyper chaotic. The sequence of hyper chaotic is sorted and depends on position of the ordered chaotic sequence the pixels of the matrix DNA encoded is jumbled. The XOR logical operation is performed for the union of the snarled matrix DNA encoded to get the intermediate encrypted. The intermediate encrypted image is decoded using DNA complementary rule (i.e. A = 10, T = 01, G = 00 and C = 11) sequence to get final cipher image.

4   Performance Analysis

In cryptographic analysis differential attacks, exhaustive attacks, and statistical attacks are performed to analyse the performance of encryption techniques. For differential attack, the "Unified Average Changed Intensity (UACI)" and "Number of Pixel Changing Rate (NPCR)" methods are used. For an exhaustive attack, the "key security" and "key space" analysis are used. The correlation coefficient analysis method is employed to verify the statistical attack. The entropy is employed to verify the error rate of a digital image.

4.1   Correlation Coefficient Analysis

The correlation coefficient is used to determine an association between the neighbouring pixels in the given digital image. The highly correlated neighbouring pixels indicate that the encryption technique is the best technique. The Pearson's correlation coefficient is shown in (6).

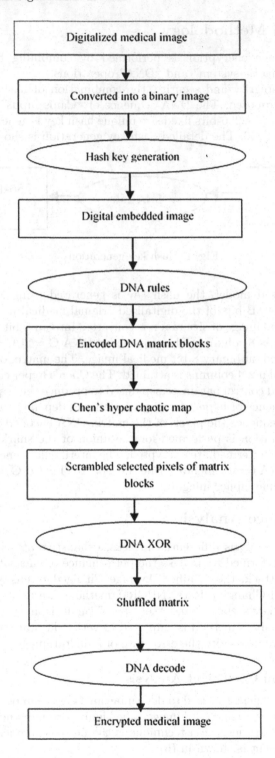

Fig. 2. Proposed system for encryption process

$$\gamma := \frac{N\Sigma I_1 I_3 - \Sigma I_1 \Sigma I_3}{\sqrt{N(\Sigma I_1^2) + (\Sigma I_1)^2}\sqrt{N(\Sigma I_3^2) + (\Sigma I_3)^2}} \tag{6}$$

where I_1 and I_3 are the gray level values of the medical digitalized image and encrypted medical image correspondingly and N is the size of an image. The +1 value of r indicates positively correlated and -1 value indicates negatively correlated [13]. A zero value indicates no correlation [14,15].

4.2   NPCR and UACI

In differential attacks, the NPCR and UACI are significant factors employed to determine the performance of encryption method for the digitalized medical image. The NPCR is defined in (7).

$$NPCR := \frac{\Sigma_{r_1 c_1} D_1(r_1 c_1)}{W_1 \times H_1} \times 100\% \tag{7}$$

where W_1 and H_1 are width and height of the digital medical image and $D_1(r_1, c_1)$ is defined in (8).

$$D_1(r_1, c_1) = \begin{cases} 0 & if \quad I_{11}(r_1, c_1) = I_3(r_1, c_1) \\ 1 & if \quad I_{11}(r_1, c_1) \neq I_3(r_1, c_1) \end{cases} \tag{8}$$

The UACI is defined in (9)

$$UACI = \frac{1}{W_1 \times H_1}\left[\Sigma_{x,y}\frac{I_{11}(r_1, c_1) - I_3(r_1, c_1)}{255} \times 100\%\right] \tag{9}$$

where I_3 and I_{11} are two ciphered images respectively obtained from an original image and one pixel value changed in an original digitalized medical image.

4.3   Entropy

The quality of the encryption algorithm is deliberated by entropy value. The entropy is the measurement of a probability distribution of grey levels throughout the image. The uniform distribution is indicated by a higher value of entropy. The higher value of entropy indicates good quality. The entropy is defined by (10).

$$E(1) = \Sigma_{i=1}^{n} R(I_i) log(R(I_i)) \tag{10}$$

where $R(I_i)$ represents the probability of distribution of gray level of the digital medical image and n represents a total number of gray levels.

5   Experimental Results

The simulation is conducted on 7th generation $Intel^\circledR$ $Core^{TM}$ i7 7500U CPU @ 2.70 GHz, 2901 MHz, 2 Core(s), 4 Logical Processor(s). The 500 medical digitalized images of different five kinds (each type 100) like Ultrasound, MRI, X-Ray, and CT are collected from "National Library of Medicine's Open Access Biomedical Images Search Engine" [16]. The ECG images are collected from ecgeducator.blogspot.com. The Matlab (R2015b) tool is employed to perform the proposed medical image encryption technique. The sample digitalized medical image is shown in Fig. 3(a). The DNA R_6 is used to conquer the encoded DNA matrix. The SHA-256 method is used to generate hash key for varying size DNA sequence of "Canis lupus" [12]. The 256-bit hash key is embedded in LSB of digitalized medical image. The initial values of the "Chen's hyper chaotic" state variables are u = 0.3, v = −0.4, w = 1.2, and p = 1 are empirically determined and control parameters e = 36, f = 3, g = 28, h = −16, and l = 0.2 are considered to produce a sequence of chaotic. The sequence of chaotic is ordered and the encoded DNA matrix pixels are jumbled based on a position of the ordered sequence. The encrypted digitalized medical image as shown in Fig. 3(b) is obtained by combining the jumbled matrix blocks. The operation of DNA XOR is intended for the union of jumbled matrix.

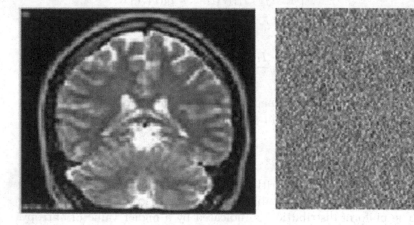

Fig. 3. (a) Sample medical image and (b) encrypted medical image

The performance analysis of the proposed system is shown in Table 1. The NPCR value is almost close to 100 i.e. 99.89 and UACI value is 32.39. The performance indicators NPCR and UACI are shows the high security analysis test for differential attacks. Entropy value is 7.90 close to 8.0 and proves the high quality of image encryption method. The correlation coefficient is nearest to 0.99 proves that the pixels are correlated. From this analysis, it is proved that it is high enough to provide security and integrity.

5.1   Key Space Analysis

In the proposed method the initial value and the system parameters of the "Chen's hyper chaotic map" are used as a secret key. The total of nine secret keys (u, v, w, p, e, f, g, h, l) and hash key is used. The size of the key is $10^{11} \times 10^{11} \times 10^{11} \times 10^{11} \times 10^{11} \times 10^{11} \times 10^{11} \times 10^{11} = 10^{88}$, if the precision is 10^{11}. The size of hash key is $256(2^8)$ bit. The space of secret key is adequate to resist exhaustive attack.

The key space should be larger than 2128(1039) to resist the brute force attack [17]. The better integrity for the digitalized medical image is provided in the proposed method by additionally considering a secret hash key of size 2^8 bits.

Table 1. Performance analysis of the proposed system

Medical image	NPCR	UACI	Entropy	Correlation coefficient
MRI image	99.88	31.57	7.98	0.987
CT image	99.77	33.23	7.76	0.992
X-ray image	99.95	32.43	7.93	0.995
Ultrasound image	99.96	32.34	7.96	0.991
ECG image	99.93	33.02	7.98	0.999

5.2   Histogram Analysis

The histogram of an image is the pictorial illustration of image pixels against the various intensity levels. The Fig. 4(a) shows the histogram of the sample medical image and Fig. 4(b) cipher medical image. From the histogram of the cipher medical image we observe that pixels are more constantly distributed, which proves that the majority of image pixel values are altered, and the technique of encryption has good confusing property.

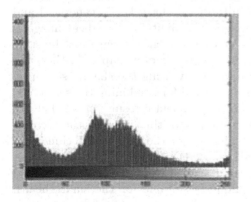 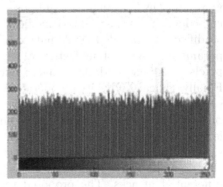

Fig. 4. (a) Histogram of sample medical image and (b) histogram of cipher medical image

5.3   Comparison Work

The proposed method is compared with the method specified in [18] shown in Table 2. In [18], the medical image is encrypted based on the intensity value of the digitalized medical image. The two different Chao's theory methods are used to shuffle the pixels of the digitalized medical image. But, in the proposed method single Chao's theory is used to shuffle the pixels. After the performance analysis results comparison is shown in Table 3. are almost equal to the method specified in [18]. But the main difference is that the proposed method takes less time and integrity is also verified.

Table 2. Comparison of the proposed system

Medical image	NPCR		UACI	
	Proposed system	Prema et al. [18]	Proposed system	Prema et al. [18]
MRI image	99.76	99.88	40.37	31.57
CT image	99.63	99.77	36.54	33.23
X-ray image	99.98	99.95	43.31	32.43
Ultrasound image	99.85	99.96	38.43	32.34

Table 3. Comparison of the performance analysis

Medical image	Correlation coefficient		Time taken	
	Proposed system	Prema et al. [18]	Proposed system	Prema et al. [18]
MRI image	0.976	0.987	40	350
CT image	0.987	0.992	41	380
X-ray image	0.990	0.995	42	400
Ultrasound image	0.979	0.991	38	330

6   Conclusion

The digitalized medical image encryption using a SHA-256 hash key, chaotic map, and DNA sequences is proposed. Initially, the hash key is generated and embedded in the digitalized medical image. The digitalized medical image is modified into encoded DNA matrix. The chaotic sequence is generated by state parameters and system factors of "Chen's hyper chaotic map". Further, the pixels of the encoded DNA matrix are jumbled by using the chaotic sequence. Finally, logical XOR bitwise operation is employed for combining the scrambled pixels of encoded DNA matrix blocks. The experimental results prove that the proposed digitalized medical image encryption algorithm improve the security level and also provides integrity. The originality of the proposed technique is to afford security and integrity using SHA-256, DNA rules, and Chen's hyper chaotic map. The crypto analysis also resists against exhaustive, statistical and differential attacks. The proposed technique is suitable for the smart health system and telemedicine applications.

References

1. Norouzi, B., Seyedzadeh, S.M., Mirzakuchaki, S., Mosavi, M.R.: A novel image encryption based on hash function with only two-round diffusion process. Multimed. Syst. **20**(1), 45–64 (2014)
2. Wang, Q., Zhang, Q., Zhou, C.: A multilevel image encryption algorithm based on chaos and DNA coding. In: Fourth International Conference on Bio-inspired Computing (BICTA 2009), 16–19 October 2009, pp. 70–74. IEEE (2009)
3. Lima, J.B., Madeiro, F., Sales, F.J.R.: Encryption of medical images based on the cosine number transform. Signal Process. Image Commun. **35**, 1–8 (2015)
4. Kanso, A., Ghebleh, M.: An efficient and robust image encryption scheme for medical applications. Commun. Nonlinear Sci. Numer. Simul. **24**(1), 98–116 (2015)
5. Kester, Q.A., Nana, L., Pascu, A.C., Gire, S., Eghan, J.M., Quaynor, N.N.: A cryptographic technique for security of medical images in health information systems. Procedia Comput. Sci. **58**, 538–543 (2015)
6. Sebastian, A., Delson, T.R.: Secure magnetic resonance image transmission and tumor detection techniques. In: 2016 International Conference on Circuit, Power and Computing Technologies (ICCPCT), pp. 1–5. IEEE (2016)
7. Vallathan, G., Devi, G.G., Kannan, A.V.: Enhanced data concealing technique to secure medical image in telemedicine applications. In: 2016 International Conference on Wireless Communications, Signal Processing and Networking (WiSPNET), pp. 186–190. IEEE (2016)
8. Al-Haj, A., Hussein, N., Abandah, G.: Combining cryptography and digital watermarking for secured transmission of medical images. In: 2016 2nd International Conference on Information Management (ICIM), pp. 40–46. IEEE (2016)
9. Anusudha, K., Venkateswaran, N., Valarmathi, J.: Secured medical image watermarking with DNA codec. Multimed. Tools Appl. **76**(2), 2911–2932 (2017)
10. Nimmy, S.F., Sarowar, M.G., Dey, N., Ashour, A.S., Santosh, K.C.: Investigation of DNA discontinuity for detecting tuberculosis. J. Ambient Intell. Humanized Comput. 1–15 (2018)
11. https://www.dreamstime.com/stock-illustration-dna-structure-nucleotide-phosphate-sugar-bases-image56366025. Accessed 22 July 2018
12. https://www.ncbi.nlm.nih.gov/nuccore/LQ820900.1. Accessed 22 July 2018
13. Vaidya, S.P., Mouli, P.C., Santosh, K.C.: Imperceptible watermark for a game-theoretic watermarking system. Int. J. Mach. Learn. Cybern. 1–17 (2018)
14. Hiremath, P.S., Akkasaligar, P.T., Badiger, S.: An optimal wavelet filter for despeckling echocardiographic images. In: International Conference Computational Intelligence and Multimedia Applications, Sivakasi, Tamilnadu, India, 13–15 December 2007, pp. 245–249 (2007)
15. Hiremath, P.S., Akkasaligar, P.T., Badiger, S.: Speckle reducing contourlet transform for medical ultrasound images. In: World Academy of Science, Engineering and Technology - Special Journal Issue, pp. 1217–1224 (2011)
16. National Library of Medicine's Open Access Biomedical Images Search Engine. https://openi.nlm.nih.gov. Accessed 11 Feb 2018
17. Wang, X., Teng, L., Qin, X.: A novel color image encryption based on chaos. Signal Process. **924**, 1101–1108 (2012)
18. Akkasaligar, P.T., Biradar, S.: Secure medical image encryption based on intensity level using Chao's theory and DNA cryptography. In: International Conference on Computational Intelligence and Computing Research, 15–17 December 2016, pp. 958–963. IEEE (2016)

A Systematic Approach for Constructing 3D MRI Brain Image over 2D Images

K. Vidhya[1], Mala V. Patil[2]($\boxtimes$), and Ravindra S. Hegadi[3]($\boxtimes$)

[1] Department of Computer Science and Engineering, VTU RRC, Belgaum, India
vidya.k2591@gmail.com
[2] Department of Computer Science, University of Horticultural Science,
Bagalkot, India
csehod15@gmail.com
[3] Department of Computer Science, Solapur University, Solapur, India
rshegadi@gmail.com

Abstract. Brain-related study must be very precise and accurate. As the brain is one of the important organ in the human body. Hence the medical image processing plays a vital role in early diagnosis and treatment planning. In the proposed paper layering mechanism is used to construct 3D brain image from several 2D Magnetic Resonance Images (MRI). In the first step, intensity-based registration technique is used to perform the registration of T1 and T2 weighted MRI images. The registered image is segmented for easy diagnosis, using gray level feature extraction and region growing segmentation methods. Finally tri-linear interpolation method is used to construct a 3D image from segmented image by matching the characteristic points of T1 and T2 weighted images. By confirming simulation results, we concluded that proposed approach is more efficient and accurate to construct 3D brain model from 2D MR images.

Keywords: Brain MRI · Registration · Segmentation ·
3D reconstruction · T1 weighted images · T2 weighted images

1 Introduction

Medical image processing plays a vital role in modern medicine for development of easy diagnosis, treatment planning systems and training simulators. As such planning system provide better pre-operative or intra-operative assistance to experts and the precise simulators can replicate exact real-world settings to provide better training environment [11]. There exist many diagnostic imaging modalities such as Computed Tomography (CT), Positron Emission Tomography (PET), Magnetic Resonance Imaging [10, 14]. Out of those MRI is the prime resource for brain disease diagnosis. In recent years many researchers are working on building 3D brain images from several 2D images. In addition to this, several studies are exists on 3D representation of brain abnormality from 2D

© Springer Nature Singapore Pte Ltd. 2019
K. C. Santosh and R. S. Hegadi (Eds.): RTIP2R 2018, CCIS 1036, pp. 154–162, 2019.
https://doi.org/10.1007/978-981-13-9184-2_14

MR images [6, 10]. However most of the existing 3D-construction techniques are times consuming because in those techniques, the data blocks are divided into cubes, each cubes are associated with eight voxels and at last they constructs a triangular mesh [6].

The brain MR image sequence can be classified into four categories: T1 weighted image, T2 weighted image, proton density and Fluid-attenuated inversion recovery (FLAIR) image. To construct 3D brain image, at first registration need to do for all four imaging sequences. Devising such registration technique is one of the challenging task. Hence in the proposed work, only T1 and T2 weighted image are considered for registration. The another reason behind the selection of T1 and T2 weighted images is, there classification is based on same criteria. That criteria is time taken from the instant when applied radio frequency signal is removed till the instant nuclei precession move out and they realign themselves with original magnetic fields. Out of these two, T1 images shows a gray and white matter and provides the excellent anatomical detail but they does not show any contrast between normal and abnormal tissues. whereas T2 weighted images lacks in anatomical detail but they gives good contrast between normal and abnormal tissues [1, 9].

In this proposed paper the reconstruction of 3D brain image is obtained by adapting a layering mechanism. The first step is registration of T1 and T2 weighted images to obtain homologous features [3, 15]. Next step is extracting extraction of brain part from cortex and at last, 3D Reconstruction from 2D MR images using tri-linear interpolation function [16, 17].

The reminder of the paper is organized as follows: Sect. 2 provides detailed information about several exiting techniques for 3D construction from 2D MR images. The detailed description of the proposed technique is discussed in Sect. 3. Section 4 contains the results of proposed technique on real patient-specific images. Section 5 gives conclusions.

2   Previous Work

There are several papers focused on 3D brain model construction from 2D MR images. In those authors used different registration, segmentation techniques to extract required portion before generation of 3D model.

Roy et al. [10] presented a 3D reconstruction of 2D brain MR slides. In this preferred system, the MRI brain slides which contain abnormalities are identified. Based on stack implementation 3D representation of abnormal slides are presented. Then the volume of the abnormality is computed. They used inter-slice interpolation, mesh generation and simplification technique to generate 3D model.

Kumar et al. [6] presented a 3D construction of a brain tumor using 2D Brain MRI. It involves the implementation of different step like pre-processing, segmentation, 3D reconstruction. Image segmentation is performed through fuzzy C mean (FCM) technique. Then mesh generation of the segmented image is performed using the marching cubes algorithm. Finally tumor is constructed using marching cubes meshing algorithm.

Lopes et al. [7] has presented tumor detection and 3D construction using 2D brain MRI. This system involves different implementations stages for identification and extraction of a tumor from the 2D MRI based on seeded region growing method. The presented method automatically selects the seed point. Along with tumor extraction, the method calculate the tumor volume to assist the radiologist.

Zahira et al. [17] presented an MRI brain images classification and reconstruction technique using depth map estimation. This referred article involves two phases namely: classification and 3D reconstruction. In the initial stage input images from the MRI dataset is pre-processed by skull stripping. In the next classification stage, skull-stripped images segmented by the watershed algorithm to extract the tumor part. Based on the attributes condensation, normal and tumor images are categorized using a neural network classifier. Finally, a 3D representation of a tumor is constructed using depth map estimation.

Thanh et al. [16] presented a 3D construction of 2D MRI cortex image. The input 2D MRI cortex images are passed to processing block. Images are pre-processed using a mean filter to eliminate the noise and image enhancement is done by histogram equalization. Otsu approach is used to segment the brain part from the cortex. Finally, the tri-linear interpolation method is used for the 3D construction of the image.

Farzana et al. [2] presented an analysis of 2D to 3D reconstruction methods in the brain MRI. In this referred method DICOM[1] images are considered as input which contains 2D brain MRI. Adaptive thresholding and k-means segmentation methods are used to extract desired part. Where as volume 3D and marching cube algorithm are used for 3D reconstruction. Based on results of the performance metrics (volumetric overlap error, relative volume difference, average symmetric surface distance, maximum symmetric surface distance, accuracy, sensitivity, specificity, and precision) authors have concluded that k-means segmentation method and marching cube algorithm for 3D model generation gives better results.

Tariq et al. [15] presented a paper on OTSU thresholding segmentation method over MRI brain images. By experimenting the data features, spatial filtering, skull stripping with other thresholding techniques, the proposed method is very helpful in segmenting the exact brain images.

Padilla et al. [9] presented NEURONAV software for image-guided surgery to Parkinson's daises. In deep brain stimulation, this software supports medical specialist. This software contains two modules. Initially, it contains planning for deep brain stimulation. This application contains many options namely: 3D image viewer, landmark medical planning and 3D brain structure visualization. NEURONAV also contains tracking of micro-electrode during surgery. This NEURONAV software is tested in specialized Colombia medical center.

Coup et al. [3] presented a registration of two different images namely: operative ultrasound and MRI. To overcome the difficulty in the registration of multimodal images, a new technique is presented namely probabilistic function which

---

[1] DICOM: Digital Imaging and Communications in Medicine.

is based on the hyperechogenic structure matching. Registration is performed by increasing the joint probability to a voxel that contains a hyperechogenic structure in two modalities.

Babu et al. [1] presented a review on automatic 3D brain image registration. This referred survey explains the problems in the structural and functional image analysis of imaging information. That analysis includes data transport, identification of boundary, volume calculation, 3D reconstruction with the display, surface rendering, analysis of shape and image overlay.

In summary, there exist lot of state of the art 3D brain reconstruction techniques from 2D MR images. Still there are many challenges need to consider such as improvement in quality of the constructed 3D MRI brain image, accurate and fully automatic registration of the image with no data loss [7, 10]. Existing methods were unable to find out the exact volume of the abnormal MRI brain images so the 3D reconstruction is not much precise [9, 17]. Adoption of effective preprocessing technique which will remove unwanted artefacts and enhance the desired contents may help to increase segmentation and visualization accuracy [12, 13].

3   Methodology

The block diagram of the proposed system is shown in Fig. 1. In the proposed system, two MR (one T1 weighted and one T2 weighted) images are taken from the database. Intensity-based registration is performed on these two input images. Then brain part extracted from the registered image using Otsu and region growing segmentation method. At last tri-linear interpolation technique is used to construct 3D brain model.

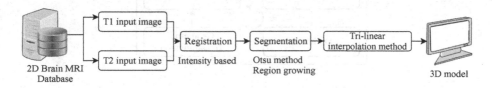

Fig. 1. Block diagram of proposed work

3.1   Image Acquisition

In the presented work, sixty brain MR images are collected from Vikram hospital, Bangalore, India. MR images are taken at every 6°. The MR images are collected in DICOM format. However, these files contain patient-specific information like a patient number, name, and age. Disclosing this information may violate the ethical laws like HIPPA[2] laws or IRB[3] protocol. So to erase the patient-specific

---

[2] HIPPA: Health Insurance Portability and Accountability Act.
[3] IRB: Institutional Review Board.

data, the MATLAB script discussed in [5] is adapted to remove patient-specific information from CT image. The script uses pre-defined function dicomanon (part of the MATLAB Image Processing Toolbox) to remove confidential medical data from a dicom file.

3.2   Image Registration

The brain MRI images are classified into a T1 and T2 weighted image which has different sensors, time intervals and various viewpoints from each other. Hence it is required to integrate the T1 and T2 weighted images to obtain a homologous features. Various image registration techniques can be used for automatic image registration. But optimization of the registered image has trade off between speed as well as accuracy [1,17]. T1 weighted image gave good contrast information of gray and white matter whereas T2 weighted image gave information on abnormal tissues. Hence registration of T1 and T2 weighted image into one homologous featured image is required. In the proposed work intensity-based registration is performed as it includes boundary identification, volume estimation, shape analysis, structural imaging that optimizes into a single image. In addition to this, the intensity-based registration technique ensures there is no loss of information in the process of registration. The flow of intensity-based registration technique is shown in Fig. 2.

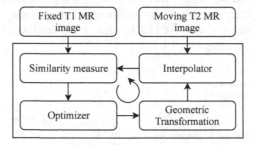

Fig. 2. Intensity-based registration

3.3   Image Segmentation

Image segmentation plays an essential role in several image processing based applications [4]. 3D model construction is no exception to this. In the proposed work registered T1 and T2 weighted MR image are segmented in order to extract the brain part from the cortex image before proceeding to 3D construction. There exist various segmentation methods like histogram-based thresholding, fusion algorithm and expection-minimization (EM) based algorithm which gives promising results. Theses methods make use of the parametric and nonparametric approach to maximize the expectation level and these are not always suitable for all types of applications [15]. The traditional segmentation technique fully depends on minimum variance threshold value, so the segmented image may give the undesired object with few lost sections. In addition to this, extracting desired information from multidimensional images is challenging task.

To overcome this challenge, a Otsu thresholding and region growing based segmentation algorithm is developed to segment the registered image of T1 and T2 weighted image. Otsu is a non-parametric approach for image segmentation and an alternative to Bayes decision rule [8]. That calculates many gray thresholds that correspond to the smallest class. Segmentation using Otsu thresholding gives the segmentation of image quickly with better accuracy results. In the case of 3D construction, choosing a threshold value for image segmentation including OTSU method is important.

After application of Otsu multi-thresholding, skull stripping and binarization is used to extract desired part from MR images. The skull stripping is used to exact brain portion. In addition to this, it is used to remove the meninges layer that comprises of the tissues that covers the brain and to obtain the remaining cerebrum region. If this is process is done manually it takes lot of time hence skull stripping for region growing method is adapted. Then the segmented image is passed to 3D model generator layer to generate 3D brain model which help doctors to come up with better diagnosis.

3.4   3D Model Reconstruction

In the proposed tri-linear interpolation algorithm is used to construct 3D brain model from registered and segmented T1, T2 MR images. The tri-linear interpolation technique calculates the approximate value between two consecutive layers. That is it is used to calculate the pixel value from the set of discrete points to construct x, y, z coordinate system. In this method each pixel of the original image is multiplied with the segmented image for the reconstruction. The model generated by tri-linear interpolation method can be rotated in any angle which is beneficial to observe different portions of brain.

4   Experimental Results

In this proposed research work the dataset of 60 MR brain images that are gathered from Vikram hospital, at every 6°, one T1 and T2 weighted image is considered. In this work T1 and T2 weighted image (shown in Fig. 3(a) and (b) respectively) is registered based on intensity-based registration as shown in Fig. 3(c). This registration technique that brings them into same sensor and helps in obtaining a homologous feature.

The registered image is segmented by using the non-parametric method OTSU algorithm, this overcomes the drawback of the automated segmentation technique, EM-based segmentation and histogram based thresholding [15], Fig. 4 represents the segmented image.

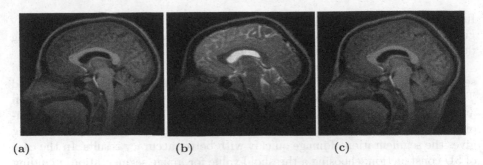

(a)   (b)   (c)

Fig. 3. (a) T1 weighted image (b) T2 weighted image (c) Registered image

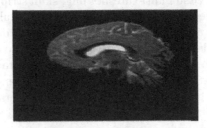

Fig. 4. Segmented image

The 3D construction is done based on tri-linear interpolation algorithm that calculates the pixel value of the images the replaces traditional techniques such as the FCM techniques. The result of tri-linear interpolation method is shown in Fig. 5. The tri-linear interpolation algorithm processes the 3D image construction that can be rotated and observed at different positions of the MRI Brain images, The obtained simulation results gives the detailing of the constructed 3D Brain represents in Fig. 6 which is helpful for easy diagnosis.

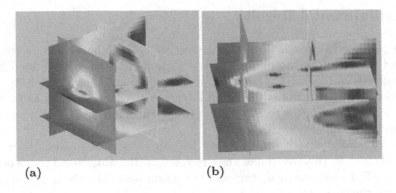

(a)   (b)

Fig. 5. Simulated interpolated images (a) vertical view (b) horizontal view

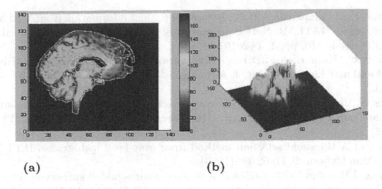

(a)   (b)

Fig. 6. Simulated 3D brain image construction

5   Conclusions

The proposed paper presents a 3D MRI Brain image construction, by taking 60 MRI Brain T1 and T2 weighted images. For detailed analysis of the simulation results obtained by using MATLAB tool. T1 and T2 weighted MR brain image are registered using intensity-based registration and then they are segmented using Otsu and region growing methods that segments the brain part for consideration and diagnosis. For 3D construction tri-linear interpolation method is used that calculates every pixel of an image and helps in construction of 3D MR image, that can easily be rotated in different angles. The simulation results obtained allows to rotate and view the 3D MR brain image in different positions thus supports doctor in early diagnosis and treatment planning. As per the comparisons made from the existing techniques, the proposed work is able to combine the T1 and T2 weighted images together for registration and segmentation ensures there is no loss of data and performs well by preserving edges as well which helps in easy construction of 3D MR Brain images. In future it is required to consider the other two classification of Brain MRI images that is Proton density layer and FLAIR layer for the 3D construction by using advanced machine learning algorithms.

References

1. Babu, G., Sivakumar, R.: Automatic 3D brain image registration-survey. Int. J. Innov. Manag. Technol. **4**(5), 502 (2013)
2. Cam, Q., Hai, N.: Analysis of 2D to 3D reconstruction techniques over brain MRI. Int. J. Innov. Res. Comput. Commun. Eng. **5**(5) (2017)
3. Coupé, P., Hellier, P., Morandi, X., Barillot, C.: 3D rigid registration of intraoperative ultrasound and preoperative MR brain images based on hyperechogenic structures. J. Biomed. Imaging **2012**, 1 (2012)
4. Hegadi, R.S., Navale, D.I., Pawar, T.D., Ruikar, D.D.: Multi feature-based classification of osteoarthritis in knee joint X-ray images. In: Medical Imaging: Artificial Intelligence, Image Recognition, and Machine Learning Techniques, chap 5. CRC Press (2019). ISBN 9780367139612

5. Hunter, E.J., Palaparthi, A.K.R.: Removing patient information from MRI and CT images using MATLAB. National Repository for Laryngeal Data Technical Memo No. 3 (version 2.0), pp. 1–4 (2015)
6. Kumar, C., Kumari, A.: 3D reconstruction of brain tumor from 2D MRIs using FCM and marching cubes. Int. J. Adv. Res. Electron. Commun. Eng. **3**(9), 970–974 (2014)
7. Lopes, S., Jayaswal, D.: A methodical approach for detection and 3D reconstruction of brain tumor in MRI. Int. J. Adv. Res. Electron. Commun. Eng. **118**(17), 37–43 (2015)
8. Otsu, N.: A threshold selection method from gray-level histograms. IEEE Trans. Syst. Man Cybern. **9**(1), 62–66 (1979)
9. Padilla, J.B., et al.: NEURONAV: a tool for image-guided surgery-application to parkinsons disease. In: Bebis, G., et al. (eds.) ISVC 2015. LNCS, vol. 9474, pp. 349–358. Springer, Cham (2015). https://doi.org/10.1007/978-3-319-27857-5_32
10. Roy, S., Sadhu, S., Bandyopadhyay, S.K.: A useful approach towards 3D representation of brain abnormality from its 2D MRI slides with a volumetric exclamation. In: 2015 Third International Conference on Computer, Communication, Control and Information Technology (C3IT), pp. 1–6. IEEE (2015)
11. Ruikar, D.D., Hegadi, R.S., Santosh, K.: A systematic review on orthopedic simulators for psycho-motor skill and surgical procedure training. J. Med. Syst. **42**(9), 168 (2018)
12. Ruikar, D.D., Santosh, K.C., Hegadi, R.S.: Automated fractured bone segmentation and labeling from CT images. J. Med. Syst. **43**(3), 60 (2019). https://doi.org/10.1007/s10916-019-1176-x
13. Ruikar, D.D., Santosh, K.C., Hegadi, R.S.: Segmentation and analysis of CT images for bone fracture detection and labeling. In: Medical Imaging: Artificial Intelligence, Image Recognition, and Machine Learning Techniques, chap 7. CRC Press (2019). ISBN 9780367139612
14. Ruikar, D.D., Sawat, D.D., Santosh, K.C., Hegadi, R.S.: 3D imaging in biomedical applications: a systematic review. In: Medical Imaging: Artificial Intelligence, Image Recognition, and Machine Learning Techniques, chap 8. CRC Press (2019). ISBN 9780367139612
15. Tavares, J.M.R.S.: Analysis of biomedical images based on automated methods of image registration. In: Bebis, G., et al. (eds.) ISVC 2014. LNCS, vol. 8887, pp. 21–30. Springer, Cham (2014). https://doi.org/10.1007/978-3-319-14249-4_3
16. Thanh, C.Q., Hai, N.T.: Trilinear interpolation algorithm for reconstruction of 3D MRI brain image. Am. J. Signal Process. **7**(1), 1–11 (2017)
17. Zahira, M.F., Sathik, M.M.: An efficient classification of MRI brain images and 3D reconstruction using depth map estimation. Adv. Comput. Sci. Technol. **10**(5), 1057–1080 (2017)

Classification of Rheumatoid Arthritis Based on Image Processing Technique

S. A. Bhisikar[1]([envelope]) and S. N. Kale[2]

[1] Department of Electronics and Telecommunication, JSPM's Rajarshi Shahu
College of Engineering, Tathawade, Pune 411033, Maharashtra, India
swatibhisikar@gmail.com
[2] Department of Applied Electronics, Sant Gadge Baba Amravati University,
Amravati 444602, Maharashtra, India
sujatankale@rediffmail.com

Abstract. Arthritis is a disabling and agonizing disease. The rapid growth of biomedical image processing techniques assists the doctor in diagnosis and treatment of the disease. In Rheumatoid Arthritis as the disease progresses, it results in reducing physical activity level of the patient. The method presented in this paper is a completely automated framework to detect and quantify joint space width. This system detects severe stage of RA that are contaminated by disease to the degree that the joint space is no longer noticeable in the X-ray image. In proposed work RA is classified in three stages Normal (Non-RA), Abnormal (RA) and Severe stage RA. Joint location accuracy achieved is 92%. 60 images were tested, Out of 60 Test images 20 images are Normal, 22 images are abnormal i.e. RA affected and 18 images are severe. SVM classifier with Radial basis function kernel is efficient compared to FFNN and k-NN as Non-RA i.e. Normal patient classification accuracy is 95%, RA classification accuracy is 70%, Severe stage classification accuracy of RA is 100%.

Keywords: Rheumatoid arthritis · Hand radiograph · Joint space width

1 Introduction

Arthritis is a form of joint disorder that involves joint inflammation of one or more joints where Greek 'Artho' means joint and 'itis' means inflammation. The early symptoms of this disease are noted in finger joints of hand. RA influences 0.5% to 1% of the old age population worldwide [19]. RA influences 0.75% of the population in India. Till today there is no established medicine for the disease, hence close checking of disease is important in medicinal treatment.

Human hand is the most complex biomechanical structure of human body. At early stage of the disease the symptoms are seen in joints of hand and wrist. RA causes joint space narrowing. Manual method of examination in hand radiograph analysis in RA is extremely tedious and time consuming for radiologists. Exact and precise analysis of hand radiographs is particularly very much needed.

© Springer Nature Singapore Pte Ltd. 2019
K. C. Santosh and R. S. Hegadi (Eds.): RTIP2R 2018, CCIS 1036, pp. 163–173, 2019.
https://doi.org/10.1007/978-981-13-9184-2_15

1.1 Hand Anatomy

There are total 14 finger bone joints of hand. Figure 1 shows hand anatomy, there are three joints in each finger except thumb. Thumb has two joints.

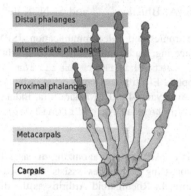

Fig. 1. Hand anatomy

Metacarpophalangeal (MCP) joint is the joint between metacarpal bones and proximal phalanges. Proximal interphalangeal (PIP) joint is between proximal and intermediate phalanx and the distal interphalangeal (DIP) joint is between intermediate and distal phalanx.

Arthritis is classified in two types Osteoarthritis and Rheumatoid arthritis. In Rheumatoid Arthritis joint space narrowing occurs in MCP and PIP joints exclusively. Segmentation is difficult because of complex hand structure. To determine diagnosis it is required to consider narrowed metacarpophalangeal, interphalangeal and distal phalangeal hand joints. In rheumatoid joint inflammation, joint space narrowing is seen mainly in MCP and PIP joints. Table 1 depicts the various finger joint name and bone name.

Table 1. Finger joint and finger bone names

Finger	Finger joint name	Finger bone name
Index, middle, ring, small	MCP	DP: Distal phalanx
	PIP	MP: Metacarpal phalanx
	DIP	PP: Proximal phalanx
Thumb	IP	DP: Distal phalanx
	MCP	PP: Proximal phalanx

To eradicate the problem, Van't Klooster created semi-automatic measurement technique for all joint margin, JSW measurement was applicable to both hands but thumb joints are not measured [1]. Wenham et al. described that in RA symptoms related to the progress of disease results in decrease in physical activity level of the patient [2]. Subramaniam et al. illustrated that computer image analysis has a very high impact on the various modalities that are available, X-ray, CT, US and MRI. These image serves vital source of information in to develop robust health care applications like visualization computer aided diagnosis and surgery simulator development [22, 23]. In arthritis X-ray images are considered to be most popular imaging technique since bones are clearly visible in X-ray and it is cost effective means amongst various modalities [3]. Langs et al. described automatic location identification of joint position and segmentation of bone contour. It evaluate result for automatic JSW measurement. It uses local linear mapping and active shape model algorithm. Texture features are used to locate the joint position on hand digital image. To identify the individual bone contour, Active shape model with statistical shape and texture model is used. As the disease progresses, joint spaces gets narrowed and bone erosions may occur [4]. Huo et al., concentrates on joint location and joint margin recognition. For the assessment, the joints have been manually delineated. For the joint margin recognition, majority of the researches trust on supervised learning, where the performance depends on choice of the training set [5]. Sharp et al. mentioned that estimation of joint space width was highly reproducible. Though erosion volume estimation was less highly reproducible [6]. Cootes, described a statistical model of appearance to images. Shape and grey-level variation learned from a training set are important parameters [7]. Sharp et al. suggested strategies of scoring joint deformities and joint space narrowing in the hands and wrists of patients with established rheumatoid arthritis were devised. The helpfulness of the scores was tested on 90 patients having X-ray of the hands and wrists 36 months or more after start of illness. Similarities were found between the extent of radiographic abnormalities or the degree of radiographic changes [8]. Bieleckia et al. presented the first phase about the computer analysis of hand X-ray digital images. The images are preprocessed followed by skeletanization of the fingers. The joints are detected and contoured. Joint widths are quantified. Likewise, while computing the minimal JSW difficulty may occur due to the sesamoid bones situated only in the thumb. Quantification of thumb joints is more complicated compared to other finger joints [9]. Huo, proposed automatic system for joint location and margin detection in RA hand radiographs. The location detection procedure depends on image features of the joint region [10]. Mahmoodi, et al. proposed a method to segment and localizes phalanx bones using active shape models. Various shape features are obtained from the segmented bone contour. A regression model and a Bayesian estimator uses these shape descriptors to evaluate skeletal growth [11]. Pietka suggested Hand bone examination can be utilized to find skeletal age. To assess bone age, parameters were extracted from phalanx analysis [12].

The diagnosis of RA is usually based on a combination of symptoms, the distribution of the affected joints, and blood and X-ray examination. There are several blood tests that play a role in diagnosing RA. As disease progresses, joint space narrowing takes place. The Rheumatoid joint space narrowing is seen only in MCP and PIP joints and changes are symmetric. In osteoarthritis only PIP and DIP joints are affected, changes are asymmetric.

2   Methodology

Measuring joint space width in a diagnostic process requires consideration of narrowed hand joints (MCP, PIP or DIP) and their symmetry.

2.1   Automatic Classification of RA

To classify RA using SVM, KNN and FFNN technique this method is proposed. In this presence of rheumatoid arthritis in patients is automatically detected. Figure 2 shows RA classification block diagram. To detect RA, Normal, Abnormal, Severe database images are first trained and 14 joints feature extracted from digital hand radiograph. Joint location is identified using second order derivatives filtering and thresholding technique. Test image database is classified using classifier based on supervised learning.

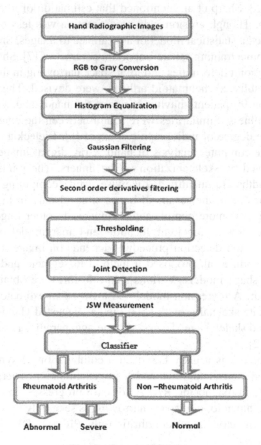

Fig. 2.   RA classification

In the classification stage, the extracted features are compared with the features stored in the data base. With this, the classifier acquires knowledge about the extracted features which aids for the classification. Based on this knowledge and figure of merit between the feature extracted from the image under analysis and stored features, the classifier assigns the normal and abnormal class of the image under analysis. As in RA severe stage, the bone ends rub together and there exists no joint space. Thus if the number of features are less than 14 then RA severe stage is detected. This is a non-invasive method of RA detection.

2.1.1 Hand Radiographic Database Images
The data consist of hand radiographic images collected from well recognized hospital. The image resolution is 2000 × 2000 pixels.

2.1.2 Image Preprocessing
In Gray level images the intensity variation is represented on gray scale. Most commonly used storage method is 8-bit storage. In 8 bit gray scale image, intensity of each pixel varies from 0 to 255, with 0 being black and 255 being white.

A color image can be represented by intensity variation using RGB to grey level conversion.

2.1.3 Histogram Equalization
Histogram equalization enhances dynamic range of grey levels. Contrast improvement of image is possible by altering image in terms of its intensity histogram [20, 21]. Adaptive histogram equalization is performed using the adapthisteq function. While histeq works on the entire image, adapthisteq operates on small regions in the image. To improve contrast and to avoid noise amplification that might be present in the image, adapthisteq is used.

2.1.4 Gaussian Filtering
First, the Gaussian filtering is applied to hand radiographic images. Gaussian filter is used to remove noise. Impulse response of Gaussian filter is a Gaussian function. A two-dimensional Gauss filter is given by following equation

$$g(x, y) = \frac{(1)}{2\pi\sigma^2} exp \frac{(-x^2 + y^2)}{\sigma^2} \tag{1}$$

Where

x: distance from the origin in the horizontal axis
y: distance from the origin in the vertical axis
σ: standard deviation

2.1.5 Second Order Derivative Filtering
The second order derivatives are used to detect horizontal and vertical edges in an image. A very popular second order operator is Laplacian operator.

$$\nabla^2 a = \frac{\partial^2 a}{\partial^2 x} + \frac{\partial^2 a}{\partial^2 y} \tag{2}$$

$$[H2x] = [H2y]^T = [1 \quad 0 \quad -1] \tag{3}$$

H2x: horizontal derivative filter
H2y: vertical derivative filter
T: transpose of matrix

The Laplacian is more sensitive operator for edge detection. Zero crossing of these operators will provide edges.

2.1.6 Thresholding

Thresholding is used to create binary images. The simplest thresholding method is to replace each pixel in an image with a black pixel if the image intensity is less than some fixed constant T, or a white pixel if the image intensity is greater than constant.
Let f(x, y) is pixel value in input image
I(x, y) is pixel value in the threshold image
then I(x, y) is written as

$$I(x, y) = 1, \ f(x, y) \geq T \tag{4}$$

$$I(x, y) = 0, \ f(x, y) < T \tag{5}$$

2.1.7 Joint Detection

First edges are detected using second order derivatives filtering technique. Double convolution is operated on the image. The joints are detected after morphological filtering. The joint space width is measured in between two points on joint. The minor axis length is considered as JSW.

2.1.8 Classifier

The classifiers like k-NN, FFNN and SVM are trained using database images and then tested the performance of each classifier. Classification results into RA and non-RA. Severe stage RA is also detected.

In these classifiers number of input feature vector is 14, since there are 14 joints in hand radiographic image. There are 15 Neurons are used with a single hidden layer and there are 2 classes as output normal and abnormal. 18 images are severe stage images. Thus there are 42 images, 70% images for training and 30% are used for testing and validation.

k-NN Classifier

k-NN classifier is a simple but very effective method of classification. k-NN is a non-parametric method of classification.

Various models can be achieved by changing value of k, where k > 1, neighbors. In this the majority result decide the outcome of the class. A higher value of k is preferred.

Feed-Forward Neural Network Classifier

The FFNN Classifier classifies the rheumatoid arthritis and non-rheumatoid arthritis person. A neural network consists of units called as neurons, these neurons are arranged in layers, which convert an input vector into output. Since there is no feedback from the next layer to the previous it is called as FFNN.

Support Vector Machine Classifier

SVM algorithm is the one that learns by example to assign labels to objects. SVM algorithm classifies data into two categories by using the best hyper-plane for separation of data points.

Non-linear SVM Kernel Functions

In non linear SVM data is separated by selecting optimum hyperplane in feature space. Various mathematical kernel functions are used for classification.

There is a class of functions C(x, y). There is a linear space S and function Φ mapping x to S such that

$$C(x, y) = \Phi(x).\Phi(y) \tag{6}$$

Polynomial kernel function for positive integer d

$$C(x, y) = (1 + <x, y>)^d \tag{7}$$

Radial Basis kernel function for positive integer σ

$$C(x, y) = exp\left(\frac{-<(x - y), (x - y)>}{(2\sigma^2)}\right) \tag{8}$$

MLP Multilayer perceptron for positive number p1 and negative number p2

$$C(x, y) = \tanh(p1 <x, y> + p2) \tag{9}$$

3 Results

All kernels are tested for estimating the performance of the system to detect RA and non-RA patient. This architecture is completely innovative approach to find Joint Space Width and classify RA. JSW is measured for each finger joint DIP, PIP and MCP. The classification is done in three types normal i.e. Non-RA, RA affected and Severe stage. We have used 60 Hand Radiographic images containing 20 normal, 22 RA and 18 severe stage images with resolution of 2000×2000 pixels.

Machine specifications are personal computer with core I5 processor with 1.6 GHz and 4 GB RAM and MATLAB R2013b in windows environment.

Table 2 shows the normal patient JSW as per standard. $\pm$ indicates the standard deviation from the respective value for normal patient.

Table 2. Standard mean joint space width of normal patient [18]

Joint	Mean JSW(mm)
DIP	0.96 ± 0.12
PIP	0.71 ± 0.33
MCP	1.66 ± 0.14
IP	1.21 ± 0.11

Features Extraction

The joint space width of all joints compare with mean JSW of normal patient. For joints having JSW is less than mean JSW of normal patient, this patient is Rheumatoid arthritis patient. From Table 3, it is observed that the joint space narrowing takes place if it is RA affected patient. MCP joint space width is highest among all finger bone joints.

Table 3. Mean JSW of rheumatoid arthritis (RA) and non-RA patient

Sr. no.	Name of joint	Mean JSW(mm) RA patient	Mean JSW(mm) Non-RA
1	Index finger-DIP	0.7	0.9
2	Middle finger-DIP	1.02	1.3
3	Ring finger-DIP	0.61	1.3
4	Little finger-DIP	0.67	1.7
5	Index finger-PIP	0.71	0.9
6	Middle finger-PIP	1.02	0.9
7	Ring finger-PIP	0.77	1.3
8	Little finger-PIP	0.88	1.6
9	Index finger-MCP	1.3	1.3
10	Middle finger-MCP	1.0	1.5
11	Ring finger-MCP	0.75	1.3
12	Little finger-MCP	1.6	1.4
13	Thumb-IP	0.8	1.2
14	Thumb-MCP	0.7	1.6

Joint location Accuracy

JLA = Number of joints located accurately/Total number of joints

Joint location accuracy for the proposed method is 92%.

In this method 60 images were tested, Out of 60 Test images 20 images are Normal, 22 images are abnormal i.e. RA affected and 18 images are severe. 14 finger joints are there in hand image, thus feature vector length for each image is 14 in Normal and abnormal case. Feature vector length is less than 14 in severe stage, as bone ends rub together and there is no joint space exists in most of the joints.

From Table 4, It is observed that in SVM, using kernel Radial basis function accuracy achieved is highest among other 4 types of kernel. Best choice of SVM kernel is rbf.

Table 4. Classifier accuracy comparison

Classifier type		Non RA accuracy %	RA accuracy%	Average accuracy %
SVM	mlp	59	70	64.5
	Linear	90.95	70	80.475
	Rbf	95	70	82.5
	Quadratic	90.9	70	80.45
	Polynomial	90.9	70	80.45
k-NN	k = 2	86.36	55	70.68
	k = 3	86.36	55	70.68
	k = 4	90.9	30	60.45
	k = 5	90.9	50	70.45
FFNN	No. of Neurons = 3	95.23	55	75.115
	No. of Neurons = 5	95.23	55	75.115
	No. of Neurons = 8	71.42	80	75.71
	No. of Neurons = 14	95.23	63	79.11
	No. of Neurons = 18	95.23	55	75.11

In k-NN classifier, as k = 3 is optimum choice, as k i.e. number of nearest neighbors increases accuracy is less. k = 2 is not considered as there must be odd number of neighbors in order to classify based on nearest neighbor comparison. In FFNN classifier with single hidden layer, it is observed that if the number of neurons are 14 then the accuracy is 79%.

Severe stage accuracy is 100%. Severe stage is detected if the number of joint features is less than 14. In severe stage joint space narrowing is such that there exist no joint space.

Figure 3 indicates that SVM using 'rbf' kernel is best for RA classification. Neural network performance is better than k-NN classifier. NN classifier accuracy is 79%, using 15 neurons, 1 hidden layer, 14 feature inputs and 2 outputs.

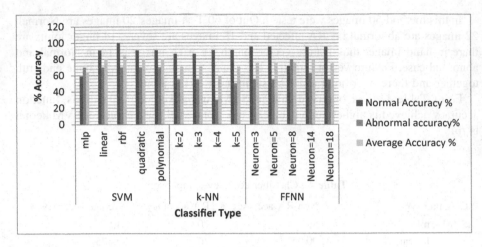

Fig. 3. Classifier comparison

Computation time is the time required to process one image is 22 s.

4 Conclusion

The proposed system is automatic and non-invasive method of RA classification. It is used to detect and measure the joint space width. Severe stage RA can be detected where the joint space is no longer observable in the X-ray image. SVM classifier with Radial basis function kernel is efficient compared to FFNN and k-NN as Non-RA i.e. Normal patient classification accuracy is 95%, RA classification accuracy is 70%, Severe stage classification accuracy of RA is 100%.

References

1. Van't Klooster, R., Hendricks, E.A., Watt, I., Kloppenburg, M., Reiber, J.H., Stoel, B.C.: Automatic quantification of osteoarthritis: validation of a new method to measure joint space width. Osteoarthr. Cartil. **16**(1), 18–25 (2008)
2. Wenham, C.Y.J., Grainger, A.J., Conaghan, P.G.: The role of imaging modalities in the diagnosis, differential diagnosis and clinical assessment of peripheral joint osteoarthritis. Osteoarthr. Cartil. **22**(10), 1692–1702 (2014)
3. Subramaniam, M., Rajini, V.: Statistical feature based classification of arthritis in knee X-ray images using local binary pattern. In: International Conference on Power, Circuits and Computation Technology (2013)
4. Georg, L., Peloshek, P., Bischof, H., Kainberger, F.: Automatic quantification of joint space narrowing and erosions in Rheumatoid arthritis. IEEE Trans. Med. Imaging **28**(1), 151–164 (2009)
5. Huo, Y., Vincken, K.L., Van der Heijde, D., De Hair, M.J., Lafeber, F.P., Viergever, M.A.: Automatic quantification of radiographic wrist joint space width of patients with Rheumatoid arthritis. IEEE Trans. Biomed. Eng. **64**(11), 2695–2703 (2017)

6. Sharp, J.T., Gardner, J.C., Bennett, E.M.: Computer-based methods for measuring joint space and estimating erosion volume in the finger and wrist joints of patients with Rheumatoid arthritis. Arthritis Rheumatol. **43**(6), 1378–1386 (2000)
7. Cootes, T.F., Edwards, G.J., Taylor, C.J.: Active appearance model. IEEE Trans. Pattern Anal. **23**(6), 681–685 (2001)
8. Sharp, J.T., Wolfe, F., Mitchell, D.M., Blochthe, D.A.: Progression of erosion and joint space narrowing scores in rheumatoid arthritis disease during the first twenty-five years of disease (2000)
9. Bieleckia, A., Korkoszb, M., Zielin, B.: Hand radiographs pre-processing, image representation in the finger regions and joint space width measurements for image interpretation. Pattern Recognit. **65**(6), 3786–3798 (2008)
10. Huo, Y., Vincken, K.L., Van der Heijde, D., De Hair, M.J., Lafeber, F.P., Viergever, M.A.: Automatic quantification of finger joint space width of patients with early Rheumatoid arthritis. IEEE Trans. Biomed. Eng. **63**(10), 2177–2186 (2016)
11. Mahmoodi, S., Sharif, B., Chester, G., Owen, J., Lee, R.: Skeletal growth estimation using radiographic image processing and analysis. IEEE Trans. Inf. Technol. **4**(4), 292–297 (2000)
12. Pietka, E., Kaabi, L., Kuo, M.L., Huang, H.K.: Feature extraction in carpal bone analysis. IEEE Trans. Med. Imaging **12**(3), 44–49 (1993)
13. Huo, Y., Vincken, K., Viergever, M., Lafeber, F.: Automatic joint detection of RA hand radiographs. In: IEEE 10th International Symposium on Biomedical Imaging (2013)
14. Vajda, S., Santosh, K.C.: A fast k-nearest neighbor classifier using unsupervised clustering. In: Santosh, K.C., Hangarge, M., Bevilacqua, V., Negi, A. (eds.) RTIP2R 2016. CCIS, vol. 709, pp. 185–193. Springer, Singapore (2017). https://doi.org/10.1007/978-981-10-4859-3_17
15. Santosh, K.C., Antani, S.: Automated chest X-ray screening: can lung region symmetry help detect pulmonary abnormalities? IEEE Trans. Med. Imaging **37**(5), 1168–1177 (2018)
16. Karagyris, A., et al.: Combination of texture and shape features to detect pulmonary abnormalities in digital chest X-rays. Int. J. Comput. Assist. Radiol. Surgery (IJCARS) **11**(1), 99–106 (2015)
17. Santosh, K.C., Vajda, S., Antani, S., et al.: Edge map analysis in chest X-rays for automatic abnormality screening. Int. J. Comput. Assist. Radiol. Surgery (IJCARS) **11**(9), 1637–1646 (2016)
18. Choi, S., Lee, G.J., Hong, S.J., Park, K.H., Urtnasan, T., Park, H.K.: Development of a joint space width measurement method based on radiographic hand images. Comput. Biol. Med. **41**, 987–998 (2011)
19. Hegadi, R.S., Navale, D.I., Pawar, T.D., Ruikar, D.D.: Multi feature-based classification of osteoarthritis in knee joint X-ray images. In: Medical Imaging: Artificial Intelligence, Image Recognition, and Machine Learning Techniques, Chap. 5. CRC Press (2019). ISBN 9780367139612
20. Ruikar, D.D., Santosh, K.C., Hegadi, R.S.: Automated fractured bone segmentation and labeling from CT images. J. Med. Syst. **43**(3), 60 (2019). https://doi.org/10.1007/s10916-019-1176-x
21. Ruikar, D.D., Santosh, K.C., Hegadi, R.S.: Segmentation and analysis of CT images for bone fracture detection and labeling. In: Medical Imaging: Artificial Intelligence, Image Recognition, and Machine Learning Techniques, Chap. 7. CRC Press (2019). ISBN 9780367139612
22. Ruikar, D.D., Hegadi, R.S., Santosh, K.C.: A systematic review on orthopedic simulators for psycho-motor skill and surgical procedure training. J. Med. Syst. **42**(9), 168 (2018)
23. Ruikar, D.D., Sawat, D.D., Santosh, K.C., Hegadi, R.S.: 3D imaging in biomedical applications: a systematic review. In: Medical Imaging: Artificial Intelligence, Image Recognition, and Machine Learning Techniques, Chap. 8. CRC Press (2019). ISBN 9780367139612

DRAODM: Diabetic Retinopathy Analysis Through Optimized Deep Learning with Multi Support Vector Machine for Classification

Emmy Bhatti[✉] and Prabhpreet Kaur

Computer Engineering and Technology, Guru Nanak Dev University,
Amritsar, India
emmybhatti@gmail.com, prabhpreet.cst@gndu.ac.in

Abstract. Diabetic retinopathy (DR) is the leading cause of avertable blindness globally. Retinal scanning of eyes is critical for examining the disease at an early stage. The concern of this study is to develop a robust mechanism to automate the process of diabetic retinopathy detection. A person suffering from DR must be referred to an ophthalmologist for an early evaluation in order to reduce the rate of vision loss hence enabling early treatment and prevention of vision loss. The proposed methodology introduces a data-driven novel algorithm using Deep learning for creating a tool for detecting DR. The approach uses DR colored fundus images and classifies them into multiple classes as stages or levels to which the eye is infected. Set of 170 colored fundus images of diabetic patients were used to train and test the model for distinguishing images into multiple classes. Entire simulation is divided into primarily two phases-firstly, the preprocessing phase where the resizing operation is performed since the input layer of the network requires predefined size. Size of 77×100 and channel of size 3 is used, indicating RGB image. Gaussian Filtering is used to tackle noise (Denoising) if any within the image. Secondly, the MSVM for segmentation and Classification phase where Multi-class SVM is used to extract critical and non-critical segments from within the training images. After extracting the features, classification is performed on the test images. Parameters used for optimization includes Accuracy, sensitivity, and specificity. Simulation is conducted in MATLAB using image processing and neural network toolbox. The proposed mechanism shows improvement in terms of classification accuracy by the margin of 5–6% which is significant, thereby enhancing the recognition rate.

Keywords: Diabetic retinopathy · Deep learning · MSVM · Accuracy · Specificity · Sensitivity

1 Introduction

Diabetic retinopathy is one of the most solemn and widely disseminated eye diseases. It is the most familiar and widespread reason for authorized blindness amongst the working-age groups of accessible nations [1]. Diabetic retinopathy happens when diabetes damages the retinal blood vessels, causing leakage of blood and fluids into the tissue. This fluid transfusion leads to the production of microaneurysms, hemorrhages,

© Springer Nature Singapore Pte Ltd. 2019
K. C. Santosh and R. S. Hegadi (Eds.): RTIP2R 2018, CCIS 1036, pp. 174–188, 2019.
https://doi.org/10.1007/978-981-13-9184-2_16

hard exudates, and cotton wool spots (a.k.a., soft exudates) [2]. Diabetic retinopathy is a silent disease and may be only detected by patients at the moment when retinal alterations have already advanced to a level where treatment proves inefficient or even impossible. Diabetes stands out amongst the most well-known reasons for visual impairment in the general population. It causes waterfall, glaucoma, and harm the veins inside the eye. Diabetic Retinopathy is an intense retinal issue implicating the adverse effects of diabetes on the retina. Around 210 million individuals everywhere throughout the world have Diabetes Mellitus; among which 10–18% of individuals are experiencing Diabetic Retinopathy.

The expanding number of diabetic retinopathy cases overall requires to strengthen the creation of instruments to determine diabetic retinopathy. Programmed recognition of diabetic retinopathy will save time and efforts. In this manner, Rubhini et al. [3] proposed a technique for the programmed discovery of microaneurysms in retinal fundus pictures. Maher et al. [4] already assessed a choice based emotionally supportive network for the programmed study of non-proliferative diabetic retinopathy. Truth be told, support vector machines were utilized by Maher et al. [4] in the computerized analysis of non-proliferative diabetic retinopathy. A few picture preprocessing strategies have additionally been projected along with the eventual classification keeping in mind the end goal to distinguish diabetic retinopathy in [5–7]. However, regardless of all these past works, the mechanized discovery of diabetic retinopathy still holds the scope for development [8].

Hence, this paper puts forward another computerized handling of retinal images with a specific end goal to help individuals recognize diabetic retinopathy in advance. The end goal of the proposed literature is to achieve the desired level of classification accuracy by reducing noise levels from within the 1–3 levels of the non-proliferative dataset where '1' indicates mild DR, '2' indicates moderate DR and '3' indicates severe or proliferative DR. Noise handling through Gaussian filtering is used at the pre-processing stage. The Gaussian filter is capable of handling noise at edges and also considered the best filters in the time domain. Resizing operation is done at the preprocessing stage thus, ensuring uniformity along the input layer for faster operation. MSVM gives multiclass segmentation and classification operation. As an output, we acquired a most extreme affectability of 94.6% and estimation capacity value of 93.8%. Heartiness as for modifications in the parameters of the calculation has also been assessed.

There are certain discrete factors leading to the worst effects of DR like the age of diabetes, poor control, etc. but these effects can be minimized or slowed down or abutted if DR is detected in the early stage of the disease. Without using automated tools, the entire process of detection is lengthy and time-consuming which is primarily due to miscommunication and delayed results causing delayed treatment and ignorance. Image processing and machine learning provides tools that could be used in order to detect the DR at an early stage. By the use of these tools, manual testing of DR can be automated.

Particulars of the proposed system are discussed in Sect. 4. The remaining paper is arranged as under: Sect. 2 provides the literature survey of techniques used to detect DR, Sect. 3 gives the phases associated with the proposed system, Sect. 4 gives the performance analysis, Sect. 5 gives conclusion and future scope followed by the references at the end.

2   Literature Survey

This section provides an in-depth survey of the various techniques used to detect diabetic retinopathy at an early stage. [5] SVM and MDA methodologies are surveyed in this paper and also their utilization for detecting diabetic retinopathy are explained. [6, 27, 28] uses the contrast enhancement, morphological filtering, and segmentation technique to detect hard exudes from the various input image. The system utilizes Contrast Limited Adaptive Histogram Equalization (CLAHE) technique [29] to improve the image and top hat transform to enhance the blood vessels. After that filtering is done and pattern recognition techniques are utilized to recognize the diseases. In [7] SVM classifier based system is utilized in to diagnose DR affected patient. It also uses test fundus images to contribute to SVM classifiers.

The technique discussed by [9] provides a mechanism to detect diabetic retinopathy through the use of Deep learning. The considerably large dataset is used for this purpose. Data-driven artificially intelligent deep learning mechanism is used to derive training and testing images for distinguishing the normal image from DR image. However, the mechanism used lacks preprocessing mechanism including noise handling within the automatic detection of and multiclass classification. Several other techniques of data mining in healthcare are surveyed by [10]. [11] discriminates between normal and abnormal tuberculosis using thoracic edge maps in terms of histograms of oriented gradients for all possible angles in the range $(0, 2\pi]$ at a different number of bins and different pyramid levels however image rotations cannot be handled with this method. [12] proposes a novel template-free geometric signature-based technique for detecting arrow annotations on biomedical images which is an important step in region-of-interest (ROI) labeling and image content study thus can help to improve datasets and their eventual mining. Algorithms associated with data mining provides the filtering mechanism in [13, 14] to ensure the better classification of the result. By analyzing discussed techniques, the best possible technique can be selected for future enhancement. The low-cost medical image processing mechanism is proposed [15]. Field programmable field array is merged along with the processor for analyzing the complex diseases like DR. [16] utilizes microaneurysm segmentation that is automatically done by using mathematical morphology although it does not work on the predefined set of directives. Sensitivity and specificity can also be further improved. However, [16] uses the multiclass SVM classifier that assures classification phase thus ensuring unwavering analysis of human observer and also presents supervised classifier that uses testing sets to obtain results. [17] proposed multilayer perceptron keeping in mind the end goal of breaking down huge information and comparing it to human services. As writing manages medicinal services of patients henceforth high level of exactness is wanted. To achieve the coveted objective correlation of SVM and multilayer perceptron on the human services informational index is made. After-effects of SVM as far as characterization is better when contrasted with multilayer perceptron. [18] classifies the DR stages using an automated system having a feature classifier. It detects the disease by extracting features using an image processing method and classifies them accordingly. [19] suggests information digging methods utilized for investigation of diabetes. Support Vector Machine (SVM) is utilized to solve the purpose. Hereditary methodology is likewise investigated for diabetic's dataset in the area of information mining. After effects of SVM acquired are better. [20] proposes five J.48 classifiers to foresee hypertension and eight different sicknesses. Expectation precision is gotten and looked at against naive Bayes approach. Results as far as J.48 are acquired to be better. [21] build up and tested a method by making a combination of shape and texture features for the classification of CXRs into two categories: TB and non-TB cases along with the SVM classifier. [22] uses a speedy method for pattern classification using a k-nearest neighbor (kNN) classifier by speeding up the distance calculation process through training data in an unsupervised manner by defining clusters. [23] proposed a mechanism to analyze various techniques to detect disease using the image and video data. Techniques which are reviewed include streaming and batch processing, big data medical analytics, selected images and video analysis and multi-model video analytic. Analyzing such a big database requires a special mechanism associated with deep learning. [24, 32] proposed a novel framework that uses convolutional neural networks (CNN) for feature extraction, where images have been scaled to a variety of different sizes by using multilevel 2D discrete Haar wavelet transform. The biggest known dataset for Indic script identification work has been used. [25] proposed a GPU based breast cancer detection mechanism. Large images processing through a single CPU is a complex task. To resolve the issue GPU based parallel computing is proposed through this literature [30].

Through the literature survey, it is concluded that the medical field is adapting the technological advancement to improve the healthcare services [31]. In addition to this, the techniques suggested requires modification in terms of pre-processing and the classification stage. The experiment corresponding to the techniques already discussed in this section along with segmentation procedure where multiple hyperplanes are required has been tackled in the proposed system.

3 Proposed System

The paper focuses on three stages of non-proliferative diabetic retinopathy namely: 1. Mild DR, 2. Moderate DR 3. Severe DR. Actually, non-proliferative category of diabetic retinopathy is the most widely recognized constituting around 80% of all detected cases. The proposed approach is divided into three stages as given below:

1. Preprocessing: Noise removal from the input image is to be done.
2. Training with deep learning: Iteratively extract the features of the image and store it into the buffer.
3. Classification: Classify the extracted feature on the basis of similarity.

3.1 Image Database

The image dataset used is DIARETDB0 consisting of 3 categories of 130 eye fundus images. Resizing operation manually as well as the automated mechanism is posted upon to fit into the input layer of the network. The images were captured and resized to 77×100 with 3 color channels (Fig. 1).

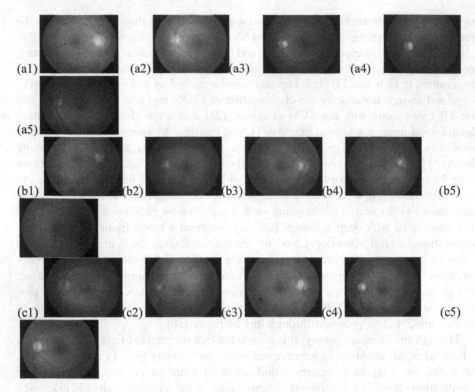

Fig. 1. (a) Mild non-retinopathy images: (b) moderate non-retinopathy images, (c) severe on-proliferative retinopathy images.

The 170 pictures are bundled in 3 sets, one for each ophthalmologic division, utilizing the PNG format. Besides, an Excel record with therapeutic conclusions for each picture is given. In this analysis, we utilized the pictures of only one ophthalmologic division including 48 pictures with mild, 48 with moderate and 34 with severe DR cases.

3.2 Pre-processing

Preprocessing mechanism used in this literature contains noise handling along with resizing operation. Noise handling is done using Gaussian filtering mechanism. This filter is capable of handling impulse noise along with smoothening operation. Equation 1 gives the operation of filtering along with smoothening.

$$G_{Smoothened_{image}} = \frac{1}{2\pi\alpha^2} e^{(a^2 + b^2)\frac{1}{2a^2}} \tag{1}$$

'α' is standard deviation, 'a' is the distance from horizontal axes and 'b' is a distance of origin from vertical axes.

After handling noise, resizing operation is done. Resizing is done to present uniform data to the input layer. Resizing is done using Eq. (2).

$$\textbf{Resized}_G = \textbf{Resize}\left(\textbf{G}_{\textbf{Smoothened}_{\text{image}}}, [\textbf{70100}]\right) \tag{2}$$

These resized images obtained is passed to the network for further processing.

3.3 Training with Deep Learning

Training operation begins by receiving the images from the pre-processing phase. To create a new network, the proposed mechanism used inbuilt layers with input layer accepting images of 77 × 100 with 3 channels. Training parameters are defined using training option command using deep learning toolbox. Train network function is used to finally train the network. The flow of network definition is given as under (Fig. 2).

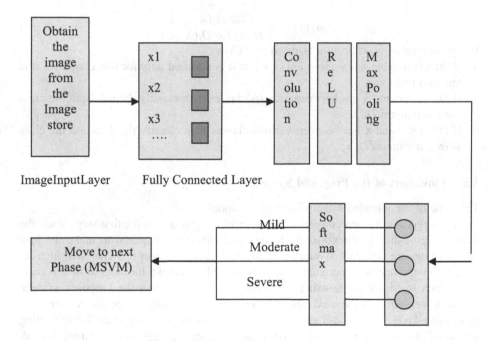

Fig. 2. The flow of the training process

3.4 Classification

To perform classification MSVM is used as the last phase. Support vector machine uses the rules-based environment to correctly reach to the solution of the given problem including outlier detection or unnecessary regions. [26] proposed this system to resolve the unclassified region. MSVM is used to realize the classification results. Optimal hyperplanes are defined to determine whether the obtained values of membership functions satisfy the hyperplane (D(x)) or not.

Satisfaction Criteria D(X) > 1.

One dimensional membership function $m_{ij}(x,y)$ is defined for determining optimal separating hyper planes $D_j(x) = 0$ as follows

1. If values of diagonal are equal (i == j)

$$m_{i_j}(x) = \begin{cases} 1 \text{ for } D_i(X) > 1 \\ D_i(x) \text{ for } D_i(X) < 1 \end{cases} \tag{3}$$

2. If values of diagonal are not equal (i ≠ j)

$$m_{i_j}(x) = \begin{cases} 1 \text{ for } D_i(X) < 1 \\ -D_i(x) \text{ for } D_i(X) > 1 \end{cases} \tag{4}$$

3. If values of diagonal are i + j > indexed diagonal elements

$$m_{i_j}(x) = \begin{cases} 0 \text{ for } D_i(X) > 1 \\ D_i(x) \text{ for } D_i(X) < 1 \end{cases} \tag{5}$$

The procedure of classification is listed as follows

4. If the pixel value x is such as Di(x) > 0 and is satisfied only for that class then it is fed into that class

5. If $D_i(x) > 0$ and x lies between various classes then classify the data into the class with maximum $D_i(x)$.

6. If $D_i(x) \leq 0$ and x lies between various classes then classify the data into the class with a minimum $D_i(x)$

3.5   Flowchart of the Proposed System

The flow of the intended system is given as under

The suggested system consists of a pre-processing phase as the first step. Since the images are of distinct categories so image smoothening is required to make the processes of disease detection symmetrical. In order to accomplish this Gaussian filtering is used. Image resizing operation is applied in order to make the images of the same size which is critical while using deep learning mechanism in the proposed system. Network definition is the next phase. The network consists of the input layer, the processing layer, and the output layer. The network is primarily used for extracting features from the image. Training process also extracts features from the images used. The features extracted from the image and image features extracted from the training phase are matched against each other. The hyperplanes defined through multi support vector machine used to decide the disease classification from the selected image. Through the proposed system it is possible to detect multiple disease stages quickly which was not possible through the existing literature (Fig. 3).

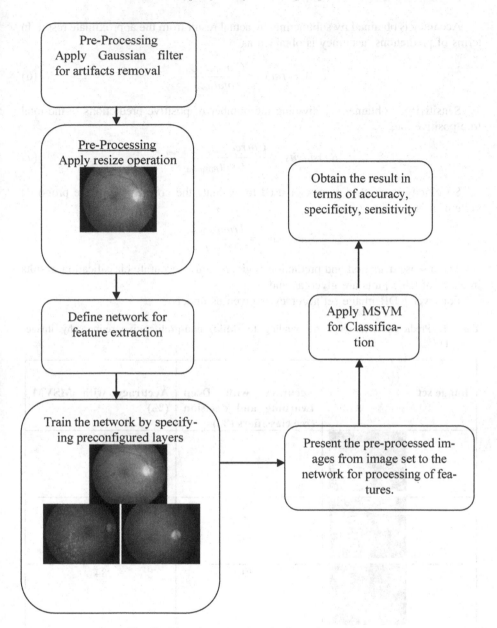

Fig. 3. Flowchart showing the flow of operation

4 Performance Analysis and Results

The performance of the system is analyzed using parameters such as accuracy, specificity, and sensitivity.

Accuracy is obtained by subtracting the actual result from the approximate result. In terms of predictions, accuracy is obtained as

$$Accuracy = \frac{Correct_{Pre}}{Total_{Pred}} \tag{6}$$

Sensitivity is obtained by dividing the number of positive predictions to the total true positive rate.

$$Sensitivity = \frac{Correct_{Positive_{predictions}}}{Total_{Positives}} \tag{7}$$

Specificity is another parameter used to evaluate the correctness of the proposed system. It is given as under

$$Specificity = \frac{True_{Negitives}}{TP + FN} \tag{8}$$

The disease detection and prediction is given though accurate classification, results in terms of table plots are given as under

For level 1 DR image set accuracy is given as under

Table 1. Predicted accuracy corresponding to (Mild) non-proliferative retinopathy images (level 1)

Image set	Accuracy with Deep Learning and decision tree classifiers (%)	Accuracy with MSVM (%)
	76	81
	78	85
	79	84
	78	83
	79	82

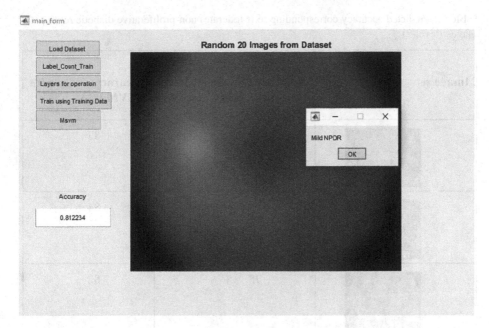

Fig. 4. Groundtruth verification and corresponding accuracy of proposed work for Mild NDPR

For level 2 retinopathy image set accuracy is given as under

For level 3 image set accuracy is given as under

Result comparison in terms of accuracy, sensitivity, and specificity are given as under (Figs. 4, 5, 6 and 7).

Results and performance analysis as indicated in the plot shows that deep learning combined with multi support vector machine yield a better result. The parametric analysis of the existing as well as proposed technique shows an average increase of 5–6% from (Existing: Accuracy = 76.66, Specificity = 77, Sensitivity = 78) to (Proposed: Accuracy = 82, Specificity = 82, Sensitivity = 82.66). Classification accuracy of the proposed system appears to be more as compared to existing techniques. Multiple class prediction mechanism shows higher accuracy hence proving the worth of study (Tables 1, 2 and 3).

Table 2. Predicted accuracy corresponding to (Moderate) non-proliferative diabetic retinopathy images (level 2)

Image set	Accuracy with Deep Learning and decision tree classifiers (%)	Accuracy with MSVM (%)
	78	82
	79	84
	78	82
	77	81
	76	82

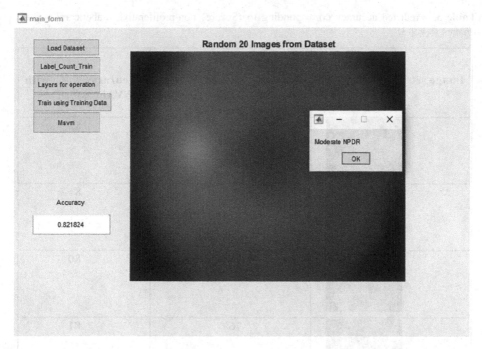

Fig. 5. Groundtruth verification and corresponding accuracy of proposed work for Moderate NDPR

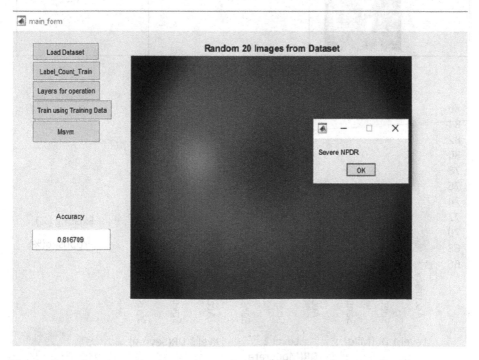

Fig. 6. Ground truth verification and corresponding accuracy of proposed work for Severe NDPR

Table 3. Predicted accuracy corresponding to (Severe) non-proliferative diabetic retinopathy images (level 3)

Image set		Accuracy with Deep Learning and decision tree classifiers (%.)	Accuracy with MSVM (%)
		78	82
		77	81
		75	80
		76	81
		78	82

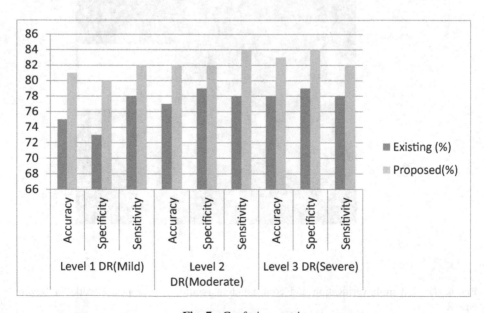

Fig. 7. Confusion matrix

5 Conclusion and Future Scope

In this paper, an automated system that utilizes multi support vector machine and deep learning mechanism for diabetic retinopathy are proposed. Pre-processing phase is critical and is well defined using noise handling and resizing operation. Obtained images are fed into the trained network for feature extraction and classification is performed using MSVM. The hybrid approach followed gives better results.

Overall enhancement in terms of accuracy is 5–6%. The main objective of this research is to check the consistency of classification accuracy when presented with larger datasets which are subsequently increased. In future, e-health system that accurately implements the steps described in Sect. 3 can be developed.

References

1. Gargeya, R., Leng, T.: Automated identification of diabetic retinopathy using deep learning. Ophthalmology **124**, 962–969 (2017)
2. Haleem, M.S., Han, L., Van Hemert, J., Li, B., Fleming, A.: Retinal area detector from Scanning Laser Ophthalmoscope (SLO) images for diagnosing retinal diseases. IEEE J. Biomed. Health Inform. **19**, 1472–1482 (2015)
3. Rubini, S.S., Kunthavai, A.: Diabetic retinopathy detection based on eigenvalues of the Hessian matrix. Procedia Comput. Sci. **47**, 311–318 (2014)
4. Maher, R., Dhopeshwarkar, M.: Automatic detection non-proliferative diabetic retinopathy using image processing techniques. J. Eng. Res. Appl. **6**, 122–127 (2016)
5. Shveta, S., Kaur, G.: Review on: detection of diabetic retinopathy using SVM and MDA. Int. J. Comput. Appl. **117**(20), 975–987 (2015)
6. Paranjpe, M.J., Kakatkar, P.M.N.: Automated diabetic retinopathy severity classification using support vector machine. Int. J. Res. Sci. Technol. **3**(3), 86–91 (2013)
7. Ramya, V.: SVM based detection for diabetic retinopathy. Int. J. Res. Sci. Innov. **V**, 11–13 (2018)
8. Adarsh, P., Jeyakumari, D.: Multiclass SVM-based automated diagnosis of diabetic retinopathy. In: International Conference on Communications Signal Processing, pp. 206–210 (2013)
9. Zhou, L., Zhao, Y., Yang, J., Yu, Q., Xu, X.: Deep multiple instance learning for automatic detection of diabetic retinopathy in retinal images. IET Image Process. **12**, 563–571 (2018)
10. Jothi, N., Rashid, N.A., Husain, W.: Data mining in healthcare - a review. Procedia Comput. Sci. **72**, 306–313 (2015)
11. Santosh, K.C., Vajda, S., Antani, S., Thoma, G.R.: Edge map analysis in chest X-rays for automatic pulmonary abnormality screening. Int. J. Comput. Assist. Radiol. Surg. **11**, 1637–1646 (2016)
12. Santosh, K.C., Wendling, L., Antani, S., Thoma, G.R.: Overlaid arrow detection for labeling regions of interest in biomedical images. IEEE Intell. Syst. **31**, 66–75 (2016)
13. Saini, M., Yadav, S., Rewri, S.: A hybrid filtering techniques for noise removal in color images. Int. J. Innov. Eng. Technol. **5**, 172–178 (2015)
14. Ma, Y., Lin, D., Zhang, B., Liu, Q., Gu, J.: A novel algorithm of image gaussian noise filtering based on PCNN time matrix. In: 2007 IEEE International Conference on Signal Processing and Communications, pp. 1499–1502. IEEE (2007)

15. Bulsara, V., Bothra, S., Sharma, P., Rao, K.M.M.: Low cost medical image processing system for rural/semi urban healthcare. In: IEEE Access, pp. 724–728 (2011)
16. Shetty, S., Kari, K.B., Rathod, J.A.: Detection of diabetic retinopathy using support vector machine (SVM). Int. J. Emerg. Technol. Comput. Sci. Electron. **23**, 207–211 (2016)
17. Naraei, P., Abhari, A., Sadeghian, A.: Application of multilayer perceptron neural networks and support vector machines in classification of healthcare data. In: Future Technologies Conference (FTC), pp. 848–852. IEEE (2016)
18. Chaudhari, V.V., Salunkhe, P.P.: Diabetic retinopathy classification using SVM classifier. Int. J. Appl. Innov. Eng. Manag. **6**, 7–11 (2017)
19. Kavakiotis, I., Tsave, O., Salifoglou, A., Maglaveras, N., Vlahavas, I., Chouvarda, I.: Machine learning and data mining methods in diabetes research. Comput. Struct. Biotechnol. J. **15**, 104–116 (2017)
20. Huang, F., Wang, S., Chan, C.: Predicting disease by using data mining based on healthcare information system. In: 2012 IEEE International Conference on Granular Computing Predict, pp. 12–15 (2012)
21. Karargyris, A., et al.: Combination of texture and shape features to detect pulmonary abnormalities in digital chest X-rays. Int. J. Comput. Assist. Radiol. Surg. **11**, 99–106 (2016)
22. Vajda, S., Santosh, K.C.: A fast k-nearest neighbor classifier using unsupervised clustering. In: Santosh, K.C., Hangarge, M., Bevilacqua, V., Negi, A. (eds.) RTIP2R 2016. CCIS, vol. 709, pp. 185–193. Springer, Singapore (2017). https://doi.org/10.1007/978-981-10-4859-3_17
23. Panayides, A.S., Pattichis, C.S., Pattichis, M.S.: The promise of big data technologies and challenges for image and video analytics in healthcare. In: 50th Asilomar Conference on Signals, Systems and Computers, pp. 1278–1282. IEEE (2016)
24. Ukil, S., Ghosh, S., Obaidullah, S., Santosh, K.C., Roy, K., Das, N.: Deep learning for word-level handwritten Indic script identification
25. Al-ayyoub, M., Alzu, S.M., Jararweh, Y., Alsmirat, M.A.: A GPU-based breast cancer detection system using single pass fuzzy C-means clustering algorithm. In: 5th International Conference on Multimedia Computing and Systems, pp. 650–654. IEEE (2016)
26. Abe, S., Inoue, T.: Fuzzy support vector machines for pattern classification. In: International Joint Conference on Neural Networks, pp. 1449–1454. IEEE (2001)
27. Ruikar, D.D., Santosh, K.C., Hegadi, R.S.: Automated fractured bone segmentation and labeling from CT images. J. Med. Syst. **43**(3), 60 (2019). https://doi.org/10.1007/s10916-019-1176-x
28. Ruikar, D.D., Santosh, K.C., Hegadi, R.S.: Segmentation and analysis of CT images for bone fracture detection and labeling. In: Medical Imaging: Artificial Intelligence, Image Recognition, and Machine Learning Techniques, Chap. 7. CRC Press (2019). ISBN 9780367139612
29. Hegadi, R.S., Navale, D.I., Pawar, T.D., Ruikar, D.D.: Multi feature-based classification of osteoarthritis in knee joint X-ray images. In: Medical Imaging: Artificial Intelligence, Image Recognition, and Machine Learning Techniques, Chap. 5. CRC Press (2019). ISBN 9780367139612
30. Ruikar, D.D., Sawat, D.D., Santosh, K.C., Hegadi, R.S.: 3D imaging in biomedical applications: a systematic review. In: Medical Imaging: Artificial Intelligence, Image Recognition, and Machine Learning Techniques, Chap. 8. CRC Press (2019). ISBN 9780367139612
31. Ruikar, D.D., Hegadi, R.S., Santosh, K.C.: A systematic review on orthopedic simulators for psycho-motor skill and surgical procedure training. J. Med. Syst. **42**(9), 168 (2018)
32. Jagtap, A.B., Hegadi, R.S.: Feature learning for offline handwritten signature verification using convolution neural network. Int. J. Technol. Hum. Interact. (IJTHI). ISSN 1548–3908

Skewness and Kurtosis of Apparent Diffusion Coefficient in Human Brain Lesions to Distinguish Benign and Malignant Using MRI

Sahan M. Vijithananda[1]([✉]), Mohan L. Jayatilake[1], Bimali S. Weerakoon[1],
P. G. S. Wathsala[1], S. Thevapriya[1], S. Thasanky[1], Tharindu D. Kalupahana[2],
and Thusitha K. Wijerathne[2]

[1] Department of Radiography/Radiotherapy, Faculty of Allied Health Sciences,
University of Peradeniya, Peradeniya 20400, Sri Lanka
deanahs@pdn.ac.lk,aquamarine.sahan@gmail.com
[2] Department of Computer Engineering, Faculty of Engineering,
University of Peradeniya, Peradeniya 20400, Sri Lanka
http://ahs.pdn.ac.lk/index.php

Abstract. The application of Diffusion Weighted Imaging (DWI) in cancer identification and discrimination is increase singly interest within last decade. DWI has significant advantages, as it does not require contrast medium and provides qualitative and quantitative information that can be helpful for lesion assessment. Therefore, this study presents the utility of skewness and kurtosis of Apparent Diffusion Coefficient (ADC) to distinguish between benign and malignant brain lesions. All the Magnetic Resonance Imaging (MRI) scans were performed with a 3 Tesla Siemens Skyra MR system using a head coil. The sample consists of six subjects with locally advanced brain lesion. The Echo-Planar Imaging pulse sequence was used to acquire axial DW MRI data with a flip angle $= 90°$, Time of Echo/Time of Repetition (TE/TR) $= 98/6400$ ms, Field of View (FOV) $= 256$ mm, matrix size $= 256 \times 256$, slice thickness of 1 mm and two levels of diffusion sensitization (b = 0 and 1000 s/mm^2). MATLAB 2014 Simulink software was used for the data analysis. The Region of Interest (ROI) the brain lesion was selected. The mean values of both the skewness and kurtosis of ADC within the ROI were determined and finally, the values were compared between benign and malignant brain lesions. The mean kurtosis and skewness of malignant and benign are 3.201, 3.738 and 0.071, 0.463 respectively. The mean kurtosis of benign is significantly high whereas mean skewness is significantly low. Therefore, there is a possibility of utilizing mean skewness and kurtosis pixel values as a potential biomarker to differentiate between benign and malignant brain lesions. ...

Keywords: ADC (Apparent Diffusion Coefficient) ·
DWI (Diffusion Weighted Imaging) · ROI (Region of Interest) ·
Brain tumor · Benign · Malignant

ⓒ Springer Nature Singapore Pte Ltd. 2019
K. C. Santosh and R. S. Hegadi (Eds.): RTIP2R 2018, CCIS 1036, pp. 189–199, 2019.
https://doi.org/10.1007/978-981-13-9184-2_17

1 Introduction

Radiological findings are established on various kinds of modern and high technological medical imaging modalities such as Computed Tomography (CT), advanced and conventional MRI, nuclear imaging, Positron Emission Tomography (PET) scanning [1, 24, 25]. In addition to this, acquiescing patient-specific data in the form medical images is the first to develop computer aided diagnosis (CAD) system [26]. The brain tumor diagnostic and therapeutic decision making system is also no exception to this. Brain tumor diagnosis system majorly uses MRI images. It is difficult to characterize neoplastic tumors due to its heterogeneity [2]. The heterogeneity can be found in spatial profiles and imaging profiles. Also there are some evidence, which prove that some radiological imaging methods do not have the ability to clearly distinguish even some lesions from healthy tissues [3, 4].

In some causes brain tumor stages, especially glyoma grades could not be differentiate by using conventional medical imaging methods. So even in this modern world which we live in, the precise way to identify a brain tumor or tumor grade would be the histo-pathological analysis which will make the patient quite uncomfortable, because most of the time is an invasive procedure which is very painful. Although this method is the only precise way, there are some limitations to have the correct diagnosis of the tumor such as errors occur in sample collection variabilities in explication.

From the past, through many scientific studies, tested the potential of using Magnetic Resonance Imaging spectroscopy as a bio-marker for identification and differentiation of brain lesions ref [5, 6]. especially they studied about the usage of decision tree algorithm with the MR spectroscopy and conventional MR imaging to differentiate brain tumor types [7]. Furthermore, perfusion and spectroscopic MRI methods were also used in some studies to identify and assess the inherent heterogeneity of cerebral tumors [8]. Also there are some researches concentrated on discrimination of secondary cerebral tumors from primaries. The studies were based on MRI diffusion tensor imaging. In this method to overcome the brain tumor identification, main investigation was the diffusivity of water molecules in peritumoral edema region [9, 10].

DWI is a significant tool to evaluate the structural characteristics of the tissues based on the Brownian motion of the water molecules. For most clinical applications of diffusion weighted imaging, the most accepted argument is that the diffusion signal vs diffusion weighting factor b decays in mono-exponential pattern. So it permits to do categorization of brain tumor types or do the definitive diagnosis of tumors [11]. In most causes the diffusion coefficient is higher than the normal brain tissues. Researchers [12–15] found a method to differentiate benign and malignant lesions (brain abscesses vs.cystic brain tumors, malignant vs desmoid tumors, meningioma, cerebellar tumors also breast, liver, and uterine tumors) by measuring the field of the value of ADC. There are several studies also can be found which were used DWI and ADC in differentiating between benign and malignant breast lesions and found out that the ADC values could differentiate benign and malignant breast lesions [16–20]. Also there are some studies found a way to differentiate the abdominal cyst stages [21, 22] by measuring means, histogram analysis, skewness and kurtosis of ADC maps within the interested volume.

In this study explorer whether there is an observable difference between skewness and kurtosis values of ADC maps of malignant and benign lesions. The study needed MRI brain image DICOM data from 3 normal subjects, 3 malignant subjects and 3 benign subjects. Also we needed the radiological report and the pathological report of each patient except healthy subjects to confirm the lesion type. From MRI image DICOM data, specially concern about t2 and t2 diffusion weighted image (DWI) data, the diffusion weighting factor (b) $b = 0$, $b = 1000$ respectively to generate ADC image for the further calculations. As a calculation material we used MATLAB 2014 Simulink software for data sorting, generate ADC images using $b = 0$ and $b = 1000$ image data, draw ROI to calculate mean ADC, skewness and kurtosis.

The hypothesis of this study was there is an observable difference of mean values of skewness and kurtosis of ADC images between malignant and benign brain lesions. The primary objective of this study is to develop a method to differentiate and identification benign and malignant brain tumors by determining skewness and kurtosis values of (ADC) images of brain lesions, which can be introduced as a primary step of brain tumor identification and differentiation in diagnostic and therapeutic radiology.

2 Materials and Methods

This prospective study was carried out using diffusion weighted MR images of the patients diagnosed as brain tumors by the Radiologist. Data was collaboratively obtained from Nawaloka Hospital, Colombo, Sri Lanka of consented diagnosed patients before undergoing Routine MRI prior to chemotherapy. The ethical approval was obtained from the Ethical Review Committee of the Faculty of Allied Health Sciences, University of Peradeniya and the informed oral consent was obtained from the participants of this study.

2.1 Data

All the MRI scan procedures were performed with a 3T Siemens Skyra MR system using a head coil. The sample consists of 9 subjects including 3 malignant, 3 benign and 3 normal brain data. The Echo-Planar Imaging (EPI) pulse sequence was used to acquire axial DW MRI data with a flip angle $= 900$, TE $= 98$ ms, TR $= 6400$ ms, (TR $=$ Time of repetition, T $=$ Time of echo), FOV $= 256$ mm, matrix size $= 256 \times 256$, slice thickness of 1 mm and two levels of diffusion sensitization ($b = 0$ and $1000\,\text{s/mm}^2$).

The software written in MATLAB 2014 Simulink (BLeDIA 1.0) was utilized to plot the ADC map (Fig. 1), select ROI (Region of Interest), and calculate the higher orders of ADC values, plot graphs (dot plots). The ADC maps were generated using $b = 0$ and $b = 1000$ images

$$ADC = \sum_{i=1}^{n} \frac{ln\left(\frac{s_i}{s_0}\right)}{b_i} \qquad (1)$$

Where i - image number, S_i − i^{th} image, S_0 - first image, n - number of images, b_i - diffusion gradient value.

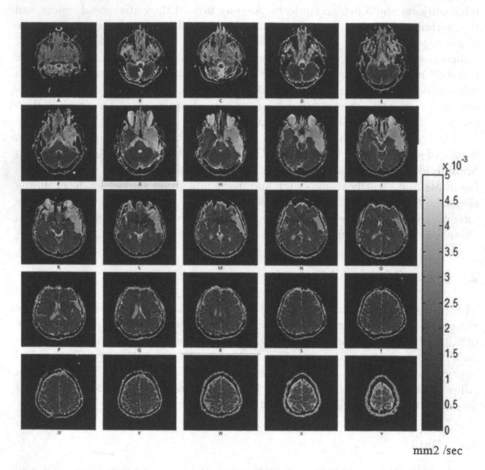

mm2 /sec

Fig. 1. The plotted ADC maps of a selected patient. ADC maps were acquired using b = 0 and b = 1000 images (A to Y) (E to P) slices including lesion areas. Units of color scale is mm²/s.

The ROI of the brain lesion was selected (Fig. 3). Then the skewness and the kurtosis of ADC within the selected ROI were calculated using BLeDIA 1.0.

On the basis of clinical radiology report and the pathological reports the ADC maps of the slices which contained the lesion areas were selected and (ROI) was drawn in each slice separately. The mean of the skewness and the kurtosis within the drawn ROI of each slice were taken using (BLeDIA 1.0).

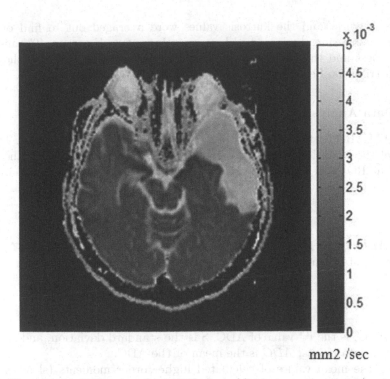

Fig. 2. The plotted ADC map of selected slice of brain tumor patient. According to the radiologist's report the patient was suspected of having malignant tumor. Units of color scale is mm^2/s.

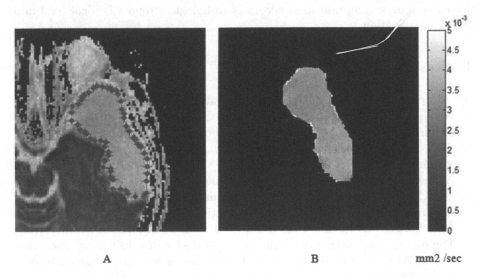

A   B   mm2 /sec

Fig. 3. The selected ROI of the ADC map in selected slice and the mask shown in the picture. (A) The zoomed image of the lesion area and the ROI. (B) The zoomed out lesion area, without background anatomy. Units of color scale is mm^2/s.

The skewness and the kurtosis values were averaged out to find out the average mean lesion skewness and kurtosis values of each patient. The mean of the skewness and the kurtosis were taken into an excel sheet to see the ADC value distribution between benign, malignant and normal brain tissues.

2.2   Data Analysis

The MATLAB 2014 Simulink software was utilized for the data analysis. The 3D (ROI) of the brain lesion was selected. The skewness and kurtosis of the ADC within the ROI were estimated using below equations

$$\text{Kurtosis} = \frac{\sum_{i=1}^{N} \left(\text{ADC}_i - \overline{\text{ADC}} \right)^4}{(N-1) \, S^4} \tag{2}$$

Where ADC_i is the ith value of ADC, S is the standard deviation, and N is the number of data points, $\overline{ADC}$ is the mean of the ADC

$$\text{Skewness} = \frac{\sum_{i=1}^{N} \left(\text{ADC}_i - \overline{\text{ADC}} \right)^3}{(N-1) \, S^3} \tag{3}$$

Where ADC_i is the ith value of ADC, S is the standard deviation, and N is the number of data points, $\overline{ADC}$ is the mean of the ADC.

Using the mean values of calculated higher order moments (skewness and kurtosis), it was compared the normal MRI brain images without lesions and compared the images of benign and malignant lesion values. In this study, the radiological report of each patient was used as a gold standard for lesion identification. The assumption used here was radiological report is ideal for brain lesion identification.

3   Results

Out of 9 patients, plotted ADC maps of one patient are shown in Fig. 1, including the lesion areas. Figure 2 shows the selected slices of the patient which includes the lesion area with the grey scale. Results from each patient were taken in to an excel sheet. Table 1 shows the skewness and kurtosis values of each slice which is including the lesion of each patient.

Average mean skewness and kurtosis values of malignant brain tissues were 0.3763, 3.2006 respectively. The average mean skewness and kurtosis values of benign brain tissues were 0.3059, 2.738 respectively. Also the average mean skewness and kurtosis values of normal brain tissues were 0.2240, 3.0905 respectively.

The overall performance outcome with average mean values of skewness and kurtosis of ADC mean of average mean ADC values of benign, malignant tumors and normal brain tissues are showed in Figs. 4 and 5 as dot plot.

Table 1. Different mean skewness and kurtosis values of ADC benign, malignant brain lesions and normal brain tissues. The chart display that the mean values of skewness and kurtosis of normal brain tissues, brain malignant tissues, brain benign tissues.

Type of lesions	Subject no.	Mean skewness value	Mean kurtosis value
Normal	1	0.2396	3.2801
	2	0.2422	3.1065
	3	0.1903	2.8848
Malignant	1	0.3664	3.6396
	2	0.3907	2.7500
	3	0.3718	3.2121
Benign	1	0.2684	2.8385
	2	0.3500	2.4662
	3	0.2995	2.9095

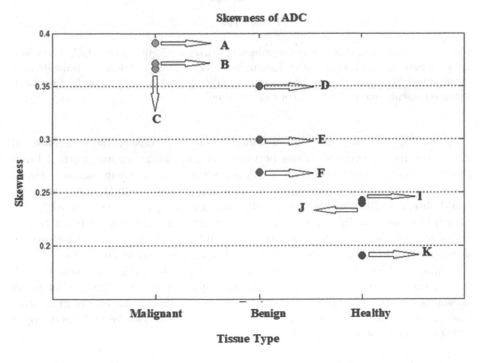

Fig. 4. Dot plot of average of mean values of skewness of calculated ADC. The green dots are indicating mean values of malignant tissues; red dots are Indicate mean values of benign tissues. Blue dots are indicating healthy brain tissues. Each letter represents one patient. (Color figure online)

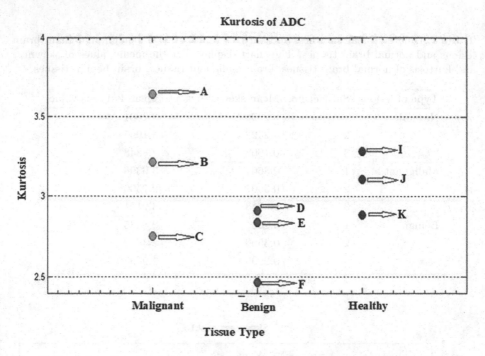

Fig. 5. Dot plot of average of mean values of kurtosis of calculated ADC. The green dots are indicate mean values of malignant brain tissues, the red dots are Indicate that mean values of benign tissues. Blue dots are indicate that healthy brain tissues. Each letter represents one patient. (Color figure online)

According to the results we obtained, there were observable differences of average of mean skewness values between benign, malignant and normal brain tissues. Also there were observable differences in average mean values between malignant and benign brain tissues. Considering average mean kurtosis values we could observe a difference between malignant and benign brain tissues. Although we could observe a difference between the malignant and normal brain tissues using average mean kurtosis values, we could not observe clear difference of average mean kurtosis values between benign and normal brain tissues.

Upper and lower margins of the mean skewness for malignant were 0.3907, 0.3664 respectively. Also in the benign the upper and lower margins of mean skewness were 0.3500, 0.2684 respectively. Upper and lower margins of mean kurtosis for malignant/ benign brain tissues were upper 3.6396, lower 2.7500/upper 2.9095, lower 2.4662 respectively.

4   Discussion

The study showed that there is an observable mean skewness and kurtosis value difference to discriminate benign and malignant brain tumors [23]. The main advantage of the results is that there is a possibility of using higher order moments as a potential bio-marker to differentiate benign and malignant brain lesions. As the results were acquired using 9 patients, further study is needed with more subjects to confirm the usage of higher order values as a bio-marker.

The accuracy and precision of the skewness and kurtosis values are significantly depend on the ROI selection of tumors as malignant tumors can have intra-tumoral heterogeneity. While drawing the ROI the necrosed areas, fluid collections cannot be avoided. The study was not able to discriminate different regions within the malignant tumor. Therefore, this study has to be improved more with the advancement of the software to draw ROI.

The other limitation of the study was being hard to discriminate between benign and normal tissues. Inadequacy of the sample size is a possible cause for this. The sample size should be increased with a wide range of benign tumor varieties to overcome this issue.

Confirmation of the benignancy or malignancy with a histopathological report of a biopsy increases the specificity and the accuracy of the study.

5   Conclusion

In this study we used 3 subjects per each category (malignant, benign, normal brain). As mentioned in the results part, there were observable differences between mean values of skewness and kurtosis between each category. The difference of average mean skewness between malignant and benign tissues were 0.0704 and difference of average mean kurtosis between malignant and benign tissues were 0.462, difference of average mean skewness between malignant and benign tissues were 0.1523, difference of average mean kurtosis between malignant and normal brain tissues were 0.11, difference of average mean skewness between normal and benign tissues were 0.082 and difference of average mean kurtosis between normal brain and benign tissues were 0.352.

The results from the study proved the hypothesis which is an observable difference among the normal, benign and malignant brain tissues according to the mean of the skewness and kurtosis values of ADC. So this method has the potential to become a one of the preliminary diagnostic procedures to identify and differentiate benign and malignant brain tumors before doing a brain biopsy.

Acknowledgments. We would like to appreciate the supervision, encouragement and guidance provided by Dr. M. L. Jayatilake and Dr. B. S. Weerakoon to make this study success, also all the staff members at the Department of Radiography/Radiotherapy, University of Peradeniya and all the staff members at the department of radiology, Nawaloka Hospital Colombo, who helped us in many ways are gratefully acknowledged.

References

1. Yousef, A., Elkharbotly, A., Settin, M., Mousa, Y.: Role of diffusion-weighted MR imaging in discrimination between the intracranial cystic masses. Egypt. J. Radiol. Nucl. Med. **45**(3), 869–875 (2014)
2. Aronen, H., et al.: Cerebral blood volume maps of gliomas: comparison with tumor grade and histologic findings. Radiology **191**(1), 41–51 (1994)
3. Krabbe, K., Gideon, P., Wagn, P., Hansen, U., Thomsen, C., Madsen, F.: MR diffusion imaging of human intracranial tumours. Neuroradiology **39**(7), 483–489 (1997)
4. Provenzale, J.M., Mukundan, S., Baroriak, D.P.: Diffusion-weighted and perfusion MR imaging for brain tumor characterization and assessment of treatment. Radiology **239**(3), 632–649 (2006)
5. Karaali, K., Bayrak, A.: Diffusion-weighted MRI: role in the differential diagnosis of the brain tumors. J. Cancer Prevent. Curr. Res. 2 (2015)
6. Cho, Y., Choi, G., Lee, S., Kim, J.: 1H-MRS metabolic patterns for distinguishing between meningiomas and other brain tumors. Magn. Reson. Imag. **21**(6), 663–672 (2003)
7. Wang, Q., Liacouras, E., Miranda, E., Kanamalla, U., Megalooikonomou, V.: Classification of brain tumors using MRI and MRS data. In: Medical Imaging 2007: Computer-Aided Diagnosis, vol. 6514, p. 65140S (2007)
8. Weber, M., et al.: Diagnostic performance of spectroscopic and perfusion MRI for distinction of brain tumors. Neurology **66**(12), 1899–1906 (2006)
9. Higano, S., et al.: Malignant astrocytic tumors: clinical importance of apparent diffusion coefficient in prediction of grade and prognosis. Radiology **241**(3), 839–846 (2006)
10. Kono, K., et al.: The role of diffusion-weighted imaging in patients with brain tumors. AJNR Am. J. Neuroradiol. **22**(6), 1081–1088 (2001)
11. Stephan, E.M., Sun, Y., Mulkern, R.V.: Diffusion imaging of brain tumors. NMR Biomed. **23**(7), 849–864 (2010)
12. Vermoolen, M., Kwee, T., Nievelstein, R.: Apparent diffusion coefficient measurements in the differentiation between benign and malignant lesions: a systematic review. Insights Imaging **3**(4), 395–409 (2012)
13. Allam, K.E., Shalaby, M.H., Moulood, I.A.: Role of diffusion weighted MRI imaging in detection of liver metakstases. Egypt. J. Hosp. Med. **69**(2), 1823–1827 (2017)
14. Rumboldt, Z., Camacho, D.L.A., Lake, D., Welsh, C.T., Castillo, M.: Apparent diffusion coefficients for differentiation of cerebellar tumors in children. AJNR Am. J. Neuroradiol. **27**(6), 1362–1369 (2006)
15. Oka, K., et al.: Usefulness of diffusion-weighted imaging for differentiating between desmoid tumors and malignant soft tissue tumors. J. Magn. Reson. Imaging **33**(1), 189–193 (2010)
16. Abdulghaffar, W., Tag-Aldeen, M.: Role of diffusion-weighted imaging (DWI) and apparent diffusion coefficient (ADC) in differentiating between benign and malignant breast lesions. Egypt. J. Radiol. Nucl. Med. **44**(4), 945–951 (2013)
17. Gupta, V.K., Liu, W., Wang, R., Ye, Y., Jiang, J.: Differentiation between benign and malignant breast lesions using ADC on diffusion-weighted imaging at 3.0 T. Open J. Radiol. **6**(1), 1 (2016)
18. Tsushima, Y., Taketomi, A., Endo, K.: Magnetic resonance (MR) differential diagnosis of breast tumors using apparent diffusion coefficient (ADC) on 1.5-T. J. Magn. Reson. Imaging **30**(2), 249–255 (2009)

19. Woodhams, R., et al.: ADC mapping of benign and malignant breast tumors. Magn. Reson. Med. Sci. 4(1), 35–42 (2005)
20. Tsushima, Y., Takahashi-Taketomi, A., Endo, K.: Magnetic resonance (MR) differential diagnosis of breast tumors using apparent diffusion coefficient (ADC) on 1.5-T. J. Magn. Reson. Imaging 30(2), 249–255 (2009)
21. Rosenkrantz, A.B., et al.: Whole-lesion diffusion metrics for assessment of bladder cancer aggressiveness. Abdom. Imaging 40(2), 327 (2015)
22. Allam, K.E., Shalaby, M.H., Moulood, I.A.: Role of diffusion weighted MRI imaging in detection of liver metakstases. J. Magn. Reson. Imaging 69(5), 249–255 (2017)
23. Delgado, A.F., et al.: Diffusion kurtosis imaging of gliomas grades II and III – a study of perilesional tumor infiltration, tumor grades and subtypes at clinical presentation. Radiol. Oncol. 51(2), 121–129 (2017)
24. Ruikar, D.D., Santosh, K.C., Hegadi, R.S.: Automated fractured bone segmentation and labeling from CT images. J. Med. Syst. 43(3), 60 (2019)
25. Ruikar, D.D., Santosh, K.C., Hegadi, R.S.: Segmentation and analysis of CT images for bone fracture detection and labeling, chap 7. In: Medical imaging: Artificial Intelligence, Image Recognition, and Machine Learning Techniques. CRC Press, Boca Raton (2019). ISBN 9780367139612
26. Ruikar, D.D., Hegadi, R.S., Santosh, K.C.: A systematic review on orthopedic simulators for psycho-motor skill and surgical procedure training. J. Med. Syst. 42(9), 168 (2018)

Segmentation of Kidney Stones in Medical Ultrasound Images

Prema T. Akkasaligar[1], Sunanda Biradar[1(✉)], and Sharan Badiger[2]

[1] Department of Computer Science and Engineering,
BLDEA's V.P. Dr. P.G. H. College of Engineering and Technology,
Vijayapur, Karnataka, India
premasb@rediffmail.com, sunanda_biradar@rediffmail.com
[2] Department of Medicine, BLDEDU's Sri. B. M. Patil Medical College
and Research Centre, Vijayapur, Karnataka, India
sharanrb@rediffmail.com

Abstract. The computer-aided diagnostic system has become an important issue in clinical diagnosis. Development of new technologies and use of various imaging modalities have raised more challenging issues. The major issue is processing and analyzing a significantly large volume of image data, to generate qualitative information for diagnosis and treatment of diseases. Medical imaging, particularly ultrasound imaging is one of the commonly used diagnostic tool by medical experts. Segmenting a region of interest in medical ultrasound image is a difficult task because of variation in object shape, orientation and image quality. In the present study, initially preprocessing of kidney ultrasound images is performed using contourlet transform and contrast enhancement using histogram equalization. The proposed method focuses on segmentation of kidney stones in preprocessed medical ultrasound images using level set method. The developed method shows better performance in segmenting renal calculi in medical ultrasound images of the kidney. The experimental results demonstrate the effectiveness of the developed software module.

Keywords: Level set segmentation · Renal calculi · Medical ultrasound image

1 Introduction

Image processing has many important applications in medical field because of the increased use of imaging modalities in medical diagnostics. Different imaging modalities assist the medical experts in their decisions by providing quantitative measures [1]. Ultrasonography (USG) is one of the oldest and widely used imaging modality in diagnosing of medical diseases. USG is most popular imaging modality. Ultrasonography is fast, needs lesser acquisition time and cheaper in cost. It can be repeated any number of times because of its harmless and low cost nature. It finds wide spread applications in diagnosing diseases associated with soft tissue organs [2]. But, it suffers from inherent speckle noise [3].

© Springer Nature Singapore Pte Ltd. 2019
K. C. Santosh and R. S. Hegadi (Eds.): RTIP2R 2018, CCIS 1036, pp. 200–208, 2019.
https://doi.org/10.1007/978-981-13-9184-2_18

Kidney stone also known as renal calculus is a most common problem faced by many people worldwide. The frequency of recurrence of kidney stone is also more. Renal calculi are presenting a significant and a challenging problem to the human population. Use of the medical imaging modalities assist in treating such problems. Kidney stone detection is a rare research in the past years. Segmentation of stone from USG images of kidney is a difficult task. So, an automated diagnostic system for kidney USG images can assist the medical experts to distinguish the kidney lesions and thus enhance the diagnosis of disease.

2 Related Work

Removal of speckle noise from USG image is one of the primary step in preprocessing prior to segmentation. Various filters such as median, Gaussian, Weiner and Gabor filters are used in noise removal [3]. The despeckling of USG image using wavelets is highlighted in [4]. Efficiency of the method is illustrated by measuring the parameters such as mean square error, peak signal to noise ratio, etc. In order to carryout segmentation and classification, various methods are used in [5]. Biomedical images become complex as they contain several annotated objects such as arrows, labels, etc. [6,7]. Annotated arrow detection is done by using a sequential classifier. Segmentation of different image layers is carried out using fuzzy binarization [8,9].

For segmentation and classification of kidney from USG images, the image is rotated initially. Angle of rotation is not unique for all images and is varying for every input image. Texture features are extracted from the rotated image. Finally, classified with the help of texture features extracted from these rotated images. Gradient vector force based segmentation is used for segmenting kidney USG images. An external force is used for avoiding improper deformation of the snake while segmenting the contours of concave nature particularly [10]. An elliptical shape is used as a common shape for kidney and Fisher Tripplets are used to know the different types of kidney images in [11]. Computerized tomography images of kidney are segmented using active contours [12]. It works on the basis of non rigid image registration. Genetic algorithms (GAs) are applied in segmentation of USG images [13].

Survey on various segmentation techniques such as region based, edge based, watershed and cluster based algorithms is done in [14]. Most of the work in literature is carried out on segmenting kidney or organ as a whole. We focus on segmentation of renal calculi portion from digital USG images of kidney. Further, analysis on segmented stone provides useful information to the medical experts.

3 Proposed Method

Segmentation of kidney stone in medical ultrasound image is proposed in this work. Proposed methodology involves several steps such as preprocessing, removal of speckle noise, segmentation of preprocessed images using level set method and finally analyzing the segmented output. The methodology is illustrated in Fig. 1.

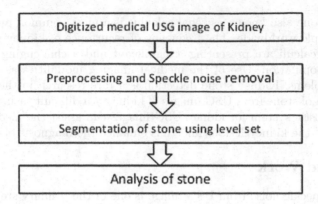

Fig. 1. Proposed method

3.1 Preprocessing of Ultrasound Kidney Images

Noise in USG images appears by the interaction between reflected waves from different individually scattered signals within the same frequency in an ultrasonography system [15]. Speckle is commonly found in the images containing organs like liver and kidney whose embedded elements are very tiny. So, it becomes difficult to determine by the wavelengths of transducers of the imaging system. Speckle is the major detraction for determination of organ lesions. It also leads to poor image quality when compared to other types of modalities such as computed tomography (CT) or magnetic resonance images (MRI). Usually most of the theories represent the speckle noise as a multiplicative noise. Acquired image I (m, n) is represented [15] as in (1).

$$I(m,n) \approx A(m,n)s(m,n) \tag{1}$$

Where A (m, n) is the actual true image and s (m, n) is the speckle noise. Dealing with multiplicative noise is more difficult than additive noise. Therefore, we have applied logarithmic transform in order to convert the multiplicative noise into additive. Contourlet transforms are found to be more suitable for despeckling of ultrasound images [15]. The method works perfectly on the images having rough contours. It works in two stages. Initially Laplacian Pyramid (LP) decomposition is performed. In this step, points of discontinuity are captured. Further, directional filter banks are used to connect these points of discontinuities. For LP decomposition, p1 levels are used. For each of these levels p2 levels of directional decompositions are applied. Thresholding is applied further. Then, inverse contourlet transform is applied to get the denoised images.

After speckle removal, contrast is enhanced to improve the quality of image using histogram equalization. Histogram equalization is one of the commonly used method for medical images [16]. The method adjusts the image, making it easy to analyze and with improved visual quality.

3.2   Segmentation Using Level-Set

Segmentation of image aims to find a collection of non-overlapping regions in an image. Presence of homogeneous intensity throughout the image is a common feature in images and becomes one of the limitations for segmentation algorithms. Particularly, region based segmentation methods work on homogeneous intensity based concept. So, the segmentation method fails in the accurate extraction of region of interest. We have used an edge based level set method for carrying out the task of segmentation.

Originally, level set method works on the concept of sign, to separate the problem image into disjoint regions. Then define a boundary using the points of continuity. Level set takes into account of fronts rather than surfaces [17]. Although the basic method of level set has several advantages, it may lead to wrong results. This is because of irregular level set function (LSF) working during deformation. This can be controlled by means of reinitialization of level set function during evolution [17,18]. Proposed work uses a signed distance function for controlling the deformation of contour. We have used an edge based level set algorithm using distance regulation term for segmentation of calculi region from renal ultrasound images specified in [19].

Initially, the input image I is smoothened with Gaussian kernel specified in (2).

$$g \triangleq \frac{1}{1 + \nabla K_\sigma * I^2} \tag{2}$$

K_σ is the Gaussian kernel function and σ is the standard deviation. Partial differentiation of edge function of smoothened image (g) is to be computed, in order to define an initial LSF $\emptyset_0$. Deformation of initial contour is based on energy function. An energy function using external energy and distance regularization term is shown in (3). External energy is calculated using (4) and term for distance regulation is calculated using (5). External energy is based on input data. μ is a constant value and its value is larger than zero. $L(\theta)$ and $A(\theta)$ use an energy coefficient α larger than zero. These functions can be defined with (6) and (7). Here, 'dl' represents delta function and 'hs' represents heaviside function. $L(\theta)$ is the energy of line integral of g along initial zero level contour. $A(\theta)$ is a weighted area of the region. Weighted area is needed to speedup the contour deformation. It effectively controls the deformation when initial contour is specified far from actual contour [19].

$$E(\theta) = \mu R_x(\theta) + E_{ex}(\theta) \tag{3}$$

$$E_{ex}(\theta) = \lambda L(\theta) + \alpha A(\theta) \tag{4}$$

$$R_x(\theta) = \int X(|\nabla \theta|) \Omega du \tag{5}$$

$$E(\theta) \triangleq \int_\Omega g.dl(\theta) \nabla \theta du \tag{6}$$

$$A(\theta) \triangleq \int_{\Omega} g.hs(-\theta)du \tag{7}$$

During deformation, if the evolved contour crosses zero points or no changes observed in consecutive iterations or if number of iterations specified initially is exceeded, then evolution of contour can be stopped.

3.3 Performance Parameters

Performance of segmentation algorithms can be measured in several ways. We have used two metrics namely dice coefficient and Jaccard index [20,21] as performance parameters.

In equations, A_1 represents segmented image of proposed algorithm. A_2 is the ground truth image.

Dice Coefficient. Dice coefficient (DC) measures the similarity between two images. We have used dice coefficient to compare the resultant segmented image with ground truth image marked by medical expert. It can be calculated using (8).

$$DC = \frac{2.|A_1 \cap A_2|}{|A_1| + |A_2|} \tag{8}$$

DC value varies between 0 and 1. For 100% accurate matching, the dice value is 1 and 0 for completely unmatched images.

Jaccard Index. The Jaccard index (JI) is also used for finding the similarity and dissimilarities between two sample data set. It is defined as the ratio of size of the intersection to the size of the union of the samples. Jacard index is also similar to that of dice coefficient. It ranges from 0 to 1. For exact match JI value is 1 and 0 for no match. It is expressed mathematically as in (9).

$$JI = \frac{|A_1 \cap A_2|}{|A_1 \cup A_2|} \tag{9}$$

4   Implementation and Results

Implementation is performed on Intel core i5 system having 4 GB RAM and 2.50 GHz processor. MATLAB 7.14 software is used for implementing the proposed method. Digital kidney USG images of various sizes and various orientations are used for implementation. The input image set is made of USG digital kidney images collected from BLDEDU's Sri. B. M. Patil Medical College and Research Centre, Vijayapur. It is prepared in consultation with the medical experts from the hospital. The images in database are captured using Phillips HD11XE USG system, curvilinear transducer with frequency ranging from 5 MHz to 7 MHz. Experimentation is carried out on entire database of 27 renal calculus USG images. The sample ultrasonic medical images of kidney stones are shown in Fig. 2(a) and (b). The speckle noise removal is performed using contourlet transform. In first stage of Laplacian pyramidal decomposition, biorthogonal filters are applied. For second stage of directional decomposition, Phoong-Kim-Vaidyanathan-Ansari (PKVA) filters are applied. We have used two levels of LP decomposition with 6 levels of directional decomposition filters. To obtain better results, hard thresholding is performed further [11]. In the preprocessing step, the despeckled image is contrast enhanced with histogram equalization to improve the image quality. Preprocessed USG images of Fig. 2(a) and (b) are shown in Fig. 3(a) and (b) respectively.

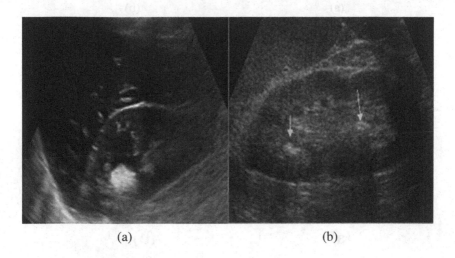

(a)   (b)

Fig. 2. Sample kidney stone US images

Segmentation of kidney stone USG image is carried out using level set method. Initial contour and final deformations of Fig. 3(a) are shown in Fig. 4(a) and (b) respectively. Final deformed contour is obtained at 205 iterations for a single stone sample USG image of kidney. Similarly, initial contour and final deformations of Fig. 3(b) are shown in Fig. 5(a) and (b) respectively. Figure 5(b) is obtained at the end of 198 iterations for a multiple stone sample USG image of kidney.

Number of stones is determined by counting number of segmented regions. The values of performance parameters for sample segmented images and number of stones are shown in Table 1. It also shows the ground truth for respective images obtained by medical expert.

Quantitative analysis of the segmented image is done by using dice coefficient and Jaccard index. Dice value calculated for an entire image set ranges between 0.614 and 0.896. An average dice coefficient obtained for an entire image set is 0.704. Jaccard index range is obtained for all the images in the data set. It ranges between 0.609 and 0.884. Average JI value obtained for an entire image

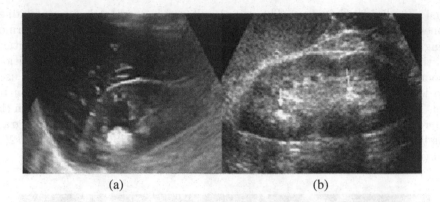

(a)   (b)

Fig. 3. Preprocessed kidney stone USG images

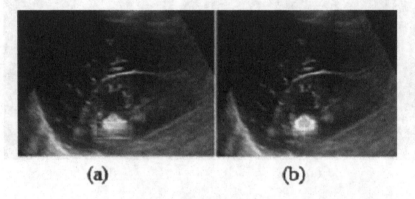

(a)   (b)

Fig. 4. (a) Initial contour (b) Finally segmented/deformed contour

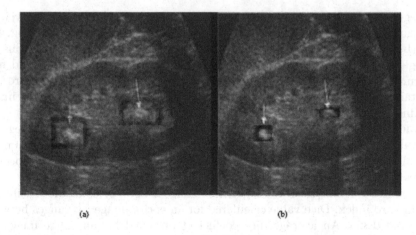

(a)   (b)

Fig. 5. (a) Initial contour (b) Finally segmented/deformed contour

Table 1. Analysis of segmented images

Sample segmented image	Ground Truth	Number of stones	DC	JI
		1	0.879	0.868
		2	0.825	0.827

set is 0.7. The results of segmentation of kidney stones in digital USG images are promising.

5   Conclusion

We have proposed a method for kidney stone segmentation in digital USG images. Preprocessing using contourlet transform and contrast enhancement using histogram equalization is carried out on clinical database. Level set method is effectively used in segmenting single as well as multiple stones in USG images. Performance of the segmentation is measured by using dice coefficient and Jaccard index. These are showing an accuracy of the implemented method. The obtained parameters can help the medical experts in their diagnosis process.

Acknowledgement. The authors are thankful to Vision Group of Science and Technology (VGST), government of Karnataka for financial support under RGS/F scheme. The authors are also thankful to Dr. Bhushita B. Lakhkar, Assistant Professor, Department of Radiology, BLDEDU's Sri. B. M. Patil Medical College and Research Centre, Vijayapur for assisting us in getting kidney USG images for preparing clinical data set for experimentation. She has also provided expert opinion for framing the ground truth. Authors also would like to thank Dr. Vinay Kundaragi, Nephrologist, Sri. B. M. Patil Medical College and Research Centre, Vijayapur for manual segmentation of USG images.

References

1. Ruikar, D.D., Hegadi, R.S., Santosh, K.C.: A systematic review on orthopedic simulators for psycho-motor skill and surgical procedure training. J. Med. Syst. **42**(9), 168 (2018)

2. Suetens, P.: Ultrasonic Imaging, Fundamentals of Medical Imaging, pp. 145–172. Cambridge University Press, Cambridge (2002)
3. Joel, T., Sivakumar, R.: Despeckling of ultrasound medical images: a survey. J. Image Graph. **1**(3), 161–166 (2013)
4. Hiremath, P.S., Akkasaligar, P.T., Sharan, B.: An optimal wavelet filter for despeckling echocardiographic images. In: International Conference on Computational Intelligence and Multimedia Applications, Sivakasi, Tamilnadu, India, 13th–15th December 2007, pp. 245–249 (2007)
5. Hafizah, W.M., Supriyanto, E.: Feature extraction of kidney ultrasound images based on intensity histogram and gray level co-occurrence matrix. In: Proceedings of IEEE Sixth Asia Modelling Symposium, pp. 115–120 (2012)
6. Santosh, K.C., Alam, N., Roy, P.P., Wendling, L., Antani, S., Thoma, G.: A simple and efficient arrowhead detection technique in biomedical images. Int. J. Pattern Recogn. Artif. Intell. **30**(5), 1657002 (2016)
7. Ruikar, D.D., Santosh, K.C., Hegadi, R.S.: Automated fractured bone segmentation and labeling from CT images. J. Med. Syst. **43**(3), 60 (2019)
8. Santosh, K.C., Roy, P.P.: Arrow detection in biomedical images using sequential classifier. Int. J. Mach. Learn. Cybern. **9**(6), 993–1006 (2018)
9. Santosh, K.C., Wendling, L., Antani, S., Thoma, G.: Overlaid arrow detection for labeling regions of interest in biomedical images. IEEE Intell. Syst. **31**(3), 66–75 (2016)
10. Kop, A.M., Hegadi, R.: Kidney segmentation from ultrasound images using gradient vector force. In: International Journal of Computer Applications Special Issue on RTIPPR, pp. 104–109 (2010)
11. Huang, J., Yang, H., Chen, Y., Tang, L.: Ultrasound kidney segmentation with a global prior shape. J. Vis. Commun. Image Represent. **24**(7), 937–943 (2013)
12. Spiegel, M., Dieter, A.H., Volker, D., Jakob, W., Joachi, H.: Segmentation of kidney using a new active shape model generation technique based on non rigid image registration. J. Comput. Med. Imaging Graph. **33**(1), 29–39 (2009)
13. Mauli, U.: Medical image segmentation using genetic algorithms. IEEE Trans. Inf. Technol. Biomed. **13**(2), 166–173 (2009)
14. Jeyakumar, V., Hasmi, M.K.: Quantitative analysis of segmentation methods on ultrasound kidney image. Int. J. Adv. Res. Comput. Commun. Eng. **2**(5), 2319–2340 (2013)
15. Hiremath, P.S., Akkasaligar, P.T., Sharan, B.: Speckle reducing contourlet transform for medical ultrasound images. World Academy of Science, Engineering and Technology Special Journal Issue, pp. 1217–1224 (2011)
16. Agarwal, T., Tiwari, M., Lamba, S.: Modified histogram based contrast enhancement using homomorphic filtering for medical images. In: IEEE International Advance Computing Conference (IACC), Gurgaon, New Delhi, India, 21st–22nd February 2014, pp. 964–968 (2014)
17. Sussman, M., Smereka, P., Osher, S.: A level set approach for computing solutions to incompressible two phase flow. J. Comput. Phys. **114**(1), 146–159 (1994)
18. Li, C., Xu, C., Gui, C., Fox, M.D.: Level set evolution without re-initialization: a new variational formulation. IEEE Trans. Imag. Process. **19**(12), 3243–3254 (2010)
19. Akkasaligar, P.T., Biradar, S.: Analysis of polycystic kidney disease in medical ultrasound images Int. J. Med. Eng. Inf. **10**(1), 49–64 (2018)
20. Cerrolaza, J.J., et al.: Quantification of kidneys from 3D ultrasound in pediatric hydronephrosis. In: IEEE International Symposium, pp. 157–160 (2015)
21. Candemir, S., et al.: Lung segmentation in chest radiographs using anatomical atlases with nonrigid registration. IEEE Trans. Med. Imag. **33**(2), 577–590 (2014)

Osteoarthritis Stages Classification to Human Joint Imagery Using Texture Analysis: A Comparative Study on Ten Texture Descriptors

Sophal Chan[(✉)] and Kwankamon Dittakan

College of Computing, Prince of Songkla University, 80 Kathu, Phuket, Thailand
{sophal.c,kwankamon.d}@phuket.psu.ac.th

Abstract. Osteoarthritis (OA) is a disease that commonly affects the joints in humans. It is the major diseases in aging and obesity society. OA has affected millions of people around the world. OA has affected 10% of Thai population in 2014. According to National Statistical Office (NSO) estimates, 31.8% of Thai population will be affected in OA by 2050. It is difficult to bring back to normal when people suffer from OA. X-ray imaging is the basic method to detect OA. There are 5 grades to differentiate the level of OA which starts from grade 0 to grade 4. This proposed work is to classify the stage of OA by applying image analysis, and classification techniques. The first step of the proposed technique is to identify the region of interest (ROI). Which regions can specify of each texture image. There are four ROIs: (i) Lateral Femur (LF), (ii) Medial Femur (MF), (iii) Lateral Tibia (LT), (iv) Medial Tibia (MT). When ROI was identified, next each ROI (sub-image) was applied with texture descriptor for extractin the feature vectors. To reduce the dimension of feature space, Correlation-based Feature Selection (CFS) approach was implemented. Finally, these selected feature vectors were used to generate the desired classifiers. The research challenge is to find which ROI can produce the better classification result. With respect to a collection of 130 images data set, the obtained result found that Medial Tibia (MT) produced the best performance with Area Under the ROC Curve (AUC) value of 0.871.

Keywords: Knee Osteoarthritis (OA) · OA classification · Texture descriptor · Osteoarthritis (OA) grading

1 Introduction

Osteoarthritis (OA) is the major diseases in elderly people. The most common of OA is knee OA [39]. OA affected over 10 million or 10% of Thai people in 2014. By 2050 the National Statistical Office (NSO) of Thailand estimated that 31.8% of total Thai population will affect by OA. The symptom of knee OA

© Springer Nature Singapore Pte Ltd. 2019
K. C. Santosh and R. S. Hegadi (Eds.): RTIP2R 2018, CCIS 1036, pp. 209–225, 2019.
https://doi.org/10.1007/978-981-13-9184-2_19

can be found out by: (i) swelling around the knee, (ii) get a pain when active of movement, (iii) stiffness in the knee, and (iv) cracking sound happened when moving. Kellgren and Lawrence [23] have classified the stage of OA into five grades: (i) Grade 0: normal case, (ii) Grade 1 or doubtful: a little sign of OA has appeared or 10% of cartilage loss, (iii) Grade 2 or mild: the joint space becomes narrowing, (iv) Grade 3 or moderate: a big sign of OA appeared or no joint space appeared, and (v) Grade 4 or severe: joint space greatly reduced and 60% of the cartilage is definitely lost. The example of OA stage images are illustrated in Fig. 1.

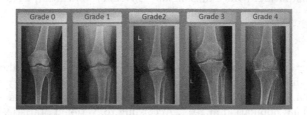

Fig. 1. Grading of OA

The patients with serious OA can significantly affect their quality of life. Thus the early detection is very important. Knee OA early detection can be done by using medical imaging technology. Medical imaging is a technology used to visualise the internal of body for clinical analysis and medical intervention [42,43]. OA detection can be applied to various types of medical imaging, for instance: magnetic resonance imaging (MRI), ultrasound (UL), computed tomography (CT), and X-ray imaging. The most typical way to detect knee OA is X-ray imaging because the cost is reasonable to everyone.

In this paper, a framework of OA stage classification is suggested by applying image texture analysis to knee X-ray images. A collection of X-ray images was obtained from two local hospitals in Thailand. The texture of the bone image was analysed using ten texture descriptors. Once the texture feature vector was extracted from the images, Correlation-based feature selection (CFS) was applied to reduce the dimension of feature vector. The classifiers were generated using various classification learning algorithms.

The proposed study offers the three advantages: (i) fast speed of the work processes, (ii) be a good sample for future work with different image modalities and (iii) can be a good framework for non-specialist researchers or new professional medical doctors to classify OA stage. Although our proposed approach offers many advantages the drawback is: manually allocate the area of interest is needed.

The remainder of the study paper is arranged as follows: Some related works are briefly represented in Sect. 2. The discussion about an overview of the framework which proposed in this study is presented in Sect. 3. The adopted image texture descriptor algorithms are pictured in Sect. 4. The detail of feature selection techniques used in this research is given in Sect. 5. As well as the brief detail of classification mechanisms used is presented in Sect. 6. The brief detail of data with respect to application domain is described in Sect. 7. The evaluation of the proposed framework is reported in Sect. 8. Lastly, the suggestions and conclusions of the future work are illustrated in Sect. 9.

2   Related Work

Medical imaging is a method to illustrate the internal human body. It has been applied to various applications of diseases detection, the examples include: osteoarthritis [31–34], brain tumor [35–37], Fracture detection and analysis [40,41] and pulmonary abnormality [27–29].

Medical image analysis plays a vital role on medical diagnosis by using image contents. There are three image contents: (i) colour, (ii) texture and (iii) shape. Most of medical imaging such as X-ray, MRI, CT-scan and ultrasound is provided an image in greyscale colour. Therefore, the texture and shape are considered in context of medical analysis. Texture has been applied in the works presented in [21,22] for osteoarthritis detection. While shape analysis was applied in the works reported in work [27,38] for pulmonary abnormality detection and brain tumors classification, respectively.

Texture is the surface of any visible object at certain scale which is one of the most important image properties which can be used for image classification or analysis. Texture can be grouped in two groups: (i) tactile and (ii) visual textures [25]. Tactile texture is the immediate tangible feel of a object surface. While visual textures are the visual impression which textures produce to human observer. Texture plays as an important mechanism for medical imaging analysis. It was adopted for pulmonary abnormalities detection from chest X-rays (CXARs) image data set in [28,29] and Tuberculosis (TB) detection in pulmonary and pleural [26].

According to the work presented in [17] reported that texture was applied for the detection of liver tumour from CT scan images. The Grey Level Co-occurrence Matrices (GLCM) texture was adopted to classify breast cancer presented in work [18]. The mass abnormalities of nodule in CT image was studied using texture in [19] have developed MAZDA method for identifying the texture of liver and kidney MRI images. The K-Gabor filter (improve from Gabor filter) was used to classify X-ray chest images presented in [20].

In the context of OA, Local binary pattern (LBP) texture descriptor was applied to detect the stage of OA on X-ray image [21]. From the work [22], it introduced texture feature include: entropy, mean, median, standard deviation, variance, and Tamura's texture features were used for OA stage classification on X-ray images. Haralick and Tamura's textures were also used for OA stage detection in [34]. In work [30] used the fractal signature as the texture feature for analysing the OA stages with X-ray image dataset. According to the study illustrated in this paper the ten texture analysis approaches, including (i) histogram feature, (ii) Local Binary Pattern (LBP), (iii) Completed LBP, (iv) Rotated LBP, (v) LBP Rotation Invariant, (vi) LBP Histogram Fourier, (vii) Local Ternary Pattern, (viii) Local Configuration Pattern, (ix) Haralick feature, and (x) Gabor filter feature descriptor are adopted. The detail of each approach is discussed in further in Sect. 4.

3   Proposed Framework

The proposed framework for generating OA stage classifier is described in this section. The framework schematic is displayed in Fig. 2.

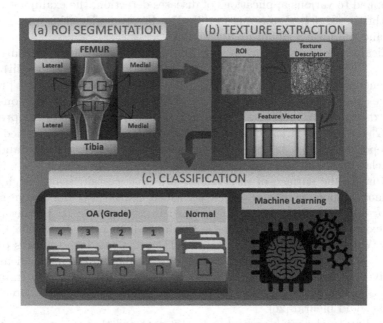

Fig. 2. The proposed framework

From the Fig. 2, the framework is consisted of three processes: (i) ROI segmentation, (ii) feature texture extracting and (iii) the classification process. The first process is the input of an X-ray imageries collection which pre-labeled using domain experts (typically Orthopaedic surgeon), for each image four sub-images are segmented including: (i) Lateral Femur (LF), (ii) Medial Femur (MF), (iii) Lateral Tibia (LT), and (iv) Medial Tibia (MT) as presented in Fig. 2(a). Once sub-images are identified, each sub-image has to represent or extract the salient features in order to facilitate in classification model generation purposes. There are 10 texture descriptors are suggested for texture extraction process with reference to the work illustrated in this paper: (i) histogram feature, (ii) Local Binary Pattern (LBP), (iii) Completed LBP (CLBP), (iv) Rotated LBP (RLBP), (v) LBP Rotation Invariant (LBP$_{ri}$), (vi) LBP Histogram Fourier (LBP-HF), (vii)

Local Ternary Pattern (LTP), (viii) Local Configuration Pattern (LCP), (ix) Haralick feature and (x) Gabor filter feature descriptor. Each texture descriptor can be then defined a feature space to which feature vector can be generated as shown in Fig. 2(b). Once a collection of texture descriptors finished, the next process will be the classification. However, prior to the classification process, it could be applied data discretization. A numeric data is converted into a set of range values; and feature selection mechanism is implemented so as to reduce the number of dimensionality in feature vector space. The Correlation-based Feature Selection (CFS) mechanism is used in this research. When the process of the feature selection is done. The feature space was reduced to which the different classifier learning methods could be applied to which the desired classifier is generated as presented in Fig. 2(c).

4   Texture Feature Descriptor

A brief detail of texture descriptors used with reference to the work illustrated in this paper is presented in this section. Texture descriptor is a technique to characterize image textures or regions. Texture descriptor can observe the region of interest in images or specific the region border.

Texture feature is one of the most important image feature which is used for image mining and image classification. Texture is a feature that give the information of intensities in an image. In this section, ten texture feature descriptors are discussed for extracting texture feature: (i) histogram feature (Subsect. 4.1), (ii) Local Binary Pattern (Subsect. 4.2), (iii) Completed LBP (Subsect. 4.3), (iv) Rotated Local Binary Pattern (Subsect. 4.4), (v) Local Binary Pattern Rotation Invariant (Subsect. 4.5), (vi) Local Binary Pattern Histogram Fourier (Subsect. 4.6), (vii) Local Ternary Pattern (Subsect. 4.7), (viii) Local Configuration Pattern (Subsect. 4.8), (ix) Haralick feature (Subsect. 4.9), and (x) Gabor filter feature descriptor (Subsect. 4.10).

4.1   Histogram Feature

The histogram feature of the grey level image is defined by state of the art histogram based feature, the histogram feature used in the work, including (i) Mean, (ii) Variance, (iii) Skewness, (iv) Kurtosis, (v) Energy, (vi) Entropy. Each feature can be defined using the Eqs. (1) to (8) respectively.

– Mean

$$\mu = \sum_{i=1}^{N}(iP(i)) \tag{1}$$

Where $P(i)$ is the probability distribution of bin i, which $P(i)$ can be written as:

$$P(i) = \frac{H(i)}{M} \tag{2}$$

$H(i)$ is the histogram function and M is the number of blocks.

- Variance

$$\sigma^2 = \sum_{i=1}^{N} (i - \mu)^2 P(i) \tag{3}$$

- Skewness

$$skew = \frac{1}{\sigma^3} \sum_{i=1}^{N} (i - \mu)^3 P(i) \tag{4}$$

Where σ is the standard deviation that can be defined as:

$$\sigma = \sqrt{\sum_{i=1}^{N} (i - \mu)^2 P(i)} \tag{5}$$

- Kurtosis

$$Kurtosis = \frac{1}{\sigma^4} \sum_{i=1}^{N} (i - \mu)^4 P(i) \tag{6}$$

- Energy

$$Energy = \sum_{i=1}^{N} [P(i)]^2 \tag{7}$$

- Entropy

$$Entropy = -\sum_{i=1}^{N} P(i) log_2 [P(i)] \tag{8}$$

4.2 Local Binary Pattern

Local Binary Pattern (LBP) [4] was used to label the pixel in the work which was applied thresholding the neighbourhood of each pixel with the output as the binary number. LBP has been considered to be a powerful feature descriptor for texture classification. LBP operator at pixel (x_c, y_c) can be illustrated by the Eq. 9 below:

$$LBP_{P,R}(x_c, y_c) = \sum_{P=0}^{P-1} S(i_P - i_c) 2^P \tag{9}$$

Where: P is the pixels, R is a radius of circle, i_c and i_p are the grave-level values of the center point in the pixel P, $S(x)$ is a function which is represented following equation:

$$S(x) = \begin{cases} 1 \text{ if } x \geq 0 \\ 0 \text{ if } x < 0 \end{cases} \tag{10}$$

4.3   Completed Local Binary Pattern

Completed Local Binary Pattern (CLBP) is of feature descriptor which applied the fundamental of LBP. In CLBP [5] a local region is defined by a center pixel and a local difference sign-magnitude transform (LDSMT). LBP has divided into two local difference sign-magnitude transform (LDSMT): (i) the difference signs (CLBP_S), and (ii) the difference magnitudes (CLBP_M).

4.4   Rotated Local Binary Pattern

Rotated Local Binary Pattern (RLBP) [6] (known as Dominant Rotated LBP [7]) is a rotation technique of LBP descriptor around the center pixel. The RLBP operator was calculated by the equation below:

$$RLBP_{P,R} = \sum_{p=0}^{P-1} S(g_p - g_c)2^{mod(p-D,P)} \tag{11}$$

Where *mood* refers to the modulus operation, g_p indicates the index of the neighbour pixel, g_p indicates the index of the center pixel, D is the dominant direction (D) in a neighbourhood that can be defined as the Eq. 12:

$$D = argmax|g_p - g_c|, \ p \in (0,1,...,P-1) \tag{12}$$

4.5   Local Binary Pattern Rotation Invariant

Local Binary Pattern rotation invariant (LBP$_{ri}$) is based on basic idea of LBP which create 2p different output values. In this work, the implementation of LBP$_{ri}$ to each pixel with 8 neighbors, LBP$_{ri}$ with 8 bin can be defined by the Eq. 13 as:

$$LBP(x,y) = \sum_{P=0}^{7} S(i_P - i_x, y)2^p \tag{13}$$

4.6   Local Binary Pattern Histogram Fourier

Local Binary Pattern Histogram Fourier (LBP-HF) considers as the rotation-invariant technique depends on LBP descriptor form. LBP-HF used Fast Fourier Transform (FFT) in static feature to calculate global features from the uniform of LBP histogram in place of calculating invariant at each pixel autonomously. In this case, LBP$_{ri}$ feature was considered as a subset of LBP-HF.

4.7   Local Ternary Pattern

Local Ternary Pattern (LTP) work on the central pixel I_c. LTP is fundamental improve from LBP operator. LTP can defined of gray-level in a zone of width

$\pm$ t. The $S(x)$ of LBP is replaced by 3-valued function $S_t(u, i_c, t)$ for LTP in the Eq. 14.

$$S_t(u, i_c, t) = \begin{cases} 1 & u \geq i_c + t \\ 0 & |u - i_c| < t \\ -1 & u \leq i_c - t \end{cases} \tag{14}$$

4.8 Local Configuration Pattern

Local Configuration Pattern (LCP) is one of rotation invariant feature description technique. LCP divided the image information architecture into two: (i) local structural information and (ii) microscopic configuration (MiC). The image information architecture includes image configuration and pixel-wise interaction relationships [8]. In addition, local structure information is directly related to the basic knowledge of LBP while MiC is used for exploring microscopic configuration information. Microscopic configuration information is defined in the Eq. 15:

$$E(a_0, ..., a_{n-1}) = |P_c - \sum_{i=0}^{n-1} a_i P_i| \tag{15}$$

Where: P_c and P_i are intensity center pixel values and neighboring pixels, $a_i (i = 0, ..., n - 1)$ are weighting parameters associated with g_i, and $E(a_0, ..., a_{n-1})$ is the reconstruction error regarding model parameters of a_i.

4.9 Haralick Feature

Haralick features considered as the common way to describe image texture, Haralick is calculated from the Gray-Level Co-occurrence Matrix (GLCM). With the basic of GLCM statistic feature, Haralicks are divided into 14 features: (i) Angular Second Moment (ASM), (ii) Contrast, (iii) Correlation, (iv) Variance, (v) Inverse Difference Moment (IDM), (vi) Sum Average, (vii) Sum Variance, (viii) Sum Entropy, (ix) Entropy, (x) Difference Variance, (xi) Difference Entropy, (xii) Information Measure of Correlation1 (IMC1), (xiii) Information Measure of Correlation 2 (IMC2), and (xiv) Maximum Correlation Coefficient (MCC).

4.10 Gabor Filter Feature

Gabor filter is feature descriptor used for texture analysis at a given location in the image. It includes the specific frequency and specific direction. Gabor filter is a part of 2D Gabor filter bank which contain 3 different parameters: (i) frequencies, (ii) orientations and (iii) smooth parameters of Gaussian envelope.

5   Correlation-Based Feature Selection

The feature selection algorithm is pointed out in this section. The Correlation-based Feature Selection (CFS) was applied in the research study to reduce the dimension of feature space for a better OA stages classification. The CFS is an algorithm used to evaluate subsets of features with an appropriate correlation measure and a heuristic search strategy [2]. The CFS algorithm has followed with the basic rule of the rule **"Good feature subsets contain features highly correlated with the classification, yet uncorrelated to each other"** [3]. CFS can be represented as the Eq. 16:

$$ CFS = \max_{S_k} \left[\frac{r_{cf_1} + r_{cf_2} + \dots + r_{cf_k}}{\sqrt{k + 2(r_{f_1 f_2} + \dots + r_{f_i f_j} + \dots + r_{f_k f_l})}} \right] \tag{16} $$

Where: S_K is the feature subset S consisting of k features, r_{cf} is the value in an average of all feature to classification correlations, r_{ff} is the value in an average of all feature to feature correlations.

6   Classification Algorithm

In this section the classification of learning algorithms are discussed. Classification algorithm is a machine learning algorithm used to classify the OA stage. It is separated into five classes, including grade 0 till grade 4. The author selected the nine famous learning methods for this study in order to know which algorithm produce the best performance in knee OA stages detection study. The nine classification learning algorithms considered with respect to this work include: (i) Decision Tree (Subsect. 6.1), (ii) Binary Split Tree (Subsect. 6.2), (iii) Bayesian Network (Subsect. 6.3), (iv) Naïve Bayes Classifier (Subsect. 6.4), (v) Average One-Dependence Estimators (Subsect. 6.5), (vi) Support Vector Machine (Subsect. 6.6), (vii) Logistic Regression (Subsect. 6.7), (viii) Sequential Minimal Optimization (Subsect. 6.8), and (ix) Back Propagation Neural Network (Subsect. 6.9).

6.1   Decision Tree

Decision tree (C4.5) is classification algorithm which the whole tree consists of three node types: (i) root node, (ii) internal nodes, and (iii) leaf nodes [9]. The parent node is known as the root node that incoming edge has not appeared in this root. Internal nodes are considered for testing on an attribute and the branch defines the output of the test. The leaf nodes are the last nodes that located at the bottom of the tree and have no outgoing edges. Leaf nodes are presented as classes for classification.

6.2   Binary Split Decision Tree

Binary split tree is one type of decision tree which each node consist of two value. While the node of decision tree can consist of multiple values. The main use of binary split tree refers to the applying in statistic dataset with skewed frequency distribution.

6.3 Bayesian Network

Bayesian Network (BN) (known as Probabilistic Networks) is a graphical prob-
ability model applied for decision making in uncertainty [10]. In addition, the
Bayesian network considers as a directed acyclic graph (DAG) and each node n
$\in$ N of BN represents a dataset attribute. BN depends on Bayes's rule which the
Bayes rule are represented in the Eq. 17 below:

$$P(C|U) = arg\ maxP(C)P(U|C) \qquad (17)$$

Where: C is the class attribute, $U = (a\_1, ..., a\_n)$ refers to unclassified test
instance.

6.4 Naïve Bayes

Naïve Bayes is considered as the technique of simple probabilistic classifier
depends on Bayes' theorem combined with strong (naive) independence assumptions [11]. Naïve Bayes has been defined as the Eq. 18 below:

$$P(C|U) = arg\ maxP(C) \prod_{i=1}^{n} P(U_i|C) \qquad (18)$$

6.5 Average One-Dependence Estimators

Average One-Dependence Estimators (AODE) is based on probabilistic classification techniques which developed from the Naïve Bayesian classifier [12] by
addressing the problem of attribute-independence. AODE can be defined as the
Eq. 19 where the probability of each class y, which y has n feature $\{x_1, ..., x_n\}$:

$$\hat{P}(y|x_1, ..., x_n) = \frac{\sum_{i:1 \leq i \leq n \wedge F(x_i) \geq m} \hat{P}(y, x_i) \prod_{j=1}^{n} \hat{P}(x_i|y, x_i)}{\sum_{y' \in Y} \sum_{i:1 \leq i \leq n \wedge F(x_i) \geq m} \hat{P}(y', x_i) \prod_{i=1}^{n} \hat{P}(x_j|y', x_i)} \qquad (19)$$

6.6 Support Vector Machine

Support Vector Machine (SVM) is a linear classifier used for the classification
task. SVM is used to separate instances of two classes in the most optimal way
by the construction of an N-dimensional hyperplane in the relationship of two
training sample classes in the feature set [13]. SVM formally divided into two
categories: (i) linear classification and (ii) non-linear classification.

6.7 Logistic Regression

Logistic regression is a statistical regression model which used on ordinary regression [14]. Logistic regression uses a logistic function to which the logistic model
is computed.

6.8   Sequential Minimal Optimisation

Sequential Minimal Optimization (SMO) is one of the developed algorithms from SVM in order to solve the quadratic programming (QP) optimization problem. Furthermore, QP has happened during the SVM training [15].

6.9   Back Propagation Neural Network

The back propagation neural network algorithm is a classification technique which is applied in neural networks to define a gradient that requires to calculate the weights feature in the network [16]. Backpropagation is a learning algorithm which detects the minimum of the error function.

7   Data Collection

In this section presents the detail of the dataset collection in the research study. A collection dataset comprising 130 X-ray images was illustrated in greyscale with DICOM (Digital Image and Communication on Medicine) format. This dataset was separated into five group as presented in Table 1 below:

Table 1. Image dataset

	Normal	Grade 1	Grade 2	Grade 3	Grade 4
No. of image	39	18	21	44	8

8   Evaluation

This section describes the evaluation of the proposed work for knee OA grading. There are 720 experiments were conducted for the evaluation of the proposed mechanism but only the most significant results are discussed. 130 X-ray images were applied in this study. Each image was taken in Postero-Anterior (PA) position. The number of image collection for training dataset in 5 grades were 39, 18, 21, 44 and 8 respectively as presented in Table 1 above. The evaluation measure parameter were used: (i) Area Under the ROC Curve (AUC), (ii) Accuracy (AC), (iii) Sensitivity (SN), (iv) Specificity (SP), (v) Precision (PR), and (vi) F-Measure (FM). The purpose of the evaluation was to declare evidence that OA grading could not be difficult detected by using the proposed technique. The three set of the experiments are discussed according to three objectives:

  i To determine the most appropriate ROIs with respect to four sub-images described in Fig. 2(a),
 ii To identify the moste appropriate feature extractor with respect to ten texture descriptors mentioned in Sect. 4, and
iii To determine the most effective on classification performance with respect to nine classification learning methods presented in Sect. 6.

Each of the above objective is discussed in detail in Subsects. 8.1 to 8.3 respectively. Ten folds Cross-Validation (TCV) was applied. To this end, all experiments were taken from the Waikato Environment for Knowledge Analysis (WEKA) machine learning workbench [24].

8.1 Region of Interests (ROIs)

This sub-section reports the first evaluation to identify the most appropriate ROIs. There are four sub-images of ROI are presented: (i) Lateral Femur (LF), (ii) Medial Femur (MF), (iii) Lateral Tibia (LT), (iv) Medial Tibia (MT). From the experiments, the LBP descriptor was adopted together with Logistic regression classifier (because experiment reported on later in this paper had presented that produced the best results). The obtained result presented in Table 2:

Table 2. Region of Interest (ROI) results.

ROI	AUC	AC	SN	SP	PR	FM
Medial Femur (MF)	0.822	0.569	0.569	0.849	0.580	0.553
Literal Femur (LF)	0.816	0.585	0.585	0.849	0.618	0.577
Medial Tibia (MT)	**0.871**	**0.654**	**0.654**	**0.907**	**0.691**	**0.658**
Lateral Tibia (LT)	0.828	0.592	0.592	0.833	0.635	0.658

From Table 2, it can be seen that the best result was obtained by applying ROI of MT, the best recorded AUC value of 0.871 and sensitivity value of 0.654. As a result, using the ROIs from Tibia gave the best two results for classification performance. It should be suggested that surgeons or clinicians who analyze X-ray images typically consider Tibia area more than Femur area.

8.2 Texture Descriptors

The second evaluation objective is reported in this sub-section in order to pick out the most adopted texture descriptors. There are ten texture feature descriptors were considered in the proposed work: (i) histogram feature, (ii) Local Binary Pattern (LBP), (iii) Completed LBP (CLBP), (iv) Rotated LBP (RLBP), (v) LBP rotation invariant (LBP_{ri}), (vi) LBP histogram Fourier (LBP-HF), (vii) Local Ternary Pattern (LTP), (viii) Local Configuration Pattern (LCP), (ix) Haralick feature, and (x) Gabor filter feature. For these set of experiments, the MT sub-image was adopted together with Logistic Regression (because it produced the best results (see Subsects. 8.1 and 8.3)). The results of the applying texture feature comparative analysis are displayed in Table 3.

From Table 3, it shows that the best performance of texture descriptor was obtained by using LBP texture descriptor with the best recorded AUC value of 0.871 and sensitivity value of 0.654. To put it briefly, using LBP and RLBP texture descriptor produce the best two results for classification performance. It should be considered that LBP should be the first choice in for learning in

Table 3. Texture descriptor results.

Texture Descriptor	AUC	AC	SN	SP	PR	FM
Histogram	0.647	0.496	0.496	0.776	0.418	0.408
LBP	**0.871**	**0.654**	**0.654**	**0.907**	**0.678**	**0.658**
CLBP	0.789	0.569	0.569	0.836	0.58	0.553
RLBP	0.817	0.585	0.585	0.856	0.578	0.559
LBP$_{ri}$	0.682	0.477	0.477	0.815	0.478	0.451
LBP-HF	0.682	0.438	0.438	0.801	0.418	0.416
LTP	0.741	0.508	0.508	0.824	0.532	0.492
LCP	0.747	0.515	0.515	0.834	0.509	0.503
Haralick	0.644	0.454	0.454	0.801	0.431	0.416
Gabor	0.772	0.523	0.523	0.833	0.537	0.514

Texture analysis. In contrast, Haralick, Histogram, LBP$_{ri}$, and LBP-HF are the texture descriptors produced the lowest performance with the AUC value of 0.644, 0.647, 0.682, and 0.682 respectively. It should be noted that Haralick, Histogram, LBP$_{ri}$, and LBP-HF the next choice after LBP and RLBP for texture analysis.

8.3 Learning Methods

This sub-section points out the last evaluation objective to identify the most appropriate learning methods. There are nine learning methods were considered (it was noted in Sect. 6): (i) Decision Tree (C4.5), (ii) Binary Split Tree, (iii) Bayesian Network, (iv) Naïve Bayes Classifier, (v) Average One-Dependence Estimators, (vi) Support Vector Machine, (vii) Logistic Regression, (viii) Sequential Minimal Optimization, and Back Propagation Neural Network. For the experiments the MT image dataset was applied because this produced the best result (see Subsect. 8.1). Again LBP texture descriptor was used due to the previous sub-section was demonstrated that this learning method made the best performance. The results of nine learning methods comparison are displayed in Table 4.

From Table 4, it illustrated that the best performance of classification among nine learning algorithms was obtained by applying Logistic Regression learning method, a best result AUC value of 0.871 and recorded sensitivity value of 0.654. To come to the point, the three methods: (i) Logistic Regression, (ii) Bayesian Network, and (iii) Naïve Bayes gave the best three results with AUC value of 0.871, 0.858, and 0.854 respectively. It should be concluded that Logistic regression, Bayesian Network, and Naïve Bayes are the first three selected for learning methods. On the contrary, using SVM, Decision Tree, and Binary Split Tree give the lowest results amount of nine methods with the recorded AUC value of 0.612, 0.628, and 0.659 respectively. It should be suggested that AVM,

Table 4. Classification algorithm results.

Machine Learning Algorithm	AUC	AC	SN	SP	PR	FM
C4.5	0.628	0.446	0.446	0.798	0.41	0.425
Binary Split Tree	0.659	0.438	0.438	0.832	0.46	0.445
AODE	0.848	0.562	0.562	0.82	0.573	0.518
Bayesian Network	0.858	0.615	0.615	0.834	0.635	0.583
Naïve Bayes	0.854	0.623	0.623	0.827	0.691	0.583
SVM	0.612	0.469	0.469	0.772	0.415	0.353
Logistic Regression	**0.871**	**0.654**	**0.654**	**0.907**	**0.671**	**0.658**
SMO	0.762	0.577	0.577	0.846	0.678	0.646
Backpropagation	0.842	0.654	0.654	0.864	0.678	0.646

Decision Tree, and Binary Split Tree are not powerful methods in case of learning on knee OA texture images.

9   Conclusion and Future Work

In short, a framework of knee OA stage detection with the operation of medical X-ray image with classification techniques was proposed. The main three findings with reference to the study in this paper are: (i) the most appropriate area to analyse OA stage was MT, (ii) in context of X-ray images LBP classification mechanism produced the best result, and (iii) the most appropriate classification learning algorithm was Logistic Regression with AUC value of 0.871. However, the framework presented here still contain a limitation of requiring manually allocate the area of interest (ROI).

The future work of this study is focused on (i) large dataset or (ii) the applying of deep learning to OA stage classification which can reduce the feature extraction process from the study.

References

1. Abraham, R., Simha, J.B., Iyengar, S.S.: Effective discretization and hybrid feature selection using naïve Bayesian classifier for medical datamining. Int. J. Comput. Intell. Res. **4** (2008)
2. Hall, M.: Correlation-based feature selection for machine learning. Methodology 21i195-i20, pp. 1–5, April 1999
3. Lewandowski, C.M., Co-investigator, N., Lewandowski, C.M.: Correlation based feature selection for discrete and numeric class machine learning. The effects of brief mindfulness intervention on acute pain experience: an examination of individual difference, vol. 1, pp. 1689–1699 (2015)
4. Ojala, T., Pietikainen, M., Harwood, D.: A comparative study of texture 710 measures with classification based on featured distributions. Pattern Recogn. **29**(1), 51–59 (1996)

5. Guo, Z., Zhang, L., Zhang, D.: A completed modeling of local binary pattern operator for texture classification. IEEE Trans. Image Proc. **19**(6), 1657–1663 (2010)
6. Mehta, R., Egiazarian, K.: Rotated local binary pattern (RLBP) rotation invariant texture descriptor. In: International Conference on Pattern Recognition Applications and Methods (ICPRAM), pp. 497–502 (2013)
7. Mehta, R., Egiazarian, K.: Dominant rotated local binary patterns (DRLBP) for texture classification. Pattern Recogn. Lett. **71**, 16–22 (2016)
8. Guo, Y., Zhao, G., Pietikainen, M.: Texture classification using a linear configuration model based descriptor. In: Proceedings of the British Machine Vision Conference, August 2016, pp. 199.1–199.10 (2011)
9. Quinlan, J.R.: Induction of decision trees. Mach. Learn. **1**(1), 81–106 (1986)
10. Friedman, N., Murphy, K., Russell, S.: Learning the structure of dynamic probabilistic networks. In: Proceedings of the Fourteenth Conference on Uncertainty in Artificial Intelligence, pp. 139–147 (1998)
11. Lowd, D., Domingos, P.: Naive Bayes models for probability estimation. In: Proceedings of the 22nd International Conference on Machine Learning ICML 2005, pp. 529–536 (2005)
12. Webb, G.I., Boughton, J.R., Wang, Z.: Not so Naive Bayes: aggregating one-dependence estimators. Mach. Learn. **58**(1), 5–24 (2005)
13. Cortes, C., Vapnik, V.: Support-vector networks. Mach. Learn. **20**(3), 273–297 (1995)
14. Wilson, J.R., Lorenz, K.A.: Modeling Binary Correlated Responses using SAS, SPSS and R. ICSA Book Series in Statistics. Springer, Cham (2015). https://doi.org/10.1007/978-3-319-23805-0
15. Platt, J.C.: Sequential minimal optimization: a fast algorithm for training support vector machines. In: Advances in Kernel Methods, pp. 185–208 (1998)
16. Kelley, H.J.: Gradient theory of optimal flight paths. ARS J. **30**(10), 947–954 (1960)
17. Wu, K., Garnier, C., Coatrieux, J.L., Shu, H.: A preliminary study of moment-based texture analysis for medical images. In: 2010 Annual International Conference of the IEEE Engineering in Medicine and Biology, Buenos Aires, pp. 5581–5584 (2010)
18. Mudigonda, N.R., Rangayyan, R., Desautels, J.E.L.: Gradient and texture analysis for the classification of mammographic masses. IEEE Trans. Med. Imaging **19**(10), 1032–1043 (2000)
19. Nugroho, H.A., Rahmawaty, M., Triyani, Y., Ardiyanto, I., Choridah, L., Indrastuti, R.: Texture analysis and classification in ultrasound medical images for determining echo pattern characteristics. In: 7th IEEE International Conference on System Engineering and Technology 2017 (ICSET), Shah Alam, pp. 23–26 (2017)
20. Humpire-Mamani, G., Traina, A.J.M., Traina, C.: k-Gabor: a new feature extraction method for medical images providing internal analysis. In: 25th IEEE International Symposium on Computer-Based Medical Systems 2012 (CBMS), Rome, pp. 1–6 (2012)
21. Kawathekar, P.P., Karande, K.J.: Severity analysis of osteoarthritis of knee joint from X-ray images: a literature review. In: 2014 International Conference on Signal Propagation and Computer Technology (ICSPCT 2014), Ajmer, pp. 648–652 (2014)
22. Kawathekar, P.P., Karande, K.J.: Use of textural and statistical features for analyzing severity of radio-graphic osteoarthritis of knee joint. In: 2015 International Conference on Information Processing (ICIP), Pune, pp. 1–4 (2015)

23. Kellgren, J.H., Lawrence, J.S.: Radiological assessment of osteo-arthrosis. Ann. Rheum. Dis. **16**(4), 494–502 (1957)
24. Hall, M.A., Frank, E., Holmes, G., Pfahringer, B., Reutemann, P., Witten, I.H.: The WEKA data mining software: an update. SIGKDD Explor. **11**, 10–18 (2009)
25. Tuceryan, M., Jain, A.K.: Texture analysis. In: The Handbook of Pattern Recognition and Computer Vision, pp. 207–248 (1998)
26. Karargyris, A., et al.: Combination of texture and shape features to detect pulmonary abnormalities in digital chest x-rays. Int. J. Comput. Assist. Radiol. Surg. **11**(1), 99–106 (2016)
27. Santosh, K.C., Vajda, S., Antani, S., Thoma, G.R.: Edge map analysis in chest x-rays for automatic pulmonary abnormality screening. Int. J. Comput. Assist. Radiol. Surg. **11**(9), 638–1646 (2016)
28. Santosh, K.C., Antani, S.: Automated chest x-ray screening: can lung region symmetry help detect pulmonary abnormalities? IEEE Trans. Med. Imaging **37**(5), 1168–1177 (2018)
29. Santosh, K.C., Wendling, L.: Angular relational signature-based chest radiograph image view classification. Med. Biol. Eng. Comput. **56**(8), 1447–1458 (2018)
30. Thomson, J., O'Neill, T., Felson, D., Cootes, T.: Automated shape and texture analysis for detection of osteoarthritis from radiographs of the knee. In: Navab, N., Hornegger, J., Wells, W.M., Frangi, Alejandro F. (eds.) MICCAI 2015. LNCS, vol. 9350, pp. 127–134. Springer, Cham (2015). https://doi.org/10.1007/978-3-319-24571-3_16
31. Wolski, M., Podsiadlo, P., Stachowiak, G.W., Lohmander, L.S., Englund, M.: Differences in trabecular bone texture between knees with and without radiographic osteoarthritis detected by directional fractal signature method. Osteoarthritis Cartilage **18**(5), 684–690 (2010)
32. Jin, C., Yang, Y., Xue, Z.J., Liu, K.M., Liu, J.: Automated analysis method for screening knee osteoarthritis using medical infrared thermography. J. Med. Biol. Eng. **33**, 471–477 (2013)
33. Gornale, S.S., Patravali, P.U., Manza, R.R.: Detection of osteoarthritis using knee x-ray image analyses: a machine vision based approach. Int. J. Comput. Appl. **145**(1), 20–26 (2016)
34. Shamir, L., Ling, S.M., Scott, W., Hochberg, M., Ferrucci, L., Goldberg, I.G.: Early detection of radiographic knee osteoarthritis using computer-aided analysis. Osteoarthritis Cartilage **17**(10), 1307–1312 (2009)
35. Ain, Q.L.G., Kazmi, S., Jaffar, A., Mirza, A.: Classification and segmentation of brain tumor using texture analysis, pp. 147–155 (2010)
36. Shree, N.V., Kumar, T.N.R.: Identification and classification of brain tumor MRI images with feature extraction using DWT and probabilistic neural network. Brain Inform. **5**(1), 23–30 (2018)
37. Zacharaki, E.I., et al.: Classification of brain tumor type and grade using MRI texture and shape in a machine learning scheme. Magn. Reson. Med. Official J. Soc. Magn. Reson. Med. **62**(5), 1609–1618 (2009)
38. Wu, P., Xie, K., Zheng, Y., Wu, C.: Brain tumors classification based on 3D shape. Adv. Future Comput. Control Syst. **2**, 277–283 (2012)
39. Hegadi, R.S., Navale, D.I., Pawar, T.D., Ruikar, D.D.: Multi feature-based classification of osteoarthritis in knee joint X-ray images. In: Medical Imaging: Artificial Intelligence, Image Recognition, and Machine Learning Techniques, chap. 5. CRC Press (2019). ISBN 9780367139612

40. Ruikar, D.D., Santosh, K.C., Hegadi, R.S.: Automated fractured bone segmentation and labeling from CT images. J. Med. Syst. (2019). https://doi.org/10.1007/s10916-019-1176-x
41. Ruikar, D.D., Santosh, K.C., Hegadi, R.S.: Segmentation and analysis of CT images for bone fracture detection and labeling. In: Medical Imaging: Artificial Intelligence, Image Recognition, and Machine Learning Techniques, chap. 7. CRC Press (2019). ISBN 9780367139612
42. Ruikar, D.D., Sawat, D.D., Santosh, K.C., Hegadi, R.S.: 3D imaging in biomedical applications: a systematic review. In: Medical Imaging: Artificial Intelligence, Image Recognition, and Machine Learning Techniques, chap. 8. CRC Press (2019). ISBN 9780367139612
43. Ruikar, D.D., Hegadi, R.S., Santosh, K.C.: A systematic review on orthopedic simulators for psycho-motor skill and surgical procedure training. J. Med. Syst. **42**(9), 168 (2018)
44. Szczypinski, P.M., Materka, A., Strzelecki, M., Kleczka, A.: MAZDA—a software package for image texture analysis. Comput. Methods Programs Biomed. **94**(1), 66–76 (2009)

Recurrent Neural Network Based Classification of Fetal Heart Rate Using Cardiotocograph

Sahana Das[1], Himadri Mukherjee[1(✉)], Sk. Md. Obaidullah[2], K. C. Santosh[3],
Kaushik Roy[1], and Chanchal Kumar Saha[4]

[1] Department of Computer Science,
West Bengal State University, Kolkata, India
`sahana.das73@gmail.com`, `himadrim027@gmail.com`, `kaushik.mrg@gmail.com`
[2] Department of Computer Science and Engineering,
Aliah University, Kolkata, India
`sk.obaidullah@gmail.com`
[3] Department of Computer Science,
The University of South Dakota, Vermillion, SD, USA
`santosh.kc@ieee.org`
[4] Biraj Mohini Matrisadan and Hospital, Kolkata, India
`chanchal1069@yahoo.com`

Abstract. Perinatal mortality and morbidity occurs mainly due to intrauterine fetal hypoxia. This can eventually lead to severe neurological damage like cerebral palsy and in extreme cases to fetal demise. It is thus necessary to monitor the fetus during intrapartum and antepartum period. Cardiotocograph (CTG) as a method of assessing the status of the fetus had been in use since the 1960s. Nowadays it is the most widely used non-invasive technique for the continuous monitoring of the fetal heart rate (FHR) and the uterine contraction pressure (UCP). Though its introduction limited the birth related problems, the accuracy of interpretation was hindered by quite a few factors. Different guidelines that are provided for the interpretation are based on crisp logic which fails to capture the inherent uncertainty present in the medical diagnosis. Misinterpretations had led to inaccurate diagnosis which resulted in many medico-legal litigations. The vagueness present in the physician's evaluation is best modeled using soft-computing based techniques. In this paper authors used the CTG dataset from UCI Irvine Machine Learning Data Repository which contains 2126 data. Re-current neural network (RNN) was used for classification of CTG using five-fold cross-validation. The result was compared with a previous work using Fuzzy Unordered Rule Induction Algorithm (FURIA) and Fuzzy Membership Function (FMF) techniques. The obtained accuracy was 98%.

Keywords: Recurrent neural network · Cardiotocograph · Fetal heart rate · FURIA

© Springer Nature Singapore Pte Ltd. 2019
K. C. Santosh and R. S. Hegadi (Eds.): RTIP2R 2018, CCIS 1036, pp. 226–234, 2019.
https://doi.org/10.1007/978-981-13-9184-2_20

1   Introduction

The problem of medical diagnosis consists of determining a patient's status from many and multifarious factors. The solution involves application of years of expertise, knowledge and intuition of the practitioners involved [1]. The situation gets more complicated if the diagnosis has to be based on indirect evidence. One such scenario is the determination of the health of the fetus from the temporal relationship of fetal heart rate (FHR) and the uterine contraction pressure (UCP) of the mother and the interdependency between the two evaluated on a real time basis especially during the intrapartum period. A plot of FHR and UCP is given in Fig. 1.

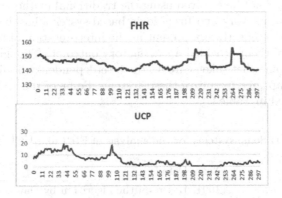

Fig. 1. Plot of FHR and UCP against time

Hypoxia can occur during labor due to the umbilical cord compression as the contraction of uterus takes place. Autonomic nervous System (ANS) of the fetus reacts to oxygen deficiency and this reaction is exhibited by the response of its two branches – Parasympathetic and Sympathetic. Parasympathetic system tends to increase the fetal heart rate (FHR) leading to tachycardia whereas the Sympathetic branch tends to decrease the heart rate leading to bradycardia. This ultimately reduces the variability of FHR. Pattern of fetal heart rate such as tachycardia (FHR > 180 bpm) for more than 15 min, reduced variability, little or no acceleration and late deceleration are some of the indications of hypoxic fetus [2].

Cardiotocograph (CTG) captures these two signals which are rich in clinical information. Though the main aim of fetal monitoring was to reduce the fetal mortality and morbidity, the incidences of perinatal death improved only marginally. According to some experts the introduction of CTG did not have much impact on the outcome of birth-related problems. Some experts even claimed that use of CTG led to unnecessary C-sections [3]. Guidelines for CTG interpretation were developed by different obstetric bodies like Federation International de Gynecologie et d'Obstetrique (FIGO) [4], National Institute of Child

Health and Development (NICHD) [5] and National Institute of Health and Care Excellence (NICE) [6]. In spite of this the problems of misinterpretation and inter and intra-observer variation persisted. This was mainly because the FHR pattern are highly complex and irregular, rendering it difficult to decipher all the information by simple visual inspection alone. Accuracy of the assessment is largely dependent on the experience and expertise of the physicians.

Initial attempts at automation were based on the classical method where crisp values of the features of FHR signal were used to classify the CTG signal into one of three classes – Normal, Suspicious and Pathological. But that failed to capture the inherent uncertainty present in any medical diagnosis. Problem arose in interpreting the pattern of FHR signals in the gray zone between what is normal and completely abnormal. Another problem was the interdependency of features that could not be resolved using the traditional method of classification.

Several neural network and fuzzy logic based systems had been proposed to address the issue of uncertainty present in the interpretation of the CTG signal. However, none of them provided a satisfactory outcome. Though the gray zone problems were somewhat taken care of, the interdependency of different features were not truly captured by any of these systems. Some of the systems developed till date are listed in Table 1.

Table 1. Some systems for the analysis of FHR and UCP signals

System	Year	Details
System 8000	1991	Based on the algorithm by Dawes and Redman proposed in 1981. Uses crisp logic [7]
NST-EXPERT	1995	Developed by Alonso-Betanzos. It is able to make diagnosis and suggest treatments [8]
Computer Aided Fetal Evaluator (CAFE)	2002	It is an upgraded version of NST-EXPERT and was developed by Alonso-Betanzos [9]
TOITU	2015	Developed by Maeda and Noguchi. Based on ANN [10]
Omniview SisPorto 3.5	2008	Developed by Ayres-de-Campos. Analyses fetal status using standard CTG parameters + ST analysis of FHR signals [11]

The authors proposed a Recurrent Neural Network (RNN) based approach to classify the status by finding the interdependency of features of the fetal heart rate and uterine contraction pressure. 2126 data sets were used, each with nine different features for the classification.

Rest of the paper is organized as follows: Sect. 2 describes RNN, Sect. 3 is the proposed model, Result and Discussion is in Sect. 4 and Sect. 5 is the Conclusion.

2   Recurrent Neural Network (RNN)

Conventionally neural network (NN) assumes that there is no interdependency between the inputs and the outputs. That is not a very good idea for the analysis of fetal heart rate because features like FHR baseline, acceleration, deceleration, FHR variability, uterine contraction pressure are dependent on each other. RNN can be thought of as having a memory that stores the information about what has been calculated so far. RNNs are feed forward network with the concept of time. Edges connecting consecutive time steps are called recurrent edges [12]. Among the several variations of RNN available we have used Long Short Time Memory (LSTM) for our work. Each node in the hidden layer is a memory cell. RNN suffers from vanishing gradient problem when error signal is back propagated through time. This is taken care of by LSTM. RNN implements LSTM in the form of weights which gradually change during training. The four components of LSTM are [13]:

- Input node stores the information about the temporal state. Activation function used is sigmoid.
- Input gate takes activation from the data point at current time, as well as from the hidden layer at time (t − 1). Its value multiplies the value of the input node.
- Forget gates remove the contents of the internal state i.e. they reset the memory of the cell.
- Internal state has a self-connected recurrent edge with fixed unit weight.
- Output gate produces a value that is the product of the internal state and the value of the output gate. Activation function used is tanh.

The current data point in our work is $x(t)^T = [f_1(t), f_2(t), ..., f_{23}(t)]^T$ where $f_i(t)$ denotes the features of FHR & UCP signals at time t. The activation function of the ith layer l at time t are $i^l(t)$, $o^l(t)$, $f^l(t)$, $c^l(t)$, $h^l(t)$ input gate, output gate, forget gate, memory cell and the output of the hidden cell respectively. They are given by the following equations:

$$i^l(t) = \sigma(W_{x_i}^l x^l(t) + W_{h_i}^l h^l(t-1) + W_{c_i}^l c^l(t-1) + b_i^l) \tag{1}$$

$$f^l(t) = \sigma(W_{x_f}^l x^l(t) + W_{h_f}^l h^l(t-1) + W_{c_f}^l c^l(t-1) + b_f^l) \tag{2}$$

$$c^l(t) = f^l(t) * c^l(t-1) + i^l(t) * tanh(W_{x_c}^l x^l(t) + W_{h_c}^l h^l(t-1) + b_f^l) \tag{3}$$

$$o^l(t) = \sigma(W_{x_o}^l x^l(t) + W_{h_o}^l h^l(t-1) + W_{c_o}^l c^l(t-1) + b_o^l) \tag{4}$$

$$h^l(t) = o^l(t) * tanh((c^l(t)) \tag{5}$$

where W terms stand for different weight matrices and b denotes the bias terms. Architecture of LSTM is shown in Fig. 2.

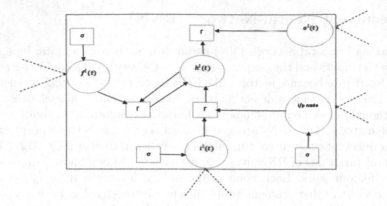

Fig. 2. Architecture of LSTM cell

3   Proposed Method

2126 data from the UCI Machine Learning Data Repository [14] were used for the proposed model. 23 features of each data were used to classify them into one of three classes: Normal (N), Suspicious (S), and Pathological (P). The results obtained were compared with the classification done by Das et al. using FURIA and fuzzy membership function. This section gives the detailed account of the proposed classification of fetal status from FHR and UCP data.

3.1   Features

The features used in the classification are given in Table 2.

Table 2. Features used in the classification

Feature	Description
LB	Baseline of FHR
AC	Accelerations
UC	Uterine contractions
DL	Light decelerations
DS	Severe decelerations
DP	Prolonged decelerations
DR	Repetitive decelerations
FM	Fetal movement
ASTV	Percentage of time with abnormal short term variability
MSTV	Mean value of short term variability

(*continued*)

Table 2. (*continued*)

Feature	Description
ALTV	Percentage of time with abnormal long term variability
MLTV	Mean value of long term variability
A	Calm sleep
B	REM sleep
C	Calm vigilance
D	Active vigilance
SH	Shift pattern
AD	Accelerative/decelerative pattern (stress situation)
DE	Decelerative pattern (vagal stimulation)
LD	Largely decelerative pattern
FS	Flat-sinusoidal pattern (pathological state)
SUSP	Suspect pattern
NSP	Normal = 1; Suspect = 2; Pathologic = 3

3.2 Proposed Architecture

Based on the size of the data and the number of features involved for each data, the authors added three dense layers after the LSTM. The LSTM is 100 dimensional and the three dense layers are 100, 50 and 8 dimensional respectively. ReLU activation function is used for the first two layers whereas for the last layer softmax activation is used. A schematic diagram of the proposed system is given in Fig. 3. Rectified Linear Unit (ReLU) is used on top of LSTM for the dense layer as it handles acceleration of gradient convergence quite efficiently and has low demanding computational time. Also it has no vanishing gradient because the derivative of ReLU is either 0 or 1.

$$f_{ReLU}(x_i) = max(0, x_i) \qquad (6)$$

The last dense layer uses softmax classifier. The softmax activation function normalizes the outputs so that probability of output of each classification label j is given by

$$P(L_t = j \mid v_t) = \frac{exp(v_t W_j)}{\sum_{i=1}^{K} exp(v_t W_i)} \qquad (7)$$

Where L_t and v_t are the label and the concatenated vector for each time step t. The most likely label is selected for output.

3.3 Cross Validation and Output

The input-output pair is (x_i^T, y_i) for the ith data point. As we have opted for 5-fold cross-validation (k = 1, 2, ..., 5) the data is divided into 5 equal parts.

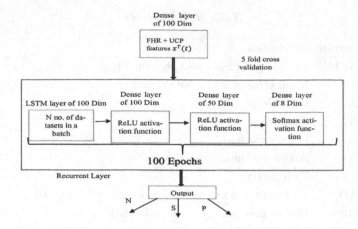

Fig. 3. Schematic representations of the proposed model

For each k = 1, 2, ..., 5 the model is fitted with subset of size λ to the rest of the k − 1 parts. The error in predicting the kth part is

$$Er_k(\lambda) = \sum_{i \epsilon k}(y_i - \alpha - x_i\beta^{-k}(\lambda))^2 \tag{8}$$

α and β are the parameters of the hyperplane to which (x_i^T, y_i) is fitted. The cross validation error or the mean squared error (MSE) to assess the fit is

$$MSE(\lambda) = \frac{1}{5}\sum_{k=1}^{5}Er_k(\lambda) \tag{9}$$

Accuracy of the model is assessed by taking the average of the accuracies of all the k cases of cross validation. The whole process is repeated for λ = 3, 4, 5, 10, 15, 20. The accepted value of λ is

$$\lambda_{final} = min(MSE_j(\lambda)) \tag{10}$$

4   Result and Discussion

The subset size λ = 5 gave an accuracy of 98%. The confusion matrix is given in Table 3. The result was compared with the result obtained by Das et al. [15] using FURIA and the fuzzy membership function (FMF) based classification. The comparison based on two parameters – Precision and Recall are shown in Table 4.

Five-fold cross validation of RNN_LSTM provided an accuracy of 98% i.e. an improvement of 3.48% over FURIA and 5.98% improvement over FMF based method. Also the Precision and Recall for all the three classes with RNN-LSTM is higher as shown in Table 4. The biggest improvement over the previous methods is the precision with which RNN can identify pathological cases.

Table 3. Confusion matrix with $\lambda = 5$

	Normal	Suspicious	Pathological
Normal	1646	9	0
Suspicious	11	282	2
Pathological	0	1	175

Table 4. Comparison between FURIA, FMF and RNN-LSTM

	FURIA		FMF		RNN-LSTM	
	Precision	Recall	Precision	Recall	Precision	Recall
Normal	98.91%	98.08%	98.49%	96.16%	99.46%	99.34%
Suspicious	91.84%	95.07%	76.74%	88%	99.59%	96.58%
Pathological	90.91%	93.02%	92.44%	96.36%	99.43%	98.87%
Accuracy	94.59%		92.14%		98%	

5   Conclusion

RNN-LSTM have been employed with success in the classification of CTG. Overall accuracy of 98% and the very high precision of recognizing pathological is promising. Using RNN an improvement of approximately 9% and 7% were obtained over FURIA and FMF respectively in detecting pathological CTGs. This is of principal importance as accurate diagnosis of fetal distress on time may prevent lifelong disability or may even prevent fetal compromise. In future we also plan to use active learning [16] and clustering-based approaches like fast KNN as proposed in [17] on a larger database and observe the performance of our system.

References

1. Yilmaz, E., Kilikçier, C.: Determination of fetal state from cardiotocogram using LSSVM with particle swarm optimization and binary decision tree. J. Comput. Math. Methods Med. **2013**, 1–8 (2013)
2. Das, S., Roy, K., Saha, C.K.: A linear time series analysis of fetal heart rate to detect the variability: measures using cardiotocography. In: Bhattacharyya, S., Das, N., Bhattacharyya, D., Mukherjee, A. (eds.) Handbook of Research on Recent Developments in Intelligent Communication Application, vol. 1, pp. 471–495. IGI Global (2017)
3. Das, S., Roy, K., Saha, C.K.: Application of FURIA in the classification of cardiotocograph. In: IEEE International Conference on Research and Development Prospects on Engineering and Technology, pp. 120–124. IEEE Press, Chennai (2013)
4. De-Campos, A., Spong, C.Y., Chandraharan, E.: FIGO consensus guidelines on intrapartum fetal monitoring. Int. J. Gynecol. Obstet. **131**(1), 13–24 (2015)

5. Macones, G.A., et al.: The 2008 National Institute of Child Health and Human Development workshop report on electronic fetal monitoring: update on definitions, interpretation, and research guidelines. J. Am. Coll. Obstet. Gynecol. **112**, 661–666 (2008)

6. Santo, S., Ayres-de-Campos, D., Santos, C., Schnettler, W., Ugwumadu, A., Garca, L.M.D.: Agreement and accuracy using the FIGO, ACOG and NICE cardiotocography interpretation guidelines. Acta Obstet. Gynecol. Scand. **96**(2), 166–175 (2017)

7. Dawes, G.S., Visser, G.H., Goodman, J.D., Redman, C.W.: Numerical analysis of the human fetal heart rate: the quality of ultrasound records. Am. J. Obstet. Gynecol. **141**(1), 43–52 (1981)

8. Alonso-Betanzos, A., Guijarro-Berdiñas, B., Moret-Bonillo, V., López-Gonzalez, S.: The NST-EXPERT project: the need to evolve. J. Artif. Intell. Med. **7**(4), 297–313 (1995)

9. Guijarro-Berdinas, B., Alonso-Betanzos, A., Fontella-Romero, O.: Intelligent analysis and pattern recognition in cardiotocographic signals using a tightly coupled hybrid system. Artif. Intell. **136**(1), 1–27 (2002)

10. Maeda, K., Noguchi, Y., Utsu, M., Nagassawa, K.: Algorithms for computerized fetal heart rate diagnosis with direct reporting. Algorithms **8**(1), 395–406 (2015)

11. de Campos, A., Sousa, P., Costa, A., Bernardes, J.: Omniview-SisPorto®3.5 - a central fetal monitoring station with online alerts based on computerized cardiotocogram+ST event analysis. J. Perinat. Med. **36**(3), 260–264 (2008)

12. A Critical Review of Recurrent Neural Networks for Sequence Learning. https://arxiv.org/abs/1506.00019. Accessed 10 Aug 2018

13. Li, J., Mohamed, A., Zweig, G., Gong, Y.: LSTM time and frequency recurrence for automatic speech recognition. In: 2015 IEEE Workshop on Automatic Speech Recognition and Understanding (ASRU), pp. 187–191. IEEE, Scottsdale (2015)

14. UCI Irvine Data Repository. http://archive.ics.uci.edu/ml/datasets/Cardiotocography

15. Das, S., Roy, K., Saha, C.K.: Fuzzy membership estimation using ANN: a case study in CTG analysis. In: Satapathy, S.C., Biswal, B.N., Udgata, S.K., Mandal, J.K. (eds.) Proceedings of the 3rd International Conference on Frontiers of Intelligent Computing: Theory and Applications (FICTA) 2014. AISC, vol. 327, pp. 221–228. Springer, Cham (2015). https://doi.org/10.1007/978-3-319-11933-5_25

16. Bouguelia, M.R., Nowaczyk, S., Santosh, K.C., Verikas, A.: Agreeing to disagree: active learning with noisy labels without crowdsourcing. Int. J. Mach. Learn. Cybern. **9**(8), 1307–1319 (2018)

17. Vajda, S., Santosh, K.C.: A fast k-nearest neighbor classifier using unsupervised clustering. In: Santosh, K.C., Hangarge, M., Bevilacqua, V., Negi, A. (eds.) RTIP2R 2016. CCIS, vol. 709, pp. 185–193. Springer, Singapore (2017). https://doi.org/10.1007/978-981-10-4859-3_17

Automatic Diagnosis of Myocardial Infarction with Left Bundle Branch Block

J. Revathi[1(✉)] and J. Anitha[2]

[1] Avinashilingam Institute for Home Science and Higher Education for Women,
Coimbatore, India
revathibmieau@gmail.com
[2] Karunya Institute of Technology and Sciences, Coimbatore, India
anithaj@karunya.edu

Abstract. The proposed study aims to reduce the risk of sudden death of patients who are affected from Myocardial infarction along with Left Bundle Branch Block. The novel method has been implemented for the automatic diagnosis of Myocardial infarction with the existence of Left Bundle Branch Block using the ECG signal. The discrete wavelet transform was performed to reduce the noise present in the ECG signal. After the denoising process, the analysis of QRS wave, S wave and T wave of ECG signal has been carried out by using detailed coefficients of discrete wavelet transform. The ECGs with Myocardial infarction and Left Bundle Branch Block were tested by using SVM and KNN classifier. The obtained results were reported by highlighting the essential feature to diagnose the Myocardial infarction along with Left Bundle Branch Block.

Keywords: Myocardial infarction · Left Bundle Branch Block ·
ST-segment

1 Introduction

In 2016, World Health Organization reported that 15.2 million deaths were due to heart disease [1]. It was a challenging issue to diagnose Myocardial Infarction (MI) along with the existence of Left Bundle Branch Block (LBBB) [2]. It was found that Myocardial Infarction with Bundle Branch Block induces high mortality rate among all other types of infarction [4]. The mortality rate can be reduced by primary percutaneous coronary intervention (PCI) and thrombolytic therapy. The primary percutaneous coronary intervention provides better result than thrombolytic therapy [3]. The ECG with Myocardial Infarction (MI) and Left Bundle Branch Block was shown in Fig. 1. Several studies were reported to diagnose myocardial infarction along with the Bundle Branch Block (BBB) based on electrocardiographic criteria. Smith et al. reported that the ST segment deviation is usually discordant to the QRS complex [5]. The author analysed the characteristics of QRS wave, T wave and ST segment of ECG signal with Left

K. C. Santosh and R. S. Hegadi (Eds.): RTIP2R 2018, CCIS 1036, pp. 235–244, 2019.
https://doi.org/10.1007/978-981-13-9184-2_21

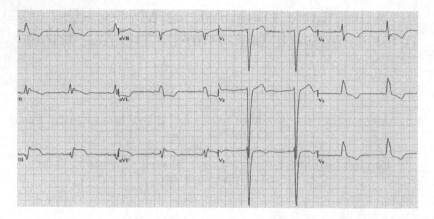

Fig. 1. ECG with Myocardial Infarction and Left Bundle Branch Block [13]

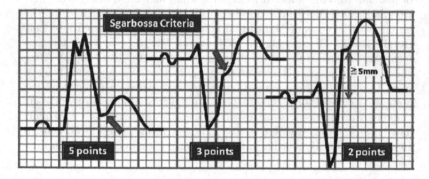

Fig. 2. Concordance and discordance of ST segment with QRS complex [16]

Bundle Branch Block (LBBB) along with MI and without MI [6]. The amplitude of QRS complex was low for the patients with myocardial infarction compared to patients without myocardial infarction [6]. The presence of ST segment depression in leads V1–V3 and the ratio between ST and S is 0.25 for the patients with Myocardial infarction [6]. There is no significance differences in the mean of QT, QTc, JT and JTc intervals between the patients with and without Myocardial Infarction [7]. Sgarbossa et al. proposed three criteria to diagnose LBBB with MI [8]. The ST segment elevation occurs in more than one lead and its amplitude is $100\,\mu V$ [8]. It is concordant with QRS wave is shown in Fig. 2. The occurrence of depression of ST segment in leads V1, V2 and V3 [8]. The ST segment discordant with QRS wave is shown in Fig. 2 and its occurred in more than one leads. Its amplitude is $500\,\mu V$. The selvester Electrocardiographic criteria was that the discordant ST segment elevation is based on the amplitude of S wave [8]. Pendell et al. studied morphology of T waves in patients who have ECG with Myocardial Infarction and LBBB. The morphology of T waves were classified into five such as (a) complete T wave is discordant with QRS wave (b) complete T wave

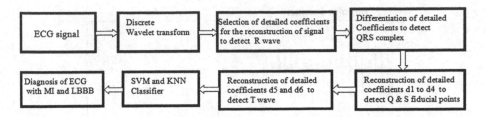

Fig. 3. Block diagram of proposed system

is concordant with QRS wave (c) early discordant and finally concordant (d) early concordant and finally discordant (e) cannot categorized because of low amplitude. It was concluded that there is no relationship between morphology of T wave and Myocardial Infarction [9].

2 Methodology

The block diagram of proposed system for an automatic detection of ECG which posses Myocardial Infarction with the existence of Left Bundle Branch Block is shown in Fig. 3. The function of each block is demonstrated in the subsequent sections.

Database. The ECG signals were downloaded from Long term ST database. It consists of 86 records obtained from 80 subjects. It shows a various events of changes in ST segment. It includes ischemic and no-ischemic ST episodes [18]. This database was constructed to analysis ST episodes in the ECG signal.

3 Preprocessing of ECG Signal

Generally the bio-signal are distorted with the noises such as artifacts, power-line interference and baseline drift. Hence, the various methods were developed recently to eliminate the noise present in the ECG signal. One of the efficient method is discrete wavelet transform [12]. In this study, the noise was eliminated by applying db6 mother wavelet at level six. The ECG signal provides information in the frequency band 0.5–100 Hz [11]. The frequency range of baseline drift and the artifacts are 0.5–1 Hz and 100–150 Hz respectively [10]. The sampling frequency of an ECG signal is 250 Hz. The signal is disintegrated into approximation coefficients and detailed coefficients at various scales. Later, the signals are reconstructed based on the coefficients obtained at level six. The results were shown in Figs. 4 and 5.

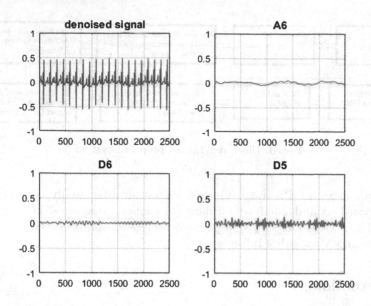

Fig. 4. Denoising of signal with approximation and detailed coefficients D6 and D5

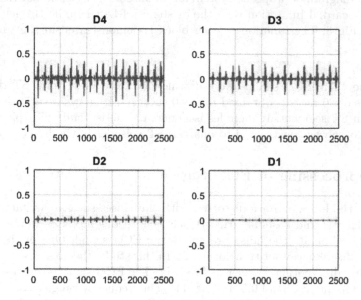

Fig. 5. Signals with detailed coefficients D1, D2, D3 and D4

3.1   Feature Extraction Technique

Feature extraction provides the information about the properties of ECG signal. The morphological and statistical features are extracted to detect an ECG waveform with the existence of MI and LBBB. These features are extracted by using discrete wavelet transform. The feature vectors are obtained based on the selection of levels of detailed coefficients for the reconstruction of an ECG wave. The wavelet transform provides better result in both time and frequency domain. This feature vectors are obtained by determining the characteristic points such as P, Q, R, S and T of an ECG signal is shown in Fig. 6. The features that are extracted from ECG waveform are P, Q, R, S, T peaks, mean of ST interval and QS interval, ratio between ST and S and also ratio between QRS and T wave. The sample features of an ECG signal with MI and LBBB is shown in Table 1.

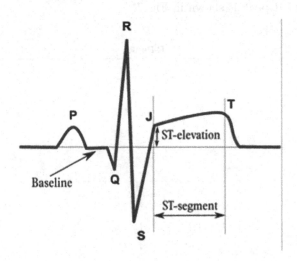

Fig. 6. Characteristic points of normal ECG signal

Table 1. Sample features of MI with LBBB ECG waveform

Record no.	QRS wave	ST segment	T wave	ST/S	QRS/T
1	0.49	0.03	0.49	0.06	1.0
2	0.49	0.02	0.49	0.102	1.0
3	0.42	0.02	0.40	0.04	0.81

Detection of QRS and R Wave. The time taken for the ventricular depolarization in the heart is defined by calculating the duration of QRS complex in ECG signal. The normal range is 40–90 ms. The presence of bundle branch block delayed the process of depolarization. So, it induces the duration of QRS complex to above 120 ms. The duration of QRS complex is calculated by applying discrete wavelet transform. It was identified that most of the energy is existed in the level 2, 3 and 4 for the sampling frequency 250 Hz. The level 2, 3 and 4 are considered for the analysis of QRS complex. The R-R interval is referred as the length of the two neighbouring QRS complex. The normal value is 0.8–0.12 s. The coefficients at level five consist of lower frequency elements. Hence it was not investigated for the analysis of R wave. The level 2 and 3 are considered for the analysis of R wave. The reconstruction of ECG wave to detect R peak is obtained from the detailed coefficients d2 and d3. To enhance QRS complex the function g(x) is implemented [17].

$$g(x) = d3 * (d2 + d4)/2^n \qquad (1)$$

The detection of R peak is shown in Fig. 7.

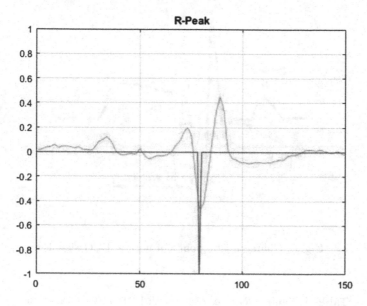

Fig. 7. Detection of R peak

Q, S and T Fiducial Points Detection. The complete QRS waves are identified by finding the fiducial points Q, S and T in the ECG signal. These points have usually low amplitude and high frequency. The energy of these points are available at low scale. The decomposition level d1 to d4 are reconstructed and it is followed by differentiation to identify Q and S point present in the signal. The T peaks are identified by the reconstruction of detailed coefficients d5 and d6 of an ECG signal. The ST segment was detected and it is shown in Fig. 8.

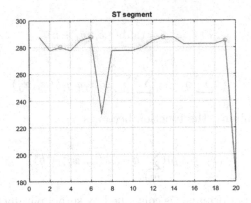

Fig. 8. Detection of ST segment

Defining Threshold to Diagnose MI with LBBB. According to Smith and Sgarbossa rule, the threshold value was defined to identify the ECG with MI and LBBB after evaluating the features of Q, R, S and T wave and ST segment for a particular record. The threshold is defined as that if the ratio between ST and R or ST and S is lesser than 0.25 and also the ratio between QRS and T wave is less than or equal to one then it was identified as MI with the existence of LBBB.

4   Classification

The ECG signal which posses MI and LBBB are classified by using SVM and KNN classifier. It was explained in detail in the following subsection.

4.1   SVM Classifier

After extracting the features from the ECG signal, a SVM classifier is used to recognize ECG with myocardial infarction with LBBB. Several studies reveals that it classifies the data accurately. In this study, the non linear SVM classifier uses kernel function for the data mapping to transformed space. The different kernel transforms includes radial bias function, polynomial and quadratic functions [15]. The kernel function is represented by the following equation [14]

$$k(\hat{x}_i, \hat{x}_j) = \hat{x}_i, \hat{x}_j \tag{2}$$

The function of Gaussian radial basis is represented by

$$k(\hat{x}_i, \hat{x}_j) = \frac{\exp \|x_i - x_j\|^2}{2 * \sigma^2} \tag{3}$$

The polynomial of a function is

$$k(\hat{x}_i, \hat{x}_j) = (\hat{x}_i, \hat{x}_j \mid 1)^d \tag{4}$$

The sigmoid is

$$k(\hat{x}_i, \hat{x}_j) = \tanh(\gamma \hat{x}_i, \hat{x}_j + \eta) \tag{5}$$

then

$$L_d = \sum_{i=1}^{l} \alpha_i - \frac{1}{2} \sum_{i=1}^{l} \sum_{i=1}^{l} \alpha_i \alpha_j y_i y_j k(\hat{x}_i, \hat{x}_j) \tag{6}$$

after completing training, the function becomes

$$f(x) = sgn(\sum_{i=1}^{l} y_i \alpha_i^* k(\hat{x}_i, \hat{x}_j) + b^*) \tag{7}$$

The SVM classifier performance is controlled by kernel parameters. The percentage of accuracy obtained is 94.7.

4.2 KNN Classifier

KNN classifier is also implemented in this study. The accuracy of classification mainly based on the K value and the distance metric type. The classification accuracy was obtained by using five fold cross validation method. The percentage of accuracy obtained to diagnose MI with LBBB is 89.5.

5 Results and Discussion

Long term ST database is employed in this work. Approximately ECG signals from 20 records were used as a training set. The sampling frequency of this signal is 250 Hz. The denoising of ECG signals were achieved by using wavelet

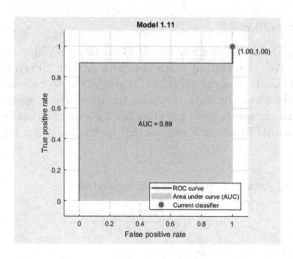

Fig. 9. ROC curve

transform. The db6 mother wavelet were employed and it is decomposed to level six and it is reconstructed based on the decomposition coefficients. The features were extracted from the signal using detailed coefficients. It was observed that the entire T wave is concordant with QRS complex. The ratio between ST and S wave are less than 0.25. The ratio between T wave and S wave are less than or equal to 1. The duration of QRS complex is also greater than 120 ms. These features were fed to the SVM classifier and KNN classifier to detect the ECGs which posses myocardial infarction and Left Bundle Branch Block (LBBB). The percentage of accuracy was 94.7 and 89.5 respectively. The ROC curve for SVM classifier is plotted in Fig. 9 and it was found that area under the curve is 0.89.

6 Conclusion

The features of ECG signals were extracted based on characteristic points such as Q, R, S and T points. The proposed system uses SVM and KNN classifier to diagnose the ECGs with Myocardial Infarction and LBBB. The 94.7 and 89.5% of accuracy was obtained. This system helps the physician to diagnose disease in early stage for further treatment.

References

1. Addison, P.S.: Wavelet transforms and the ECG: a review. Physiol. Meas. **26**(5), R155–R159 (2005)
2. Edhouse, J.A., Sakr, M., Angus, J., Morris, F.P.: Suspected myocardial infarction and left bundle branch block: electrocardiographic indicators of acute ischaemia. Emerg. Med. J. **16**, 331–335 (1999)
3. Widimsky, P.: Acute myocardial infarction due to the left main coronary artery occlusion: electrocardiographic patterns, angiographic findings, revascularization and in-hospital outcomes. Cor et Vasa **54**(1), e3–e7 (2012)
4. Terkelsen, C.J., et al.: Mortality rates in patients with ST-elevation vs. non-ST-elevation acute myocardial infarction: observations from an unselected cohort. Eur. J. **26**(1), 18–26 (2004)
5. Di Marco, A., et al.: Assessment of Smith algorithms for the diagnosis of acute myocardial infarction in the presence of left bundle branch block. Revista Española de Cardiología (Engl. Ed.) **70**(7), 559–566 (2017)
6. Dodd, K.W., Elm, K.D., Smith, S.W.: Comparison of the QRS complex, ST-segment, and T-wave among patients with left bundle branch block with and without acute myocardial infarction. J. Emerg. Med. **51**(1), 1–8 (2016)
7. Dodda, K.W., Elm, K.D., Dodd, E.M., Smith, S.W.: Among patients with left bundle branch block, T-wave peak to T-wave end time is prolonged in the presence of acute coronary occlusion. Int. J. Cardiol. **236**, 1–4 (2017)
8. Gregg, R.E., Helfenbein, E.D., Babaeizadeh, S.: New ST-segment elevation myocardial infarction criteria for left bundle branch block based on QRS area. J. Electrocardiol. **46**, 528–534 (2013)
9. Meyers, H.P., Jaffa, E., Smith, S.W., Drake, W., Limkakeng Jr., A.T.: Evaluation of T-wave morphology in patients with left bundle branch block and suspected acute coronary syndrome. J. Emerg. Med. **51**, 229–237 (2016)

10. Martis, R.J., Acharya, U.R., Lim, C.M., Suri, J.S.: Characterization of ECG beats from cardiac arrhythmia using discrete cosine transform in PCA framework. Knowl.-Based Syst. **45**, 76–82 (2013)
11. Ai, D., Yang, J., Wang, Z., Fan, J., Ai, C., Wang, Y.: Fast multi-scale feature fusion for ECG heartbeat classification. EURASIP J. Adv. Signal Process. **2015**, 46 (2015)
12. Zidelmala, Z., Amirou, A., Ould Abdeslam, D., Merckle, J.: ECG beat classification using a cost sensitive classifier. Comput. Methods Programs Biomed. **111**, 570–577 (2013)
13. https://www.ecgguru.com/ecg/left-bundle-branch-block-acute-inferior-wall-mi
14. Khazaee, A., Ebrahimzadeh, A.: Classification of electrocardiogram signals with support vector machines and genetic algorithms using power spectral features. Biomed. Signal Process. Control **5**, 252–263 (2010)
15. Martis, R.J., Acharya, U.R., Min, L.C.: ECG beat classification using PCA, LDA, ICA and discrete wavelet transform. Biomed. Signal Process. Control **8**(5), 437–448 (2013)
16. Cai, Q., et al.: The left bundle-branch block puzzle in the 2013 ST-elevation myocardial infarction guideline: from falsely declaring emergency to denying reperfusion in a high-risk population. Are the Sgarbossa criteria ready for prime time? Am. Heart J. **166**(3), 409–413 (2013)
17. https://shodhganga.inflibnet.ac.in/bitstream/10603/174847/11/11_chapter%202.pdf
18. Jager, F.: Mark: long-term ST database: a reference for the development and evaluation of automated ischaemia detectors and for the study of the dynamics of myocardial ischaemia. Med. Biol. Eng. Comput. **41**(2), 172–182 (2003)

Exudates Detection from Digital Fundus Images Using GLCM Features with Decision Tree Classifier

Parashuram Bannigidad and Asmita Deshpande[✉]

Department of Computer Science, Rani Channamma University, Belagavi,
Karnataka, India
parashurambannigidad@gmail.com, asd_bca@yahoo.com

Abstract. Diabetes affects a number of human organs, the most common organ being the human eye. Diabetic Retinopathy, Glaucoma, Macular Edema are some of the common ophthalmic disorders found in diabetic patients. Ophthalmologists diagnose Diabetic Retinopathy in a digital fundus image with the presence of exudates. The proposed algorithm consolidates morphological operations for blood vessel removal, segmentation and optic disc removal followed by exudates detection. In this experiment the GLCM features are extracted. These features enhance the detection of affected regions in a retinal image as it depicts how frequently various combinations of gray levels co-exist in an image section. This experiment also explores the use of SVM, k-NN and Decision tree classifiers to distinguish between diseased and healthy retinal images. It is observed from experimentation that the Decision tree classifier yields best results of classifying exudates in digital fundus images. The PPV of the proposed algorithm with decision tree classifier is 100% for DIARETDB0, 97.6% for DIARETDB1, 97% for e-Ophtha EX and 100% for Messidor databases. The sensitivity of the proposed algorithm is 100% for DIARETDB0, 100% for DIARETDB1, 91.6% for e-Ophtha EX and 94.5% for Messidor databases. The proposed algorithm also exhibits 100%, 97.6%, 94.5% and 94.5% accuracy values for DIARETDB0, DIARETDB1, e-Ophtha EX and Messidor databases, respectively using GLCM features with decision tree classifier. Thus, Decision tree classifier is proposed as robust means for detecting exudates and examining the presence of Diabetic Retinopathy in digital fundus images.

Keywords: Diabetic retinopathy · Fundus images · Blood vessels ·
Morphological operations · Optic disc · GLCM features · Exudates ·
Decision tree classifier

© Springer Nature Singapore Pte Ltd. 2019
K. C. Santosh and R. S. Hegadi (Eds.): RTIP2R 2018, CCIS 1036, pp. 245–257, 2019.
https://doi.org/10.1007/978-981-13-9184-2_22

1 Introduction

Diabetes is a lifestyle disorder that affects 422 million people globally. It deteriorates the functioning of a number of human organs particularly the human eye. Diabetic patients develop ophthalmic complications such as Diabetic Retinopathy, Glaucoma and Macular Edema. Such diseases can be diagnosed by ophthalmologists by examining the affected portion of retina. A non-mydriatic fundus camera is nowadays commonly employed to diagnose these disorders. The fundus camera aids in documenting the interior portion of the eye giving a view of fovea, optic disc, macula and retina. In the early stages of Diabetic Retinopathy microaneurysms start appearing as small red dots caused by swellings of the capillaries. As the disease progresses exudates appear on the retina as a result of breakdown of the vessels, leaking serum proteins or lipids. Exudates can be seen as irregularly shaped yellow lesions which can be present anywhere in the fundus image of a diabetic patient. Figure 1 depicts a digital fundus image with exudates.

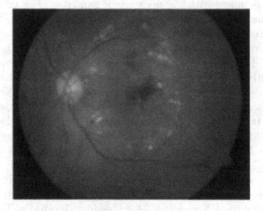

Fig. 1. Fundus image with exudates

Most of the researchers have explored image processing techniques to detect exudates in a fundus image. Chen et al. [1] have explored segmentation technique based on histogram and morphological operations for detection of exudates and obtained 90% PPV and 94% specificity. Omar et al. [4] obtained 96.73% accuracy and 98.68% sensitivity while identifying exudates using texture features, Local Binary Pattern (LBP) variants and ANN classifier. Akram et al. [5] have used filter banks to extract candidates and a Bayesian classifier to detect exudates. They have obtained sensitivity 93.7% and PPV 97.54%. Prakash and Selvathi [6] have proposed a histogram based technique and mathematical morphological operations to detect hard and soft exudates. They have achieved 94.6% specificity and 96.7% sensitivity. Mukherjee and Dutta [7] have worked on a data set of 130 images and obtained 94.98% sensitivity. Pan and Bing-Kun [8] have suggested preprocessing the green channel and histogram based thresholding. They have also used Fuzzy C-means technique to obtain 87.5% PPV and 84.8% sensitivity. Liu et al. [9] have explored a unique segmentation strategy based on location and used random forest classifier to achieve 75% PPV and 76% sensitivity. Rekhi et al. [10] worked on adaptive thresholding with SVM classifier to achieve 90% accuracy. Gupta and Karandikar [11] proposed an approach to detect exudates based on morphological operations. They have extracted GLCM and Splat features and obtained 87% sensitivity and 88% accuracy. Somkuwar et al. [12] devised a mechanism for classifying exudates and non exudates regions using 6 dimensional intensity based features and Euclidean Distance classifier. They have obtained an average accuracy of 96.08% and 95% for e-Ophtha and Messidor databases respectively. Zhang et al. [13] have detected exudates using mathematical morphology and contextual feature set. They have proposed the use of random forest classifier and obtained AUC 0.95. Bannigidad and Deshpande [17] have proposed a combination of morphological operations, Gaussian filtering, Mean filtering and CLAHE for enhancing the fundus images. Bannigidad and Deshpande [22] explored segmentation based on thresholding along with texture features and k-NN classifier to extract exudates in fundus images. Santosh et al. [23] have detected arrow annotations in biomedical images using fuzzy techniques. Zohora and Santosh [24] have proposed the use of circular Hough transform for identification of foreign objects of circular nature from X-ray images. Santosh and Zohora [25] have used edge detection along with morphological operations for segmentation of circular elements in chest radiographs. Zohora et al. [26] have suggested the use of normalized cross correlation and unsupervised clustering to detect circular elements.

2   Materials and Methods

For the purpose of implementation, the digital fundus images are obtained from publicly available databases, namely; DIARETDB0 [15], DIARETDB1 [20], e-Ophtha EX [19] and Messidor [21]. DIARETDB0 is a fundus database with 130 images captured with 50° field-of-view and can be used by the research community to test their algorithms. 110 of these images show symptoms of Diabetic Retinopathy whereas 20 images are normal. The fundus images from DIARETDB1 database are captured with 50° field-of view. It consists of 5 normal images and 84 images containing signs of diabetic retinopathy. e-Ophtha EX is a freely available database of color fundus images having 47 images with exudates and 35 normal images. These images have a size ranging between 1440×960 pixels and 2544×1696 pixels. The Messidor is a digital fundus image database having 1200 images captured at 45° field-of-view with the help of a color video 3CCD camera on a Topcon TRC NW6 non-mydriatic retinograph. These images are represented with 8 bits per color plane at 2240×1488, 2304×1536 or 1440×960 pixels.

3   Proposed Method

A multi-stage methodology is followed for detection and identification of exudates present in a fundus image. The multistage process initially performs preprocessing, blood vessel removal, segmentation, optic disc removal, and exudates detection followed by feature extraction and classification. The proposed method is depicted in the flow diagram shown in the below Fig. 2.

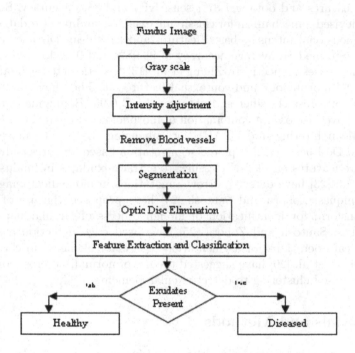

Fig. 2. Flow diagram of the proposed method

3.1   Preprocessing

The objective of preprocessing a digital fundus image is to reduce the uneven illumination and highlight finer details in the fundus like blood vessels, lesions and exudates. The centre of a fundus image inherently has more contrast whereas the periphery exhibits less contrast [17]. During the preprocessing stage, the input image is resized to 576 × 720 Standard size and converted into grayscale. The contrast between foreground and background is enhanced by mapping the intensities to higher levels. Figure 3(a) shows the grayscale image and Fig. 3(b) shows the contrast adjusted image. In order to highlight the edges in the fundus image, canny edge detection is used [14]. This is followed by detection and masking of outer circular border of the fundus image. Now thresholding is applied and the image is binarized.

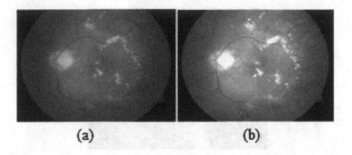

(a)   (b)

Fig. 3. (a) Grayscale image (b) Contrast adjusted image

3.2 Blood Vessel Removal

In order to highlight the affected portion of retina comprising of exudates it is necessary to detect the entire vasculature of blood vessels and eliminate it. The vessel path is detected and closed using morphological operations. Morphological opening and closing operations are applied by selecting a disc shaped structuring element in the following manner:

$$A(x,y) = (I \oplus S)(x,y) = maxI(x - x', y - y') + S(x',y') \tag{1}$$

$$B(x,y) = (I \ominus S)(x,y) = minI(x + x', y + y') - S(x',y') \tag{2}$$

$$In = A(x,y) - B(x,y) \tag{3}$$

where, I(x,y) is the image obtained after external border masking during pre-processing and $S(x',y')$ is the structuring element. These morphological opening and closing operations enhance the size of the exudates and eliminate smaller components like blood vessels. Next a ball shaped structuring element is defined and once again morphological closing operation is applied to eliminate blood vessels completely. The image obtained after removal of blood vessels is shown in Fig. 4.

3.3 Segmentation and Optic Disc Removal

Fundus images are characterized by variations in intensity as well as variations in foreground and background texture. Hence, segmentation based on local thresholding is proposed. Column wise neighborhood operation is performed on image In. The image is then transformed to binary and a predefined threshold is used to extract the lesions.

$$I(x,y) = 1 \ if \ p(x,y) > T \tag{4}$$

$$I(x,y) = 0 \ if \ p(x,y) < T \tag{5}$$

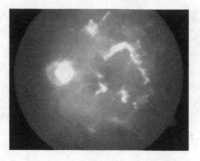

Fig. 4. Image after blood vessel removal

where, T = 0.45. In the grayscale image In, pixels with intensity p(x, y) that are in the range 0.45–1 is converted as white and the pixels that are in the range 0.45–0 is converted to black. Exudates and Optic disc are the brighter regions of the retina and hence represented as white pixels in the binarized image.

Optic disc is the brightest spot on the retina. It is a normal circular anatomical structure present near the left or right border of the retina. Since the exact location of the optic disc is not fixed, it is necessary to identify and eliminate it. To find the location of the optic disc, the largest connected component in the segmented image is computed [18] by retrieving the largest value of the columns of image. A circular mask is then computed having mesh grid vectors x and y and the medians of row and column coordinate vectors.

$$Im = \sqrt{(x - mc)^2 - (y - mr)^2} \tag{6}$$

This is followed by subtracting the mask from segmented image Ixy to completely eliminate optic disc from the fundus image.

$$Ind = I(x, y) - Im \tag{7}$$

Next the outer circular border is eliminated from Ind to obtain image Inc. Finally a morphological closing operation extracts the precise shape of exudates.

$$Iex = f(Inc) \tag{8}$$

Further, CLAHE is applied to adjust the contrast and distinguish between exudates and non exudates region. Φ(Iex) complements the image and transforms white pixels to black and vice versa. Ψ(Iex) is used for inversion highlighting all dark feature pixels. Next a 'AND' operation on preceding two images filters all the white pixels present in the connected region of the exudates.

$$Ie = AND(\Phi(Iex), \Psi(Iex)) \tag{9}$$

The segmented image after the optic disc elimination is then shown in Fig. 5.

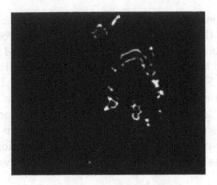

Fig. 5. Image after segmentation and optic disc removal

4 Feature Extraction and Classification

The proposed method explores Gray Level Co-occurrence Matrix (GLCM) [4] for classification of exudates. The GLCM and associated texture feature calculations enhance accurate detection of exudates in a retinal image as it depicts how frequently various compositions of gray levels co-exist in an image. The GLCM provides a value of variation in intensity at a particular pixel. In other words, texture feature provide insights into the texture of a pixel of interest. These features are particularly useful in applications that involve automatic extraction of features that distinguish a normal tissue from an abnormal tissue. The proposed technique uses the following feature set:

- Energy – It is the sum of squared elements in the GLCM.

$$Energy = \sum_{i,j=0}^{N-1} P(ij)^2 \tag{10}$$

- Contrast – The contrast attribute is a measure of local intensity variations in a GLCM matrix.

$$Contrast = \sum_{i,j=0}^{N-1} (i-j)^2 \tag{11}$$

- Homogeneity – It is a measure of uniformity in gray levels of an image.

$$Homogeneity = \sum_{i,j=0}^{N-1} Pij/1 + (i-j)^2 \tag{12}$$

- Correlation – Given a pair of pixels, this feature measures the probability of occurrence for the joint pixel pairs.

$$Correlation = \sum_{i,j=0}^{N-1} P(ij)(i-\mu)(j-\mu)/\sigma^2 \tag{13}$$

where, Pij is the element in the GLCM matrix, N is the number of gray levels in the image and μ is the GLCM mean. In this study, three different supervised learning classifiers, namely, Support Vector Machine (SVM), k-Nearest Neighbor (k-NN) and Decision tree are used to analyze the digital fundus images for presence of exudates. SVM is a supervised leaning algorithm that is commonly used for classification and regression analysis. It builds a model that classifies the images as either healthy or diseased. Another important non-parametric method that can be used for classification fundus images is k-nearest neighbour. Decision tree classification is based on tree representation wherein every leaf node represents a class label and the internal nodes represent attributes.

5   Experimental Results and Discussion

For the purpose of experimentation, digital fundus images were obtained from publicly available database DIARETDB0, DIARETDB1, e-Ophtha EX and Messidor which are treated as standardized databases and many researchers have worked on it. In all, 114 images from DIARETDB0 database, 89 from DIARETDB1, 82 from e-Ophtha EX and 74 from Messidor database are tested in this experiment. The proposed algorithm is coded in MATLAB R2015b. The Fig. 6(a) shows sample original digital fundus images; a(i) DIARETDB0, a(ii) DIARETDB1, a(iii) e-Ophtha EX and a(iv) Messidor databases. The segmented images are shown in Fig. 6(b). The classification of images from DIARETDB0, DIARETDB1, e-Ophtha EX, and Messidor databases based on the clinical test parameters using proposed algorithm with Support Vector Machine, k-NN and Decision tree classifiers are shown in the Table 1.

Table 1. Clinical test parameters for DIARETDB0, DIARETDB1, e-Ophtha EX and Messidor databases.

Database name	Digital fundus images			SVM classifier				k-NN classifier				Decision tree classifier			
	Diseased	Healthy	Total	TP	FP	FN	TN	TP	FP	FN	TN	TP	FP	FN	TN
DIARETDB0	110	04	114	110	00	04	00	109	01	04	00	110	00	00	04
DIARETDB1	85	04	89	85	00	04	00	85	00	03	01	83	02	00	04
e-Ophtha EX	68	14	82	68	00	14	00	67	01	12	02	66	02	06	08
Messidor	68	06	74	68	00	06	00	68	00	06	00	68	00	05	01

The proposed algorithm is evaluated based on the following performance evaluation measures [16]:

– Sensitivity (Se): It ratio of True Positives (TP) correctly identified from a given set of images.

$$Se = TP/(TP + FN) \qquad (14)$$

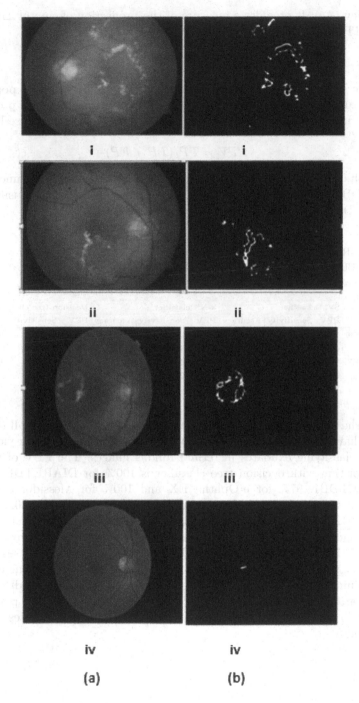

Fig. 6. (a) Sample original fundus images; a(i) DIARETDB0; a(ii) DIARETDB1; a(iii) e-Ophtha EX; a(iv) Messidor database (b) Segmented images of (a)

- Accuracy (Ac): It provides a measure of the proportion of sum of True Positives (TP) and True Negatives (TN) to the total population.

$$Ac = TN + TP/(TP + FP + TN + FN) \qquad (15)$$

- Positive Predictive Value (PPV): It indicates the percentage of population that actually shows the symptoms of the disease. It measures the proportion of True Positives (TP) to the sum of True Positives (TP) and False Positives (FP).

$$PPV = TP/(TP + FP) \qquad (16)$$

The results of clinical test statistics based on the test parameters on DIARETDB0, DIARETDB1, e-Ophtha EX and Messidor databases using SVM, k-NN and Decision tree classifiers are given in the Table 2.

Table 2. Results of clinical test statistics based on the test parameters on DIARETDB0, DIARETDB1, e-Ophtha EX, Messidor databases using SVM, k-NN and decision tree classifiers.

Database	SVM classifier			k-NN clasiifier			Decision tree classifier		
	PPV	Sensitivity	Accuracy	PPV	Sensitivity	Accuracy	PPV	Sensitivity	Accuracy
DIARETDB0	100%	96.4%	96.4%	99%	99%	95.6%	100%	100%	100%
DIARETDB1	100%	95.5%	95.5%	100%	96.5%	96.6%	97.6%	100%	97.6%
e-Ophtha EX	100%	82.9%	82.9%	98.5%	85.8%	84.1%	97%	91.6%	94.5%
Messidor	100%	91.8%	91.8%	100%	91.8%	91.8%	100%	94.5%	94.4%

From the above Table 2, it can be clearly observed that though all the three classifiers have yielded encouraging results, the Decision tree classifier yields best results of classifying exudates in retinal fundus images. The PPV of the proposed algorithm with decision tree classifier is 100% for DIARETDB0, 97.6% for DIARETDB1, 97% for e-Ophtha EX and 100% for Messidor databases. The sensitivity of the proposed algorithm is 100% for DIARETDB0, 100% for DIARETDB1, 91.6% for e-Ophtha EX and 94.5% for Messidor databases. The proposed algorithm also exhibits 100%, 94.5%, 94.5% and 97.6%, accuracy values for DIARETDB0, e-Ophtha EX, Messidor and DIARETDB1databases, respectively. Thus, we propose the use of Decision tree classifier for detecting exudates and examining the presence of Diabetic Retinopathy in digital fundus images. The comparative results of clinical test statistics based on the test parameters on DIARETDB0, DIARETDB1, e-Ophtha EX and Messidor databases are given in the Table 3.

Table 3. Comparative results of clinical test statistics based on the test parameters on DIARETDB0, DIARETDB1, e-Ophtha EX and Messidor databases.

Database	Authors	PPV	Sensitivity	Accuracy
DIARETDB0	Omar et al. [4]		98.68%	
	Garcia et al. [2]	85.7%	95.9%	
	Lin and Bing-Kun [8]	87.5%	84.8%	
	Akram et al. [5]	97.5%	93.7%	
	Proposed method	100%	96.4%	96.4%
DIARETDB1	Shilpa and Nagbhushan [18]	89.13%	100%	
	Akyol et al. [3]	88.46%	93.27%	
	Chen et al. [1]	90%	94%	
	Proposed method	100%	95.5%	95.5%
e-OphthaEX	Zhang et al. [13]	0.95(AUC)		
	Liu et al. [9]		76%	75%
	Somkuwar et al. [12]			96.08%
	Proposed method	100%	82.9%	82.9%
Messidor	Prakash et al. [6]		96.1%	
	Rekhi et al. [10]		76%	90%
	Gupta and Karandikar [11]		87%	88%
	Proposed method	100%	91.8%	91.8%

6 Conclusion

This paper consolidates a multistage approach for exudates detection that includes morphological operations for blood vessel removal, segmentation, optic disk removal and exudates detection. In this experiment the GLCM features are extracted. This experiment also explores the use of SVM, k-NN and Decision tree classifiers to distinguish between diseased and healthy retinal images. It is observed from experimentation that the Decision tree classifier yields best results of classifying exudates in digital fundus images. The PPV of the proposed algorithm with decision tree classifier is 100% for DIARETDB0, 97.6% for DIARETDB1, 97% for e-Ophtha EX and 100% for Messidor databases. The sensitivity of the proposed algorithm is 100% for DIARETDB0, 100% for DIARETDB1, 91.6% for e-Ophtha EX and 94.5% for Messidor databases. The proposed algorithm also exhibits 100%, 97.6%, 94.5% and 94.5% accuracy values for DIARETDB0, DIARETDB1, e-Ophtha EX and Messidor databases, respectively. The promising values of performance evaluation measures indicate the robustness of the proposed algorithm.

Acknowledgement. The authors are indebted to Dr. Uttara Deshpande, Ophthalmologist, Lions NAB Hospital Miraj, Maharashtra for providing valuable insights in the area of fundus image analysis and visualization. The authors are also grateful to the developers of DIARETDB0, DIARETDB1, e-Ophtha EX and Messidor databases.

References

1. Chen, X., Bu, W., Wu, X., Dai, B., Teng, Y.: A novel method for automatic hard exudates detection in color retinal images. In: Proceedings of International Conference on Machine Learning and Cybernetics, ICMLC 2011, China, pp. 1175–1181 (2011)
2. Garcia, M., Valverde, C., Lopez, M.I., Poza, J., Hornero, R.: Comparison of logistic regression and neural network classifiers in the detection of hard exudates in retinal images. In: 2013 Proceedings of International Conference on Engineering in Medicine and Biology Society Japan, pp. 5891–5894 (2013)
3. Akyol, K., Sen, B., Bayir, S., Cakmak, H.B.: Assessing the importance of features for detection of hard exudates in retinal images. Turk. J. Electr. Eng. Comput. Sci. **25**(2), 1223–1227 (2017)
4. Omar, M., Khelifi, F., Tahir., M.A.: Detection and classification of retinal fundus images exudates using region based multiscale LBP texture approach. In: 2016 Proceedings of IEEE International Conference on Control, Decision and Information Technologies, pp. 227–232 (2016)
5. Akram, M.U., Tariq, A., Anjum, M.A., Javed, M.Y.: Automated detection of exudates in colored retinal images for diagnosis of diabetic retinopathy. Appl. Opt. **51**(20), 4858–4866 (2012)
6. Prakasha, N.B., Selvathi, D.: An efficient approach for detecting exudates in diabetic retinopathy images. Biomed. Res. Spec. Issue, S414–S418 (2016)
7. Mukherjee, N., Dutta, H.S.: A new approach for color distorted region removal in diabetic retinopathy detection. In: Gupta, S., Bag, S., Ganguly, K., Sarkar, I., Biswas, P. (eds.) Advancements of Medical Electronics. LNB, pp. 85–97. Springer, New Delhi (2015). https://doi.org/10.1007/978-81-322-2256-9_9
8. Lin, P., Bing-Kun, Z.: An effective approach to detect hard exudates in color retinal image. In: Qian, Z., Cao, L., Su, W., Wang, T., Yang, H. (eds.) Recent Advances in Computer Science and Information Engineering. LNEE, vol. 124, pp. 541–546. Springer, Heidelberg (2012). https://doi.org/10.1007/978-3-642-25781-0_80
9. Liu, Q., et al.: A location-to-segmentation strategy for automatic exudates segmentation in colour retinal fundus images. Comput. Med. Imaging Graph. 78–86 (2017)
10. Rekhi, R., Issac, A., Dutta, M., Travieso, C.M.: Automated classification of exudates from digital fundus images. In: Proceedings of International Conference and Workshop on Bioinspired Intelligence (2017)
11. Gupta, S., Karandikar, A.M.: Diagnosis of diabetic retinopathy using machine learning. Int. J. Res. Dev. **3**(2), 2–6 (2015)
12. Somkuwar, A.C., Patil, T.G., Patankar, S.S., Kulkarni, J.V.: Intensity features based classification of hard exudates in retinal images. In: 2015 Proceedings of Annual IEEE India Conference, pp. 1–5 (2015)
13. Zhang, X., et al.: Exudate detection in color retinal images for mass screening of diabetic retinopathy. Med. Image Anal. **18**(7), 1026–1043 (2014)
14. Gonzalez, R.C., Woods, R.E.: Digital Image Processing, 3rd edn. Prentice Hall, Upper Saddle River (2014)
15. DIARETDB0. http://www.it.lut.fi/project/imageret/diaretdb0. Accessed 4 Oct 2018

16. Wikipedia. https://en.wikipedia.org/wiki/Sensitivity_and_specificity. Accessed Oct 2018
17. Bannigidad, P., Deshpande, A.: A hybrid approach for digital fundus images using image enhancement techniques. Int. J. Comput. Eng. Appl. **XII**(I), 122–131 (2017)
18. Shilpa, B.V., Nagabhushan, T.N.: An ensemble approach to detect exudates in digital fundus images. In: Proceedings of IEEE Second International Conference on Cognitive Computing and Information Processing (2016)
19. Decencière, E., et al.: Machine learning and image processing methods for teleophthalmology. IRBM **34**(2), 196–203 (2013)
20. DIARETDB1. http://www.it.lut.fi/project/imageret/diaretdb1/. Accessed 4 Oct 2018
21. Decencière, E., et al.: Feedback on a publicly distributed image database: the Messidor database. Image Anal. Stereol. **33**(3), 231–234 (2014)
22. Bannigidad, P., Deshpande, A.: A multistage approach for exudates detection in fundus images using texture features with k-NN classifier. Int. J. Adv. Res. Comput. Sci. **9**(1), 1–5 (2018)
23. Santosh, K.C., Wendling, L., Antani, S., Thoma, G.R.: Overlaid arrow detection for labeling regions of interest in biomedical images. IEEE Intell. Syst. **31**(3), 66–75 (2016)
24. Zohora, F.T., Santosh, K.C.: Circular foreign object detection in chest X-ray images. In: Santosh, K., Hangarge, M., Bevilacqua, V., Negi, A. (eds.) RTIP2R 2016. CCIS, vol. 709, pp. 391–401. Springer, Singapore (2017). https://doi.org/10.1007/978-981-10-4859-3_35
25. Zohora, F.T., Santosh, K.C.: Foreign circuar element detection in chest X-rays for effective automated pulmonary abnormality screening. Int. J. Comput. Vis. Process. **7**(2), 36–49 (2017)
26. Zohora, F.T., Antani, S., Santosh, K.C.: Circle like foreign element detection in chest X-ray using normalized cross-correlation and unsupervised clustering. In: Procceedings of SPIE 10574, Medical Imaging (2018)

WT and PDE Approach for Forest Species Recognition in Macroscopic Images

Rohini A. Bhusnurmath[1]($\boxtimes$) and P. S. Hiremath[2]

[1] Department of Computer Science, Akkamahadevi Women's University,
Vijayapura 586105, Karnataka, India
rohiniabmath@gmail.com
[2] Department of Computer Science (MCA), KLE Technological University,
BVBCET Campus, Hubli 580031, Karnataka, India
hiremathps53@yahoo.com

Abstract. The paper addresses the industrial problem of automatic recognition of forest species using wavelet transform (WT) and partial differential equation (PDE) approach. Presently, the task of identification is done manually by human experts, which is a tedious and time consuming process. Hence, there is a need to develop the automatic classification of forest species using computer vision techniques. This work is extremely difficult since forest species are richly comparable in visual look. The proposed approach would be useful in automatic classification of forest species and will save the illegal trading of endangered species. The proposed method obtains directional information using wavelet transform. Anisotropic diffusion, a PDE based filtering, is applied on this directional information to obtain textural component. Discriminant features are obtained by applying linear discriminant analysis (LDA) to the statistical features computed from textural component. Classification is performed using k-NN. The experimentation has been done using benchmark dataset of forest species. The proposed approach exhibits classification accuracy of 98.37% with reduced features.

Keywords: Image processing · Pattern recognition ·
Wavelet transform · Forest species · Discriminant features ·
Partial differential equation · Anisotropic diffusion

1 Introduction

Automatic classification of industrial products has driven a great attention due to its emergent need. Computer vision algorithms can solve this problem very efficiently. The efforts of human experts in the industry employed for this purpose are subjective and often prone to errors. The industry focuses on ways to prevent frauds at the cost of huge capital. There are chances that wood traders are mixing a cheaper species with noble one. Identification of wood timber and log is

© Springer Nature Singapore Pte Ltd. 2019
K. C. Santosh and R. S. Hegadi (Eds.): RTIP2R 2018, CCIS 1036, pp. 258–269, 2019.
https://doi.org/10.1007/978-981-13-9184-2_23

a tough task, since it is devoid of its leaves, fruits and flowers when brought outside the forest [8,9]. The thrust areas like wood identification and the problem of forest species recognition need to be addressed using computer vision expert system [1–10] for the sake of precision. Cavalin et al. [1] introduced a method for the reduction of testing cost with comparable classification accuracy using adaptive multi-level framework. Reduction in resolution is done for cost reduction at feature extraction level. Hafemann et al. [10] used convolution neural networks for texture classification and obtained interesting results on macroscopic dataset. It was observed that models can detect edges and color based features and holes in wood. Forest species recognition using gray level co-occurrence matrix (GLCM) and neural network classifier is implemented in Tou et al. [2–4] and Khalid et al. [5]. Filho et al. [8] obtained better recognition rate by using divide and conquer method. Filho et al. [9] extracted fifteen texture descriptors and also considered combinations of these descriptors for classification purpose. It was observed that the combination of descriptors and classifiers improved the accuracy. The intra class variability is achieved as compared to the single classifier and completed local binary pattern (CLBP) descriptors. Many other techniques for forest species recognition have been adopted in the literature that include Gabor filters [6], local binary patterns [7], local phase quantization [9].

In [11], the robustness of multiresolution local directional binary pattern approach is demonstrated on the four different benchmark datasets and this method is applied on the industrial application of engineered wood recognition which exhibited the improved recognition rate with reduced computational time. In [12,19], a robust approach based on wavelet transform and partial different equation for anisotropic diffusion (AD) is proposed. The approach is tested on three benchmark datasets and it is demonstrated that better classification accuracy with reduced computational cost is achieved. The recognition system based on color texture for industrial application is implemented in [13]. Further, industrial applications using anisotropic diffusion are demonstrated in [14]. Vajda and Santosh [20] have proposed a fast k-NN classifier. Use of radon transform for texture classification is attempted by Santosh and Lamiroy [21] using dynamic time warping. Many statistical features for texture classification have been used in the literature [22–30] and tested with different classifiers.

The task of forest species recognition in a macroscopic image of forest area is a challenging one, since the wood textures are similar in nature. This paper proposes a method based on wavelet transform (WT) and partial differential equation (PDE) of anisotropic diffusion for texture classification problem with reference to forest species recognition.

The contributions of the paper are: first, to investigate the applicability of the WT and PDE based algorithm on a industrial application, specifically forest species recognition; second, to investigate the effect of first and second order texture descriptors on forest species identification; third, to investigate the effectiveness of texture descriptors considered in combinations; and, fourth, to enhance the classification accuracy.

2   Proposed Methodology

The methodology extracts the features from the macroscopic images of forest species as follows:

- Apply Haar wavelet to the texture image of forest species to yield directional information.
- AD is applied on directional information to obtain the texture approximation.
- Texture approximation is utilized to obtain statistical features. Then, discriminant features are obtained by applying linear discriminant analysis (LDA).
- Classification is done using k-NN.

The steps are briefed below.

2.1   Wavelet Transform

The implementation and theoretical details of wavelet based algorithms are discussed in [15,16]. Time dependent signal analysis and two dimensional image analysis at different resolutions is facilitated by wavelet transform.

2.2   Partial Differential Equation for Anisotropic Diffusion

It is a non linear partial differential equation model for anisotropic diffusion (AD) process. AD is a powerful tool in image enhancement and edge detection. The more description of AD is given in the Appendix A1.

2.3   Features Extracted

The different feature sets of first order statistics (FV1) and second order statistics (FV2–FV9), that are used for texture representation, are tabulated in the Table 1. The details of the features, that are obtained for experimentation, are presented in the Appendix A2.

Besides the individual features FV1 to FV9, different combinations are also considered for the experimentation.

Table 1. The first (FV1) and second (FV2–FV9) order statistics used for experimentation [19].

Feature set	No. of features	Feature set	No. of features
FV1	5	FV6 [26,27]	6
FV2 [23]	7 features × 4 angles = 28 number	FV7 [28]	4
FV3 [22,30]	5	FV8 [29]	2
FV4 [24]	5	FV9	5
FV5 [25]	4		

Table 2. Different combinations of feature sets.

Name of feature set	Combination of feature set	Name of feature set	Combination of feature set
FV10	FV1+FV3	FV21	FV1+FV3+FV5
FV11	FV1+FV3+FV4	FV22	FV1+FV3+FV6
FV12	FV1+FV3+FV4+FV5	FV23	FV1+FV4+FV5
FV13	FV1+FV3+FV4+FV5+FV6	FV24	FV3+FV4+FV5
FV14	FV1+FV3+FV4+FV5+FV6+FV7	FV25	FV6+FV7
FV15	FV1+FV3+FV4+FV5+FV6+FV7+FV8	FV26	FV4+FV5
FV16	FV1+FV3+FV4+FV5+FV6+FV7+FV8+FV9	FV27	FV3+FV4
FV17	FV1+FV3+FV5+FV6	FV28	FV5+FV6
FV18	FV3+FV4+FV5+FV6	FV29	FV8+FV9
FV19	FV1+FV4+FV5+FV6	FV30	FV2+FV4
FV20	FV6+FV7+FV8+FV9		

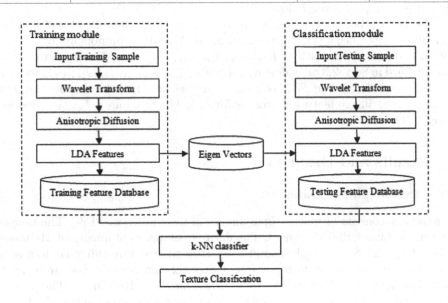

Fig. 1. The outline of proposed WT and PDE method

2.4   Texture Classification

An important characteristic of texture classification is to develop texture features that discriminates arbitrary textures with a higher degree of accuracy. The outline of proposed texture classification approach is depicted in the Fig. 1.

Feature Extraction Procedure: The procedure for feature extraction is given below:

Read the input sample image I. The 1-level Haar WT is applied to I. The directional components, namely, diagonal (D), horizontal (H), vertical (V) are obtained. D, H and V components are used to obtain texture approximations I_{Dtxr}, I_{Htxr}, I_{Vtxr}, using AD. Features listed in Tables 1 and 2 are computed using texture components and are stored in feature vector F. The F is stored in the feature database, with class label. The above procedure is repeated for the entire input samples and set of training feature (TF) is obtained. Discriminant feature set (TFLDA) is obtained by subjecting TF to LDA. TFLDA is then used for texture classification.

The feature extraction procedure is executed up to different numbers of diffusion steps (t) and is followed by extraction of various features as discussed in the Sect. 2.3.

Testing Procedure: Texture classification is done using k-NN [18]. The strategy of two-fold experimentation carried out five times for random sets. The testing procedure is given below:

The 1-level Haar WT is applied to test sample I_{test}. The D, H and V components are obtained. AD is performed on H, V and D components of I_{test} to obtain texture approximations $I_{testDtxr}$, $I_{testHtxr}$, $I_{testVtxr}$. The statistical features defined in Sect. 2.3 are calculated using $I_{testDtxr}$, $I_{testHtxr}$, $I_{testVtxr}$ to form feature vector F_{test}. The F_{test} is used to project on TFLDA components. The $F_{testLDA}$ (test image feature vector) is formed. The test image I_{test} is classified using k-NN classifier.

3   Results and Discussion

3.1   Texture Image Dataset

The dataset consists of macroscopic images of the timber wood [9]. The images are acquired using digital camera. The dataset comprises of images of 41 classes of size 3264×2448 pixels. The samples per class change from thirty-six to ninety nine. For the experimentation purpose, thirty six samples per class are considered. The detailed procedure to capture the images is given in [9]. The sample texture images of forest species used for experimentation are shown in the Fig. 2.

The imperative properties of the dataset, that is considered herein, are recorded in the Table 3. The non-overlapping training and testing sets are randomly divided to ensure unbiased texture classification. For training, half of the sub images are considered and the remaining half of sub images are taken for testing.

3.2   Results and Discussion

The experimentation is performed on Intel(R) Core(TM) i3-2330M @ 2.20 GHz having 4 GB RAM, using MATLAB7.9 software. The empirical values of anisotropic diffusion parameters are: k (conduction coefficient) = 60 and λ (lambda) = 0.25 (Appendix A1).

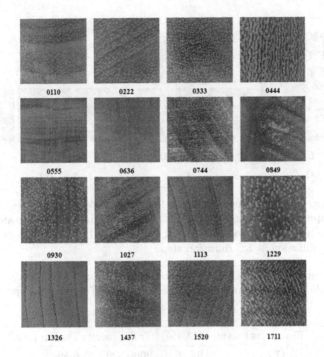

Fig. 2. Sample macroscopic images from the forest species dataset.

Table 3. Essential dataset properties.

Dataset name	Number of classes	Images per class	No. of sub images for each class	Total sub images	Image size (pixels)	Sub image size (pixels)	Image format
Forest species dataset [9]	41	36	576	23616	512×512	128×128	jpg

The Fig. 3 shows an input sample image of forest species dataset and the resultant images obtained after applying WT and PDE of anisotropic diffusion.

Feature sets exhibiting classification accuracy (CA) greater than 90% are listed in the Table 4. The corresponding training time (TrTm), testing time (TstTm) and optimal diffusion step (DS) are also presented in the Table 4.

From the Table 4, it is observed that these ten feature sets appear in combinations perform better than feature sets taken separately, but not in combination.

The summary of classification accuracy (%) attained by the proposed method and that by other methods on macroscopic forest species database are tabulated in the Table 5.

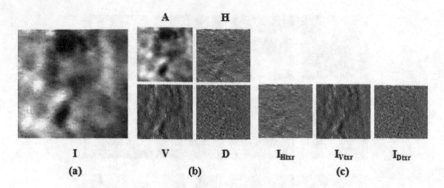

Fig. 3. A macroscopic forest species image I of size 128 × 128 and its resultant images: (a) original input texture image I (0110), (b) 1 - level Haar wavelet transform of the image I, (c) texture approximation I_{Htxr}, I_{Vtxr}, I_{Dtxr} of H, V, D components obtained by anisotropic diffusion at 8th diffusion step.

Table 4. Feature sets exhibiting CA greater than 90% on forest species dataset and the corresponding training time (TrTm), testing time (TstTm) and optimal diffusion step (DS).

Sl. No.	Feature set name	CA (%)	TrTm (sec.)	TstTm (sec.)	DS
1	FV17	94.76	2069.529	0.592	10
2	FV18	98.37	2251.867	0.517	8
3	FV19	95.24	2096.887	0.592	9
5	FV22	94.63	1326.887	0.492	7
6	FV11	91.59	1647.920	0.441	10
7	FV12	94.21	1702.164	0.534	5
8	FV13	95.06	1538.903	0.725	3
9	FV14	93.54	1741.837	0.845	5
10	FV15	91.46	2121.854	0.915	9

From the Table 5, it is observed that a direct comparison of the results is not possible as different datasets are used in every approach. However, it depicts the snapshot of research work in this area. It is observed that proposed method performs better than the methods dedicated to field of industrial application. Further, it is noted that by virtue of the WT and anisotropic diffusion based approach, there is considerable saving in computational time cost involved in feature extraction and classification as compared to that in other methods. It has been demonstrated in [12] that the proposed approach yielded better classification results at reduced computational cost when experimented on three different benchmark datasets.

Table 5. Summary of classification accuracy (CA) attained by proposed method and methods on macroscopic forest species database.

Method	CA (%)	No. of classes	Images/class	Features/approach	Classifier
Tou et al. [2]	72.0	5	360/5	GLCM	Neural network
Tou et al. [3]	72.8	5	360/5	GLCM, 1DGLCM	Neural network, K-NN
Khalid et al. [5]	95.0	20	1949/20	GLCM	Neural network
Tou et al. [4]	85.0	6	600/6	GLCM, Gabor	k-NN
Yusof et al. [6]	90.3	30	3000/30	Gabor, GLCM	Neural network
Nasirzadeh et al. [7]	96.6	37	3700/37	LBPU2, LBPH, F	Neural network
Filho et al. [8]	80.8	22	1270/22	Color, GLCM	Neural network
Filho et al. [9]	97.77	41	1025/41	CLBP + chromatic features	Neural network
Hafemann et al. [10]	95.77	41	2942/41	Convolution NN	SVM
Cavalin et al. [1]	96.48	41	–	Adaptive multi-level framework	6 different classifiers
Proposed method	**98.37**	41	**23616/41**	**WT + PDE**	**k-NN**

4   Conclusions

The present work focuses on using macroscopic images for automatic recognition of forest species. The novelty is to extract features using wavelet transform and PDE based approach using k-NN classifier. The proposed method is experimented for different combination of feature sets and the optimal set of features to perform the classification task is obtained. The most dominant feature sets are FV1, FV3, FV4 and FV5. The experimental results demonstrate that the proposed PDE based method exhibit better classification accuracy of 98.37% on macroscopic forest species dataset at reduced computational cost.

Acknowledgments. The authors thank reviewers for suggestions and critical comments.

Appendix A1. Anisotropic Diffusion

A major drawback of the linear diffusion is its uniform filtering of local signal features and noise. This problem was addressed by Perona and Malik [17]. He proposed a nonlinear diffusion process, where diffusion can take place with a variable diffusion in order to control the smoothing effects. A discrete form of generalized diffusion equation [17,31] is given by the Eq. (A1):

$$I_s^{t+\Delta t} = I_s^t + \frac{\lambda}{|\overline{\eta_s}|} \sum_{p \in \overline{\eta_s}} g\left(\nabla I_{s,p}^t\right) \nabla I_{s,p}^t \tag{A1}$$

where I_s^t is the discretely sampled image, s denotes the pixel position in a discrete two-dimensional grid, and $0 \leq \lambda \leq 1/4$ is a scalar that controls the numerical stability, $\overline{\eta_s}$ is the number of pixels in the window (usually four, except at the image boundaries). Each new image in the family is computed by applying this equation to the previous image. Consequently, anisotropic diffusion is an iterative process where a relatively simple set of computations is used to compute each successive image in the family. The choice of the anisotropic diffusion function plays a significant role in preserving image edges. Generally, as proposed in [17], it is chosen as either

$$g(x) = \frac{1}{1 + (\frac{x}{k})^2} \tag{A2}$$

or

$$g(x) = exp\left\lfloor -(x/k)^2 \right\rfloor \tag{A3}$$

where g(x) is the gradient magnitude defined in the Eqs. (A2) and (A3), k is a threshold for the gradient magnitude and it determines the extent to which edges are to be preserved during the anisotropic diffusion process. The first equation favors wide regions over smaller ones, while the latter favors high contrast edges over low contrast ones.

Appendix A2. Features Extracted

The two types of texture feature measures are considered for feature extraction, namely, first and second order statistics [19]. The first order texture feature statistics are independent of neighboring pixel relationships. The second order statistics considers relationship between neighboring pixels. The details of the first order statistics (FV1) and second order statistics (FV2–FV9), that are used for experimentation, are given the Table 6.

Table 6. The details of first order (FV1) and second order statistics (FV2–FV9) considered for experimentation [19].

Feature set	Description	Features	No. of features
FV1	First order statistics	Median, mean, standard deviation, skewness and kurtosis	5
FV2	Haralick features [23]	Entropy, homogeneity, contrast, energy, maximum probability, cluster shade and cluster prominence	7 features × 4 angles = 28 number
FV3	Gray level difference statistics [22,30]	Contrast, homogeneity, energy, entropy and mean	5
FV4	Neighborhood gray tone difference matrix [24]	Busyness, complexity, coarseness, contrast and texture strength	5
FV5	Statistical feature matrix [25]	Coarseness, contrast, period and roughness	4
FV6	Law's texture energy measures [26,27]	Six texture energy measures	6
FV7	Fractal dimension texture analysis [28]	Roughness of a surface	4
FV8	Fourier power spectrum [29]	Radial sum and angular sum	2
FV9	Shape	Size (x,y), area, perimeter and $perimeter^2/area$	5

References

1. Cavalin, P.R., Kapp, M.N., Oliveira, L.S.: Multi-scale forest species recognition systems for reduced cost (2017). arXiv:1709.04056v1 [cs.CV]
2. Tou, J.Y., Lau, P.Y., Tay, Y.H.: Computer vision based wood recognition system. In: Proceedings of International Workshop on Advanced Image Technology, pp. 197–202 (2007)
3. Tou, J.Y., Tay, Y.H., Lau, P.Y.: One-dimensional grey-level co-occurrence matrices for texture classification. In: International Symposium on Information Technology, pp. 1–6 (2008)
4. Tou, J.Y., Tay, Y.H., Lau, P.Y.: A comparative study for texture classification techniques on wood recognition problem. In: Proceeding of the 5th International Conference on Natural Computation, pp. 8–12 (2009)
5. Khalid, M., Lee, E.L.Y., Yusof, R., Nadaraj, M.: Design of an intelligent wood species recognition system. IJSSST **9**(3), 9–19 (2008)

6. Yusof, R., Rosli, N.R., Khalid, M.: Using Gabor filters as image multiplier for tropical wood species recognition system. In: 12th International Conference on Computer Modelling and Simulation, pp. 284–289 (2010)
7. Nasirzadeh, M., Khazael, A.A., Khalid, M.B.: Woods recognition system based on local binary pattern. In: 2nd International Conference on Computational Intelligence, Communication Systems and Networks, pp. 308–313 (2010)
8. Paula Filho, P.L., Oliveira, L.S., Britto, A.S., Sabourin, R.: Forest species recognition using color-based features. In: Proceedings of the 20th International Conference on Pattern Recognition, pp. 4178–4181 (2010)
9. Paula Filho, P.L., Oliveira, L.S., Nisgoski, S., Britto, A.S.: Forest species recognition using macroscopic images. Mach. Vis. Appl. **25**(4), 1019–1031 (2014)
10. Hafemann, L.G., Oliveira, L.S., Cavalin, P.: Forest species recognition using deep convolutional neural networks. In: Proceedings of 22nd International Conference on Pattern Recognition (ICPR), pp. 1103–1107 (2014)
11. Hiremath, P.S., Bhusnurmath, R.A.: Multiresolution LDBP descriptors for texture classification using anisotropic diffusion with an application to wood texture analysis. Pattern Recogn. Lett. **89**, 8–17 (2017). https://doi.org/10.1016/j.patrec.2017.01.015
12. Hiremath, P.S., Bhusnurmath, R.A.: Texture classification using PDE approach and wavelet transform. Pattern Recogn. Image Anal. **27**(3), 473–479 (2017). https://doi.org/10.1134/S1054661817030154
13. Hiremath, P.S., Bhusnurmath, R.A.: Industrial applications of colour texture classification based on anisotropic diffusion. In: Santosh, K.C., Hangarge, M., Bevilacqua, V., Negi, A. (eds.) RTIP2R 2016. CCIS, vol. 709, pp. 293–304. Springer, Singapore (2017). https://doi.org/10.1007/978-981-10-4859-3_27
14. Hiremath, P.S., Bhusnurmath, R.A.: Performance analysis of anisotropic diffusion based colour texture descriptors in industrial applications. Int. J. Comput. Vis. Image Process. **7**(2), 50–63 (2017). https://doi.org/10.4018/IJCVIP.2017040104
15. Daubechies, I.: Ten Lectures on Wavelets. SIAM, Philadelphia (1992)
16. Mallat, S.G.: A theory for multiresolution signal decomposition: the wavelet representation. IEEE Trans. Pattern Anal. Mach. Intell. **11**, 674–693 (1989)
17. Perona, P., Malik, J.: Scale-space and edge detection using anisotropic diffusion. IEEE Trans. Pattern Anal. Mach. Intell. **12**(7), 629–639 (1990)
18. Duda, R.O., Hart, P.E., Stork, D.G.: Pattern Classification. Wiley, New York (2001)
19. Bhusnurmath, R.A., Hiremath, P.S.: LDA based discriminant features for texture classification using WT and PDE approach. Cogn. Comput. Inf. Process. **801**, 191–200 (2018). https://doi.org/10.1007/978-981-10-9059-2_18
20. Vajda, S., Santosh, K.C.: A fast k-nearest neighbor classifier using unsupervised clustering. In: Santosh, K.C., Hangarge, M., Bevilacqua, V., Negi, A. (eds.) RTIP2R 2016. CCIS, vol. 709, pp. 185–193. Springer, Singapore (2017). https://doi.org/10.1007/978-981-10-4859-3_17
21. Santosh, K.C., Lamiroy, B.: DTW-radon-based shape descriptor for pattern recognition. Int. J. Pattern Recogn. Artif. Intell. (IJPRAI) **27**(3), 30 (2013). https://doi.org/10.1142/S0218001413500080
22. Weszka, J.S., Dyer, C.R., Rosenfeld, A.: A comparative study of texture measures for terrain classification. IEEE Trans. Syst. Man Cybern. **6**(4), 269–285 (1976)
23. Haralick, R.M., Shanmugam, K., Dinstein, I.: Textural features for image classification. IEEE Trans. Syst. Man Cybern. **3**(6), 610–621 (1973)
24. Amadasun, M., King, R.: Texural features corresponding to texural properties. IEEE Trans. Syst. Man Cybern. **19**(5), 1264–1274 (1989)

25. Wu, C.M., Chen, Y.C.: Statistical feature matrix for texture analysis. CVGIP Graph. Models Image Process. **54**(5), 407–419 (1992)
26. Laws, K.I.: Rapid texture identification. In: SPIE, vol. 238, pp. 376–380 (1980)
27. Haralick, R.M., Shapiro, L.G.: Computer and Robot Vision, vol. 1. Addison-Wesley, Boston (1992)
28. Mandelbrot, B.B.: The Fractal Geometry of Nature. Freeman, San Francisco (1982)
29. Rosenfeld, A., Weszka, J.: Picture recognition. In: Fu, K. (ed.) Digital Pattern Recognition, pp. 135–166. Springer, Berlin (1980). https://doi.org/10.1007/978-3-642-67740-3
30. Aggarwal, N., Agrawal, R.K.: First and second order statistics features for classification of magnetic resonance brain images. J. Signal Inf. Process. **3**, 146–153 (2012). https://doi.org/10.4236/jsip.2012.32019
31. Tsiotsios, C., Petrou, M.: On the choice of the parameters for anisotropic diffusion in image processing. Pattern Recogn. **46**(5), 1369–1381 (2012). https://doi.org/10.1016/j.patcog.2012.11.012

Diabetes Detection Using Principal Component Analysis and Neural Networks

R. Haritha[1]([✉]), D. Sureshbabu[2], and P. Sammulal[3]

[1] Computer Science Engineering, Sree Chaitanya Institute of Technological Sciences, Karimnagar, India
haritha.ravula@gmail.com
[2] Computer Science and Engineering, Kakatiya Government College, Hanamkonda, India
sureshd123@gmail.com
[3] Computer Science and Engineering, Jawaharlal Nehru Technological University, Hyderabad, India
sammulalporika@gmail.com

Abstract. Data mining is a growing discipline in the medical field that aims to extract knowledge relevant large amounts of data. It uses tools from statistics, artificial intelligence, and optimization techniques, etc. This paper present the detection of diabetes on the basis of data taken form UCI repository (PIMA), with help of neural network and principal component analysis. Data training and testing perform according to k fold verification and NN based approach yields 99% of accuracy. Further PCA NN approach is proposed for dimension reduction techniques and it gives accuracy 98.7% marginally low from NN based approach.

Keywords: UCI · NN · PCA · PIMA

1 Introduction

1.1 Expert System

An expert system is a decision support tool capable of reproducing the cognitive mechanisms of an expert or of a group of experts. It consists of 3 parts; a fact base; a rule base and an inference engine. So, an expert system is a software able to answer questions by reasoning from facts and known rules [1]. The fields of interrogation to be filled by the user can be more or less numerous and the concepts displayed in these fields will be treated by the search engine of the software by using the Boolean language "AND", "OR", "NOT".

1.2 Diabetes Diagnostic Assistance

Types of diabetes: There are several types of diabetes in the world. The most famous are:

© Springer Nature Singapore Pte Ltd. 2019
K. C. Santosh and R. S. Hegadi (Eds.): RTIP2R 2018, CCIS 1036, pp. 270–285, 2019.
https://doi.org/10.1007/978-981-13-9184-2_24

(1) *Type 1 diabetes* (Known as "insulin-dependent diabetes") usually affects children and adults under 30/40 (young subjects). The onset of this type of diabetes is brutal. Affected individuals lose weight due to their diabetes and are treated for life with insulin therapy.

(2) *Type 2 diabetes* (Known as "non-insulin-dependent diabetes") is also called diabetes of maturity and in contrast to type 1 diabetes appears very insidiously in people generally older (typically >40 years) and overweight [2].

Type 2 diabetes is much more prevalent than type 1 diabetes. It accounts for almost 90% of diabetes worldwide. Reports of type 2 diabetes in children - formerly rare - are on the rise. In some countries, diabetes accounts for almost half of new cases diagnosed in children and adolescents [2].

1.3 Classification Systems

These are systems based on classification methods. These classification methods are intended to identify the classes to which objects belong from certain descriptive features. These types of systems apply to a large number of human activities and are particularly suited to the problem of automated decision-making [3, 4]. The use of computers for the realization of this classification becomes more and more frequent. Even though the expert's decision is the most important factor in the diagnosis, classification systems provide substantial assistance as they reduce errors due to fatigue and the time required for diagnosis. As part of the classification, Artificial Neural Networks (ANN) have been widely proven in the scientific and industrial community [5–12].

1.4 Problem Located in Classification Systems

The presence of redundant attributes or highly noisy attributes in the databases, performance of the system may decline. This extreme situation risks leading to a classification without any real interest for the user [13]. This requires the use of variable selection techniques which aim to select or extract an optimal subset of the most relevant characteristics or parameters to make a better learning and ensure a good performance of the classification system.

2 Proposed Methodology

2.1 System Model

Diabetes is increasing internationally and is fourth among chronic diseases in our country and around the world. Its causes are complex but are largely due to rapid increases in the incidence of overweight, obesity and sedentary lifestyle. Although diabetes has serious consequences for the human body, a large proportion of its cases and complications could be prevented by good glycaemic control and early diagnosis. This requires the use of a diagnostic support system to facilitate decision-making and minimize uncertainty about the current or future state of the patient (Fig. 1)

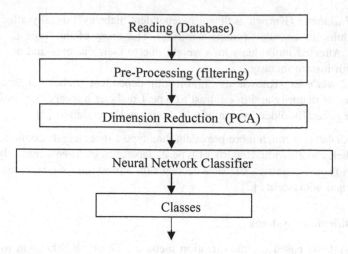

Fig. 1. Generalized architecture of diabetic detection using neural network

2.2 Dimension Reduction

The PCA is a method of exploratory data analysis: from a set of N observations characterized by m initial quantitative variables, we try to condense the representation of the data while conserving at best their global organization. For this, we represent the data on k new variables (the principal components, k < m) obtained as linear combinations of the initial variables, keeping as much variance as possible.

The purpose of principal component analysis (PCA) is to transform an X matrix of p variables in another Y matrix of wrong virtual variables ordered from highest to lowest variance.

Let n be patterns of dimension p, that is [14]:

$$X^* = \left[x_{ij}\right]_{i=1,p}^{j=1,n} \tag{1}$$

We can also represent the information "centered" on its average, and obtain its average and variance:

$$X^* = \left[x_{ij} - \bar{x}_i\right]_{i=1,p}^{j=1,n} \tag{2}$$

$$\bar{x} = \frac{1}{n}\sum\nolimits_{j=1}^{n} x_{ij} \tag{3}$$

$$\sigma = \sqrt{\frac{1}{n-1}\sum\nolimits_{j=1}^{n} \left(x_{ij} - \bar{x}_i\right)^2} \tag{4}$$

$$Z = \left[\frac{x_{ij} - \bar{x}_i}{\sigma_i}\right]_{i=1,p}^{j=1,n} \tag{5}$$

The covariance matrix is defined, as:

$$S = \frac{XX^T}{n-1} = (S_{ij})_{i=1,p}^{j=1,p} \tag{6}$$

Fulfilling the matrix S the following properties:

- S is a symmetric matrix.
- S is defined as NO negative (non-negative eigen values).
- The trace of S is equal to the Inertia of the points with respect to the origin.

The first factorial axis U_1 relative to the points under study is when this axis maximizes the inertia explained.

$$F_{kj} = Z_j^t U_k \tag{7}$$

$$F_k = \{F_{kj}\}^{j=1,n} = Z\bar{U}_k \tag{8}$$

Thus the selection of factorial axes is carried out in order of "relevance" in terms of contribution of information (eigenvalues), so each axis that is determined must contribute less and less information.

2.3   Classification

Neural networks have been widely used in the field of classification because of the simplicity of their reasoning and their learning performance inspired by human reasoning. To make a good learning, the model of the neural network will be chosen in an experimental way. It depends on the number of hidden layers, neurons in each layer, number of inputs and outputs. To have good results, the chosen model can have a very complex architecture. It possible to select the set of most relevant variables to make the right learning and at the same time to choose the best model of the neural network while ensuring its simplicity.

Artificial Neural Networks
A network of artificial neurons is a model of calculation whose design is very schematically inspired by the functioning of biological neurons. The neural networks are generally optimized by probabilistic learning methods. They are placed on the one hand in the family of statistical applications, and on the other hand in the family of artificial intelligence methods to which they provide a perceptual mechanism independent of the programmer's own ideas.

The Neuron: The Perceptron of Rosenblatt [15] transposes the behaviour of neurons into an equation integrating the great principles observed in nature. From a stimulus represented by an observation vector $x \in R_n$, each component x_i, Neural networks and variable selection $i \in [1, n]$ is multiplied by a connection weight w_i. The sum of these weighted inputs is then made by adding a bias θ (or activation threshold). For reasons of convenience of notation, this bias is transformed into a new output neuron 1 with a weight link w_0 such that $\theta = w_0$. Finally, the sum (or activation state) is passed in a non-linear staircase function f (or threshold function). In the end the output of a neuron is written in a synthetic way (Eq. (9)):

$$y = f\left(\sum_{i=-w_i x_i}^{n}\right) \tag{9}$$

Perceptron alone can be seen as a discrimination function between two classes for a classification problem. It partitions the input space into two regions with a linear decision boundary. With well-studied weights, this linear surface can represent logical functions like AND, OR, and NOT. The Perceptron cannot simulate the exclusive OR (XOR) because in this case, the decision surface is nonlinear. Criticism of the case of XOR by [16] has also provoked a temporary but historic disaffection for the Perceptron. Although a computer neuron is not a perfect model of its biological version, it remains none the less close experimentally observed phenomena. Moreover, aggregated in network, the yet limited capacities of the artificial neuron produce very interesting results as well at the purely functional level as at the level of the modelization. Several algorithms have been proposed to determine w_i weights, starting with the experiment-based [15] method, the least squares technique [17] and finally the gradient-descent techniques as in the case of Multi-Layer Perceptron (MLP).

Computer units are no longer called Perceptron but more simply neurons or nodes. In addition to the layered topology, the main difference with the version of [15] is the use of differentiable and nonlinear activation functions such as the sigmoid, also called the logistic function.

The idea of such a topology is old and it took a number of years to see the appearance of algorithms to calculate the weight of such a network in particular because of the introduction of hidden layers. Proposed for the first time by [18] in 1994, the use of the retro propagation of the error gradient in multi-layered systems will again be brought to the forefront in 1986 by [19], and simultaneously, under a similar name, at [20] during his paper.

These networks are often fully connected, which means that each neuron in a layer i is connected to all the neurons in the $i + 1$ layer. On the other hand, in a classical scheme, the neurons of the same layer are never connected to each other.

MLPs are essentially used for two purposes: partitioning a shape space for classification problems and approximating functions. Unlike the Perceptron of [15], the MLP can represent any Boolean function with n variables, although some may require an exponential number of neurons in the hidden layers. Due to the non-linearity of the sigmoid as an activation function, the separation boundaries are better adapted to each class in the case of a classification problem. This property is also found in the case of the function approximation that produces continuous and smooth curves at a time.

MLPs have interesting mathematical properties. Many of them are valid for networks with only two hidden layers, which testifies to the potential power of MLPs. It should be noted that these properties are rarely constructive in the sense or that it is shown that a certain number of neurons is sufficient to perform a task, the property gives no information on the topology to choose in order to solve the problem. The majority of the properties are proven without the assumption of the use of the sigmoid, it is enough simply that the function of activation is bounded (raised and minuted), increasing and continuous.

The difficulty of using this network lies in the fact that it is necessary to determine its topology, it is a question of defining the number of neurons of the different layers as well as their interconnections.

If the number of hidden neurons is too small, the learning algorithm will not be able to construct an intermediate representation of the problem that is linearly separable and some of the examples will not be learned correctly. Conversely, if this number is too high, there is a risk of learning the problem by heart: the network perfectly recognizes the learning examples but will give poor results on new data that it did not see during the study.

Learning: The best-known approach for learning a MLP is the gradient descent technique. Indeed, the use of differentiable activation functions makes it possible to use this technique that is both simple to implement and above all very computationally efficient.

We will use in the rest of this section the following notations:

- P the number of shapes in the learning base.
- x_p, $p \in [1, P]$ the form $n°p$ of the learning base.
- L the number of network layers (including the input and output layer).
- N_l, $l \in [0, L-1]$ the number of neurons in layer $n°l$.
- $o_{l,j}$, the calculated output of neuron $n°j$ in layer $n°l$.
- $d_j(x_p)$ the component $n°j$ of the expected output for the form x_p.
- $w_{l,j,i}$ the weight of the connection between neuron $n°j$ in layer $l-1$ and neuron $n°j$ in layer l.
- f is the activation function.

The output of any neuron is given by (10):

$$o_{l,j} = f\left(\sum_{i=0}^{N_{l-1}} w_{l,j,i} o_{l-1,i}\right) \tag{10}$$

The cost function E to be minimized in the case of a learning is a measure of the error between the desired output for a shape and the output calculated by the network. The error on a form p is generally quantified by a quadratic error $E_p(w)$:

$$E_p(w) = \frac{1}{2}\sum_{q=1}^{N_L} \left(O_{L,q}(x_p) - d_q(x_p)\right)^2 \tag{11}$$

The error for the set of forms $E_p(w)$ is therefore:

$$E(w) = \sum_{p=1}^{P} E_p(w) \tag{12}$$

The problem boils down to:

$$\min E(w) \tag{13}$$

To solve this type of problem, a classic optimization technique resulting from operational research consists in determining by successive iterations the values of the parameter w. It consists of using an existing point w_0 and making it move in the direction of the anti-gradient. The new point obtained by the translation $w \rightarrow w + \mu x_p \sigma_p$ has a smaller value for the objective function. The parameter μ is a positive step called in this case no learning and σ is the gradient of the error. The translation operation is repeated until a satisfactory solution is obtained. By using the retro propagation of the gradient of the error, the summary of the progress of the method is given by (Algorithm 1).

Algorithm 1
Learning a MLP by gradient back propagation
1: Random initialization of network weights

2: repeat for each sample of the learning base do - Propagate the sample in the network

   - Calculates error on the output layer

   - Propagation of the error on the lower layers - Weight adjustment

   end
   Update the total error until Stop criterion

Although the error is minimized locally, the technique converges to a minimum and gives good practical results. In most cases, few problems due to local minima are encountered. However, it persists two problems that we encounter in a real application which are on the one hand the slowness of the convergence if μ is badly chosen and on the other hand the possible risk of converging towards a local and not global minimum of the surface error.

The main defect of this method is a relatively long convergence time which depends on different parameters such as the initialization at the instant $t = 0$ of the synaptic weights or the initial value of the parameter μ. Nevertheless, it gives good experimental results.

In an implementation of the error propagation retroactivity algorithm, it is also difficult to determine when the weight adjustment of the MLP should be completed. Several stop criteria are used: the iterations stop when the norm of the gradient is close to zero (the weights then vary only very little), or else as soon as the error at the exit is below a certain threshold.

The first criterion is more interesting mathematically because it corresponds to the stabilization of the solution in a minimum, the second is closer to real (interpretable) criteria of good correlation between calculated solution and expected solution. In the latter case, if the problem studied concerns a classification task, we can consider that the learning ends when all the forms are classified, which makes it possible to dispense with the determination of the error rate not to be used exceed.

In practice, we go from this last stopping criterion to a second which takes into account a maximum number of iterations not to be crossed. Indeed, it is not guaranteed that the network can classify all forms, even with an infinite number of iterations. The combination of the two conditions makes it possible to obtain a correct solution in a reasonable time.

3 Simulation Results

3.1 PIMA Database

The tests of the proposed method are carried out on the basis of Pima Indians Diabetes data [22]. The dataset was chosen from the UCI repository that conducts a study of 768 Pima Indian women (500 non-diabetic 268 Diabetics). These same women who stopped their migrations in Arizona (USA) adopting a Westernized way of life develop diabetes in almost 50% of cases. The diagnosis is a variable binary value "class" that allows to know if the patient shows signs of diabetes according to the criteria of the World Health Organization. The eight clinical descriptors are:

1. Npreg: number of pregnancies.
2. Glu: concentration of plasma glucose.
3. BP: diastolic blood pressure, (mmHg).
4. SKIN: triceps skin fold thickness, (mm).
5. Insulin: insulin dose, (mu U/ml).
6. BMI: body mass index, (weight in kg/(height in m2).
7. DPF: Diabetes pedigree function (heredity).
8. Age: age (Year).

3.2 Analysis of Database Data

Table 1 contains information on the parameters taken into consideration.

Table 1. Descriptor information in the database

Attribute name	Min/Max	Standard deviation	Segregated
Npreg	0/17	3.37	17
Glu	0/199	31.973	136
BP	0/122	19.356	47
Skin	0/99	15.952	51
Insu	0/846	115.244	186
Bmi	0/67.1	7.884	248
Ped	0.078/2.42	0.331	517
Age	21/81	11.76	52

3.3 Evaluation Criteria

Data classification performance was evaluated by calculating true positives (TPs), true negatives (TNs), false positives (FPs) and false negatives (FNs), percent sensitivity (S_e), specificity (S_p) and the classification rate (TC), their respective definitions are as follows:

- VP: diabetic classified diabetic.
- VN: non-diabetic classified non-diabetic.
- FP: non-diabetic classified diabetic.
- FN: diabetic classified as non-diabetic.

Sensitivity is the ability to give a positive result when the disease is present. It is calculated by:

$$S_e = \frac{VP}{VP + FN} \tag{14}$$

Specificity is the ability to give a negative result when the disease is absent. It is calculated by:

$$S_p = \frac{VN}{VN + FP} \tag{15}$$

Classification rate is the percentage of correctly classified examples. It is calculated by:

$$TC = \frac{VP + VN}{VN + VN + FP + FN} \tag{16}$$

3.4 Results and Interpretation

Experimentation 1

In the first experiment we used a multilayer perceptron with the topology [8: 6: 1] which has as input the eight parameters of the base PIMA: with a learning step = 0.5 (Fig. 2).

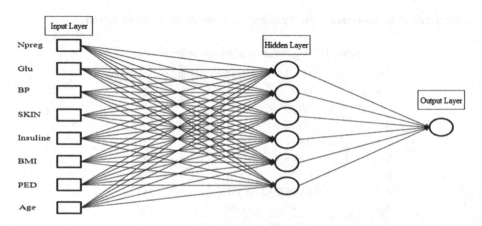

Fig. 2. Architecture used in experimentation

The results of this experiment show that the specificity of the system is very high which means that the system has made a good apprenticeship for the negative data. So when a patient is non-diabetic our model detects it very successfully.

On the other hand the sensitivity of the system is very weak which means that the system has made a bad recognition of the positive data. So many diabetic patients have been recognized as non-diabetic. This can generate a major risk for the health of the patient.

With these performances, we can say that the model gave an average classification rate and a good specificity. On the other hand it gave a weak sensitivity. What remains a disadvantage to study (Fig. 3).

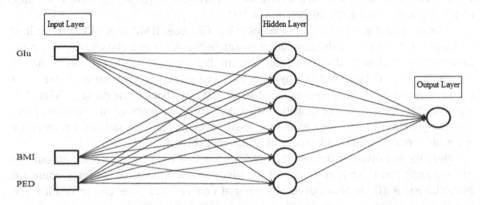

Fig. 3. Confusion matrix plot for NN based method

Here the confusion matrix of Neural network shows that out of 78 there are 77 outputs gives correct result while rest 1 gives wrong output, so accuracy count for Non-diabetic is 98.7%.

Now out of 229 cases of Diabetic 227 gives correct result while rest 2 gives wrong result, so accuracy count for Diabetic is 99.1%. So exact accuracy count drawn for both Diabetic and Non-Diabetic are 99.0%.

$$\text{False Positive Rate} = \frac{FP}{FP+TN} = \frac{1}{1+77} = 1.3\%$$

$$\text{Specificity} = \frac{TN}{FP+TN} = \frac{77}{1+77} = 98.7\%$$

$$\text{True Positive Rate} = \frac{TP}{TP+FN} = \frac{227}{227+2} = 99.1\%$$

$$\text{Miss} = \frac{\text{FN}}{\text{TP} + \text{FN}} = \frac{2}{227 + 2} = 0.9\%$$

$$\text{Rate Error} = \frac{\text{FN} + \text{FP}}{\text{TOTAL}} = \frac{2 + 1}{307} = 1.0\%$$

$$\text{Rate Accuracy} = \frac{\text{TN} + \text{TP}}{\text{TOTAL}} = \frac{77 + 227}{307} = 99.0\%$$

Experiment 2:

In order to improve the performance of the previous experiment, we applied the PCA method to the same architecture of Experiment 1 [8: 6: 1] with a learning step = 0.5 and a maximum of iteration = 200.

This method has selected the variables (Glu: Glucose, BMI: mass, PED: Heredity) as variables that have the discriminating power between the two classes. Coming back to the nature of diabetes disease we can confirm that this method has actually found the most relevant variables: that is why the change in glucose concentration is the most used parameter for the diagnosis of this disease and if we back to the causes of diabetes, we note that the majority of diabetic men suffer from the problem of overweight and that the genetic factor is responsible for most cases of pancreas failures which is the main abnormality of a development of a diabetes (Fig. 4).

Here the confusion matrix of PCA-NN shows that out of 36 there are 36 outputs gives correct result while rest none of the gives wrong output, so accuracy count for Non-diabetic is 100%. Now out of 113 cases of Diabetic 111 gives correct result while rest 2 gives wrong result, so accuracy count for Diabetic is 98.2%. So exact accuracy count drawn for both Diabetic and Non-Diabetic are 98.7%.

$$\text{False Positive Rate} = \frac{\text{FP}}{\text{FP} + \text{TN}} = \frac{0}{0 + 36} = 0\%$$

$$\text{Specificity} = \frac{\text{TN}}{\text{FP} + \text{TN}} = \frac{36}{1 + 36} = 100\%$$

$$\text{True Positive Rate} = \frac{\text{TP}}{\text{TP} + \text{FN}} = \frac{111}{111 + 2} = 98.2\%$$

$$\text{Miss} = \frac{\text{FN}}{\text{TP} + \text{FN}} = \frac{2}{111 + 2} = 1.8\%$$

$$\text{Rate Error} = \frac{\text{FN} + \text{FP}}{\text{TOTAL}} = \frac{2 + 0}{149} = 1.3\%$$

$$\text{Rate Accuracy} = \frac{\text{TN} + \text{TP}}{\text{TOTAL}} = \frac{36 + 118}{149} = 98.7\%$$

We also note a slight decrease in its specificity. Table 2 summarizes the results obtained by the two experiments:

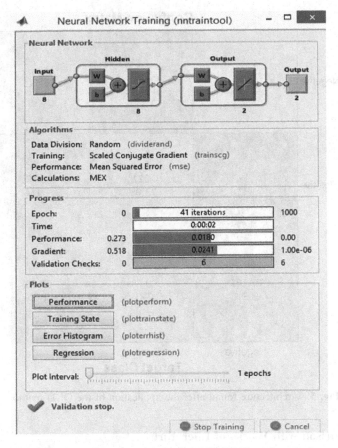

Fig. 4. Confusion matrix plot for PCA-NN based method

Table 2. Performance table

Method	Number of variables	Error rate	Specificity	Accuracy
NN (Proposed)	8	1%	%	99.0%
PCA+NN (Propose)	3	1.3%	100%	98.7%

Comparing the results of the two experiments, we note that the proposed method decreased the number of input variables by more than 60% and that the number of connections decreased by more than 55%, which gave a good optimization of the architecture of the model and a strong improvement of the classifier performances (Fig. 5).

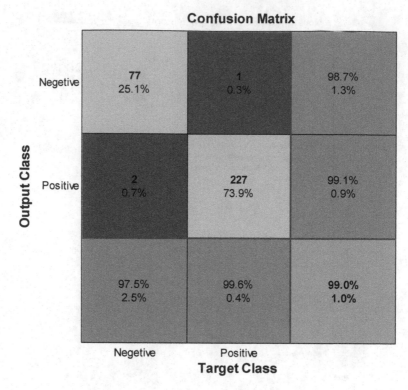

Fig. 5. Architecture found after the application of the OCD method

3.5 Comparison with Works of Literature

In order to situate the performance of the proposed approach, we carried out a comparative study between the results obtained and those of the work already done in this field (studied in the state of the art) with the PIMA database. Table 3 summarizes the comparison with the other works:

Table 3. Comparison table with literature works

Name of the method	Type of method	Number of variables	Accuracy
FCBF+SVM [23]	Filter	4	77.99%
GR+RBF [24]	Filter	5	86.46%
GR+MLP [24]	Filter	5	78.21%
CAFS+MLP [25]	Wrapper	6	76.18%
GA+SVM [26]	Wrapper	4	81.50%
TS1+MLP [27]	Wrapper	4	79.55%
OCD+MLP	Embedded	3	83.59%
NN (Proposed)	Neural network	8	99.0%
PCA+NN (Propose)	PCA and neural network	3	98.7%

After several researches on this problematic we can confirm that the method proposed is the first method of the type Embedded applied on the PIMA database and the Number of variables found by this method is the smallest number of variables selected so far. We also note that the classification rate is the best among the methods that use the MLPs and even among the other methods except in the work of Karegowda et al. [24] they found a better rate but using the RBF as a classifier. From the Table 3, it is clear that the proposed work outperforms other works (Figs. 6 and 7).

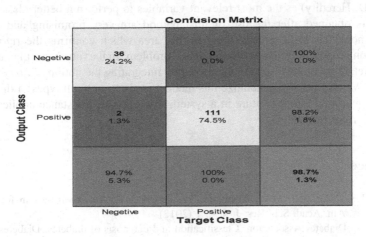

Fig. 6. Neural network training

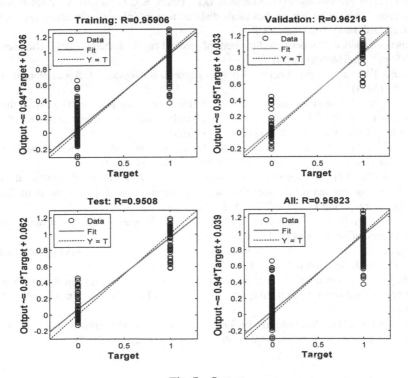

Fig. 7. Output

4 Conclusion

We presented a variable selection method that confirmed its performance in the test performed and that gave an interesting improvement of the model error, sensitivity and classification rate while optimizing its architecture. After comparing the results obtained with work in the literature, we noticed that the results found are comparable or better than the other results. This method selected the variables (Glu: Glucose, BMI: mass, PED: Heredity) as the most relevant variables to perform a better classification. The results obtained after the use of this method are very promising and are well located among the work already done in this area which confirms the rigor of the contribution proposed for the resolution of our problem. In the future, we plan to ensure the interpretability of the results of the model by integrating the notion of fuzzy with the classifier. We also want to generalize this modest application to all types of diseases in order to integrate it into the future in a system of diagnostic assistance applicable in a hospital or a medical office.

Reference

1. Zeki, T.S., Malakooti, M.V., Ataeipoor, Y., Tabibi, S.T.: An expert system for diabetes diagnosis. Am. Acad. Sch. Res. J. **4**(5), 1 (2012)
2. American Diabetes Association: Classification and diagnosis of diabetes. Diabetes Care **38** (Suppl. 1), S8–S16 (2015)
3. Lavery, L.A., Armstrong, D.G., Murdoch, D.P., Peters, E.J., Lipsky, B.A.: Validation of the infectious diseases society of America's diabetic foot infection classification system. Clin. Infect. Dis. **44**(4), 562–565 (2007)
4. American Diabetes Association: Standards of medical care in diabetes—2014. Diabetes Care **37**(Suppl. 1), S14–S80 (2014)
5. Amato, F., et al.: Artificial neural networks in medical diagnosis. J. Appl. Biomed. **11**(2), 47–58 (2013)
6. Jayalakshmi, T., Santhakumaran, A.: A novel classification method for diagnosis of diabetes mellitus using artificial neural networks. In: 2010 International Conference on Data Storage and Data Engineering (DSDE), pp. 159–163. IEEE, February 2010
7. Ahmadlou, M., Adeli, H.: Enhanced probabilistic neural network with local decision circles: a robust classifier. Integr. Comput.-Aided Eng. **17**(3), 197–210 (2010)
8. Karegowda, A.G., Manjunath, A.S., Jayaram, M.A.: Application of genetic algorithm optimized neural network connection weights for medical diagnosis of pima Indians diabetes. Int. J. Soft Comput. **2**(2), 15–23 (2011)
9. Iyer, A., Jeyalatha, S. Sumbaly, R.: Diagnosis of diabetes using classification mining techniques, arXiv preprint (2015). arXiv:1502.03774
10. Durairaj, M., Kalaiselvi, G.: Prediction of diabetes using soft computing techniques-a survey. Int. J. Sci. Technol. Res. **4**(3), 190–192 (2015)
11. Erkaymaz, O., Ozer, M.: Impact of small-world network topology on the conventional artificial neural network for the diagnosis of diabetes. Chaos, Solitons Fractals **83**, 178–185 (2016)
12. Erkaymaz, O., Ozer, M., Perc, M.: Performance of small-world feedforward neural networks for the diagnosis of diabetes. Appl. Math. Comput. **311**, 22–28 (2017)

13. Mahajan, A., Kumar, S. Bansal, R.: Diagnosis of diabetes mellitus using PCA and genetically optimized neural network. In: 2017 International Conference on Computing, Communication and Automation (ICCCA), pp. 334–338. IEEE, May 2017
14. Mangathayaru, N., Mathura Bai, B., Srikanth, P.: Clustering and classification of effective diabetes diagnosis: computational intelligence techniques using PCA with kNN. In: Satapathy, S.C., Joshi, A. (eds.) ICTIS 2017. SIST, vol. 83, pp. 426–440. Springer, Cham (2018). https://doi.org/10.1007/978-3-319-63673-3_52
15. Rosenblatt, F.: The perceptron: a probabilistic model for information storage and organization in the brain. Psychol. Rev. **65**(6), 386 (1958)
16. Minsky, M., Papert, S.A., Bottou, L.: Perceptrons: An Introduction to Computational Geometry. MIT press, Cambridge (2017)
17. Widrow, B., Hoff, M.E.: Adaptive switching circuits, No. TR-1553–1. Stanford Univ CA Stanford Electronics Labs (1960)
18. Werbos, P.J.: The Roots of Backpropagation: from Ordered Derivatives to Neural Networks and Political Forecasting, vol. 1. Wiley, Hoboken (1994)
19. Rumelhalt, D.E.: Learning internal representations by error propagation. Parallel Distrib. process. **1**, 318–362 (1986)
20. Le Cun, Y.: Learning process in an asymmetric threshold network. In: Bienenstock, E., Soulié, F.F., Weisbuch, G. (eds.) Disordered Systems and Biological Organization. NATO ASI Series (Series F: Computer and Systems Sciences), vol. 20, pp. 233–240. Springer, Heidelberg (1986). https://doi.org/10.1007/978-3-642-82657-3_24
21. Tetko, I.V., Livingstone, D.J., Luik, A.I.: Neural network studies. 1. comparison of overfitting and overtraining. J. Chem. Inf. Comput. Sci. **35**(5), 826–833 (1995)
22. Pima, A.F., Asuncion, A.: Pima Indians Diabetes Data Set. UCI Machine Learning Repository, University of California, Irvine, School of Information and Computer Sciences (2010). http://archive.ics.uci.edu/ml
23. Balakrishnan, S., Narayanaswamy, R.: Feature selection using fcbf in type ii diabetes databases. Int. J. Comput. Internet Manag. **17**(1), 50–58 (2009)
24. Karegowda, A.G., Manjunath, A.S., Jayaram, M.A.: Comparative study of attribute selection using gain ratio and correlation based feature selection. Int. J. Inf. Technol. Knowl. Manag. **2** (2), 271–277 (2010)
25. Kabir, M.M., Islam, M.M., Murase, K.: A new wrapper feature selection approach using neural network. Neurocomputing **73**(16–18), 3273–3283 (2010)
26. Huang, C.L., Wang, C.J.: A GA-based feature selection and parameters optimization for support vector machines. Expert Syst. Appl. **31**(2), 231–240 (2006)
27. Wang, Y., Li, L., Ni, J., Huang, S.: Feature selection using tabu search with long-term memories and probabilistic neural networks. Pattern Recogn. Lett. **30**(7), 661–670 (2009)

Microaneurysm Detection in Diabetic Retinopathy Using Genetic Algorithm and SVM Classification Techniques

Nitta Gnaneswara Rao[1], S. Deva Kumar[1(✉)], T. Sravani[2],
N. Ramakrishnaiah[3], and V. Rama Krishna S[4]

[1] Department of CSE, VFSTR University, Guntur, Andhra Pradesh, India
gnani.nitta@gmail.com, deva.248@gmail.com
[2] TCS, Hyderabad, India
[3] Department of CSE, JNTUK Kakinada, Kakinada, Andhra Pradesh, India
[4] Department of CSE, VFSTR University, Guntur, Andhra Pradesh, India

Abstract. Diabetics is the common disease faced by many of the people in India. It can be detected through microaneurysm. Using genetic algorithm and SVM classification techniques, sores are viewed as the most punctual indications of diabetic retinopathy. The diabetic retinopathy is an infection caused by diabetes and is considered as the significant reason for visual impairment in working age populace. The proposed technique depends on numerical morphology and comprises in expelling parts of retinal life systems to achieve the sores. This strategy comprises of four phases: (a) Image acquisition; (b) Preprocessing; (c) Extracting the features using genetic algorithm; (d) Classification. The exactness of the strategy was obtained using genetic algorithm and SVM classifier is also applied to identify whether the person is diabetic or non diabetic patient.

Keywords: Diabetic retinopathy · Genetic algorithm ·
SVM classifier · Microaneurysms

1 Introduction

Granting to International Diabetes Federation (IDF), in 2011, 366 million people were recognized with diabetes. It is anticipated that by 2030 the count will increase to 522 million. The risk increased through diabetes in any type of eye disease and occurs blindness is known as Diabetic Retinopathy (DR). DR means causing damage to the blood vessels in the retina. Broadly speaking, if not treated there is gradual vision loss. According to the survey since for 20 years, who has diabetes more than 75% of people, will hold some type of DR. These complexities can be forestalled if the analysis is done properly and regularly at least formerly a year. It is hard to identify the symptoms in the patients, who have DR, So that it would be late for giving effective treatment. Thus, for spotting and early medical involvement is vital. Normally, ophthalmologists observe

© Springer Nature Singapore Pte Ltd. 2019
K. C. Santosh and R. S. Hegadi (Eds.): RTIP2R 2018, CCIS 1036, pp. 286–293, 2019.
https://doi.org/10.1007/978-981-13-9184-2_25

DR based on characteristics, such as hemorrhages, Microaneurysms (MAs), texture and the expanses of the blood vessels. MAs are the first scientific sign to illustrate the diabetic retinopathy and specify it as red lesions. Agreeing to a written report, declared in [1,2], half of 205,000 ophthalmologists are globally established in areas such as India, China, Russia, Brazil and USA [3,4]. It is likewise noted that in 23 countries for one million patients there have less than one ophthalmologist; 30 countries with less than four; 48 with lesser than 25 for every million citizens; 74 countries with less than 100. Only 18 have more than 100 ophthalmologists for every one thousand people. In summation, it is anticipated that the count of ophthalmologists will grow just 2%, while the diabetic patients will be 54% in 2030. For examination of each eye it takes 15 to 30 min. For this reason the work mainly concentrated on algorithm evolution and that algorithm is used to detect Microaneurysms in eye images. The finding of diabetic retinopathy is made after an unmistakable examination of the retina with an ophthalmoscope [5]. The retinal master depends upon two or three tests to screen the improvement of the illness and to settle on choices for the fitting treatment [6]. This paper includes the part extraction using genetic computation and SVM portrayal [7,8]. In this paper we have to detect Microaneurysms for three steps i.e. first step is applying preprocessing for the given retinal image, second step is extraction of features from the preprocessed image and last step is applied Support Vector Machine classification. A method for red lesion detection is represented in [9]. At this stage, to improve the image digital Curve lets transformation is applied and by manipulating the coefficients of transformation the lesions are detached, and then these red lesions are named as Candidates. In [10], MAs detection is gained over preprocessing the pixels. Detection is applied to the set of attributes like elevation; shape and size of the picture elements are measured. In paper [11] detection of MAs is done with coarse segmentation by using morphological operations and fine segmentation by using the SVM classifier. The problem for MAs detection is formed in [12]; the problem is finding regions of interest (blobs). To characterize these blobs many region descriptors are presented. In [13], there are two approaches for detection of MAs. They are: (1) an approach based on visibility and spatial location the MA candidates are brought away. (2) An approach of adaptive weighting which is based on spatial location and contrast the MAs are extracted. The results exhibited in this approach are based on selection of outcomes and individual detectors paper [14] introduces a singular method for MAs detection, which is based on multi overlapping and (RT) Radon Transform. On the preprocessing stage, to take out the background top-hat transform and averaging filter is applied [15]. In processing stage the total image is separated into sub-icons, and on every sub-image RT is applied for masking [16]. After masking and detection of ONH and blood vessels MAs are detected. And for performance evaluation 3 databases are applied. They are: (ROC) Retinopathy Online Challenge database, Mashhad database and Local database. For the above determined causes, there is a necessity to implement an automated technique of pre-diagnosis to attain a quick valuation of the retinal images and to designate if there are any kind of lesion that need to be diagnosed with the expert. Furthermore, the procedure of recognition and categorization of lesions ought to be humble to simplify its succeeding mass distribution. The interpretation of the pre-diagnosis, primary and introductory diagnosis would profit the procedures that necessitate fewer evaluating assets. This paper outlined as follows. Section 2 describes the methodology of this paper to detect Microaneurysms, Sect. 3 describes the result analysis and Sect. 4 describes the conclusion of the paper.

2 Methodology

Diabetic retinopathy happens when the blood vessels of the retina in the back piece of the eye are harmed. Harms because of little vessels would be known as smaller scale vascular malady while harms because of the courses would be large scale vascular disease. This can be identified by the following stages is depicted in Fig. 1.

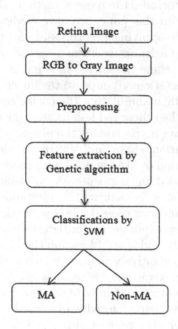

Fig. 1. Architecture of proposed methodology

Step 1: Image Acquisition
The gained shading pictures of retina from Drive datasets individually are utilized for enlistment and distinguishing proof. These shading pictures are converted into grayscale first and then apply the operations.

Step 2: Preprocessing

Application of effective preprocessing technique is important to improve image quality and make it suitable further processing [17,18]. The proposed method and the lesion detection algorithm's performance depends on the quality of the retinal images. For identifying the pathetic character of the images there are so many elements. They are noisy, low contrast, non-uniform illumination, diffusion and variation in light reflection, differences in cameras and difference in retinal chrome. For identifying the pathetic character of the images there are so many elements. They are noisy, low contrast, non-uniform illumination, diffusion and variation in light reflection, differences in cameras and difference in retinal chrome. To slim down the image variations and to amend the image quality the Preprocessing stage is really important. Median filter is applied to green plane images to decrease the noise before a (CLAHE) Contrast Limited Adaptive Histogram Equalization is applied [19]. Median filter is a nonlinear method which is used for reducing the noise. As random bit error occurs in the communication channel the noise will also be existed. In this process, adaptive histogram equalization is used and then the image is separated into regions and then histogram equalization is enforced on every region. To visible the hidden things in the image gray values is used. Then in the consecutive step, for contrast enhancement and to avoid over saturation of the same areas in retinal image CLAHE is applied. To slim down the image variations and to amend the image quality the Preprocessing stage is really important.

Step 3: Feature Extraction

(i) Feature extraction through genetic algorithm

Genetic Algorithm is applied on the fundus images for feature extraction and improve the accuracy. Genetic Algorithms (GAs) are stochastic hunt techniques in view of the standards of normal hereditary frameworks. They play out a pursuit in giving an ideal answer for assessment (fitness) function of an improvement problem. GA deal at the same time with numerous arrangements and utilize just the wellness work esteems. While taking care of a streamlining issue using GAs, every arrangement is coded as a string of limited length over a limited letters in order A. Each string or chromosome is considered as a person. A gathering of M (M is limited) such people is known as a population. GAs begins with a haphazardly created populace of size M. When it is applied on the retina to detect the Microaneurysms. The genetic algorithm is implemented using four different steps for image enhancement.

   i. Initialization: The individual pixels are randomly selected from the image known as population.
   ii. Fitness Function: The Fitness is evaluated for the individual pixels and its value is determined.
   iii. Genetic Operations: The selected individuals having the fitness function more than the threshold value are taken for genetic crossover and mutation operations.
   iv. Termination Criteria: The steps from 1 to 3 are iterated for given number of times as per the trial and error approach. The enhanced pixel obtained from

this approach are considered as the initial centroid and Euclidean distance for the nearby pixels is computed for the clustering the images.

Step 4: Classification

For image classification in general (or) for DR we are using (SVM) Support Vector Machine classifier, which is a supervised learning method. Support Vector Machines are useful to avoid the over segmentation of the retinal images that minimizes the blurriness of an image and means of selection the essential features that are mostly useful for classification the shapes in the retinal images. The considered retinal images have different shapes for blood vessels, microaneurysms and red lesions. These shaped are detected effectively by means of SVM. The hyper plane is used in decision boundary which isolates the negative and the positive data points, and this isolation should be maximal. The classifier equation is given below:

$$\bullet \, f(x) = W^T \mathrm{x} + \mathrm{b}$$

In the above equation "w" is the weight vector of the Hyper plane and "b" is the bias.

3   Experiment Results and Discussions

By using this methodology, we are taken by considering Drive datasets of retinal images. The retinal size of this dataset is $565 * 584$ pixels with .tif format. The result of the proposed system can be shown from Figs. 2, 3, 4, 5 and 6. Figure 2 shows that the retinal images can be obtained from Drive dataset. Figure 3 generates the preprocessing result. The features extraction of preprocessed images can be extracted through Genetic algorithm are shown in Fig. 4 and these features are classified using SVM classifier as illustrated in Fig. 5. Figure 6 represents the recognition accuracy of diabetic and non diabetic individual by considering their taken retinal images from dataset. The accuracy of proposed system is verified with taken different samples of retinal images. It is observed that from Fig. 6 obtained with high accuracy in identifying the diabetic person using genetic and SVM algorithm.

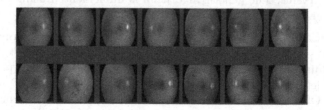

Fig. 2. Enrolled images of user

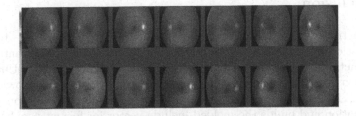

Fig. 3. Gray color images of retina

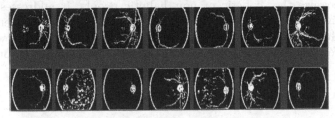

Fig. 4. Preprocessing using Daughman's algorithm

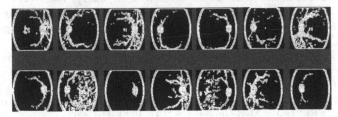

Fig. 5. Feature extraction using genetic algorithm

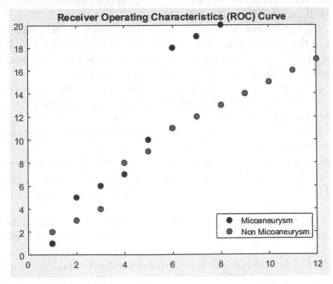

Fig. 6. ROC curve depicting diabetic and non diabetic patients

4   Conclusion

Eye biometric has the high steadiness contrasted with other biometric characteristics. This paper outlines the portrayal and acknowledgment of highlight extraction of retina through hereditary calculation. The retinal highlights are extricated through hereditary calculation in proposed acknowledgment framework. The more profound portrayal with hereditary calculation denoises the information and builds the melded include vector for format age. SVM classifier is expert to improve the acknowledgment precision of proposed framework to recognize the diabetic or non-diabetic people. It addresses the non-comprehensiveness and caricaturing assaults by limiting the equivalent mistake rate. Convolution neural systems can be joined on proposed acknowledgment framework for proficient portrayal and reproduction of highlight layouts.

References

1. Lazar, I., Hajdu, A.: Retinal microaneurysm detection through local rotating cross section profile analysis. IEEE Trans. Med. Imaging 32(2), 400–7 (2013)
2. Sinthanayothin, C., Boyce, J.F., Williamson, T.H., Cook, H.L., Mensah, E., Lal, S.: Automated detection of diabetic retinopathy on digital fundus images. Diabet. Med. 19, 105–12 (2002)
3. Júnior, S.B., Welfer, D.: Automatic detection of microaneurysms and hemorrhages in color eye fundus images. Int. J. Comput. Sci. Inf. Technol. 5(5), 21–37 (2013)
4. Quellec, G., Lamard, M., Josselin, P.M., Cazuguel, G., Cochener, B., Roux, C.: Optimal wavelet transform for the detection of microaneurysms in retina photographs. IEEE Trans. Med. Imaging 27(9), 1230–41 (2008)
5. Niemeijer, M., Van Ginneken, B., Staal, J., Suttorp-Schulten, M.S., Abramoff, M.D.: Automatic detection of red lesions in digital color fundus photographs. IEEE Trans. Med. Imaging 24(5), 584–592 (2005)
6. Aimmanee, P., Uyyanonvara, B., Jitpakdee, P.: A survey on hemorrhage detection in diabetic retinopathy retinal images. In: Proceedings of the 9th International Conference on Electrical Engineering, Electronics, Computer, Telecommunications and Information Technology (ECTI-CON), pp. 1–4 (2012)
7. Bosman, P.A.N., Thierens, D.: Linkage information processing in distribution estimation algorithms. In: Proceedings of the 1st Annual Conference on Genetic and Evolutionary Computation, vol. 1, pp. 60–67 (1999)
8. Cormen, T.H., Leiserson, C.E., Rivest, R.L.: Introduction to Algorithms. McGraw-Hill Book Company, New York (1990)
9. Esmaeili, M., Rabbani, H., Dehnavi, A.M., Dehghani, A.: New curvelet transform based method for extraction of red lesions in digital color retinal images. In: Proceedings of the 17th International Conference on Image Processing, pp. 4093–4096 (2010)
10. Bae, J.P., Kim, K.G., Kang, H.C., Jeong, C.B., Park, K.H., Hwang, J.M.: A study on hemorrhage detection using a hybrid method in fundus images. J. Digit. Imaging 24(3), 394–404 (2011)
11. Lazar, I., Hajdu, A.: Retinal microaneurysm detection through local rotating cross section profile analysis. IEEE Trans. Med. Imaging 32(2), 400–7 (2012)

12. Sopharak, A., Uyyanonvara, B., Barman, S.: Simple hybrid method for fine microaneurysm detection from non-dilated diabetic retinopathy retinal images. Comput. Med. Imaging Graph. **37**(5–6), 394–402 (2013)
13. Adal, K.M., Sidibé, D., Ali, S., Chaum, E., Karnowski, T.P., Mériaudeau, F.: Automate detection of microaneurysms using scale-adapted blob analysis and semi-supervised learning. Comput. Methods Programs Biomed. **114**, 1–10 (2014)
14. Antal, B., Hajdu, A.: Improving microaneurysm detection in color fundus images by using context-aware approaches. Comput. Med. Imaging Graph. **37**(5–6), 403–408 (2013)
15. Javidi, M., Pourreza, H.R., Harati, A.: Vessel segmentation and microaneurysm detection using discriminative dictionary learning and sparse representation. Int. J. Comput. Methods Programs Biomed. **139**, 93–108 (2016)
16. Rao, N.G., Sajja, V.R., Kumar, S.D., Rao, M.V.: An improved IHBM using smoothing projections. Int. J. Control Theory Appl. **8**(1), 326–335 (2015)
17. Ruikar, D.D., Santosh, K.C., Hegadi, R.S.: Automated fractured bone segmentation and labeling from CT images. J. Med. Syst. **43**(3), 60 (2019). https://doi.org/10.1007/s10916-019-1176-x
18. Ruikar, D.D., Santosh, K.C., Hegadi, R.S.: Segmentation and analysis of CT images for bone fracture detection and labeling (Chap. 7). In: Medical Imaging: Artificial Intelligence, Image Recognition, and Machine Learning Techniques. CRC Press (2019). ISBN: 9780367139612
19. Hegadi, R.S., Navale, D.I., Pawar, T.D., Ruikar, D.D.: Multi feature-based classification of osteoarthritis in knee joint X-ray images (Chapt. 5). In: Medical Imaging: Artificial Intelligence, Image Recognition, and Machine Learning Techniques. CRC Press (2019). ISBN: 9780367139612

Compressive Sensing for Three-Dimensional Brain Magnetic Resonance Imaging

Selrina D'souza[1], H. Anitha[1(✉)], and Karunakar Kotegar[2]

[1] Department of Electronics and Communication, Manipal Institute of Technology,
Manipal Academy of Higher Education, Udupi, Karnataka, India
selrina18@gmail.com, anitha.h@manipal.edu
[2] Department of Computer Application, Manipal Institute of Technology,
Manipal Academy of Higher Education, Udupi, Karnataka, India
karunakar.ak@manipal.edu

Abstract. Three dimensional (3D) Magnetic Resonance Imaging (MRI) reconstructions depend heavily on the imaging speed. Magnetic Resonance (MR) images consist of large volume of redundant and sparse data. Therefore, the need to reduce this data without degrading the image information. In Fourier Domain, sparse nature of MR images enables image reconstruction with fewer Fourier coefficients. Fourier Transform (FT) maps the image into the frequency domain using fixed and same size window throughout the analysis. In our paper, a method to perform compressive sensing for MR image is presented. Anisotropic filtering using Active Contour Modelling is performed to smoothen the image in order to preserve edge information. MR image is converted into Fourier Domain using Discrete Fourier Transform (DFT). l1 and l2 reconstruction algorithms are used to reconstruct the images using minimum coefficients that have maximum information.

Keywords: Magnetic Resonance Imaging (MRI) ·
Compressive Sensing (CS) · Discrete Fourier Transform (DFT)

1 Introduction

1.1 Magnetic Resonance Imaging

MRI creates two dimensional (2D) images of the body, on the basis of hydrogen atom's magnetic fields present in the body. Under the influence of a strong uniform magnetic field, atoms align themselves along the applied field which gives rise to small net magnetization. These excited atoms generate a small RF field called echo due to equilibrium. Spatial frequency space or k-space are used to record the measurements. Signals are recovered from these echoes and images are formed of the scanned object. Using sophisticated computer calculation, these 2D images obtained are joined together to produce a 3D model of the object scanned [8]. This is called 3D-MRI reconstruction.

© Springer Nature Singapore Pte Ltd. 2019
K. C. Santosh and R. S. Hegadi (Eds.): RTIP2R 2018, CCIS 1036, pp. 294–302, 2019.
https://doi.org/10.1007/978-981-13-9184-2_26

1.2   MRI-MEG/MRI-EEG Co-registration

In MRI-MEG/MRI-EEG co-registration, human brain functional data obtained from electroencephalography (EEG) or magnetoencephalography (MEG) techniques are combined with the brain structural data obtained from MRI techniques. Which results in head model for activated area visualization and source localization. A surface matching technique is used to perform co-registration, it aligns EEG/MEG and MRI co-ordinate. This technique is used to identify activated brain regions, or epileptic foci for surgical and diagnostic applications.

1.3   Drawbacks of MRI

Acquisition of image set for 3D imaging brain study takes 30 min–40 min and sometimes approaching to an hour. This leads to reduced MRI throughput and imaging speed. Traditionally, human brain MRI data sets were manually analyzed. However, with advances in technology MR imaging lead to higher resolution images giving enormous data which was difficult to interpret manually. This lead to the evolution of sophisticated computer tools to extract relevant clinical information of brain MRI data. In contrast to the high-resolution MR image, the tools used are low-resolution methods with high computational complexity, fairly giving the good performance. Combining low-level analysis techniques with high complexities is a task and limits the clinical application of such computer-based tool. Hence, there is a need to reduce the data for processing, which in turn will reduce the complexity of algorithms thereby increasing MRI registration process speed. One such method to do so is based on sparsity property by the virtue of which some of the data values in space convert into zeros in an appropriate transform domain and the image or signal is reconstructed from the rest of the data values without any much loss of information.

2   Literature Review

MRI-MEG co-registration consist of mapping neuro-physiological functions of brain contained in the MEG data to MRI data, this registration of MEG-MRI data is done based on Powell method [10]. This fusion helps to localize the origination of neurophysiological functions of the brain, thereby enabling to study the spatiotemporal aspects of these processes. This co-registration may give large errors due to motion artefacts. Wherein, the digitized head shape from MEG was co-registered with a head skin surface of MRI.

It is possible to speed up the co-registration process as MR images are sparse either in space or in a transform domain. This is done by considering only those coefficients that have maximum information and discarding the others [5]. This is done by converting MR images into sparse domain and performing compressive sensing. The information on boundaries of soft tissues and organs is very sparse. Hence with compressive sensing, a signal or an image can be reconstructed from a very limited number of samples less than Nyquist criteria [1,3]. This enables the reconstruction of very high-quality images with a subset of the acquired samples that consist maximum information thereby significantly reducing the MRI co-registration time, scan duration and the time patient spends inside the magnet [7]. Moreover, increasing the throughput of a clinic.

To perform compressive sensing, we first need to convert the MR image into a sparse domain [11]. DFT is the most common sparsifying transform and is widely applied in a variety of academic and industrial fields [2,4]. However, it is not better at representing object boundaries, curves and edges present in the image. Hence, DFT gives sparse representation only for smooth image [9]. Therefore, it is suboptimal to use 2D-DFT for non-smoothened images. This limitation encourages to use a method, which makes MR image smoother and then get a sparser image. Anisotropic filters are proved to have higher smoothing quality than isotropic filters [6].

3   Methodology

3.1   MRI Image

The MRI signal acquisition is slow process. The MRI signal obtained gives the direction of magnetization and the spatial positions of these magnetization within the patient's body. A collection of serial data frames gives a single MR image (Fig. 1).

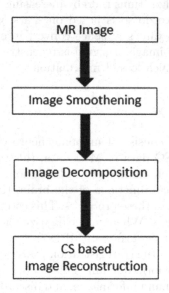

Fig. 1. Block diagram of the proposed methodology

3.2   Image Smoothening

Image smoothening is performed using Anisotropic filters. It reduces image noise without degrading edges or lines present in the images.

$$\frac{dI}{dt} = div(c(x,y,t) \bigtriangledown I) = \bigtriangledown c \bigtriangledown I + c(x,y,t) \bigtriangleup I \qquad (1)$$

I is unfiltered image. div(), $\bigtriangleup$ and $\bigtriangledown$ denotes divergence, laplacian and gradient operator and c(x, y, t) is the coefficient of diffusion, it determines the rate of diffusion and is a function of image gradient which in turn controls the amount of edge preservation in the image.

3.3   Image Decomposition

Images are often compressible with perceptible loss of information. The smoothened MR image is subjected to sparsifying transform to obtain a matrix of sparse coefficients. An inverse transform is applied to generate an image from minimum number of coefficients, this allows the straightforward use of a Transform using an efficient algorithm. Commonly used sparsifying transform for brain MRI is Discrete Fourier Transform (DFT).

3.3.1   Discrete Fourier Transform. To decompose the MR image, DFT is used. In 1D-DFT, signal is deconstructed into waves that are infinitely long like cosine waves and sinewaves. In 2D-DFT imaging, each row in sparse coefficient matrix corresponds to the echo data signal. 2D-DFT is obtained by computing 1D-DFT along the row and then along the column of the MR image.

3.4   Compressive Sensing (CS)

The CS is used to perform image reconstruction using less number of samples, thereby compressing the image. Basic requirements of CS are that a signal or an image should be sparse. This sparsity can be mathematically expressed as:

$$x = \sum_{i=1}^{N} S_i \Psi \qquad (2)$$

x is a discrete signal expressed in terms of orthogonal basis $[\Psi]$ and S_i are the coefficients of x.

$$x = \Psi S \qquad (3)$$

Above equation is the matrix form

x is k-sparse when k entries of S are non-zeroes and remaining (N-k) entries are zero. Thus enabling to exactly reconstruct x from these non-zero entries.

$$y = \Phi x \qquad (4)$$

Above equation is used to measure N coefficients of x and store them in vector y. $[\Phi]$ is a M × N measurement matrix. Now we can write,

$$y = \Phi x = \Phi \Psi S = \Theta S \tag{5}$$

Θ is a M × N measurement/dictionary matrix.

The result is obtained by recovering x from a significantly less number of measurements contained in y. This may be done by solving below equations

$$||\widehat{x}||_{l_1} = \sum_{i=1}^{N} |\widehat{x}| \ , then \ min \ ||\widehat{x}||_{l_1} \ subjected \ to \ \Phi \widehat{x} = y \tag{6}$$

$$\widehat{x} = arg \ min||y - \Psi x||_{l_2}^2 + \lambda ||\Psi x||_{l_1} \tag{7}$$

x and $\widehat{x}$ are signal of interest and reconstructed signal respectively, Φ is the acquisition matrix of size $N \times M$; y is the captured data, λ is a controlling parameter and Ψ is the sparsifying transform and k is the regularization parameter. The objective function here is the l_1-norm minimization given in Eq. 6 and the l_2-norm constraint given in Eq. 7 enforces data consistency. In words, out of all the potential sparse solutions, the equation selects one solution that is compressible.

4   Results

Figure 2 shows the cross-sectional MR image data obtained from young adult of OASIS-1 database obtained from Open Access Series of Imaging Studies (OASIS) http://www.oasis-brains.org/, with dimension 176 by 208 pixels. OASIS-1 dataset consists of a cross-sectional collection of 416 subjects aged 18 to 30.

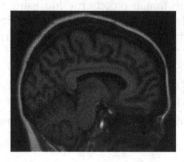

Fig. 2. The orginal MR image obtained from OASIS http://www.oasis-brains.org/

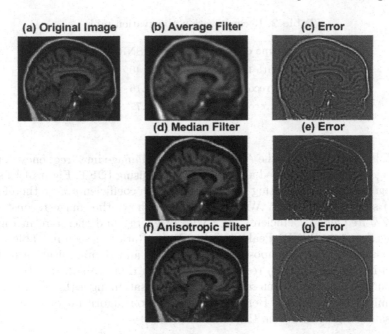

Fig. 3. Filtered image. (a) Original Image (b) Average Filtered Image (c) Absolute error associated with average filtering (d) Median Filtered Image (e) Absolute error associated with median filtering (f) Anisotropic Filtered Image (g) Absolute Error associated with Anisotropic filtering

Figure 3 shows the filtered images. Anistropic filter out performs Average filter and Median filter, as shown in Fig. 3(f). It retains the image edge details and does not blur the image. On the other hand, Average filter blurs the image and Median filter completely looses the edge information, as shown in Fig. 3(b) and (d) respectively. PSNR of the filtered images are listed in Table 1.

Table 1. Filtering observation table

Type of filter	PSNR
Average filter	18.40
Median filter	22.74
Anisotropic filter	27.45

Table 2. Reconstruction obesrvation table

Type of reconstruction	PSNR
With thresholding	1.40
l1 reconstruction	4.16
l2 reconstruction	2.77

DFT is used to convert the Anisotropic filtered image into frequency domain. The image is then compressed and reconstructed using IDFT. Figure 4(b) shows the image reconstructed using all the frequency coefficients and therefore in Fig. 4(c) shows a zero error. Whereas Fig. 4(d) shows the image reconstructed using 62% frequency coefficients with 90% of energy and therefore in Fig. 4(e) the image has 1.40 PSNR (Peak Signal to Noise Ratio) as seen in Table 1.

Reconstruction of decomposed image using l2 and l1 minimisation methods is shown in Fig. 5(b) and (c) respectively. As seen from Table 2 results and also from visual means, it can been seen that l1 minimisation algorithm over performs l2 minimisation algorithm. Hence, l1 minimisation algorithm overperforms l2 minimisation algorithm when the data is sparse.

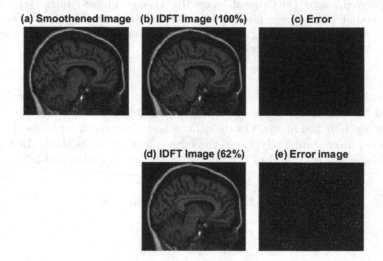

(a) Smoothened Image   (b) IDFT Image (100%)   (c) Error

(d) IDFT Image (62%)   (e) Error image

Fig. 4. Fourier Transformed image and reconstructed image using thresholding method. (a) Anisotropic Smoothened Image (b) IDFT of the image considering all the coefficients (c) Absolute error associated with (b). (d) IDFT of the image considering 62% coefficients with 90% energy (e) Absolute error associated with (d).

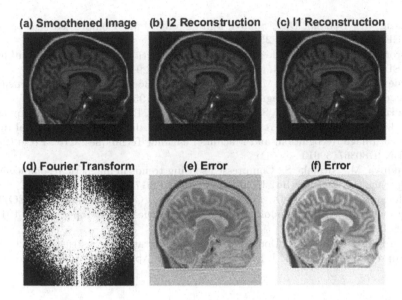

Fig. 5. Fourier Transformed Image and reconstructed image using compressive sensing. (a) Anisotropic Smoothened Image (b) l2 reconstruction (c) l1 reconstruction (d) Fourier transform of the image (frequency domain representation) (e) Absolute error associated with (b). (f) Absolute error associated with (c)

5   Conclusions

Our method exploited the sparse nature of 2D MR image using DFT. Obtained the least number of coefficients with maximum information and reconstructed the image using these minimum coefficients. Thereby reducing redundant data in the MR image reconstruction by about 30%.

Acknowledgements. MRI data was downloaded from Open Access Series of Imaging Studies (OASIS) http://www.oasis-brains.org/ for making the neuroimaging datasets freely available to the scientific community. The codes were implement on MATLAB R2017b.

References

1. Candes, E., Tao, T.: Near optimal signal recovery from random projections: universal encoding strategies. IEEE Trans. Inf. Theory **52**(1) (2006)
2. Chavez-Roman, H., Ponomaryov, V.: Super resolution image generation using wavelet domain interpolation with edge extraction via a sparse representation. IEEE Geosci. Remote Sens. Lett. **11**(10), 1777–1781 (2014)
3. Donoho, D.: Compressed sensing. IEEE Trans. Inf. Theory **52**(4), 1289–1306 (2006)
4. Fang, L., Wu, L., Zhang, Y.: A novel demodulation system based on continuous wavelet transform. Math. Probl. Eng. **9** (2015)

5. Lustig, M., Donoho, D.L., Santos, J.M., Pauly, J.M.: Compressed sensing MRI. IEEE Sign. Process. Mag. **25**(2), 72 (2008)
6. Perona, P., Malik, J.: Scale-space and edge detection using anisotropic diffusion. IEEE Trans. Pattern Anal. Mach. Intell. **12**(7), 629–639 (2017)
7. Pruessmann, K., Weiger, M., Scheidegger, M., Boesiger, P.: SENSE: sensitivity encoding for fast MRI. Magn. Reson. Med **42**(5), 952–962 (1999)
8. Ruikar, D.D., Sawat, D.D., Santosh, K.C., Hegadi, R.S.: 3D imaging in biomedical applications: a systematic review (Chap 8). In: Medical Imaging: Artificial Intelligence, Image Recognition, and Machine Learning Techniques. CRC Press (2019). ISBN 9780367139612
9. Sandilya, M., Nirmala, S.: Compressed sensing trends in magnetic resonance imaging. Eng. Sci. Technol. Int. J. **204**, 1342–1352 (2017)
10. Schwartz, D., Lemoine, D., Poiseau, E., Barillot, C.: Registration of MEG/EEG data with 3D MRI: methodology and precision issues. Brain Topogr. **9**(2), 101–116 (1996)
11. Zhang, Y., Wu, L.B.S.P., Dong, Z.: A two-level iterative reconstruction method for compressed sensing MRI. J. Electromagn. Waves Appl. **25**(8–9), 1081–1091 (2011)

Segmentation of Lungs from Chest X Rays Using Firefly Optimized Fuzzy C-Means and Level Set Algorithm

Ebenezer Jangam[1(✉)] and A. C. S. Rao[2]

[1] Vignan's Foundation for Science Technology and Research,
Guntur, Andhra Pradesh, India
ebenezer.jangam@gmail.com
[2] IIT (ISM) Dhanbad, Dhanbad, Jharkhand, India

Abstract. Segmentation of lungs from chest x ray is a non trivial task required as a preprocessing step for detection of different diseases like cardiomelagy, tuberculosis, pneumonia. High accuracy in segmentation of lung results in high accuracy of detection of diseases from lungs. For the past four decades multiple techniques were proposed for automatic segmentation of lungs. In this paper, we propose a hybrid segmentation technique based on firefly optimized fuzzy c-means clustering algorithm. The output of the fuzzy c-means is given to level set to finalize the segmentation of the lungs. The performance of the proposed technique is evaluated using two public chest x ray datasets: JRST and Montgomery County. JRST contains 247 chest x-rays and MC dataset contains 138 chest x-rays. The Jaccard coefficient for the proposed segmentation technique is 95.1 which is on par with the state of art segmentation techniques.

1 Introduction

With the discovery of x ray [15] in 1895, there is a revolution in the field of diagnostics. With the invention of the modern digital computer in late 1940s, attempts were made to make computers perform tasks which need human intelligence for the completion. In 1960s researchers published articles about radiology report analysis using computer [7]. In 1970s, focus was upon the detection of abnormalities in chest x-ray using a computer.

The traditional chest analysis is the most prevalent radiological procedure, making up a minimum of a third of all exams in the radiology division. Moreover, Pulmonary diseases like pneumonia, tuberculosis, emphysema and lung cancer can be screened based on the chest radiograph [32]. But, computerized interpretation of a chest radiograph is extremely challenging due to presence of superimposed anatomical structures. The complexity of computerized analysis of chest x-ray along with their prevalence in radiology department is the main reason for the researchers to concentrate on the development of computer algorithms to assist radiologists in reading chest images.

© Springer Nature Singapore Pte Ltd. 2019
K. C. Santosh and R. S. Hegadi (Eds.): RTIP2R 2018, CCIS 1036, pp. 303–311, 2019.
https://doi.org/10.1007/978-981-13-9184-2_27

Segmentation of lungs is the essential step in the Computer Aided Detection (CADe) and Computer Aided Diagnosis (CADx) [19,22] of chest x-ray images. It is the basic step performed in the automatic tuberculosis screening [31]. It is the part of automatic pneumonia screening. It is used as the first step to detect cardiomegaly (enlargement of heart). It is the preliminary step in lung nodule detection from chest x-rays using computer algorithms. To find out the abnormalities in the lungs from chest x-rays, the primary task is to delineate the lungs from the chest x-rays [12,17,18,23,25].

Researchers proposed a wide range of techniques for automatic lung segmentation, rotation [24] and foreign object detection [37]. Some of the techniques used for lung segmentation are thresholding, region growing [21], connected component [20], neural networks, active contours, level set [8], pixel classification, structured edge detection, adversarial networks, graph cuts, game theory. Although there are a variety of existing techniques for lung segmentation, researchers are investigating for the novel techniques which can result in high accuracy. Higher the accuracy in segmentation of the lungs, higher is the accuracy in classification and detection of diseases like cardiomegaly, pneumonia and other lung related diseases. In recent times, hybrid techniques are being investigated to segment the lung regions with high accuracy.

We propose a hybrid lung segmentation technique which is the combination of fuzzy c-means and level set algorithm. The performance of the proposed technique is compared with the existing techniques. The accurate detection of lung boundary results in high accuracy in estimating heart boundaries and thereby increases the accuracy in the detection of cardiomegaly (enlargement of heart).

The content of the paper is organized in the following manner. Section 2 contains details about the proposed hybrid lung segmentation technique. Section 3 has the information about the public datasets and performance metrics used for evaluation of proposed lung segmentation technique. The performance of proposed technique is compared with existing techniques in Sect. 4. Segmented lung boundaries are used for cardiomegaly detection in Sect. 4. Section 5 concludes the paper by giving possible future directions.

2   Lung Field Segmentation Using Fuzzy C-Means and Level Set Algorithm

2.1   Fuzzy C-Means Clustering

Fuzzy c-means clustering is one of the popular algorithms used for segmentation of medical images [1,5,13]. In fuzzy clustering, centroid and scope of each cluster are estimated by minimising the cost function. The origin of FCM is K-means algorithm where N objects are assigned to N classes ($K \leq N$) with minimum value of cost function.

In the context of medical images, N is total number of pixels in the image. These N pixels should be assigned to one of the clusters. Centre of the cluster should be finalized. According to FCM clustering, the pixels nearer to the centroid will receive high values and the pixel far away from the centroid will receive low values.

2.2   Level Set Algorithm

When level set methods are used for image segmentation, an image is segmented based on boundaries which change dynamically. When fuzzy clustering is used for segmentation, it employs pixel based classification [27]. Level set algorithm embeds active contours into a time dependent PDE function f(t, x, y). The zero level set C(t) is used to approximate the active contours.

$$\begin{cases} \phi(t,x,y) < 0(x,y) \text{ is inside } C(t) \\ \phi(t,x,y) = 0(x,y) \text{ is at } C(t) \\ \phi(t,x,y) > 0(x,y) \text{ is outside } C(t) \end{cases} \tag{1}$$

The following equation determines the evolution of ϕ.

$$\begin{cases} \frac{\partial \phi}{\partial t} + |F \bigtriangledown \phi| = 0 \\ \phi(0,x,y) = \phi_0(x,y) \end{cases} \tag{2}$$

where $\bigtriangledown \phi$ denotes the normal direction, $\phi_0(x,y)$ is the initial contour and F represents the comprehensive forces [13].

2.3   Hybrid Algorithm for Lung Segmentation

Hybrid algorithm for lung segmentation uses the combination of spatial fuzzy clustering and level set algorithm to achieve better accuracy. After preprocessing, the fuzzy clustering algorithm is used to classify the pixels which belong to region of interest. The outcome of fuzzy clustering algorithm is utilised by the level set segmentation algorithm for proper initialization. The outcome of fuzzy clustering is also useful to estimate the controlling parameters of level set segmentation. In other words, Level set segmentation algorithm finds out the contours of interest in a medical image from the outcome of fuzzy clustering algorithm.

Let R_k be the region of interest which is obtained from the fuzzy c-means clustering algorithm. Then the level set function can be initialized using the following equation.

$$\phi_0(x,y) = -4\epsilon(0.5 - B_k) \tag{3}$$

where ϵ is a constant regulating Dirac function. B_k is a binary image obtained from the equation $B_k = R_k \geq b_0$, where b_0 is an adjustable threshold.

The parameters controlling the level set segmentation are selected to get optimal results by using iterative method. The initial values are chosen according to the method outlined in [13].

3   Datasets and Performance Metrics

3.1   Public Datasets of Chest X-Ray for Segmentation and Disease Detection

The following are the public datasets available for segmentation of chest X rays.

- JSRT/SCR dataset [33]
- MC dataset [11]

The following dataset contains chest x-rays labelled with multiple diseases.

- Chest x-ray 14 dataset [33]

SCR Dataset. JSRT dataset is a Chest X ray image database of 247 chest radiographs [30]. JSRT is the public dataset for lung nodule detection. SCR dataset was made public in order to promote comparision of techniques proposed for segmentation of anatomical structures [33]. SCR dataset is the most common dataset used to evaluate segmentation of anatomic structures (lungs, heart, clavicles) in a CXR as shown in Table 2.

Montgomery County Dataset. Montgomery County (MC) dataset contains 138 PA CXRs in this dataset which are collected under TB control programme. 80 CXRs are considered to be normal and 58 are abnormal with manifestations of TB [11]. Manual segmentation on images of MC dataset was performed and binary lung masks were generated. Montgomery dataset was primarily made available for tuberculosis screening but it is useful for segmentation of lung fields.

Chest X-Ray8 Dataset. Chest X-ray 8 dataset is a massive dataset released publicly in 2017 [35] for detection of multiple diseases. Initially, it was a dataset of Chest X-ray images with eight different disease labels and no finding label. Along with images and labels, there was other information about patients' visit, gender, etc. We selected 150 chest x-rays with label cardiomegaly and 150 chest x-rays which are normal with no labels. A customized dataset with 300 chest x-rays is chosen for evaluation of technique for the detection of cardiomegaly (Table 1).

Table 1. Public datasets used for evaluation

Data set	Purpose	Source	Available at	Cardinality
SCR-JSRT [33]	Segmentation of heart, clavicles and lungs	Image Sciences Institute	JSRT Website	247
Montgomery [11]	Segmentation of lungs	USNLM	NLM	138
Chest X-ray14 [35]	Multi label classification	NIH	NIH	112, 120

3.2   Performance Metrics for Segmentation of Lungs

The Jaccard similarity coefficient Ω and Dice coefficient are the commonly used metrics to measure the performance of a segmentation technique.

$$\Omega = \frac{|S \cap GT|}{|S \cup GT|} = \frac{|TP|}{|FP| + |TP| + |FN|} \tag{4}$$

where TP (true positives) is the count of pixels which are classified correctly, FP (false positives) is the count of pixels which are identified as part of the object but they belong to background in reality, and FN (false negatives) are the pixels which are classified as background but the fact is that they belong to the object.

$$DSC = \frac{|S \cap GT|}{|S| + |GT|} = \frac{2|TP|}{2|TP| + |FP| + |FN|} \tag{5}$$

Level set algorithm used the spatial fuzzy c-means clustering algorithm to improve the segmentation accuracy. The proposed technique was evaluated using JSRT SCR dataset and MC dataset.

Once the lungs are segmented, cardiothoracic ratio can be calculated by estimating the heart region. Accuracy of prediction of cardiomegaly is comparable to the state of art techniques.

4 Comparision of Performance of the Proposed Lung Field Segmentation Technique

4.1 Comparision of Performance of the Proposed Lung Field Segmentation Technique on JSRT SCR Dataset

Using proposed hybrid technique, lung region is extracted from all the images in JSRT SCR dataset. Average values are computed. The results are compared with the other lung segmentation techniques in Table 2. Our proposed segmentation technique has recorded an overlap of 95.6 ± 1.5 and DSC of 97.6 ± 1.2 which is better than human observer.

Highest accuracy is 96.3 ± 1.2 when lower order adaptive region growing technique [4] is used. Human observer accuracy is calculated as 94.6 ± 1.8 and more than half of the segmentation techniques generated an accuracy more than human observer.

4.2 Comparision of Performance of Proposed Lung Field Segmentation Technique on Montgomery County Dataset

Using proposed hybrid technique, lung region is extracted from all the images in MC dataset. Average values are computed. The results are compared with the other lung segmentation techniques in Table 3. Proposed hybrid technique has recorded a overlap of 93.5 ± 2.1 and DSC of 95.8 ± 1.5 which are comparable to the performance of state of art techniques listed in Table 3.

Only a few segmentation techniques are evaluated using Montgomery County [3,4,6]. Lower order region growing approach has reported high accuracy of 96.6 ± 1.8 as shown in Table 3. SCAN Technique has recorded an accuracy of 91.4 ± 0.61 with MC data set against 94.7 ± 0.4 using JSRT SCR dataset.

Table 2. Comparision of performance of lung field segmentation techniques on JSRT SCR Dataset

Method	Jaccard overlap	DSC	Time in sec
Lower order adaptive region growing [4]	96.3 ± 1.2	98.3 ± 0.7	3–20
Multitask FCN [34]	95.9 ± 1.7	-	-
Proposed method	95.6 ± 1.5	97.6 ± 1.2	25–30
Structured edge detector UCM [36]	95.2 ± 1.8	97.5 ± 1.0	<0.1
SIFT Flow [3]	95.4 ± 1.5	96.7 ± 0.8	20–25
Landmark based segmentation [9]	95.3 ± 2.0	-	5.3
MISCP [26]	93.0 ± 4.5	-	13–28
Hybrid voting [33]	94.9 ± 2.0	-	>34
Hierarchical lung segmentation [28]	94.6 ± 1.9	97.2 ± 1.0	35.2
SCAN [6]	94.7 ± 0.4	-	0.84
Human observer [33]	94.6 ± 1.8	-	-
Game theoretic framework [10]	94.6 ± 2.2	-	38
InvertedNet [16]	94.6	97.2	7.7
Pixel Classification post processed [33]	94.5 ± 2.2	-	30
SSM and SAM [14]	93.09 ± 2.1	-	40–80
MSCCP [26]	93.0	-	-
ASM tuned [33]	92.7 ± 3.2	-	1
ASM-SIFT [29]	92.0 ± 3.1	-	75
AAM whiskers [33]	91.3 ± 3.2	-	3
Graph cut [2]	91.0 ± 3.7	-	-

Table 3. Comparision of performance of lung field segmentation techniques on Montgomery County Dataset

Method	Jaccard overlap	DSC	Time in sec
Lower order region growing [4]	96.6 ± 1.8	97.8 ± 0.5	-
SIFT Flow [3]	94.1 ± 3.4	96.0 ± 1.8	
Proposed method	93.5 ± 2.1	95.8 ± 1.5	-
SCAN [6]	91.4 ± 0.61	-	0.84

5 Conclusion and Future Scope

Using Fuzzy c-means clustering for initialization of parameters in Level set method yielded an average overlap is 95.6 for JRST dataset, which is on par with the state of art segmentation techniques. For MC dataset the overlap is 95.8, which is on par with the state of art segmentation techniques used for lung segmentation from chest x rays. The proposed hybrid lung segmentation method is used to estimate heart boundaries. Cardio Thoracic Ratio is computed from heart boundaries and the overall accuracy is 97.0% for cardiomegaly detection. Age related information is also considered to predict the cardiomegaly with high accuracy.

Optimization techniques can be applied to improve the results further as a future work. Investigation is needed to compare the performance of different lung segmentation techniques with datasets of different sizes.

References

1. Cai, W., Chen, S., Zhang, D.: Fast and robust fuzzy c-means clustering algorithms incorporating local information for image segmentation. Pattern Recogn. **40**(3), 825–838 (2007)
2. Candemir, S., Jaeger, S., Palaniappan, K., Antani, S., Thoma, G.: Graph-cut based automatic lung boundary detection in chest radiographs. In: IEEE Healthcare Technology Conference: Translational Engineering in Health and Medicine, pp. 31–34 (2012)
3. Candemir, S., et al.: Lung segmentation in chest radiographs using anatomical atlases with nonrigid registration. IEEE Trans. Med. Imaging **33**(2), 577–590 (2014)
4. Chondro, P., Yao, C.Y., Ruan, S.J., Chien, L.C.: Low order adaptive region growing for lung segmentation on plain chest radiographs. Neurocomputing **275**, 1002–1011 (2018)
5. Chuang, K.S., Tzeng, H.L., Chen, S., Wu, J., Chen, T.J.: Fuzzy c-means clustering with spatial information for image segmentation. Comput. Med. Imaging Graph. **30**(1), 9–15 (2006)
6. Dai, W., et al.: Scan: structure correcting adversarial network for organ segmentation in chest x-rays. arXiv preprint arXiv:1703.08770 (2017)
7. Giger, M.L., Chan, H.P., Boone, J.: Anniversary paper: history and status of cad and quantitative image analysis: the role of medical physics and AAPM. Med. Phys. **35**(12), 5799–5820 (2008)
8. Hegadi, R.S., Navale, D.I., Pawar, T.D., Ruikar, D.D.: Multi feature-based classification of osteoarthritis in knee joint x-ray images (Chap. 5). In: Medical Imaging: Artificial Intelligence, Image Recognition, and Machine Learning Techniques. CRC Press (2019). ISBN 9780367139612
9. Ibragimov, B., Likar, B., Pernuš, F., Vrtovec, T.: Accurate landmark-based segmentation by incorporating landmark misdetections. In: 2016 IEEE 13th International Symposium on Biomedical Imaging (ISBI), pp. 1072–1075. IEEE (2016)
10. Ibragimov, B., Likar, B., Pernus, F., et al.: A game-theoretic framework for landmark-based image segmentation. IEEE Trans. Med. Imaging **31**(9), 1761–1776 (2012)
11. Jaeger, S., Candemir, S., Antani, S., Wáng, Y.X.J., Lu, P.X., Thoma, G.: Two public chest x-ray datasets for computer-aided screening of pulmonary diseases. Quant. Imaging Med. Surg. **4**(6), 475 (2014)
12. Karargyris, A., et al.: Combination of texture and shape features to detect pulmonary abnormalities in digital chest x-rays. Int. J. Comput. Assist. Radiol. Surg. **11**(1), 99–106 (2016)

13. Li, B.N., Chui, C.K., Chang, S., Ong, S.H.: Integrating spatial fuzzy clustering with level set methods for automated medical image segmentation. Comput. Biol. Med. **41**(1), 1–10 (2011)
14. Li, X., Luo, S., Hu, Q., Li, J., Wang, D., Chiong, F.: Automatic lung field segmentation in x-ray radiographs using statistical shape and appearance models. J. Med. Imaging Health Inform. **6**(2), 338–348 (2016)
15. Mould, R.F.: A Century of X-rays and Radioactivity in Medicine: with Emphasis on Photographic Records of the Early Years. CRC Press, Boca Raton (1993)
16. Novikov, A.A., Lenis, D., Major, D., et al.: Fully convolutional architectures for multi-class segmentation in chest radiographs. IEEE Trans. Med. Imaging **37**(8), 1865–1876 (2018)
17. Rao, N.G., Kumar, V.V., Krishna, V.V.: Texture based image indexing and retrieval. Int. J. Comput. Sci. Network Secur. **9**(5), 206–210 (2009)
18. Rao, N.G., Sravani, T., Kumar, V.V.: OCRM: optimal cost region matching similarity measure for region based image retrieval. Int. J. Multimedia Ubiquit. Eng. **9**(4), 327 (2014)
19. Ruikar, D.D., Hegadi, R.S., Santosh, K.C.: A systematic review on orthopedic simulators for psycho-motor skill and surgical procedure training. J. Med. Syst. **42**(9), 168 (2018)
20. Ruikar, D.D., Santosh, K.C., Hegadi, R.S.: Automated fractured bone segmentation and labeling from CT images. J. Med. Syst. **43**(3), 60 (2019). https://doi.org/10.1007/s10916-019-1176-x
21. Ruikar, D.D., Santosh, K.C., Hegadi, R.S.: Segmentation and analysis of CT images for bone fracture detection and labeling (Chap. 7). In: Medical Imaging: Artificial Intelligence, Image Recognition, and Machine Learning Techniques. CRC Press (2019). ISBN 9780367139612
22. Ruikar, D.D., Sawat, D.D., Santosh, K.C., Hegadi, R.S.: 3D imaging in biomedical applications: a systematic review (Chap. 8). In: Medical Imaging: Artificial Intelligence, Image Recognition, and Machine Learning Techniques. CRC Press (2019). ISBN 9780367139612
23. Santosh, K., Antani, S.: Automated chest X-ray screening: can lung region symmetry help detect pulmonary abnormalities? IEEE Trans. Med. Imaging **37**(5), 1168–1177 (2018)
24. Santosh, K., et al.: Automatically detecting rotation in chest radiographs using principal riborientation measure for quality control. Int. J. Pattern Recogn. Artifi. Intell. **29**(02), 1557001 (2015)
25. Santosh, K., Vajda, S., Antani, S., Thoma, G.R.: Edge map analysis in chest x-rays for automatic pulmonary abnormality screening. Int. J. Comput. Assist. Radiol. Surg. **11**(9), 1637–1646 (2016)
26. Seghers, D., Loeckx, D., Maes, F., Vandermeulen, D., Suetens, P.: Minimal shape and intensity cost path segmentation. IEEE Trans. Med. Imaging **26**(8), 1115–1129 (2007)
27. Sethian, J.A., et al.: Level set methods and fast marching methods. J. Comput. Inform. Technol. **11**(1), 1–2 (2003)
28. Shao, Y., Gao, Y., Guo, Y., Shi, Y., Yang, X., Shen, D.: Hierarchical lung field segmentation with joint shape and appearance sparse learning. IEEE Trans. Med. Imaging **33**(9), 1761–1780 (2014)
29. Shi, Y., et al.: Segmenting lung fields in serial chest radiographs using both population-based and patient-specific shape statistics. IEEE Trans. Med. Imaging **27**(4), 481–494 (2008)

30. Shiraishi, J., et al.: Development of a digital image database for chest radiographs with and without a lung nodule: receiver operating characteristic analysis of radiologists' detection of pulmonary nodules. Am. J. Roentgenol. **174**(1), 71–74 (2000)
31. Vajda, S., et al.: Feature selection for automatic tuberculosis screening in frontal chest radiographs. J. Med. Syst. **42**(8), 146 (2018)
32. Van Ginneken, B., Romeny, B.T.H., Viergever, M.A.: Computer-aided diagnosis in chest radiography: a survey. IEEE Trans. Med. Imaging **20**(12), 1228–1241 (2001)
33. Van Ginneken, B., Stegmann, M.B., Loog, M.: Segmentation of anatomical structures in chest radiographs using supervised methods: a comparative study on a public database. Med. Image Anal. **10**(1), 19–40 (2006)
34. Wang, C.: Segmentation of multiple structures in chest radiographs using multitask fully convolutional networks. In: Sharma, P., Bianchi, F.M. (eds.) SCIA 2017. LNCS, vol. 10270, pp. 282–289. Springer, Cham (2017). https://doi.org/10.1007/978-3-319-59129-2_24
35. Wang, X., Peng, Y., Lu, L., Lu, Z., Bagheri, M., Summers, R.M.: ChestX-ray8: Hospital-scale chest x-ray database and benchmarks on weakly-supervised classification and localization of common thorax diseases. In: 2017 IEEE Conference on Computer Vision and Pattern Recognition (CVPR), pp. 3462–3471. IEEE (2017)
36. Yang, W., et al.: Lung field segmentation in chest radiographs from boundary maps by a structured edge detector. IEEE J. Biomed. Health Inform. **22**(3), 842–851 (2018)
37. Zohora, F.T., Antani, S., Santosh, K.: Circle-like foreign element detection in chest x-rays using normalized cross-correlation and unsupervised clustering. In: Medical Imaging 2018: Image Processing, vol. 10574, p. 105741V. International Society for Optics and Photonics (2018)

Detection and Classfication of Non-proliferative Diabetic Retinopathy Using Retinal Images

D. B. Mule[1]($\boxtimes$), S. S. Chowhan[2]($\boxtimes$), and D. R. Somwanshi[3]($\boxtimes$)

[1] Department of Computer Science, STNMS Latur, Latur, Maharashtra, India
deepa.101@rediffmail.com
[2] Symbiosis Institute of Computer Science and Research (SICSR),
Symbiosis International (Deemed University), Pune, India
drschowhan@gmail.com
[3] Department of Computer Science, College of Computer Science and Information
Technology (COCSIT), Latur, Maharashtra, India
somwanshi1234@gmail.com

Abstract. Diabetic retinopathy is one of the general eye disease observed in diabetic patients and which is the most important reason of vision loss. Cataracts and glaucoma are some other eye diseases found in diabetic patients due to complication in diabetes. Diabetic Retinopathy is shown generally because of the impairment of retinal blood vessels in the patients suffering from diabetic patients. So there is a need of recognition of Diabetic Retinopathy in early stage of diabetes is required and also to safeguard or to save the diabetic patient from early vision loss and for that it is extreme necessary to detect and extract the retinal image and retinal blood vessels from retinal image. Automated segmentation and recognition system for retinal images is proposed in this paper to determine the intensity of diabetes. Segmentation of retinal image from eye image, segmentation of blood vessels from retinal images, segmentation of microaneurysms and hardexudates are performed more accurately. Proposed technique uses image processing techniques and support vector machine for detection and classification of non-proliferative diabetic retinopathy grades like normal, mild, moderate and severe. The proposed method is tested on STARE dataset and results are compared using parameters of sensitivity, specificity, and accuracy, and found high accuracy of 96.32 which can help to detect and prevent diabetic retinopathy.

Keywords: Diabetic retinopathy · STARE · Machine learning ·
Support vector machine

1 Introductions

Detection and Classification of retinal images and detection of blood vessels in retinal images are the most imperative part in different types of medical

© Springer Nature Singapore Pte Ltd. 2019
K. C. Santosh and R. S. Hegadi (Eds.): RTIP2R 2018, CCIS 1036, pp. 312–320, 2019.
https://doi.org/10.1007/978-981-13-9184-2_28

diagnoses process. Width and color identification of blood vessels, reflectivity and tortuosity, vessels and irregular branching of blood vessels in retinal images and the count of vessels of a definite width in as image are used to analyze information for diagnosis process. When the amount of blood vessels in an image are huge, or when a huge amount of images are acquired, then the manual description of the vessels is monotonous or even difficult. Detection and classification of micro-aneurysms and hard-exudates is also essential for more perfect analysis of diabetic retinopathy, and which need to be efficiently accomplished. Diabetes affects almost every one and produces numbers of complications such as vision loss, heart failure and stroke. Numbers of such eye problems of diabetic eye diseases have to face by the people with diabetes because of complication in diabetes. Cataracts and glaucoma are the two most common problems seen in the patients having diabetes, and the other disease's affect on the retina is the main cause of vision loss [1]. Diabetic Retinopathy is mainly caused by complication in diabetes in diabetic patients because of that abnormality in the retina is seen and in the worst case loss of sight or severe vision loss is occurred [1]. The main reason behind the Diabetic retinopathy is diabetes and due to that blood vessels in retina are changes: the thin, light-sensitive inner lining are seen and such changes are mainly observed at back side of the eye and are known as diabetic retinopathy [2]. There are no such symptoms in the early stages of diabetes but the number and severity mostly increase as the time passes. In most of the patients diabetic changes in the retina are seen after approximately 20 years of diabetes [2]. The blurred or unclear vision, floaters and flashes and finally the loss of vision are some of the common symptoms of the diabetic retinopathy [3]. Retinal images plays important role in detection of diabetic retinopathy in diabetic patients. Changes in the blood vessels of the retina can be observed with the help of retinal images. Due to the changes in blood vessels, blood leaks from the vessels and they grow fragile new vessels. Due the damage of nerve cell, impaired vision problem can occur and because of changes in verve cell blurred vision may be observed. If this problem is untreated, hemorrhage can be seen and detachement of retina is can also be observed. The main cause of blindness in the United States Diabetic is only due to the diabetic retinopathy. Following are some symptoms of Diabetic retinopathy.

- Unclear image appears in front of eye
- Abrupt vision loss
- Rings around lights may appear.
- Shady spots or blinking lights

Performing the mass screening of diabetes patients will result in a large number of images that need to be examined. The cost of manual examination is prohibiting the implementation of screening on a large scale. A possible solution could be the development of an automated screening system for retinal images [1]. Such a system should be able to distinguish between affected retinal images and normal retinal images. This will significantly reduce the workload for the ophthalmologists as they have to examine only those images diagnosed by the system as possibly abnormal [4]. Eye screening is a key part of diabetes care

[5]. If the patient suffers from diabetes, then the patient's eyes are at risk of damage from diabetic retinopathy, a condition that can lead to sight loss if it's not treated. Diabetic retinopathy is one of the most common causes of sight loss among people of working age [6,7]. Screening is a way of detecting the condition early before patient notice any changes to his vision. Diabetic retinopathy is categorized into two types: proliferative and non-proliferative diabetic retinopathy. Non-proliferative diabetic retinopathy is the most common type of diabetic retinopathy and measured in grade such as normal, mild, moderate and severe. This paper proposes a technique for detection, segmentation of microaneurysms, hardexudates, and blood vessels, and classification of non-proliferative diabetic retinopathy using support vector machine. The proposed method is tested on STARE dataset and results are compared using parameters of sensitivity, specificity, and accuracy, and found high accuracy of 96.32% which can help to detect and prevent diabetic retinopathy.

2   Related Work

Stall et al. [4] works ridge based vessels segmentation and uses color retinal images. Their method suggests automatic screening of diabetic retinopathy and their algorithms is based on extraction of image ridges, they also uses the k-NN classifier for feature vector classification. Tariq et al. [5] work on background and noise segmentation of color retinal images. They present a novel approach for segmentation of colored retinal images, coarse segmentation and background segmentation used for this. Siva Sundhara Raja et al. [6] works on detection of blood vessels in retinal images, they use anisotropic diffusion filter and binary morphological operation and achieved 98.08% accuracy using DRIVE dataset. Karegowda et al. [7] works on exudates detection in retinal images using Artificial Neural Networks. Decision tree and GA-CFS method are used as input to the BPN model for detection of exudates in retinal images. Mendonça and Campilho [8] suggested vessels segmentation algorithm, there algorithms uses centerline of the vessel then the filtering process for vessels. Multiscale morphological enhancement technique was used for enhancement of the contrast of the blood vessels. The authors achieved 96.33% accuracy in DRIVE dataset and 95.79% accuracy in STARE dataset. Wilcoxon matched pairs testing algorithm was used to prove the accuracy of their results. Palomera-Perez et al. [9] have used region developing algorithm based on feature mining for the segmentation of blood vessels. The domain partitioning based parallelism was used to group the vessels. The authors achieved 92.5% accuracy in DRIVE dataset and 92.6% accuracy in STARE dataset. Fraz et al. [10] uses vessels segmentation method based on collaborative classification. The gradient vector field and Gabor transform were constructed and uses collaborative classification for feature selection. The authors achieved 72.62% sensitivity, 97.64% specificity, and 95.11% accuracy in STARE dataset and 74.06% sensitivity, 98.07% specificity, and 94.8% accuracy in DRIVE dataset.

3   Datasets

Different dataset are available for the proposed study, these are mainly DRIVE, STARE, Diaretdb0 and Diaretdb1 are the publicly available dataset for detection of Diabetic retinopathy, among that we have used STARE and DRIVE Dataset for this study.

3.1   STARE Dataset

This dataset [11] comprises 20 eye fundus color images captured with a Top Con TRV-50 fundus camera at 35°. FOV. The images were digitalized to 700 × 605 pixels. This dataset is observed by different observes and it consists of two groups of manual segmentation. Performance is computed with the segmentations of the first observer as ground truth. DRIVE dataset also contains manually blood vessel segmentation.

3.2   DRIVE Dataset

This dataset [12] contains 400 images of patients suffers from diabets and whose age in the range of 25 to 90 years from that 400 images we have selected 40 images from that 33 does not show the any indication of Diabetic Retinopathy (DR) and remaining 7 shows the signs DR. Images from the database are compressed with JPEG technology. Canon CR5 CCD camera is used to acquire the images with the 45° field of view (FOV). Color level of each available images is 8 bit and resolution of each image is 768 × 584. The diameter of each image is around 540 pixel and FOV is circular. A mask of each image is also available in the database and images are cropped for FOV.

4   Methodology

The proposed methodology is as shown in Fig. 1. Image enhancement operation is performed on image which is taken as input and also highlighted area affected by lesion in retinal image. After that image segmentation operation is performed. Here segmentation is categorized into four steps these are namely: Noise Segmentation, background segmentation, vessels segmentation and microaneurysms and hard-exudates segmentation. Noise Segmentation is performed to remove the noise from retinal image and similarly Background segmentation is performed to remove background image from the retinal image. Noise Segmentation and Background segmentation is mainly performed because these area are actually not needed and removing these areas means no need to process unwanted area means it reduces the processing time, finally optic disc of retinal image is separated and vessel are extracted after the noise and background segmentation steps. The detailed algorithm is presented as below:

1. Acquiring the input retinal image.
2. Resizing the image to clearly identify the vessels.
3. Extracting green component from RGB image.
4. Complement and apply adaptive histogram.
5. Apply morphological opening.
6. Extracting optic disc.
7. Perform background and Noise Segmentation.
8. Perform filtering and image adjustment.
9. Perform Microaneurysms and hardexudates Segmentation.
10. Apply gray level threshold and convert image into binary image.
11. For vessels segmentation Single-level discrete 2-D wavelet transform is need to apply.
12. A last step is to highlight the blood vessels and detecting the lesion.

Images are acquired from STARE dataset as discussed above. Most of the images in database are affected by diabetic retinopathy. After acquiring the image, image size is changed for representing retinal image with the vessels are marked. The

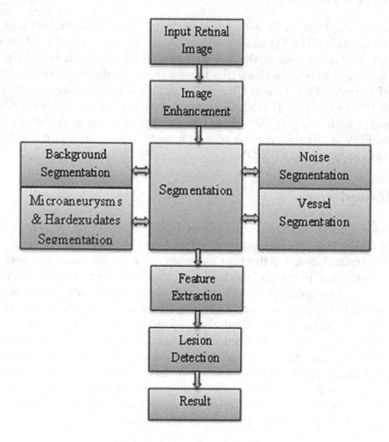

Fig. 1. Proposed methodology

green channel image from RGB image is separated, then applied the complemetation operation on image and finally adaptive histogram of the image is calculated. To represents the specific detailed portion from the image morphological opening operation is used. Optic disc is the joining point of the all blood vessels in retina so removing or eliminating the optic disc is performed. Removing of optic disc is also needed to identify some vessels that are crosses through it. Therefore the area of optic disc is removed before the vessels are segmented. The morphological opening operation is used for detection and segmentation of the optic disc and that is performed using the green channel of original retinal image. To remove the noise from the image, 2-Dimensional median filter is used. And finally to remove the background image again Morphological opening operation is used. Gray level threshold is calculated to convert the image in to binary image. Finally single level discrete 2-D wavelet transform is used to segment the boundaries of vessels in retinal image After that microaneurysms and exudates are segmented. Microaneurysms and exudates indentification is required and which is most essential factor in diagnosis of diabetic retiopathy. Microaneurysms is simply the are abnormal protions in the blood vessels, which looks like a tiny and oval shape dots appear in portion of the blood vessels. Hard exudates generally are the unmatched yellow and white colored intra-retinal dots which are varies between trivial flecks to superior acnes and which might developed into trinkets called as circinates. For the detection and segmentation of Microaneurysms, disc-based dilation operation and Single-level discrete 2-D wavelet transformation is applied then tracing region boundaries in binary image is performed. For Detection of hard exudates, the green component is extracted from the RGB image and a threshold-based binarization depending on the standard deviation of the green component is applied. The grades of nonproliferative diabetic retinopathy (NPDR) are categorized with the help of Support Vector Machine (SVM) based techniques. Grades in the non-proliferative diabetic retinopathy are characterized into: normal, mild, moderate and severe. These grades are provided by medical experts and are based on the following criteria: Normal if (μA $= 0$ and H $= 0$), Mild NPDR if (($0 < \mu$A $<= 5$) and (H $= 0$)), Moderate NPDR if (($5 < \mu$A < 15 or $0 < $ H < 5) and (VN $= 0$)) and Severe NPDR if ((μA $>= 15$) or (H $>= 5$) or (NV $= 1$)). In the above expression μA is called the number of Micro-aneurysms, H is called the number of Hemorrhages and Presence of neovascularization is denoted by NV.

5   Result

For the non-proliferative diabetic retinopathy grade classification, the proposed method selects the four features these are: density of blood vessel, possibility of total number of microaneurysms, definite number of microaneurysms and density of hard exudates. The implementation of the proposed method is performed in Matlab 2010. The detection and classification performance of the method proposed for non-proliferative diabetic retinopathy grades detection is tested using STARE and DRIVE Dataset. Total of 80 images were tested from these

database. Based on human eye observation weather it is the accurate segmentation, or poor segmentation can be decided by comparing the already extracted
images from the above databases. Segmentation results of this technique are
presented in Table 1.

In Table 1, row one, shows the original retinal image, vessels segmented image
and original images with vessels are marked, in next row again shows original image, Green Component image and Microaneurysms Segmented Image, in
last row again shows original image, Green Component image after Morphological operation and hard exudates segmented image. Finally the support vector
machine classifier is trained with the four features that we have extracted. The
following table, Table 1 shows the accuracy, sensibility and specificity. The per-

Table 1. Original retinal image, vessels segmented image and original images with
vessels are marked.

Original Image	Vessels Segmented Image	Original Image with Vessels are Marked
Original Image	Green Component image	Microaneurysms Segmented Image
Original Image	Green Component image after Morphological operation	hard exudates segmented Image

formance of the SVM Classifier for the classification of grades of non-proliferative diabetic retinopathy is shows in Table 2:

Table 2. Accuracy, sensibility and specificity of SVM classifier

Metric	Support vector machine
Accuracy	95.32%
Sensibility	96.33%
Specificity	97.32%

Table 3. Performance of the SVM classifier for the classification of grades of non-proliferative diabetic retiopathy

Metric	Grade0 (normal)	Grade1 (mild)	Grade2 (moderate)	Grade3 (severe)	Average
Accuracy	81.06%	93.02%	86.06%	88.05%	87.05%
Sensibility	96.07%	2.02%	22.23%	81.25%	50.39%
Specificity	72.03%	99.98%	93.25%	92.23%	89.37%

6   Conclusion

This study presents the effective and accurate method for the detection and segmentation of blood vessels, microaneurysms, the optic-disc, and hard exudates that are more beneficial to decide the strength of diabetes. The method uses image processing techniques and support vector machine for detection and classification of grades of non-proliferative diabetic retinopathy such as severe, moderate, mild and normal. The features that are collected demonstrate that, these are more suitable for the classification of grades of non-proliferative diabetic retinopathy. The proposed method is tested on STARE and DRIVE dataset and results are compared using parameters of sensitivity, specificity, and accuracy, and found high accuracy of 96.32% which can help to detect and prevent diabetic retinopathy. This method can be improved in future for the detection of soft exudates, edema, and macula, and for using the application of texture analysis in order to improve the accuracy and sensibility of the method presented (Table 3).

References

1. Archer, D.B.: Diabetic retinopathy: some cellular, molecular and therapeutic considerations. Eye **13**, 497–523 (1999)
2. Antal, B., Hajdu, A.: An ensemble-based system for microaneurysm detection and diabetic retinopathy grading. IEEE Trans. Biomed. Eng. **59**(6), 1720–1726 (2012)

3. Fong, D.S., Aiello, L., Gardner, T.W., et al.: Retinopathy in diabetes. Diab. Care **27**(Suppl. 1), S84–S87 (2004)
4. Stall, J., Viergever, M.A.: Ridge based vessel segmentation in color image of retina. IEEE Trans. Med. Imaging **23**(4), 501–509 (2004)
5. Tariq, A., Akram, M.U.: An automated system for colored retinal image background and noise segmentation. In: IEEE Symposium on Industrial Electronics and Applications, vol. 1, pp. 423–427 (2010)
6. Siva Sundhara Raja, D., Vasuki, S.: Automatic detection of blood vessels in retinal images for diabetic retinopathy diagnosis. In: Computational and Mathematical Methods in Medicine, vol. 2015, pp. 1–12. Hindawi Publishing Corporation (2014)
7. Karegowda, A.G., Nasiha, A., Jayaram, M.A.: Exudates detection in retinal images using back propagation neural network. Int. J. Comput. Appl. **25**(3), 25–31 (2011)
8. Mendonça, A.M., Campilho, A.: Segmentation of retinal blood vessels by combining the detection of centerlines and morphological reconstruction. IEEE Trans. Med. Imaging **25**(9), 1200–1213 (2006)
9. Palomera-Perez, M.A., Martinez-Perez, M.E., Benitez-Perez, H., Ortega-Arjona, J.L.: Parallel multiscale blood vessel segmentation. IEEE Trans. Biomed. Eng. **59**(9), 2538–2548 (2012)
10. Fraz, M.M., Remagnino, P., Hoppe, A., et al.: An ensemble classification-based approach applied to retinal blood vessel segmentation. IEEE Trans. Biomed. Eng. **59**(9), 2538–2548 (2012)
11. Goldbaum, M.D.M.: STARE dataset website. Clemson University, Clemson (1975). http://www.ces.clemson.edu
12. Staal, J.J., Abramoff, M.D., Niemeijer, M., Viergever, M.A., van Ginneken, B.: Digital retinal image for vessel extraction (DRIVE) database. Image Sciences Institute, University Medical Center Utrecht, Utrecht (2004). http://www.isi.uu.nl/Research/Databases/DRIVE/

Public Datasets and Techniques for Segmentation of Anatomical Structures from Chest X-Rays: Comparitive Study, Current Trends and Future Directions

Ebenezer Jangam[1]([✉]) and A. C. S. Rao[2]

[1] Vignan's Foundation for Science Technology and Research,
Guntur, Andhra Pradesh, India
ebenezer.jangam@gmail.com
[2] IIT (ISM) Dhanbad, Dhanbad, Jharkhand, India

Abstract. Segmentation of anatomical structures from chest x ray has an increasing importance in the past four decades and researchers have proposed various techniques and evaluated them using different datasets. In order to evaluate and compare a proposed technique, it is necessary to have knowledge about public datasets available. In this survey, properties and characteristics of different public chest x ray datasets available for segmentation of anatomical structures are studied. Different approaches for segmentation of anatomical structures (lung, heart, clavicles) are summarized. Segmentation techniques for each anatomical structure for a given dataset are compared and analyzed. The paper outlines the issues where further research can be focused.

1 Introduction

With the discovery of x ray [15] in 1895, there is a revolution in the field of diagnostics. With the invention of the modern digital computer in late 1940s, attempts were made to make computers perform tasks which need human intelligence for the completion. In 1960s researchers published articles about radiology report analysis using computer [8]. In 1970s, focus was upon the detection of abnormalities in chest x-ray using a computer.

The traditional chest analysis is the most prevalent radiological procedure, making up a minimum of a third of all exams in a typical radiology division. Moreover, Pulmonary diseases like pneumonia, tuberculosis [20,21], emphysema and lung cancer can be screened based on the chest radiograph [26]. But, computerized interpretation of a chest radiograph is extremely challenging due to presence of superimposed anatomical structures. The complexity of computerized analysis of chest x-ray along with their prevalence in radiology department is the main reason for the researchers to concentrate on the development of computer algorithms to assist radiologists in reading chest images.

Researchers have developed a variety of algorithms for computer aided analysis of medical images (X-ray, computed tomography, for instance) [17].

© Springer Nature Singapore Pte Ltd. 2019
K. C. Santosh and R. S. Hegadi (Eds.): RTIP2R 2018, CCIS 1036, pp. 321–331, 2019.
https://doi.org/10.1007/978-981-13-9184-2_29

Segmentation of organs (like lung, heart, clavicles) has been regarded as one of most important problems in computer aided diagnostics applications [18,19]. Higher the accuracy in segmentation of the anatomical structures, higher is the accuracy in classification and detection of diseases like cardiomegaly, pneumonia and other lung related diseases.

One of the major problems faced by the researchers was the lack of public chest x-ray datasets which can act as benchmark for the comparison of performance of different techniques proposed. Performance of an algorithm was evaluated on customized x-ray data sets for about three decades from 1970s to late 1990s. In 2000, a public dataset [25] from JSRT was made available to researchers. A few more public datasets were made available which can act as benchmark for the evaluation of proposed algorithms.

Although, in recent years, a few more public datasets [7,9,12,27,29] of chest x-ray are dedicated, the information about the recent datasets is not available in any of the existing surveys according to our knowledge. Authors in [14] have focused on different segmentation techniques on chest x-ray datasets but the recent techniques are not included. Therefore, the focus of this survey is on the public datasets suited for segmentation of anatomical structures from chest x-rays. The use of publicly available datasets for evaluation of a given approach has two main advantages. First advantage is that the time and resources can be saved as new chest x-ray data set need not be obtained and researchers can spend their efforts on development of their algorithms and implementations. Second advantage is the use of common datasets enables comparison of performance of different approaches proposed for a given task [4].

The scope of the survey is public chest x-ray datasets for segmentation of anatomical structures. All the techniques that are evaluated using a specific dataset are compared in terms of corresponding performance metrics. Section 2 gives description about three public datasets available for segmentation of anatomical structures. Section 3 gives details about commonly used performance metrics for segmentation of anatomical structures. Section 4 compares different techniques based on the common data set used for evaluation. Section 5 concludes the paper by outlining some of the observations which are helpful for future work.

2   Public Datasets of Chest X-Ray for Segmentation of Anatomical Structures

The following are the public datasets available for segmentation of anatomical structures (lung, heart and clavicles).

- JSRT/SCR for lung segmentation, heart segmentation and clavicle segmentation [27]
- MC dataset for lung field segmentation [12]
- CRASS dataset for lung field segmentation [9]

Some datasets like Montgomery County (MC) can be use for multiple purposes. It can be used for lung field segmentation and tuberculosis screening.

2.1 SCR Dataset

JSRT in cooperation with Japanese Radiological Society has developed a Chest X ray image database of 247 chest radiographs with and without nodule. The images are collected from thirteen distinct institutions in Japan and 1 in the USA in 1988 and made it as a public dataset [25]. Out of 247 images, 154 CXR images have lung nodules, while 93 are actually normal with no nodules. JSRT is the only public dataset available for lung nodule detection (Figs. 1 and 2).

ISI, University Medical Centre Utrecht, The Netherlands has established SCR dataset [27] in order to promote comparision of techniques proposed for segmentation of lung regions, the heart and the clavicles [27]. For each image from JSRT dataset, the borders of both lungs, the heart, and both the clavicles were stored in files with .pfs extension. Individual anatomic structures are stored with .gif extension [27]. SCR dataset is the most common dataset used in studies related to segmentation of anatomic structures (lungs, heart, clavicles) in a CXR as shown in Table 2. Sample masks are shown in the Fig. 3.

Fig. 1. Sample clavicle segmentation masks for images in SCR dataset

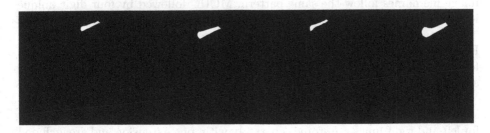

Fig. 2. Sample clavicle segmentation masks for images in SCR dataset

2.2 CRASS Dataset

CRASS dataset was collected from African region where tuberculosis is prevalent. It contain a set of 548 PA chest radiographs acquired from adults of age greater than 15 years. Out of 548 images, 333 are abnormal and 225 are normal. Among 333 abnormal images, 220 are abnormal at upper lung area near the

Fig. 3. Heart segmentation masks for images in SCR dataset

clavicle. Among 548 images, 299 are marked as training set and the remaining 249 images are considered as test set. The main purpose of CRASS dataset is to form a benchmark for clavicle segmentation.

Researchers have proposed different techniques for clavicle segmentation and evaluated on CRASS dataset as shown in the Table 5. Human observers performed better than all other techniques [9]. Better techniques for clavicle segmentation need to be developed.

2.3   Montgomery County Dataset

U.S. National Library of Medicine (USNLM) and the Department of Health and Human Services, MC, MD, USA has collected Montgomery County (MC) dataset. There are 138 PA CXRs in this dataset which are collected under TB control programme. 80 CXRs are considered to be normal and 58 are abnormal with manifestations of TB [12].

All images are deidentified and are available in DICOM format. The spatial resolution of the CXR images is either 4020 by 4892 or 4892 by 4020 pixels. All image file names follow the same pattern: MCUC followed by four digit unique identifier. For each CXR, corresponding clinical readings are stored in a file with .txt extension. Clinical reading comprises of age, gender and lung abnormality. For example, a clinical reading of a CXR in the MC appears in the following form: Patient's Sex: M Patient's Age: 031Y Cavitary nodular infiltrate in RUL; active TB.

Manual segmentation on images of MC dataset was performed under the supervision of a radiologist and binary lung masks were generated. Mask images for left and right lungs are stored separately with .png extension and are included in seperate folders in the dataset [12]. Montgomery dataset was primarily made available for tuberculosis screening but it is useful for segmentation of lung fields. Table 6 gives different techniques and their performance when MC dataset is used. Lower order region growing technique [5] achieved higher accuracy 96.6 ± 1.8 when compared to other techniques. Segmentation techniques should be evaluated on multiple datasets (SCR and MC) to achieve better insight about their performance.

Table 1. Public datasets for segmentation of anatomical structures

Data Set	Purpose	Source	Available at	Cardinality
SCR-JSRT [27]	Segmentation of heart, clavicles and lungs	Image Sciences Institute	JSRT Website	247
Montgomery [12]	Segmentation of lungs	USNLM	NLM	138
CRASS [9]	Segmentation of clavicles	USNLM	Grand challenge	548

3   Performance Metrics for Segmentation of Anatomical Structures

There are different ways to measure the performance of Segmentation technique but the final decision whether the segmentation is sufficiently accurate or not is determined by the requirements of the target application. In general, the problem of segmentation is considered as a relation between lung and background. Most of the research papers consider classical accuracy, sensitivity, and specificity as performance metrics (Table 1).

$$accuracy = \frac{N_{TP} + N_{TN}}{N_{TP} + N_{TN} + N_{FP} + N_{FN}} \tag{1}$$

$$sensitivity = \frac{N_{TP}}{N_{TP} + N_{FN}} \tag{2}$$

$$specificity = \frac{N_{TN}}{N_{TN} + N_{FP}} \tag{3}$$

N_{TP} denotes the true positive portion and it is equivalent to the portion of image identified correctly as lung region, N_{TN} denotes the true negative portion of the image which is equivalent to the portion of image correctly identified as background region, N_{FP} denotes the false positive portion and it is equivalent to the part of the image incorrectly classified as lung region, and N_{FN} is the false negative fraction which is same as the part of the image incorrectly classified as background region.

The Jaccard similarity coefficient is the overlap measure. It is the measured as the coincidence between the ground truth (GT) and the estimated segmentation mask (S) over all pixels in the image.

$$\Omega = \frac{|S \cap GT|}{|S \cup GT|} = \frac{|TP|}{|FP| + |TP| + |FN|} \tag{4}$$

where TP (true positives) is the count of pixels which are classified correctly, FP (false positives) is the number of pixels which are identified as part of the object but they belong to background in reality, and FN (false negatives) are the pixels which are identified as background but are in actually part of the object.

Dice coefficient is the metric to measure intersection between the GT and S as given below.

$$DSC = \frac{|S \cap GT|}{|S| + |GT|} = \frac{2|TP|}{2|TP| + |FP| + |FN|} \qquad (5)$$

Average contour distance (ACD) is the average distance between the segmentation boundary S and the ground truth boundary GT [3].

4   Comparitive Study of Segmentation Techniques for Each Dataset

4.1   Comparision of Performance of Lung Field Segmentation Techniques on JSRT SCR Dataset

SCR dataset was used to evaluate the performance of different lung segmentation techniques as shown in Table 2. Highest accuracy is 96.3 ± 1.2 when lower order adaptive region growing technique [5] is used. Human observer accuracy is calculated as 94.6 ± 1.8 and more than half of the segmentation techniques generated an accuracy more than human observer. Accuracy could be improved further and execution time could be decreased.

Table 2. Comparision of performance of lung field segmentation techniques on JSRT SCR dataset

Method	Jaccard overlap	DSC	Time in sec
Lower order adaptive region growing [5]	96.3 ± 1.2	98.3 ± 0.7	3–20
Multitask FCN [28]	95.9 ± 1.7	-	-
Structured edge detector UCM [30]	95.2 ± 1.8	97.5 ± 1.0	<0.1
SIFT flow [3]	95.4 ± 1.5	96.7 ± 0.8	20–25
Landmark based segmentation [10]	95.3 ± 2.0	-	5.3
MISCP [22]	93.0 ± 4.5	-	13–28
Hybrid voting [27]	94.9 ± 2.0	-	>34
Hierarchical lung segmentation [23]	94.6 ± 1.9	97.2 ± 1.0	35.2
SCAN [6]	94.7 ± 0.4	-	0.84
Human observer [27]	94.6 ± 1.8	-	-
Game theoritic framework [11]	94.6 ± 2.2	-	38
InvertedNet [16]	94.6	97.2	7.7
Pixel classification post processed [27]	94.5 ± 2.2	-	30
SSM and SAM [13]	93.09 ± 2.1	-	40–80
MSCCP [22]	93.0	-	-
ASM tuned [27]	92.7 ± 3.2	-	1
ASM-SIFT [24]	92.0 ± 3.1	-	75
AAM whiskers [27]	91.3 ± 3.2	-	3
Graph cut [2]	91.0 ± 3.7	-	-

4.2 Comparision of Performance of Heart Segmentation Techniques on JSRT SCR Dataset

Segmentation of heart from a given chest x-ray is a challenging task as it is difficult to extract the heart region exactly. In spite of the complexity, various techniques were proposed and evaluated on JSRT SCR dataset. Most of them have low accuracy when compared to human observer as shown in Table 3. Highest accuracy 89.9 ± 4.4 was achieved by using Fully Convolutional Networks [28].

Table 3. Comparision of performance of heart segmentation techniques on JSRT SCR dataset

Method	Jaccard overlap	DSC	Time in sec
Multitask FCN [28]	89.9 ± 4.4	-	-
InvertedNet with ELU [16]	88.2	93.7	-
SIFT Flow [3]	95.4 ± 1.5	96.7 ± 0.8	
Human observer [27]	87.8 ± 5.4	-	-
Hybrid voting [27]	86.0 ± 5.6	-	-
SCAN [6]	86.6 ± 1.2	-	-
Hybrid ASM/PC [27]	83.6 ± 8.2	-	-
Hybrid AAM/PC [27]	82.7 ± 8.4	-	-
AAM whiskers BFGS [27]	81.4 ± 7.0	-	-
Pixel classification postprocessed [27]	82.4.5 ± 7.7	-	-
Sparse shape composition [23]	94.6 ± 1.9	97.2 ± 1.0	35.2
Pixel Classification [27]	81.1 ± 7.7	-	-
ASM tuned [27]	81.4 ± 7.6	-	-
ASM default [27]	77.5 ± 13.5	-	-
Mean shape [27]	64.3 ± 14.7	-	-

4.3 Comparision of Performance of Clavicle Segmentation Techniques on JSRT SCR Dataset

Clavicle segmentation is the most challenging task as it is very difficult to seperate the clavicles from a given chest x-ray. Even though automated techniques were proposed, none of them performed better than human observer as shown in Table 4. Maximum accuracy achieved was 89.6 ± 3.7 by the human observer.

4.4 Comparision of Performance of Clavicle Segmentation Techniques on CRASS Dataset

Clavicle segmentation is quite challenging but researchers have addressed the problem by adopting pixel classification based methods, HDAP, Fully Convolution Networks and Active Shape Model. None of the techniques have resulted in better accuracy than human observer as shown in Table 5.

Table 4. Comparision of performance of clavicle segmentation techniques on JSRT SCR Dataset

Method	Jaccard overlap	DSC	Time in sec
Human observer [27]	89.6 ± 3.7	-	-
InvertedNet with ELU [16]	86.8	92.9	-
Multitask FCN [28]	86.3 ± 4.5	-	-
ASM tuned [27]	73.4 ± 13.7	-	-
Hybrid voting [27]	73.6 ± 10.6	-	-
ASM default [27]	69.0 ± 14.3	-	-
Hybrid AAM/PC [27]	66.3 ± 15.7	-	-
AAM whiskers BFGS [27]	62.5 ± 17.1	-	-
Hybrid ASM/PC [27]	61.3 ± 20.6	-	-
Pixel classification postprocessed [27]	61.5 ± 12.3	-	-
Pixel classification [27]	61.8 ± 10.0	-	-
AAM default [27]	50.5 ± 23.4	-	-
Mean Shape [27]	30.3 ± 21.4	-	-

4.5 Comparision of Performance of Lung Field Segmentation Techniques on Montgomery County Dataset

Only a few segmentation techniques are evaluated using Montgomery County Dataset [3,5,6]. Lower order region growing approach has reported high accuracy of 96.6 ± 1.8 as shown in Table 6. SCAN Technique has recorded an accuracy of 91.4 ± 0.61 with MC data set against 94.7 ± 0.4 using JSRT SCR dataset.

5   Conclusion and Future Scope

Lung field segmentation has attracted attention from most of the researchers and some of the techniques have attained an accuracy more than the accuracy of human observer. Segmentation of other anatomical structures heart and clavicles was not focused much during the last four decades. The accuracies reported in the automatic segmentation of heart and clavicles were not encouraging due to the reason that medical applications demand an accuracy more than the accuracy of human observer.

Another observation results from the fact that most of the researchers have used JSRT SCR dataset alone for the evaluation of the performance of the technique proposed. It is advisable to evaluate the performance of the proposed technique using all the available datasets to have a better insight.

Eventhough CRASS and JSRT datasets are available for clavicle segmentation, segmentation of clavicle remains as a challenging task. Better techniques should be proposed to increase the accuracy of clavicle segmentation.

Table 5. Comparision of performance of clavicle segmentation techniques on CRASS dataset

Method	Jaccard overlap	DSC	Time in sec
Third observer [9]	93.0 ± 4.0	-	-
Second observer [9]	93.0 ± 4.0	-	-
HDAP No shape [9]	86.0 ± 1.0	-	-
HDAP [9]	85.0 ± 1.0	-	-
HDAP No borde [9]	80.0 ± 1.0	-	-
HAP [9]	77.0 ± 1.0	-	-
PC postprocessed [9]	73.0 ± 1.1	-	-
SFCN-ML [1]	70.47	81.40	-
ASM default [9]	69.0 ± 1.9	-	-

Table 6. Comparision of performance of lung field segmentation techniques on Montgomery County Dataset

Method	Jaccard overlap	DSC	Time in sec
Lower order region growing [5]	96.6 ± 1.8	97.8 ± 0.5	-
SIFT flow [3]	94.1 ± 3.4	96.0 ± 1.8	
SCAN [6]	91.4 ± 0.61	-	0.84

As massive datasets of chest x-rays are available, deep learning techniques could play a major role in automatic multiple disease detection.

Paediatric chest x-ray datasets are needed to analyze and process the chest diseases related to children. Hence more paediatric pubic datasets are needed for evaluation of segmentation and disease detection techniques.

References

1. Bi, L., Kim, J., Kumar, A., Fulham, M., Feng, D.: Stacked fully convolutional networks with multi-channel learning: application to medical image segmentation. Vis. Comput. **33**(6–8), 1061–1071 (2017)
2. Candemir, S., Jaeger, S., Palaniappan, K., Antani, S., Thoma, G.: Graph-cut based automatic lung boundary detection in chest radiographs. In: IEEE Healthcare Technology Conference: Translational Engineering in Health and Medicine, pp. 31–34 (2012)
3. Candemir, S., et al.: Lung segmentation in chest radiographs using anatomical atlases with nonrigid registration. IEEE Trans. Med. Imaging **33**(2), 577–590 (2014)
4. Chaquet, J.M., Carmona, E.J., Fernández-Caballero, A.: A survey of video datasets for human action and activity recognition. Comput. Vis. Image Underst. **117**(6), 633–659 (2013)

5. Chondro, P., Yao, C.Y., Ruan, S.J., Chien, L.C.: Low order adaptive region growing for lung segmentation on plain chest radiographs. Neurocomputing **275**, 1002–1011 (2018)
6. Dai, W., et al.: Scan: structure correcting adversarial network for organ segmentation in chest x-rays. arXiv preprint arXiv:1703.08770 (2017)
7. Demner-Fushman, D., et al.: Preparing a collection of radiology examinations for distribution and retrieval. J. Am. Med. Inform. Assoc. **23**(2), 304–310 (2015)
8. Giger, M.L., Chan, H.P., Boone, J.: Anniversary paper: history and status of cad and quantitative image analysis: the role of medical physics and AAPM. Med. Phys. **35**(12), 5799–5820 (2008)
9. Hogeweg, L., Sánchez, C.I., de Jong, P.A., Maduskar, P., van Ginneken, B.: Clavicle segmentation in chest radiographs. Med. Image Anal. **16**(8), 1490–1502 (2012)
10. Ibragimov, B., Likar, B., Pernuš, F., Vrtovec, T.: Accurate landmark-based segmentation by incorporating landmark misdetections. In: 2016 IEEE 13th International Symposium on Biomedical Imaging (ISBI), pp. 1072–1075. IEEE (2016)
11. Ibragimov, B., Likar, B., Pernus, F., et al.: A game-theoretic framework for landmark-based image segmentation. IEEE Trans. Med. Imaging **31**(9), 1761–1776 (2012)
12. Jaeger, S., Candemir, S., Antani, S., Wáng, Y.X.J., Lu, P.X., Thoma, G.: Two public chest x-ray datasets for computer-aided screening of pulmonary diseases. Quant. Imaging Med. Surg. **4**(6), 475 (2014)
13. Li, X., Luo, S., Hu, Q., Li, J., Wang, D., Chiong, F.: Automatic lung field segmentation in x-ray radiographs using statistical shape and appearance models. J. Med. Imaging Health Inform. **6**(2), 338–348 (2016)
14. Mittal, A., Hooda, R., Sofat, S.: Lung field segmentation in chest radiographs: a historical review, current status, and expectations from deep learning. IET Image Process. **11**(11), 937–952 (2017)
15. Mould, R.F.: A Century of X-rays and Radioactivity in Medicine: with Emphasis on Photographic Records of the Early Years. CRC Press, Boca Raton (1993)
16. Novikov, A.A., Lenis, D., Major, D., et al.: Fully convolutional architectures for multi-class segmentation in chest radiographs. IEEE Trans. Med. Imaging **37**(8), 1865–1876 (2018)
17. Ruikar, D.D., Hegadi, R.S., Santosh, K.C.: A systematic review on orthopedic simulators for psycho-motor skill and surgical procedure training. J. Med. Syst. **42**(9), 168 (2018)
18. Ruikar, D.D., Santosh, K.C., Hegadi, R.S.: Automated fractured bone segmentation and labeling from CT images. J. Med. Syst. **43**(3), 60 (2019). https://doi.org/10.1007/s10916-019-1176-x
19. Ruikar, D.D., Santosh, K.C., Hegadi, R.S.: Segmentation and analysis of CT images for bone fracture detection and labeling (chap. 7). In: Medical imaging: Artificial Intelligence, Image Recognition, and Machine Learning Techniques. CRC Press, Boca Raton (2019). ISBN 9780367139612
20. Santosh, K., Antani, S.: Automated chest x-ray screening: can lung region symmetry help detect pulmonary abnormalities? IEEE Trans. Med. Imaging **37**(5), 1168–1177 (2018)
21. Santosh, K., Vajda, S., Antani, S., Thoma, G.R.: Edge map analysis in chest x-rays for automatic pulmonary abnormality screening. Int. J. Comput. Assist. Radiol. Surg. **11**(9), 1637–1646 (2016)
22. Seghers, D., Loeckx, D., Maes, F., Vandermeulen, D., Suetens, P.: Minimal shape and intensity cost path segmentation. IEEE Trans. Med. Imaging **26**(8), 1115–1129 (2007)

23. Shao, Y., Gao, Y., Guo, Y., Shi, Y., Yang, X., Shen, D.: Hierarchical lung field segmentation with joint shape and appearance sparse learning. IEEE Trans. Med. Imaging **33**(9), 1761–1780 (2014)
24. Shi, Y., et al.: Segmenting lung fields in serial chest radiographs using both population-based and patient-specific shape statistics. IEEE Trans. Med. Imaging **27**(4), 481–494 (2008)
25. Shiraishi, J., et al.: Development of a digital image database for chest radiographs with and without a lung nodule: receiver operating characteristic analysis of radiologists' detection of pulmonary nodules. Am. J. Roentgenol. **174**(1), 71–74 (2000)
26. Van Ginneken, B., Romeny, B.T.H., Viergever, M.A.: Computer-aided diagnosis in chest radiography: a survey. IEEE Trans. Med. Imaging **20**(12), 1228–1241 (2001)
27. Van Ginneken, B., Stegmann, M.B., Loog, M.: Segmentation of anatomical structures in chest radiographs using supervised methods: a comparative study on a public database. Med. Image Anal. **10**(1), 19–40 (2006)
28. Wang, C.: Segmentation of multiple structures in chest radiographs using multi-task fully convolutional networks. In: Sharma, P., Bianchi, F.M. (eds.) SCIA 2017. LNCS, vol. 10270, pp. 282–289. Springer, Cham (2017). https://doi.org/10.1007/978-3-319-59129-2_24
29. Wang, X., Peng, Y., Lu, L., Lu, Z., Bagheri, M., Summers, R.M.: ChestX-ray8: hospital-scale chest x-ray database and benchmarks on weakly-supervised classification and localization of common thorax diseases. In: 2017 IEEE Conference on Computer Vision and Pattern Recognition (CVPR), pp. 3462–3471. IEEE (2017)
30. Yang, W., et al.: Lung field segmentation in chest radiographs from boundary maps by a structured edge detector. IEEE J. Biomed. Health Inform. **22**(3), 842–851 (2018)

An Imperceptible Secure Transfer of Medical Images for Telemedicine Applications

B. Madhu[1(✉)] and Ganga Holi[2(✉)]

[1] Department of CSE, Dr. Ambedkar Institute of Technology, Bangalore, India
bmadhu.cs@gmail.com
[2] Department of ISE, Global Academy of Technology, Bangalore, India
gangaholi@gmail.com

Abstract. Medical image transfer is an important task to transfer the patient's medical observations to the doctors at remote areas to know patient condition through the Internet/open network for expert opinion. But information transmission through unsecured channel raises confidentiality problems for any image data. This paper addresses the issue of confidentiality for medical images. The proposed technique uses a hybrid combination of Stationary Wavelet Transformation (SWT) and Singular Value Decomposition of the original medical image for the embedding of Patient watermark into it. The robustness is improved by verifying the peak signal to noise ratio to avoid wrong diagnosis of medical conditions. The robust performance of the proposed scheme is evaluated for known common processing attacks such as Salt and pepper, Gaussian, Mean and Median Filters, Histogram Equalization etc. The method is robust for embedded watermark at the acceptable visual quality of the watermarked image.

Keywords: Stationary Wavelet Transformation (SWT) ·
Telemedicine · Singular Value Decomposition (SVD) ·
Mammogram images · Correlation coefficient ·
Peak signal to noise ratio

1 Introduction

As the usage of the internet is growing, the medical field is exposing from static sharing to dynamic sharing of medical images improving the efficiency of diagnostics [12,13]. Telemedicine uses an electronic media to diagnose patient clinically by exchanging medical information. The number of applications are growing now a days like E-Mails, Wireless tools, Video Conferencing, and Telecommunication technology. Electronic-health (E-Health) includes patients consultations using patient portals, nursing call centers, medical image transmission, video conferencing are all coming under telemedicine. Telemedicine works on demand basis and by the requirement of the patient. Watermarking [9,10] of medical images

K. C. Santosh and R. S. Hegadi (Eds.): RTIP2R 2018, CCIS 1036, pp. 332–340, 2019.
https://doi.org/10.1007/978-981-13-9184-2_30

is very important as the medical images [20] have information which belongs to life changing scenarios of human subject. The rest of this paper is organized as follows. In Sect. 2 work done so far has been discussed. Section 3 Concepts of stationary wavelet transformation and singular value decomposition followed by proposed method have been discussed. In Sect. 4, the experimental results of the proposed scheme are shown. Finally, the conclusions are given in Sect. 5.

2   Literature Review

An image is considered as an important data type for the transmission of information. There is a necessity to provide security for such images as it carries sensitive information like medical image. The importance of generating, storing and accessing the medical images is explained [1]. The concept of storing the medical images over the cloud was implemented [2]. The major problem of sharing the medical images over the internet is the probability of changing it in an unnoticeable manner. So aspects of privacy have become an important issue [3] and various policies were suggested for privacy issue. The changes in the medical image will impact diagnostic reports of the patients [4]. The above problem can be solved using cryptographic and watermarking methods. Cryptographic methods are effective, but will take much higher time for providing security for larger images. Watermarking is a technique of hiding watermark over the cover image, and ensuring image integrity. Discrete Wavelet Transformation (DWT) [7,14] for the transmission of Medical images has been proposed to provide copyright protection of it [15,16]. Combination of Discrete Cosine Transformation (DCT) and DWT watermarking techniques has been used [5] for transmission of medical images with division between the Region of Interest and Non Region of Interest. The ROI based reversible watermarking and RONI based robust watermarking has been proposed [6]. Two level Contourlet Transform for Computed Tomography (CT) and MRI images has been proposed and a good PSNR is achieved [8]. A Blind Medical Image watermarking for medical image as cover image and Lena image as a watermark with Lifting wavelet transformation has been proposed by taking computer tomography (CT), ultrasound (US) images for Telemedicine applications [11]. PSNR and Correlation coefficient values are good by taking X-ray, CT Scan and MRI images with DWT and DCT technique [17]. Performance of watermarking can be improved by taking hybrid combination of transformation techniques.

3   Stationary Wavelet Transformation (SWT)

SWT algorithm is very simple and is close to DWT. SWT is a transformation works very similar to DWT, but keep time invariance property known as Undecimated-DWT. The stationary wavelet transform (SWT) is a wavelet transformation designed to overcome the translation invariance property of discrete wavelet transformation [12]. The property of invariance can be achieved by upsampling the filters by inserting zeros between the filter coefficients and avoiding the down-sampling of decimated algorithm. The SWT is well known for its inherent redundancy scheme as the output of each level of SWT contains the same number of samples as the input. So for a decomposition of N levels, there is a redundancy of N in the wavelet coefficients.

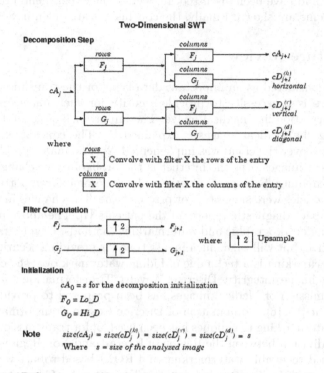

Fig. 1. 2D Stationary Wavelet Transformation (Courtesy Math works)

Figure 1 2D Stationary Wavelet Transformation (Courtesy Math works) depicts the digital implementation of SWT for 2D transformation. Apply appropriate high and low pass filters to the data at each level to produce two sequences at each level. Advantages of SWT are

1. SWT has an evident advantage than DWT when the requirement of real time signal denoising is not high, and it can offer more precise flow information.

2. SWT is translation invariant, even if the signal is shifted, the obtained coefficients will not change.

3. SWT performs better in denoising, edge detecting and image fusion and break down point detection.

4. DWT is only suitable for implementing for discrete signals of images whose size is power of 2 but SWT can be applied to any arbitrary size of images.

4   Singular Value Decomposition (SVD)

SVD [11] is an attractive algebraic transform for image processing. SVD has prominent properties in imaging and have technique for handling matrices (sets of equations) that do not have an inverse. SVD is robust and reliable orthogonal matrix decomposition method. This includes square matrices whose determinant is zero and all rectangular matrices. The Singular Value Decomposition is considered as a good mathematical tool in linear algebra and can be divided into real or complex matrix. The Singular values are referred with the diagonal entries of the matrix. The left-singular vectors and right-singular vectors are well known as columns of the matrix. The singular value decomposition (SVD) is a matrix factorizations given by Eq. (1) If A is an $m \times n$ matrix, then A is a product of three factors:

$$A = U\varSigma V^*, \tag{1}$$

where U is an orthogonal $m \times m$ matrix, V is an orthogonal $n \times n$ matrix, V^* is the transpose of V, and $\varSigma$ is an $m \times n$ matrix that has all zeros except for its diagonal entries, which are nonnegative real numbers.

Advantages of SVD are

  1. Reconstruction of the image is easy.
  2. Cost is less.

The Singular values are used for the implementation because it is less affected if general image processing is performed and the modification of the largest coefficients would cause a greater measure of image degradation.

5   Proposed Method

An Image Watermarking with the hybrid combination of SWT and SVD is proposed to embed watermark in the Medical images for Telemedicine applications. The method is evaluated by dividing the original image and watermark image into coefficients and selecting the sub bands which are less sensitive to the system. The changes in low frequency components will reflect in the cover image, so take high frequency components to provide security for an image. Input images are taken from MIAS Dataset, and watermark image is standard Lena image. The process of embedding and extraction will be explained in Sects. 5.1 and 5.2 respectively.

5.1   Process of Embedding

The steps required to embed a watermark into a mammogram image can be listed as below.

Step 1: Input mammogram image of size (1024×1024) as input image (I).

Step 2: Apply Stationary Wavelet Transformations (SWT) with Symlet 2. After applying SWT to the input image it will be divided into four frequency bands LL, LH, HL, and HH respectively.

Step 3: Apply Singular Value Decomposition to high band coefficients. Apply SVD to HH subband.

Step 4: Input Lena image as watermark (W), repeat the above steps.

Step 5: Append Input Image with Watermark using the scaling factor.

$$WM = I + a * W;$$

Step 6: Inverse SWT to get Watermarked Image.

The output of the above process is watermarked image which is ready for the transmission from sender.

5.2   Process of Extraction

The steps required to extract a watermark from a watermarked Mammogram image can be listed as below.

Step 1: Input watermarked mammogram image (WM).

Step 2: Apply Stationary Wavelet Transformations (SWT) with Symlet 2 to watermarked image to get LL, LH, HL and HH bands.

Step 3: Apply Singular Value Decomposition to high band coefficients. Apply SVD to HH sub bands.

Step 4: Repeat the same steps to the input image (I).

Step 5: Subtract watermarked with input image using the scaling factor.

$$W = (WM - I)/a;$$

Step 6: Inverse SWT to get watermark.

The output of the above process is watermark image to verify the integrity of the input image at the receiver (Table 1). The experiment is conducted over 100 image samples and tabulated for 10 images by considering the scaling factor as 0.5.

Table 1. PSNR values without attacks

Host image	Alpha	Time	CC for watermark	CC for host	PSNR watermark	PSNR host
mbd001	0.5	4.315	1	1	73.89	80.37
mbd002	0.5	4.315	1	1	76.47	76.91
mbd003	0.5	4.315	1	1	73.25	81.56
mbd004	0.5	4.315	1	1	74.30	79.94
mbd005	0.5	4.315	1	1	72.2	85.45
mbd006	0.5	4.315	1	1	72.84	83.42
mbd007	0.5	4.315	1	1	72.86	83.09
mbd008	0.5	4.315	1	1	72.97	82.92
mbd009	0.5	4.315	1	1	72.77	83.39
mbd010	0.5	4.315	1	1	73.89	80.37

6 Performance Parameters

The results of the proposed work is evaluated by using peak signal to noise ratio (PSNR) and Correlation Coefficient (CC). PSNR is calculated for the input image and watermarked image as in Eq. (2).

$$PSNR = 10 \log_{10} \frac{255 \times 255}{MSE} \tag{2}$$

where MSE is mean square error between two images.

The Correlation Coefficient is calculated between the original medical image to the watermarked medical image Eq. (3).

$$CC = \frac{\sum_{i=1}^{n}(x_i - \overline{x})(y_i - \overline{y})}{\sqrt{\sum_{i=1}^{n}(x_i - \overline{x})^2(y_i - \overline{y})^2}} \tag{3}$$

7 Experimental Results

The algorithm is executed in MATLAB R2013a version. Experiment is conducted to test over 322 Mias dataset [21] with size 1024×1024. The algorithm is implemented by taking the Mammogram images as cover images and Lena image as watermark image. Watermark is inserted into the Medical Image with the scaling factor 0.5. Table 2 depicts the sample input image and watermark image taken for the implementation of the proposed algorithm. Table 3 indicates watermarked image, recovered watermark with attacks and without attacks. Various attacks have been tested on the watermarked image with Gaussian noise and Salt and pepper noise of 0.01. Default parameters for Histogram Equalization, Intensity Adjustment, Mean and Median Filters. Average time taken for implementing 10 mammogram images are 4.38 s, average PSNR of original watermark and recovered watermark is 70 DB, PSNR of original mammogram and watermarked image is above 70 DB and Correlation Coefficient between host and watermarked images are proved to be 1.

The algorithm is checked for both the watermark and for the input image. The visual quality of the watermark remains unchanged at the extraction step. The watermarked mammogram image quality is checked at the end of embedding

Table 2. Host image and watermark images

Host Image	Watermark Image

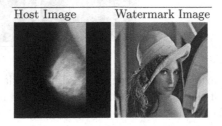

Table 3. Resultant watermarked and extracted watermark images under various attacks

Index	Parameter	Watermarked	EWatermark
1	Without Attacks		
2	Gaussion Noise		
3	Hist Equalization		
4	Intensity Adjustment		
5	Mean		
6	Median		

step. Comparison of various methods with the different dataset has been given in Table 4. The state of the art indicates that our method is better in providing imperceptibility for the medical images.

Table 4. State of art comparison of existing methods to solve medical image security

Ref. method	Images	Methodology	Size	PSNR host
[5]	MRI, CT scan, Ultrasound images	DWT+DCT	512 × 512	51
[6]	X-ray, MRI_brain, CT Scan_head, MRI_head	ROI, NROI using SS	512 × 512, 256 × 256	44
[17]	MRI, CTScan, X-ray	DWT+DCT	512 × 512	51
[18]	Ultrasound, MRI, CT, mammogram, X-ray	DWT+DCT+SVD	512 × 512	56
[19]	Mega colon, abdominal	DWT+DCT+SVD	512 × 512	58.03
[Proposed Method]	Mammogram	SWT+SVD	1024 × 1024	79.04

8   Conclusion

The proposed novel method for watermarking of medical images provides lossless procedure for embedding and extraction of the watermark. The input images are Mammogram Images and watermark image is Lena image. The Efficiency of the proposed method can be verified by calculating PSNR of Original Watermark and Extracted Watermark. The stability of the proposed method has been verified against Gaussian noise, salt and pepper, mean and median filter, histogram attacks. The PSNR value of the suggested method is 70 DB which is good to visualize the image data. The future work can be extended by applying it to RGB image datasets.

References

1. Ruff, E.: System and method for generating storing and accessing secured medical imagery. United States Patent Application Publication. Pub. No.: US 2017/0277831 A1, pp. 1–4, 18 September 2017
2. Deshmukh, P.: Design of cloud security in the EHR for Indian healthcare services. J. King Saud Univ. Comput. Inf. Sci. **29**(3), 281–287 (2017)
3. Haasa, S., Wohlgemuthb, S., Echizenb, I., Soneharab, N., Müller, G.: Aspects of privacy for electronic health records. Int. J. Med. Inform. **80**(2), e26–e31 (2011)
4. Singh, A., Dutta, M.K.: Imperceptible watermarking for security of fundus images in tele-ophthalmology of retina diseases. Int. J. Med. Inform. **108**, 110–124 (2017). ISSN 1386–5056
5. Sharma, A., Singh, A.K., Ghrera, S.P.: Secure hybrid robust watermarking technique for medical images. In: 4th International Conference on Eco-Friendly Computing and Communication Systems, ICECCS 2015 (2015). Procedia Comput. Sci. (ScienceDirect) **70**, 778–784, 1877–0509
6. Maitya, H.K., Maityb, S.P.: Joint robust and reversible - watermarking for medical images. Sci. Direct Procedia Technol. **6**, 275–282 (2012). 2212–0173

7. Thanki, R., Borra, S.: A color image steganography in hybrid FRT-DWT domain. J. Inf. Secur. Appl. **40**, 92–102 (2018). 2214–2126
8. Mousami, V.M., Mousami, P.T.: An efficient contourlet based multiple watermarking scheme for health information system. Int. J. Comput. Sci. Eng. Res. Pap. **6**(2), 61–65 (2018). E-ISSN 2347–2693
9. Liu, X., Lou, J., Wang, Y., Du, J., Zou, B., Chen, Y.: Discriminative and robust zero-watermarking scheme based on completed local binary pattern for authentication and copyright identification of medical images. In: Proceedings of SPIE 10579, Medical Imaging 2018: Imaging Informatics for Healthcare, Research, and Applications (2018)
10. Dinesh Kumar, G., Teja, D.P., Reddy, S.S., Sasikaladevi, N.: An efficient watermarking technique for biometric images. In: 7th International Conference on Advances in Computing & Communications (ICACC 2017). Proc. Comput. Sci. **115**, 423–430 (2017). ISSN: 1877-0509
11. Venkatram, N., Reddy, L.S.S., Kishore, P.V.V.: Blind medical image watermarking with LWT-SVD for telemedicine applications. WSEAS Trans. Sig. Process. **10**, 200–300 (2014). E-ISSN 2224–3488
12. Madhu, B., Holi, G., Srikanta Murthy, K.: Secure digital image watermarking technique using stationary wavelet transform and singular value decomposition based on scaling factor. In: International Conference on Electrical, Electronics, Computers, Communication, Mechanical and Computing (EECCMC), 28th and 29th January 2018 (2018). IEEE Conference Proceedings, pp. 84–88. ISBN: CFP18037-PRT:978-1-5386-4303-7
13. Madhu, B., Holi, G., Murthy, K.S.: An overview of image security techiques. Int. J. Comput. Appl. **154**(6), 37–46 (2016)
14. Venkatram, N., Reddy, L.S.S., Kishore, V.: DWT-BAT based medical image watermarking for telemedicine applications. J. Eng. Technol. **2**, 18–37 (2014)
15. Jaiswal, S., Agrawal, H.: Encryption and secure transmission of telemedicinal image in watermarking using DWT HAAR Wavelet Algorithm. Int. J. Recent Innov. Trends Comput. Commun. **3**(5), 3273–3277 (2015). ISSN 2321–8169
16. Pratibha, K., Bhatia, S.K.: Security in telemedicine using DWT-CDCS. Int. J. Comput. Appl. **136**(7), 11–15 (2016). ISSN 0975–8887
17. Mehto, A., Mehra, N.: Adaptive lossless medical image watermarking algorithm based on DCT and DWT. In: International Conference on Information Security and Privacy (ICISP 2015) (2016). Procedia Comput. Sci. (ScienceDirect) 88–94 (2016)
18. Swaraja, K.: A hybrid secure watermarking technique in telemedicine. Int. J. Eng. Technol. (IJET) **9**(3), 265–270 (2017). ISSN (Print) 2319–8613, ISSN (Online) 0975–4024
19. Assini, I., Badri, A., Safi, K., Sahel, A., Baghdad, A.: A robust hybrid watermarking technique for securing medical image. Int. J. Intell. Eng. Syst. **11**(3), 169–176 (2018)
20. Abd-Eldayem, M.M.: A proposed security technique based on watermarking and encryption for digital imaging and communications in Medicine. Egypt. Inform. J. **14**(1), 1–13 (2013)
21. Suckling, J.: The mammographic image analysis society digital mammogram database exerpta medica. In: International Congress Series vol. 1069, pp. 375-378 (1994). http://peipa.essex.ac.uk/info/mias.html

Preprocessing and Segmentation of Retina Images for Blood Vessel Extraction

Ambaji S. Jadhav$^{(\boxtimes)}$ and Pushpa B. Patil

Dr. P. G. H. College of Engineering and Technology, Vijayapur, India
asjadhavec@gmail.com

Abstract. In every field there is use of technology as part of in medical field lot of analysis is done using images. Processing of images will give good analysis when there is no noise or very less noise is present so processing retina images also gives correct results so we used different filters to find out which filter is suitable for pre-processing of retina images available in DRIVE database by computing mean square error (MSE) and Peak signal to noise ratio (PSNR) for different noises. After pre-processing images are segmented using discrete wavelet transform (DWT) and extracted blood vessel pixels are computed and compared with first observer result available in data base and results are very close to manual segmentation which is given DRIVE database.

Keywords: Blood vessels · Pre-processing · Optic disc · Retina image

1 Introduction

Every year the number of people suffering from diabetes is increasing but the medical experts available for correct diagnosis is very less so there need for computerized diagnosis which can reduce the time of experts for detection of diabetic retinopathy. Researchers are worried for the image enhancement since it could be significantly improve the visual perception and as numerical attributes of the image that has a direct impact on the accuracy of analysis by medical experts. To remove noise effects, it is necessary to remove or reduce the noise with preserving image details during the pre-processing that help the correct segmentation of parameters like blood vessels, exudates, optic disc etc which are indicators of diabetic progress in the person. In present work wiener filtering is used as linear filter and median filtering as a non-linear filter. Image Enhancement basically includes noise reduction from the given image can contain noises like Gaussian, salt and pepper.

1.1 Related Work

Retina images acquired by a special camera called "fundus camera" are sometimes low contrast or uneven distribution of light can create strong challenges

© Springer Nature Singapore Pte Ltd. 2019
K. C. Santosh and R. S. Hegadi (Eds.): RTIP2R 2018, CCIS 1036, pp. 341–348, 2019.
https://doi.org/10.1007/978-981-13-9184-2_31

in the process detection of sign of diabetic retinopathy related parameters like blood vessel extraction, optic disc detection, exudates detection and microaneurysm detection etc. which can provide good information about the healthiness of retina. Razban et al. [1], Presented a method for isolating blood vessels from their background content in image acquired by fundus camera that helps in understanding the condition of functioning of retina part of the eye. They used Gabor wavelet to separating background content and mathematical morphology for removing blood vessels. Nisha et al. [2], Proposed a new method that finds the process related to velocity of blood flow in human retina cells which very necessary for providing nutrients to retina cells. This is also called as "major temporal arcade" that helps in detection of blood vessels. In this they used Gabor filter and Hough transform for blood vessel detection. Tejaswi et al. [3], Used matched filter for segmentation of blood vessels from retina images. Neha Gupta and Aarti [4], Implemented a method to detect blood vessels using combination of different filters so as to reduce the possible noise level present in the image and helps for more accurate segmentation of blood vessel network in turn that provide information regarding status of blood vessel functionality. Salazar-Gonzalez et al. [5], Presented a new method called "graph cut technique" for segmentation of blood vascular network from retina image that estimates the location of blood vessels based on graph theory with intensity as a major information source and determined the unusual distribution of small and branched blood veins. Annie Edel Quinn and Gokula Krishnan [6], Developed a method which that uses contrast adjustment over the input image and curvelet transform approach as a next step. They separated green channel from RGB retina image and then contrast adjustment is done over the input image after that curvelet transform is operated to detect edges which form the vessels of retina. Deepa and Mymoon Zuviriya [7], Presented a method for abnormal blood vessel (Thick vessels and extra branches created) detection from retina image using features of Gray level features and moment invariants features and then these features used for classification of pixels belonging blood vessel or not belonging blood vessel. Support vector machine (SVM) is used here as a classifier. Chrastek et al. [8], Segmentation of blood vessels is performed to remove them from retina image later which the optic disc is extracted from retina image captured using fundus camera that provide details required for analysis of retina for determination of diabetes effect. Here blood vessels are separated using "distance map algorithm" and then optic disc is separated using morphological operation, Hough transform and active contour that provide status of optic disc where blood vessels originate in retina. Lowell et al. [9], Segmentation of blood vessels is performed using "template matching algorithm" along with directionally sensitive gradients that is effective in determining blood vessels in different direction throughout the retina which serve as nutrients providing network for retina cells. Welfer et al. [10,11], Extraction of optic disc from retina images is done using adaptive morphological operation and "watershed transform". Here watershed transform is used for marking and detecting optic disc boundary in the retina and morphological operation is used blood vessel segmentation from remaining objects of retina.

2   Proposed Methodology

The algorithm for different preprocessing methods is given below: 1. Read input image. 2. Convert RGB to color. 3. Add different noises to input image. 4. The input become noisy image. 5. Apply different filters. 6. Compute mean square error (MSE) and peak signal to noise ratio (PSNR) for all filters. 7. After preprocessing apply discrete wavelet transform to extract blood vessels. 8. Compute blood vessel pixels and compare with manual segmentation. The processing of images is required to in order to avoid or reduce non-uniform illumination which introduced in process of capturing of retina images. The process of noise removal helps in the task of segmentation of image contents that in turn helps in the process of analysis and makes the image appropriate for automated detection [12,13]. Figure 1 shows the flowchart for proposed work.

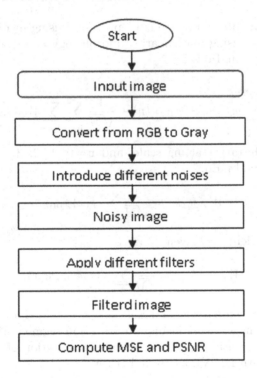

Fig. 1. Flowchart for pre-processing

2.1   Preprocessing Methods

Different filters work effectively on different noises and type of images so it is important to do analysis of filters with different type of images to identify which filter works well for different noise. In this we have considered four noises namely Salt and Pepper noise, Speckle noise, Poisson noise and Gaussian noise. Once pre-processing is carried out its effectiveness is measured listed in Tables 1, 2, 3, 4, and 5 of results and discussion section. It is found that from above tables median filter can suppress the noise from retina images of DRIVE database. Using median filter for preprocessing and discrete wavelet transform (DWT) is used for segmentation the blood vessel pixels are computed tabulated in Table 6.

A **Filtering:** Filtering of image is performed using different masks with different size like 3×3, 5×5 and with different valued coefficients arranged in various manners. The mask is placed on image such that the center of the mask coincides with first pixel and for median filter all pixels overlapped by mask coefficients are arranged in ascending order and center value is found to be median value then that first pixel is replaced by median value. In next step mask is moved to second pixel in image so that mask center overlaps with it and previous procedure is repeated. This process is performed for all the pixels in the image.

B **Segmentation:** After an image is preprocessed using median filter segmentation is carried out using mathematical morphology and discrete wavelet and results are tabulated in Table 6.

$$f(n) = \frac{1}{\sqrt{M}} \sum_k w_\phi(j_0, k) \varphi_{j0,k} \, k(n) + \frac{1}{\sqrt{M}} \sum_{j=j_0}^{\infty} \sum_k W_\psi(j, k) \psi_{j,k}(n) \qquad (1)$$

where j_0 is an arbitrary starting scale, and $n = 0, 1, 2, 3 \ldots m$. Approximate coefficients are given by Eq. 2

$$W_\phi(j_0, k) = \frac{1}{\sqrt{M}} \sum_x f(x) \phi_{j0,k} \qquad (2)$$

The detailed coefficients are given by Eq. 3

$$W_\phi(j_0, k_0) = \frac{1}{\sqrt{M}} \sum_x f(x) \phi_{j0,k} \, 0(x) \qquad (3)$$

Table 6, shows the comparison of pixels count obtained by segmentation of DRIVE database images using median filtering and segmentation using wavelet transform technique with respect to manual segmentation of DRIVE database images provided with DRIVE as ground truth.

2.2 Performance Measures

For all segmented images how many pixels are belonging to blood vessels and how many pixels are belonging to background is estimated and compared with manual segmented images and their pixels count is considered for performance measurement. Mean Square Error (MSE): Definition: The MSE is the cumulative squared error between the compressed or reconstructed and the original image. In statistics, the mean squared error (MSE) or mean squared deviation

(MSD) of an estimator measures the average of the squares of the errors or deviations - that is, the difference between the estimator and what is estimated. A lower value of MSE means lesser value. Peak Signal to Noise Ratio (PSNR1): Definition: "It is the ratio between the maximum possible power of a signal and the power of corrupting noise that affects the fidelity of its representation or PSNR is a measure of peak error." The Mean Square Error (MSE) and the Peak Signal to Noise Ratio (PSNR) are the two error metrics used to compare image compression quality. The MSE represents the cumulative squared error between the compressed and the original image, where as PSNR represents a measure of the peak error. High PSNR means good image quality and less error introduced to in the image. Once pre-processing is carried out its effectiveness is measured listed in above Tables 1, 2, 3, 4, and 5. It is found that from above tables median filter can sup-press the noise from retina images of DRIVE database. Using median filter for pre-processing and discrete wavelet transform (DWT) is used for segmentation the blood vessel pixels are computed tabulated in Table 6.

3   Results and Discussion

The pre-processing of retina images helps in segmentation of images that leads to extraction of diabetic characteristics from available database images. The results obtained by the proposed methodology shows that the median filter is good preprocessing filter for DRIVE database images since it produces results our experiments, with known number of images in each category and number of category in the database. Table 6 shows the comparison of percentage of pixels classified as blood vessel pixels and percentage of pixels classified as non blood vessel pixels that indicates accuracy of segmentation process .

Table 1. MSE and PSNR values for mean filter

Sl. No	Types of noise image	Average MSE	Average PSNR
1	Salt and pepper noise	7.6504	36.90
2	Speckle noise	9.6207	35.96
3	Poisson noise	8.9772	39.78
4	Gaussian noise	11.187	32.99

Table 2. MSE and PSNR values for median filter

Sl. No	Types of noise image	Average MSE	Average PSNR
1	Salt and pepper noise	6.6504	39.90
2	Speckle noise	6.6207	39.96
3	Poisson noise	6.9772	39.78
4	Gaussian noise	6.187	39.99

Table 3. MSE and PSNR values for arithmetic mean filter

Sl. No	Types of noise image	Average MSE	Average PSNR
1	Salt and pepper noise	39.45	43.98
2	Speckle noise	36.83	46.30
3	Poisson noise	36.68	38.99
4	Gaussian noise	38.11	36.59

Table 4. MSE and PSNR values for SVD decomposition filter

Sl. No	Types of noise image	Average MSE	Average PSNR
1	Salt and pepper noise	121.6959	27.72
2	Speckle noise	122.74	26.07
3	Poisson noise	122.81	27.366
4	Gaussian noise	118.67	27.12

Table 5. MSE and PSNR values for weighted mean filter

Sl. No	Types of noise image	Average MSE	Average PSNR
1	Salt and pepper noise	97.99	28.38
2	Speckle noise	98.44	28.49
3	Poisson noise	99.24	29.41
4	Gaussian noise	95.42	28.42

Table 6. Comparison of extracted blood vessel pixels by proposed method and manual segmentation.

Test images (DRIVE DB)	No. of pixels obtained	No. of pixels (1st observer or manual)	Diff. in percentage
Image1	41220	38419	7.2
Image2	42135	38457	9.41
Image3	47056	38480	22.23
Image4	39344	38514	2.14
Image5	36061	38480	6.28
Image6	39812	38501	3.40
Image7	42811	38404	11.41
Image8	48020	38429	24.91
Image9	41351	38470	7.41
Image10	30674	38463	20.20
Image11	39879	38460	3.61

(*continued*)

Table 6. (*continued*)

Test images (DRIVE DB)	No. of pixels obtained	No. of pixels (1st observer or manual)	Diff. in percentage
Image12	43155	38458	12.21
Image13	33769	38448	12.15
Image14	43779	38421	13.91
Image15	39596	38410	3.01
Image16	37434	38481	2.72
Image17	44873	38414	16.80
Image18	37618	38434	0.47
Image19	38036	38461	1.10
Image20	38130	38414	0.71

4 Conclusion

In this paper pre-processing of retina images to help in segmentation is presented, segmentation of images that leads to extraction of diabetic characteristics from available database. The results obtained by the proposed methodology shows that the median filter gives good pre-processing for DRIVE database since it produced low value of mean square error and high peak signal to noise ratio. We have made an attempt to study image enhancement by using linear and non linear filtering technique. Also salt and pepper noise and Gaussian noise added into image after applying median filtered, it was observed that salt and pepper noise and Gaussian noise reduction was better than wiener filter. The median filter perform better than wiener filter, it is not only better for noise reduction also remove the blurred effect in image After pre-processing the blood vessels are segmented and extracted blood veins pixels are counted they shows the result which is very close to manual segmentation process which is available in DRIVE database.

References

1. Razban, A., Nooshyar, M., Mahjoory, K.: Segmentation of retinal blood vessels by means of 2D Gabor wavelet and fuzzy mathematical morphology. In: IEEE Conference ICSPIS 2016, Amirkabir University of Technology, Tehran, Iran, 14–15 December 2016
2. Nisha, K.L., Sreelekha, G., Savithri, S.P., Mohanachandran, P., Vinekar, A.: Fusion of structure adaptive filtering and mathematical morphology for vessel segmentation in fundus images of infants with retinopathy of prematurity. In: IEEE 30th Canadian Conference (CCECE) on Electrical and Computer Engineering (2017)
3. Prakash, T.D., Rajashekar, D., Srinivasa, G.: Comparison of algorithms for segmentation of blood vessels in fundus images. In: IEEE 2nd International Conference on Applied and Theoretical Computing and Communication Technology (iCATccT) (2016)

4. Gupta, N., Aart: Performance evaluation of retinal vessel segmentation using a combination of filters. In: IEEE 2nd International Conference on Next Generation Computing Technologies 2016 (NGCT-2016), Dehradun, India, 14–16 October 2016
5. Salazar-Gonzalez, A., Kaba, D., Li, Y., Liu, X.: Segmentation of the blood vessels and optic disk in retinal images. IEEE J. Biomed. Health Inform. 18(6), 1874–1886 (2014)
6. Annie Edel Quinn, E., Gokula Krishnan, K.: Retinal blood vessel segmentation using curvelet transform and morphological reconstruction. In: IEEE International Conference on Emerging Trends in Computing, Communication and Nanotechnology (ICECCN 2013) (2013)
7. Deepa, M., Mymoon Zuviriya, N.: Comparative analysis on supervised classification techniques for segmentation and detecting abnormal blood vessels in retinal images. In: IEEE International Conference on Emerging Trends in Science, Engineering and Technology (ICETS) (2012)
8. Chrastek, R., et al.: Automated segmentation of the optic nerve head for diagnosis of glaucoma. Med. Image Anal. 9(1), 297–314 (2005)
9. Lowell, J., et al.: Optic nerve head segmentation. IEEE Trans. Med. Imaging 23(2), 256–264 (2004)
10. Welfer, D., Scharcanski, J., Kitamura, C.M., Ludwig, D.P., Marinho, D.: Segmentation of the optic disk in color eye fundus images using an adaptive morphological approach. Comput. Biol. Med. 40(1), 124–137 (2010)
11. Ruikar, D.D., Santosh, K.C., Hegadi, R.S.: Automated fractured bone segmentation and labeling from CT images. J. Med. Syst. 43(3), 60 (2019). https://doi.org/10.1007/s10916-019-1176-x
12. Ruikar, D.D., Santosh, K.C., Hegadi, R.S.: Segmentation and analysis of CT images for bone fracture detection and labeling. In: Medical Imaging: Artificial Intelligence, Image Recognition, and Machine Learning Techniques, chap. 7. CRC Press (2019). ISBN 9780367139612
13. Hegadi, R.S., Navale, D.I., Pawar, T.D., Ruikar, D.D.: Multi feature-based classification of osteoarthritis in knee joint X-ray images. In: Medical Imaging: Artificial Intelligence, Image Recognition, and Machine Learning Techniques, chap. 5. CRC Press (2019). ISBN 9780367139612

Design a Novel Detection for Maculopathy Using Weightage KNN Classification

Chetan Pattebahadur$^{(\boxtimes)}$, Ramesh Manza$^{(\boxtimes)}$, and Anupriya Kamble$^{(\boxtimes)}$

Department of Computer Science and IT, Dr. Babasaheb Ambedkar Marathwada
University, Aurangabad, Maharashtra, India
chetu358@gmail.com, manzaramesh@gmail.com, anupriya.k.145@gmail.com

Abstract. Diabetic Maculopathy is damage to the macula. Scientifically also known as a pathological disorder. It's a very serious upshot of diabetes. Maculopathy early detection is very important as it causes blindness and is irreversible if proper treatment is taken. The present study deals with the design of a novel detection technique for early diagnosis of the diabetic maculopathy. For that Digital image processing have been used and for extracting the feature wavelet Filter has been used. Weighted KNN classification technique for grading of image i.e. mild, moderate and sever on the standard fundus images. The blood vessel extraction of the retina is first preferred because when macula starts getting affected by diabetes at the same time some abnormal blood vessel is created which is known as neovascularization. This Neovascularization cause's blindness because the retina gets nourishes with the blood vessels that's why blood vessel extraction is very important. Diabetic Maculopathy is one of the complications of diabetes mellitus that is considered as the major cause of vision loss among people around the world. It results from the leakage of fluid rich in fat and cholesterol from the damaged retinal vasculature. Accumulation of these fluids called exudates near the center of the retina. Development of diabetic Maculopathy is slow and silent, very frequently without any symptoms in the early stages. If Maculopathy is not detected in the early stage then the damage of the macula or visual field is irreversible and can lead to blindness. Therefore, compulsory regular screening of diabetic eye will help to identify the Maculopathy at the initial stage and reduce the risk of severe vision loss. Digital screening of Maculopathy results in the generation of a large number of retinal images to be Manually analyzed by an expert [2]. This often leads to observer fatigue and increase in the time taken for diagnosis. Non clinically significant (NCSME) and clinically significant (CSME) are two types of maculopathy stages.

Keywords: Macula · Blood vessel · Digital image processing

© Springer Nature Singapore Pte Ltd. 2019
K. C. Santosh and R. S. Hegadi (Eds.): RTIP2R 2018, CCIS 1036, pp. 349–360, 2019.
https://doi.org/10.1007/978-981-13-9184-2_32

1   Introduction

Diabetes. Diabetes is the long term disorder, when body cannot increase the insulin or the percentage of sugar is high than normal in the body that time we say it is a diabetic mellitus. Normal person sugar range is between 70 to 99 mg/dl and diabetes patient sugar range is more than 99 mg/dl means it may be between 100 to 126 mg/dl. Long term diabetes may affect on human organ like kidney, heart, eye, nerves, and blood vessels. Diabetes mellitus mainly are divided in two types Type I and Type II. In Type I body cannot produce Insulin and is generally diagnosed in children and young people. In which patient are given an insulin injection [1]. In Type II it does not respond to insulin. This type of diabetes is diagnosed in middle age or aged people. Both types are chronically lead high blood sugar levels. Both increases the risk of complications. Diabetes symptom are like frequent urination, drinking a lot of water, want to eat more means feeling hungry more than usual, feeling very fatigued, having a blurry vision because it affected on the human organ. When blood sugar goes high in the body that i.e. it gets affected a blood vessel causing leakage of fluid rich in fat and cholesterol from the damaged retinal vasculature. Accumulation of these fluids called exudates near the center of the retina if these types of results are shown in the retina or if a center of retina means that macula gets affected, and results in blindness it is called as diabetic maculopathy.

Diabetic Maculopathy. Diabetic Maculopathy is one of the complications of diabetes mellitus that is considered as the major cause of vision loss among people around the world. It results from the leakage of fluid rich in fat and cholesterol from the damaged retinal vasculature. Accumulation of these fluids called exudates near the center of the retina. Development of diabetic Maculopathy is slow and silent, very frequently without any symptoms in the early stages. If Maculopathy is not detected in the early stage then the damage of the macula or visual field is irreversible and can lead to blindness. Therefore, compulsory regular screening of diabetic eye will help to identify the Maculopathy at the initial stage and reduce the risk of severe vision loss. Digital screening of Maculopathy results in the generation of a large number of retinal images to be Manually analyzed by an expert [2]. This often leads to observer fatigue and increase in the time taken for diagnosis. Non clinically significant (NCSME) and clinically significant (CSME) are two types of maculopathy stages.

2   Methodology

The present study deals with the design of a novel detection technique for early diagnosis of the diabetic maculopathy. For that we used standard fundus image database which is STARE [4], DRIVE [5], DIRETDB0 [6], and DIRETDB1 [7] in which from STARE 35 images were taken, from DRIVE 20 images, that from DIRECTDB0 25 images and from DIRECTDB1 20 images were taken, these database are publically available in open sorce, and then Digital image processing have been used and for feature extraction we used wavelet Filter. Weighted KNN classification technique which is a supervised type of classification technique

(a)Normal Fundus Image	(b)Abnormal fundus image
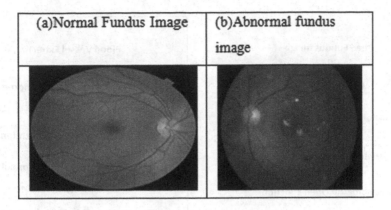	

Fig. 1. Normal and abnormal fundus image.

have been used for grading of image i.e. mild, moderate and severe on the standard fundus images. The blood vessel extraction of the retina is first preferred because when macula starts getting affected by diabetes at the same time some abnormal blood vessel is created which is known as neovascularization. This neovascularization causes blindness because the retina gets nourishes with the blood vessels that's why blood vessel extraction is very important. For the extraction of Retinal blood vessel, the funds images are taken from a standardized database. From which, firstly, the extraction of green from Red Green Blue (RGB) image is done. After that secondly, the extraction of green channel, intensity transformation function has been applied to enhance the funds image. Then in third step, histogram equalization was applied on intensity transformed image because it highlighted the blood vessel. And then in the fourth step, morphological open function is applied for thinning the blood vessel. In fifth step, median filter is applied to remove noise which appeared when we used morphological open function. Then in the sixth step, threshold is applied for extraction of blood vessel and then in the seventh step apply Sym4 wavelet and extract the blood vessel. After the whole precess Area, Diameter, Length, Thickness and Mean Diameter is calculated. Normal blood vessel diameter are >25 mm through that value we predict the normal and abnormal images. Then apply Weighted KNN classification for grading the disease mild, moderate and severe (Fig. 1) .

2.1 Green Channel

In this process green channel plays an important role in following line figure showing red, green and blue channel with its histogram.

2.2 Red Channel

$$r = \frac{R}{(R + G + B)} \tag{1}$$

Here Red channel is viewed in which R, G, and B means Red Green and Blue respectively (Fig. 2).

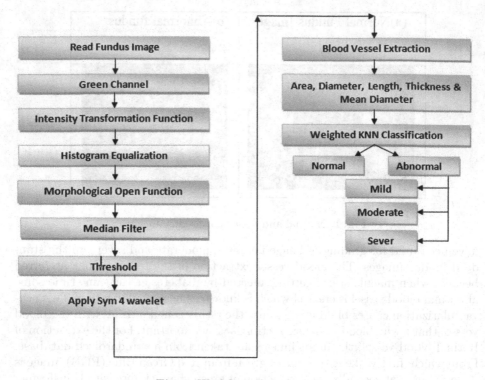

Fig. 2. Workflow of methodology

2.3 Green Channel

$$g = \frac{G}{(R + G + B)} \tag{2}$$

Here Green channel is viewed in which R, G, and B means that Red Green and Blue respectively.

2.4 Blue Channel

$$b = \frac{B}{(R + G + B)} \tag{3}$$

2.5 Intensity Transformation Function

The intensity transformation function is very easiest technique in image processing. If the need of an image is light lighter and dark darker at that time, it changes the intensity values, intensity transformation function which increase

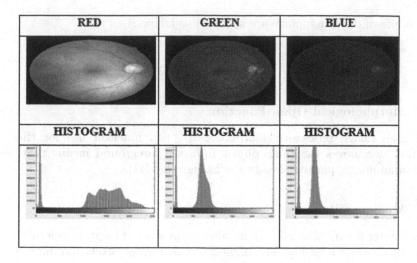

Fig. 3. Red, Green and Blue channel with histogram. (Color figure online)

contrast on certain value [15,16]. [8] The values of pixel pre and post processing are denoted to f(x, y) and g(x, y) (Fig. 3).

$$g(x,y) = Tf(x,y) \tag{4}$$

T for transformation of pixel value from f(x, y) into pixel (x, y). Input image is a f(x, y) and g(x, y) is output or processed image. [9] here we used intensity transformation for highlighting the retinal blood vessel.

2.6 Histogram Equalization

Histogram equalization produces output image with same allocation of pixel intensity means that the histogram of the output image is compress and increase systematically [10].

Here ps(s) and pd(d) is the image probability density functions. In following line we have histogram equalization of the image.

$$u = T(s)$$
$$T(s) = \int_0^s ps(x)dx \tag{5}$$

The histogram equalization image is acquired by a same transformation function as follows:

$$v = Q(d)$$
$$Q(d) = \int_0^d pd(x)dx \tag{6}$$

The values of d for the image are acquired as follows:

$$d = Q^{-1}[u]$$
$$Q^{-1}[u] = Q^{-1}T[s] \tag{7}$$

2.7 Morphological Open Function

In Digital image processing morphological open function use for the noise removal, it removes the small object from the foreground means that bright pixel of an image, placing then in the background [11].

2.8 Median Filter

Median filter use median value in its filtering process. It improves picture clarity improving, reduce least high and highest intensity value pixels and impulse noise [12].

2.9 Threshold

Thresholding is the way of partitioning an image foreground into background. This is a type of image segmentation that converts the grayscale image into binary image, it's very effective in an image with high level of contrast [13].

2.10 Wavelet

For the extraction of Diabetic maculopathy lesion we use here Wavelet filter. Here we used Symlet 4 wavelet. Wavelet performs a mathematical operation like image compression and decompose a signal into single representation and showing signal information. Wavelet can reduce noise and compress data and done many other operations. Wavelet computes the approximation coefficients matrix and details coefficients matrices i.e. horizontal, vertical, and diagonal, respectively. Here, inverse wavelet i.e. reconstructed image gives good result. The reconstructed or inverse wavelet have approximation coefficients matrix X, based on approximation matrix CA and details matrices CH, CV, and CD. In that proposed study we used sym4 wavelet.

2.11 Blood Vessel Extraction

In blood vessel extraction we calculate the Area, Diameter, Length, Thickness and Mean Diameter. By following formulas.

2.11.1 Area

$$Area = \pi(r2) \tag{8}$$

2.11.2 Diameter

$$Diameter = \sqrt{\frac{Area}{\pi}} \qquad (9)$$

2.11.3 Length

$$Length = \frac{Area}{2} \qquad (10)$$

2.11.4 Thickness

$$Thickness = \frac{Area}{Length} \qquad (11)$$

2.11.5 Mean Diameter

$$Mean = \frac{M}{c} \qquad (12)$$

2.12 Weighted KNN Classification

Weighted KNN is the supervise classification technique which classify the data very easily [17]. Nominal Classes prediction is the main goal of classification version of weighted KNN and this classification works with the majority of the nearest neighbors and can be formulated by:

$$\hat{P}n\left(\omega_i|x\right) = \frac{p_n\left(x, w_i\right) \cdot w_i}{\sum_{i=1}^{M} p_n\left(x, w_i\right) \cdot w_i} = \frac{k_i \cdot w_1}{k \cdot w} \qquad (13)$$

where w works as a function of the distance between the ith nearest neighbor and the test point. The distances on which the search for the nearest neighbors is based in the first step, have to be transformed into similarity measures which can be used as weights [14].

2.13 Experimentation and Result

The fundus images are taken from the standard database and extract the blood vessel in following line see the experiment on fundus image. In above experiment it can be seen that the abnormal blood vessel are born and some vessel are going to damage and some are already bleeding here blood vessel extraction experiment is done (Fig. 4).

Following table shows the blood vessel Area, Diameter, Length, Thickness And Mean Diameter in statistical format. Then using Weighted KNN classification, classification and grading has been performed. Here we only focused on Retinal blood vessel diameter if its statistical value is 25 so its is a normal retina if its greter than or less than 25 so its is a diabetic maculopathy retina. We also predict the grading range for mild moderate and sever (Fig. 5).

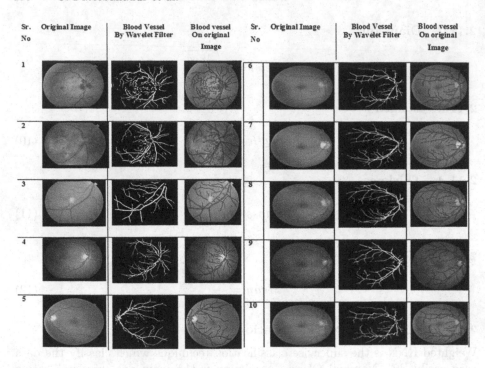

Fig. 4. Extraction of retinal blood vessels.

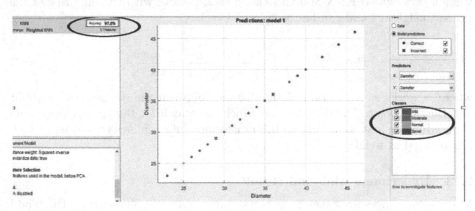

Fig. 5. Weightage KNN classification result. (Color figure online)

Classification and Grading. For the Classification and grading we used Weighted KNN supervise technique and we got the 97% accuracy on 100 fundus image, we also used a Support Vector Machine, Decision Tree, Linear Regression etc. classifier used but Weighted KNN is giving to much good result than other, thats why here we used Weighted KNN. In following figure we can see classifier accuracy and grading. In above figure we draw a black circle on result and grad-

Table 1. Statistical parameters of retinal blood vessels by symlet wavelet

Sr. no.	Area	Diameter	Length	Thickness	Mean diameter
1.	98.34875	32	49	2	20
2.	64.38625	26	32	2	20
3.	70.17125	27	35	2	20
4.	52.41125	23	26	2	20
5.	97.68875	31	49	2	20
6.	84.62875	29	42	2	20
7.	85.27375	29	43	2	20
8.	126.6763	36	63	2	20
9.	88.97	30	44	2	20
10.	87.79	30	44	2	20
11.	96.29375	31	48	2	2
12.	83.54625	29	42	2	20
13.	123.9963	35	62	2	20
14.	81.38375	29	41	2	20
15.	95.16125	31	48	2	20
16.	155.4675	40	78	2	20
17.	76.77875	28	38	2	20
18.	72.97	27	36	2	20
19.	83.04	29	42	2	20
20.	105.565	33	53	2	20
21.	82.82	29	41	2	20
22.	78.49375	28	39	2	20
23.	63.6525	25	32	2	20
24.	67.83125	26	34	2	20
25.	128.42	36	64	2	20
26.	96.39375	31	48	2	20
27.	84.655	29	42	2	20
28.	89.3725	30	45	2	20
29.	74.04375	27	37	2	20
30.	101.405	32	51	2	20
31.	122.345	35	61	2	20
32.	86.69	30	43	2	20
33.	126.6763	36	63	2	20
34.	117.2775	34	59	2	20
35.	109.3588	33	55	2	20
36.	141.445	38	71	2	20
37.	88.5375	30	44	2	20
38.	81.415	29	41	2	20

(*continued*)

Table 1. (*continued*)

Sr. no.	Area	Diameter	Length	Thickness	Mean diameter
39.	120.1463	35	60	2	20
40.	126.1963	36	63	2	20
41.	83.85	29	42	2	20
42.	127.305	36	64	2	20
43.	84.725	29	42	2	20
44.	76.4925	28	38	2	20
45.	128.9163	36	64	2	20
46.	88.2875	30	44	2	20
47.	89.47375	30	45	2	20
48.	100.9788	32	50	2	20
49.	62.41875	25	31	2	20
50.	70.72625	27	35	2	20
Bitmap sr. no.	Area	Diameter	Length	Thickness	Mean diameter
51.	83.08	29	42	2	20
52.	90.35	30	45	2	20
53.	100.2288	32	50	2	20
54.	83.41625	29	42	2	20
55.	65.0575	26	33	2	20
56.	85.315	29	43	2	20
57.	89.675	30	45	2	20
58.	70.795	27	35	2	20
59.	91.33	30	46	2	20
60.	82.89625	29	41	2	20
61.	79.0975	28	40	2	20
62.	94.30125	31	47	2	20
63.	81.91875	29	41	2	20
64.	80.125	28	40	2	20
65.	194.8813	44	97	2	20
66.	110.1225	33	55	2	20
67.	96.805	31	48	2	20
68.	80.63875	29	40	2	20
69.	90.43	30	45	2	20
70.	88.14	30	44	2	20
71.	62.565	25	31	2	20
72.	70.415	27	35	2	20
73.	123.75	35	62	2	20
74.	114.0988	34	57	2	20
75.	72.09125	27	36	2	20
76.	151.3663	39	76	2	20
77.	119.585	35	60	2	20

(*continued*)

Table 1. (*continued*)

Sr. no.	Area	Diameter	Length	Thickness	Mean diameter
78.	106.9825	33	53	2	20
79.	90.62375	30	45	2	20
80.	210.0325	46	105	2	20
81.	91.52125	30	46	2	20
82.	124.7875	36	62	2	20
83.	73.8575	27	37	2	20
84.	56.23875	24	28	2	20
85.	60.6425	25	30	2	20
86.	99.4975	32	50	2	20
87.	140.5138	38	70	2	20
88.	77.16125	28	39	2	20
89.	95.79375	31	48	2	20
90.	82.03875	29	41	2	20
91.	61.25125	25	31	2	20
92.	71.53875	27	36	2	20
93.	93.17375	31	47	2	20
94.	121.285	35	61	2	20
95.	174.755	42	87	2	20
96.	67.6075	26	34	2	20
97.	194.8813	44	97	2	20
98.	110.1225	33	55	2	20
99.	96.805	31	48	2	20
100.	80.63875	29	40	2	20

ing, first circle showing the accuracy of result and second showing the grading. Blue color showing the Mild, orange showing Moderate, Yellow showing Normal and Purple showing Sever of Diabetic maculopathy grading. PCA is disable there. In that Weighted KNN classification we used five fold cross validation it also called K-fold validation and get the 97% good result on database (Table 1).

3 Conclusion and Discussion

Blood vessel extraction is very important for maculopathy as it creates new abnormal blood vessels by which it not only damages the macula but increases high risk of vision loss. So to reduce this risk by Weighted KNN classification blood vessel have been extracted. In which area, diameter, length etc. has been extracted. And by calculating the diameter we can came to the conclusion that less than and greater than 25 mm. diameter the blood vessel is abnormal.

Weighted KNN classification gave 97% accuracy on standard database. As 97% is a good accuracy rate, the research is useful for ophthalmologist.

References

1. American Diabetes Association: American Diabetes Association Copyright 1995–2018 [Internet]. http://www.diabetes.org/diabetes-basics/type-1/
2. Shodhganga.   http://shodhganga.inflibnet.ac.in/bitstream/10603/3095/14/14_chapter%206.pdf
3. Noronha, K., Nayak, K.P.: Automated diagnosis of diabetes maculopathy: a survey. J. Med. Imaging Health Inform. **3**, 1–8 (2013)
4. Structured Analysis of the Retina. http://cecas.clemson.edu/~ahoover/stare/
5. Staal, J.J., Abramoff, M.D., Niemeijer, M., Viergever, M.A., van Ginneken, B.: Ridge based vessel segmentation in color images of the retina. IEEE Trans. Med. Imaging **23**, 501–509 (2004)
6. DIARETDB0 - Standard Diabetic Retinopathy Database. http://www.it.lut.fi/project/imageret/diaretdb0/
7. DIARETDB1 - IMAGERET. http://www.it.lut.fi/project/imageret/diaretdb1/
8. CS425 Lab: Intensity Transformations and Spatial Filtering. http://www.cs.uregina.ca/Links/class-info/425/Lab3/
9. Kushwaha, S., Rabindra, K.S.: Study and analysis of various image enhancement method using MATLAB. Int. J. Comput. Sci. Eng. (IJCSE) **3**(1), 15–20 (2015)
10. Dixit, M., Singh, R.P.: Histogram equalization a strong technique for image enhancement. Int. J. Signal Process. Image Process. Pattern Recogn. **8**(8), 345–352 (2015)
11. Gonzalez, R.C., Woods, R.E.: Digital Image Processing, 3rd edn. (2008). ISBN 9789332570320. OCLC 979415531
12. Garg, P.K., Verma, P., Bhardwaj, A.: A survey paper on various median filtering techniques for noise removal from digital images. Am. Int. J. Res. Formal Appl. Nat. Sci. (2014). ISSN (Print) 2328-3777, ISSN (Online) 2328-3785, ISSN (CD-ROM) 2328-3793
13. Mathworks. https://in.mathworks.com/discovery/image-thresholding.html
14. Kozak, K., Kozak, M., Stapor, K.: Weighted k-nearest-neighbor techniques for high throughput screening data. Int. J. Chem. Mol. Eng. **1**(12), 155 (2007). World Academy of Science, Engineering and Technology
15. Ruikar, D.D., Santosh, K.C., Hegadi, R.S.: Automated fractured bone segmentation and labeling from CT images. J. Med. Syst. **43**(3), 60 (2019). https://doi.org/10.1007/s10916-019-1176-x
16. Ruikar, D.D., Santosh, K.C., Hegadi, R.S.: Segmentation and analysis of CT images for bone fracture detection and labeling. In: Medical Imaging: Artificial Intelligence, Image Recognition, and Machine Learning Techniques, Chap. 7. CRC Press (2019). ISBN 9780367139612
17. Hegadi, R.S., Navale, D.I., Pawar, T.D., Ruikar, D.D.: Multi feature-based classification of osteoarthritis in knee joint X-ray images. In: Medical Imaging: Artificial Intelligence, Image Recognition, and Machine Learning Techniques, Chap. 5. CRC Press (2019). ISBN 9780367139612

Histopathological Image Classification: Defying Deep Architectures on Complex Data

Suvidha Tripathi[✉] and Satish Singh

Indian Institute of Information Technology Allahabad,
Devghat, Jhalwa, Allahabad 211015, U. P., India
suvitri24@gmail.com, sk.singh@iiita.ac.in

Abstract. Automatic analysis of medical images is a challenging research which requires both the skill of a pathologist and computer vision knowledge to develop efficient systems. In this work, we have taken up the task of classifying different types of cell nuclei in histopathological Colon Cancer Images. We aim to show the relevance and effect of a complex histopathological dataset on the performance of current deep learning architectures. We have experimented with pre-trained (on ImageNet) AlexNet, VGG16, and VGG19 architectures and applied transfer learning approach to train these architectures. On the basis of the results obtained on the Histopathological image dataset, while using fine tuned AlexNet, VGG16, and VGG19 architectures; the suitability of using pure architectures is somehow questionable and these state of the art algorithms straightaway cannot be used for the sophisticated classification of very complex cancer tissue dataset. Comparative evaluation of the above state of the art methods have been done and the possibility of devising hybrid deep architectures is investigated thereof.

Keywords: Transfer learning · Traditional features ·
Neural networks · Deep learning · Histopathology · Nuclei classification

1 Introduction

Traditional Technique of extracting features from a set of images is becoming obsolete and as a result, works claiming to use it are being taken lightly unless the authors make a thorough comparison with deep architectures. However, this is also true that no one has actually been able to come up with any such hand-crafted or object level feature descriptors that has outperformed deep neural networks. One thing that gets remarkably get overlooked is the type of diversity of dataset that is being used to produce high accuracies in this domain. Histopathological or medical data, in general, is one such dataset that has been challenging researchers around the world due to their heterogeneity and uncertain nature of patterns. In this work, we have taken some of the very famous

© Springer Nature Singapore Pte Ltd. 2019
K. C. Santosh and R. S. Hegadi (Eds.): RTIP2R 2018, CCIS 1036, pp. 361–370, 2019.
https://doi.org/10.1007/978-981-13-9184-2_33

deep learning methods to do the classification task using transfer learning technique and highlight the fact that even the most efficient networks may fail to improve the classification accuracy when the dataset involved is highly complex in nature. Starting from AlexNet [1], VGG16, and VGG19 [2], three of the most used architectures due to their low computational requirements and high performance, are used in this experimental study.

The histopathological dataset used here is a set of different types of nuclei found in colon cancer i.e., fibroblasts, inflammatory, lymphocytes, neutrophils, eosinophils, etc. The presence of each type of these cell nuclei indicates the nature of cancer [4]. This information is crucial for the pathologists to diagnose the severity and type of cancer. The factors that influence the decision of diagnosis are size, structure, density, chromatin texture and intensity (depends on the staining dye) of the nuclei present in the affected tissues [5]. These factors change with the type of nuclei and hence, pathologists need to know beforehand the type to make a conscious decision. The properties of this dataset are crucial since such type of datasets changes with the type of staining technique used to stain the nuclei and stroma of the tissues. Hence, different staining technique gives different color and texture features. So to summarize, color and texture features play a very important role in classifying different structures, even more than the shape and size features.

With deep learning architectures we can very easily extract the features after each layer but, lack of interpretability restricts us to know the actual type of features extracted in the process. Therefore, it is difficult to firmly establish whether the quality of features deep learning network is extracting will give good classification performance.

In past years, much work has been done on histopathological images for nuclei classification using handcrafted feature descriptors such as morphological, texture, shape and color features. Shape representation through DTW-Radon based descriptors by authors in [3] and [6] establised a rotation and scaling invariant lossless transform for detecting various shape properties in several numeral, character and symbol datasets. Their methods could be used to detect shape features in nuclei datasets for classification purposes but, since the dataset is very huge and contains large number of samples, their method would incur huge computational cost. Liu et al. in [7] tested the various types of features that can be extracted from images and through feature selection methods they found out the most relevant features for cell nuclei classification. However they did not mention the kind of dataset they used for extracting features, therefore it is hard to say if their findings are universal for all types of datasets. Authors in [8] studied various nuclei classification methods on different types of cancer kinds such as prostate, breast, renal clear cell and renal papillary cell cancer. They showed that the classification methods gave different accuracies on each one of these datasets. Their results proved that there cannot be one definitive method that would give better results across all types of cancer. They also used deep learning methods like LeNet, EncoderNet, Color-EncoderNet on their datasets but only Color-EncoderNet was able to give best results among the

10 methods they tested on 3 out of 4 datasets. These studies hint that even using the deep learning framework does not guarantee good results in the case of complex histopathological images. Other effective methods such as [10] and [11] to detect candidate region of interests and extracting features for further processing of biomedical images use hand-crafted feature descriptors and local variations within the images. These methods are effective for small datasets that have greater inter and less intra class separability. But, in our case, where the dataset is complex and has less inter class separability, relying only on hand-crafted features is not a feasible approach. To prove this, we have tested few state of the art algorithms on hand-crafted feature descriptors and compared the obtained results with deep learning methods. Deep learning method used in [12] would have been an initial approach for classification but their method use grayscale images and also the dataset used is not from biomedical domain. So, the complexity and feature relation is highly deviated from our intensity and color centric RGB histopathological dataset.

2   Experiments

We have taken some of the very recent deep architectures and trained our dataset on them to find out their performance.

2.1   Dataset

Image dataset from which nuclei points are extracted as patches is taken from [9]. The dataset came with annotated nuclei and their location in the data. We prepared our own data points using the method in [14]. From each image in the dataset as shown in Fig. 1, the nuclei present in this image were annotated by pathologists. Annotated nuclei center pixel coordinates were recorded for each of the images along with their corresponding labels. Using this recorded information about all the nuclei, total 22444 nuclei samples of height and width 27 around the center pixel coordinate, with RGB color channels, were collected in a folder. 22444 nuclei were segregated into four classes viz. Epithelial nuclei, Inflammatory nuclei, Fibroblast nuclei and miscellaneous other types as the fourth category. The number of samples in each class affect the final results by a great margin. In our dataset, class 1 i.e. epithelial class has total 7,722 nuclei, class 2 (inflammatory nuclei) has 5,712 samples, class 3 (fibroblast nuclei) has 6,971 class points and the miscellaneous category has mixed type data of total 2,039 sample points. These raw nuclei images were then divided into train and test set. 70% of the samples from each class were taken as input for training and the rest 30% of the samples were used in testing. However, The input size of each image in our dataset had to be resized to $224 \times 224 \times 3$ since, this is the size that the AlexNet, VGG16 and VGG19 architectures take as input. Figure 2 shows the sample nuclei dataset.

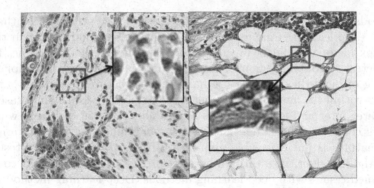

Fig. 1. 500×500 H&E stained histology image samples of colorectal adenocarcinomas

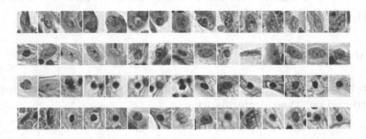

Fig. 2. Example of nuclei dataset. Row 1: epithelial nuclei, Row 2: inflammatory nuclei, Row 3: fibroblasts, Row 4: miscellaneous

2.2 AlexNet

AlexNet by Krizhevsky et al. [1] is the very first architecture inspired from LeCun et al. [15] which gained popularity after 2012 ImageNet challenge. It has 5 convolutional layers followed by 3 fully connected layers. We divided the dataset into 7:3 ratio for training and testing. Initially, we kept the learning rate incremental, starting from 0.01 and increased it up to 0.00001 i.e. 1e−05. we observed the minibatch accuracy very low and the overall accuracy on pretrained AlexNet was observed 0. So, incremental learning rate did not work with our dataset. Hence, we kept the learning rate constant.

2.3 VGG16

We investigated the effect of increasing depth of convolutional layers by testing the performance of VGG16 [2] on our dataset accuracy. The number of parameters increases with the depth and hence the computation requirements. We trained our dataset using the pre-trained model because learned features are often transferable to different data and then it also takes less training time as compared to the experiment where the model is trained from the scratch [13].

Training any deep learning architecture from scratch is not feasible for both accuracy and time performance since the network has to learn again the trivial features like edges and lines which becomes a redundant task if the accuracy does not improve as the training progress. Using the concept of transfer learning helps propagate the generic features through the model. Only the features specific to the dataset are learned through model training.

2.4   VGG19

VGG19 [2] has more depth than VGG16 i.e. 19 convolutional layers and hence, improved performance. Working on this theory we trained VGG19 on our dataset and made few observations included in Sect. 3.

Apart from transfer learning, random changes in batch size and number of epochs were performed to select the optimal hyperparameters. We selected the batch size of 300 and trained the architectures for 100 number of epochs.

3   Results and Discussions

We have evaluated our classifier performance using Precision (or Positive Predictive Value PPV), Recall (or True Positive Rate TPR), F1 score, Accuracy and time taken by three architectures. Accuracy and time comparison among three architectures are shown in Table 1. It is observed that with deep architectures large batches can be parallelized across many machines, reducing training time significantly. Also, large batch size reduces the number of parameter updates required to train a model which in turn results in reduced model training times. Therefore, we kept the batch size high. To establish our design choice of a large batch size we did random batch size changes, starting from 64. We noticed no change in accuracy but, the time required to train each batch increased by 100%. Earlier, the time for each epoch, in case of 300 batch size, was around 20 min, which increased to 42 min when the batch size was reduced to just 64 and number of epochs to 30. This happens when lower batch size takes a number of iterations to do the weight update due to more number of computations. So, It was more feasible to train our dataset with a larger batch size considering the time efficiency. The recent article by authors in [21] have studied the effect of increasing batch sizes on ImageNet and CIFAR10 datasets using recent state of the art deep learning algorithms like ResNet and Inception-ResNet-V2. They confirmed that the large batch size reduces the training times significantly and are better than decaying learning rate when the effect on accuracy is not significant. Figures 3a, b and c are the ROC curves of three networks. Each figure has four curves representing four classes of nuclei i.e., Epithelial, Fibroblast, Inflammatory and miscellaneous. We have also compared our deep learning architecture performance with handcrafted descriptors we used in [14] to measure their retrieval performance on our dataset. Comparison Table 5 clearly outlines the fact the handcrafted descriptors are clearly no match to deep learning algorithms since

Table 1. Accuracy and time comparison

Architectures	Accuracy (%)	Time (secs)
Alexnet	72.68	2564.69
VGG16	73.89	9331.79
VGG19	73.54	59976.00

there is a huge difference in classification metrics. While the same descriptors performed better in retrieving CT, MRI, and ultrasound images such as in [16–18], they performed very poorly on our dataset when we used the same feature subset for classification. It is important to note that the feature descriptors specially designed to retrieve medical images in [16–18] performed even poorer than the ones that were designed for retrieving colored images [19,20]. So, it establishes the fact the color information is an important feature in case of histopathological images. Handcrafted features that work on grayscale images will not give optimum performance in such datasets.

Table 2. Confusion matrix of AlexNet

Output class	Target class				
	1	2	3	4	Precision
1	2017	261	145	93	80.9%
2	152	1104	280	138	65.9%
3	52	198	1524	233	75.9%
4	6	151	142	148	33.1%
Recall	90.9%	64.4%	72.9%	24.2%	72.5%

We made following observations from the results we obtained.

1. ROC curves are shown in Fig. 3 shows the performance of each architecture AlexNet (Fig. 3a), VGG16 (Fig. 3b), and VGG19 (Fig. 3c). To compare the differences among these curves we took True Positive Rate (TPR) value at 90% in all three curves and noticed the corresponding False Positive Rate (FPR) with respect to each class. FPR value should be minimized with respect to each class. In case of class 1 (Epithelial) minimum, FPR is given by VGG19 and maximum FPR is by AlexNet whereas for class 4 (Miscellaneous) minimum and maximum FPR is given by VGG16 and VGG19 respectively. For Inflammatory nuclei category, FPR is almost similar in all three methods and in the class of Fibroblast nuclei, VGG16 gives the minimum FPR and VGG19 outputs maximum FPR. After analyzing the three ROC curves, we inferred that there is no unique pattern to declare the best classifier for all 4 classes. They show different patterns with respect to each class. This difference in

patterns may become a problem when determining the best classifier among the three. However, due to an imbalance in the data samples, it is expected that the fourth class which has the least number of samples will perform the worst. This gives the clue to the best classifier question, which is, the classification method that performs the best with minority class should be the best classifier. Here VGG16 has the minimum FPR with minority class. Hence, VGG16 is the best classifier among the three. This is also reflected in the classification metrics Table 5.

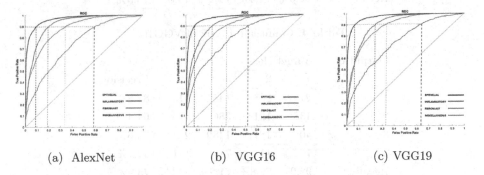

(a) AlexNet   (b) VGG16   (c) VGG19

Fig. 3. ROC curves of AlexNet, VGG16 and VGG19

2. if we compare our results with ImageNet dataset accuracies on these networks, that is AlexNet has top 1 accuracy of 56.1% [1] and top 5 accuracies of 80%, VGG16 has top 1 and top 5 accuracies of 70.6% and 89.9%, and for VGG19 it is 68% and 85.5% respectively [2], we see that there is a significant improvement of atleast 12% in top 1 accuracy and 6% increase when comparing top 5 accuracies of AlexNet and VGG19.
3. Hence, by observation of accuracy changes among datasets, we can very certainly say that our dataset was indeed difficult to classify for these architectures.
4. We also made observations among class wise accuracy, and uniformly we noticed from confusion matrices that class 1 i.e. Epithelial nuclei scored the best with highest percentage of 84.9% in case of VGG19 (Table 2). Class 2 (Inflammatory nuclei) second with the highest percentage of 65.9% in AlexNet (Table 3), class 3 (Fibroblasts) third with highest 76.6% in VGG16 (Table 4) and miscellaneous nuclei in class 4 scored the least accuracy among all three architectures with best value of 42.7% in VGG16 (Table 4).
5. This variation in accuracies reflect on the structure of the nuclei in the database. Miscellaneous nuclei contained all other small groups of nuclei found in colon cancer, hence this class did not have any particular pattern in majority. Therefore, the classifier could not make the best decision for this class.
6. We observed from Table 5 that despite VGG19 having the deepest network, did not perform better than VGG16. But, it is however not a very significant improvement. VGG16 is only 1% more sensitive (recall) than VGG19

Table 3. Confusion matrix VGG19

Output class	Target class				
	1	2	3	4	Precision
1	2119	191	117	68	84.9%
2	145	1224	411	128	64.2%
3	42	191	1488	295	73.8%
4	11	108	75	121	38.4%
Recall	91.5%	71.4%	71.2%	19.8%	73.5%

Table 4. Confusion matrix of VGG16

Output class	Target class				
	1	2	3	4	Precision
1	2132	208	145	91	82.8%
2	138	1214	386	146	64.4%
3	40	181	1489	234	76.6%
4	7	111	71	141	42.7%
Recall	92.0%	70.8%	71.2%	23.0%	73.9%

Table 5. Comparison between methods through performance parameters

Method	Precision	Recall	F1-score	Accuracy
LBDP	38.80%	31.45%	34.74%	40.7%
LCOD	48.27%	38.00%	42.52%	46.20%
LWP	38.06%	31.05%	34.20%	39.30%
LDEP	37.30%	30.15%	33.34%	39.10%
RSHD	49.30%	36.65%	42.04%	44.10%
AlexNet	63.95%	63.10%	63.52%	72.50%
VGG16	**66.62%**	**64.25%**	**65.41%**	**73.90%**
VGG19	65.34%	63.47%	64.39%	73.50%

(Table 5). Also, when we look at the time took by VGG16 and VGG19 from Table 1 for training, VGG19 took 6 times more time than VGG16. Hence, if we have to choose between VGG16 and VGG19, VGG16 becomes the better choice both in terms of accuracy and time.

7. From the comparison of the handcrafted and deep learning architectures in Table 5, it is trivial to deduce that deep architectures performed better than handcrafted descriptors used in this study.

4   Conclusion

Through this experimental work, our objective was to establish that, the state of the art deep learning networks perform better than handcrafted features but may not produce great results for all kinds of datasets such as the Histopathological data whereas, AlexNet, VGG16, and VGG19 produces classification accuracy better in ImageNet dataset as mentioned in point 2 of Sect. 3. Histopathological data is highly complex and incomprehensible to the non-experts. Without the consultation of domain expertise of the experienced pathologists, one can never be sure of the nature of the objects present in the images. Hence, proper classification of the images is a complex task even for humans for such datasets. Deep learning algorithms do not address the dataset heterogeneity problem and their performance in different domains of data. Therefore, with our work, we have tried to reflect on the fact that otherwise widely used deep learning algorithms used for classifying histopathological data are not the best feasible methodology alone. Hence, only handcrafted or only deep learning architectures are not enough for classifying complex histopathological data. Their combination shall be exploited to achieve the better performance.

Acknowledgment. This research was carried out in Indian Institute of Information Technology, Allahabad and supported by Ministry of Human Resource and Development, Government of India. We are also grateful to the NVIDIA corporation for supporting our research in this area. Currently, we are using a donated TITANX(PASCAL) GPU with 3584 CUDA cores to train models for this research work.

References

1. Alex, K., Ilya, S., Geoffrey, H.: Image net classification with deep convolutional neural network. In: NIPS (2012)
2. Simonyan, K., Zisserman, A.: Very deep convolutional networks for large-scale image recognition. arXiv preprint arXiv:1409.1556 (2014)
3. Santosh, K.C., Lamiroy, B., Wendling, L.: DTW-radon-based shape descriptor for pattern recognition. Int. J. Pattern Recogn. Artif. Intell. **27**(03), 1350008 (2013)
4. https://www.cancerresearchuk.org/what-is-cancer/how-cancer-starts/types-of-cancer
5. Zink, D., Fischer, A.H., Nickerson, J.A.: Nuclear structure in cancer cells. Nat. Rev. Cancer **4**, 677–687 (2004). https://doi.org/10.1038/nrc1430
6. Santosh, K.C., Lamiroy, Bart, Wendling, Laurent: DTW for matching radon features: a pattern recognition and retrieval method. In: Blanc-Talon, Jacques, Kleihorst, Richard, Philips, Wilfried, Popescu, Dan, Scheunders, Paul (eds.) ACIVS 2011. LNCS, vol. 6915, pp. 249–260. Springer, Heidelberg (2011). https://doi.org/10.1007/978-3-642-23687-7_23. Inria-00617287
7. Liu, S., Mundra, P.A., Rajapakse, J.C.: Features for cells and nuclei classification. In: 2011 Annual International Conference of the IEEE Engineering in Medicine and Biology Society, Boston, MA, pp. 6601–6604 (2011). https://doi.org/10.1109/IEMBS.2011.6091628

8. Singh, M., Zeng, Z., Kalaw, E.M., Giron, D.M., Chong, K.-T., Lee, H.K.: A study of nuclei classification methods in histopathological images. In: Chen, Y.W., Tanaka, S., Howlett, R., Jain, L.C. (eds.) InMed 2017. SIST, vol. 71, pp. 78–88. Springer, Cham (2018). https://doi.org/10.1007/978-3-319-59397-5_9

9. Sirinukunwattana, K., Raza, S.E.A., Tsang, Y.W., Snead, D.R.J., Cree, I.A., Rajpoot, N.M.: Locality sensitive deep learning for detectionand classification of nuclei in routine colon cancer histology images. IEEE Trans. Med. Imaging **35**(5), 1196–1206 (2016)

10. Santosh, K.C., Wendling, L., Antani, S., Thoma, G.R.: Overlaid arrow detection for labeling regions of interest in biomedical images. IEEE Intell. Syst. **31**(3), 66–75 (2016). https://doi.org/10.1109/MIS.2016.24

11. Ravi, M., Hegadi, R.S.: Detection of Glomerulosclerosis in diabetic nephropathy using contour-based segmentation. In: International Conference on Advanced Computing Technologies and Applications ICACTA (2015)

12. Ukil, S., Ghosh, S., Obaidullah, S.M., Santosh, K.C., Roy, K., Das, N.: Deep learning for word-level handwritten Indic script identification, arXiv preprint arXiv:1801.01627

13. Yosinski, J., Clune, J., Bengio, Y., Lipson, H.: How transferable are features in deep neural networks? In: Advances in Neural Information Processing Systems 27 (NIPS 2014). NIPS Foundation (2014)

14. Tripathi, S., Mishra, S., Singh, S.K.: Routine colon cancer detection using local image descriptors. In: IEEE Region 10 Conference (TENCON), Singapore 2016, pp. 2062–2065 (2016). https://doi.org/10.1109/TEN-CON.2016.7848388

15. LeCun, Y., et al.: Backpropagation applied to handwritten zip code recognition. Neural Comput. **1**(4), 541–551 (1989)

16. Dubey, S.R., Singh, S.K., Singh, R.K.: Local diagonal extrema pattern: a new and efficient feature descriptor for CT image retrieval. IEEE Signal Process. Lett. **22**(9), 1215–1219 (2015)

17. Dubey, S.R., Singh, S.K., Singh, R.K.: Local bit-plane decoded pattern: a novel feature descriptor for biomedical image retrieval. IEEE J. Biomed. Health Inform. **20**(4), 1139–1147 (2015)

18. Dubey, S.R., Singh, S.K., Singh, R.K.: Local wavelet pattern: a new feature descriptor for image retrieval in medical CT databases. IEEE Trans. Image Process. **24**(12), 5892–5903 (2015)

19. Dubey, S.R., Singh, S.K., Singh, R.K.: Rotation and scale invariant hybrid image descriptor and retrieval. Comput. Electr. Eng. **46**, 288–302 (2015)

20. Dubey, S.R., Singh, S.K., Singh, R.K.: Local neighbourhood-based robust colour occurrence descriptor for colour image retrieval. IET Image Process. **9**(7), 578–586 (2015)

21. Smith, S.L., Kindermans, P.J., Ying, C., Le, Q.V.: Don't decay the learning rate, increase the batch size. In: ICLR 2018, arXiv preprint arXiv:1711.00489

Biometrics and Applications

The Quantification of Human Facial Expression Using Trapezoidal Fuzzy Membership Function

M. R. Dileep[1(✉)] and Ajit Danti[2]

[1] Department of Computer Science and Applications, St. Aloysius College (Autonomous), Mangaluru, Karnataka, India
dileep.kurunimakki@gmail.com
[2] Department of Computer Science and Engineering,
Christ (Deemed to be University), Bangalore, Karnataka, India
ajit.danti@christuniversity.in

Abstract. Fuzzy Inference System is an interesting approach. Major benefit of the FIS is, it permits the natural narration in linguistic terms of tribulations that can be resolved rather than in requisites of associations between accurate arithmetical points. This helps, handling with the complicated systems in easy way, is the major motive why fuzzy system is broadly incorporated in practice. In the present research paper, an effective approach is proposed that quantifies the human facial expression using Mamdani implication based fuzzy logic system. The recent principle engages in retrieving arithmetical values from person's face and feed them to a fuzzy classifier. Fuzzification and Defuzzification process issues trapezoidal fuzzy membership function for input as well as output. The diverse characteristic of this method is its effortlessness and maximum correctness. Experimental outcome on Image dataset depicts excellent accomplishment of the proposed methodology. In this paper, a legitimate procedure proposed for quantification of human facial expression from the features of the face by means of Mamdani type fuzzy inference system, which is proficient to set up a convenient membership association involving the various dimensions of the happy expression. Values representing features of the face are fed to a Mamdani-type fuzzy classifier. This system recognizes three levels of same happy expression namely Normal, Bit Smiley and Loud Laugh. The total output expressions for this proposed scheme is three. Another discrete element of the proposed methodology is the membership method model of expression outcome which stands on various surveys and readings of psychology.

Keywords: Fuzzy rule · Quantification of expression ·
Membership function

1 Introduction

Human facial appearance classification is the most interesting subject in the field of biometric, which is having a vast choice of applications namely Business,

K. C. Santosh and R. S. Hegadi (Eds.): RTIP2R 2018, CCIS 1036, pp. 373–380, 2019.
https://doi.org/10.1007/978-981-13-9184-2_34

Managerial, Organizational, Cultural contexts and so on. For this aspire, automatic detection of facial expression of the human faces is most active and hot area of research in machine vision and learning inside the most recent few years. Expressions on the face are produced by the variations in the face muscles that create features of face viz raising eyebrows, stretching corner lips, opening eyes and so on. Facial expression classification process generally extracts the facial expression arguments from the still picture of the face. This technique is known as "Quantification" of expression. These retrieved features are inputted to the FIS for the quantification of human facial expression by defining the range of expression with their membership function. In this paper, a complete system for quantification of facial expression in different range of happy faces is proposed. The heart of this proposed scheme is the Mamdani type fuzzy inference system which is employed to quantify the human facial expression from facial features of the face. Due to the complexity of facial expression, the quantification of human facial expression will turn out to be a topic with great challenges. In this paper, a Mamdani based fuzzy inference technique applied for quantification of human facial expression in the human face. Introduced a scheme of expression recognition depends on DWT-PCA/LDA, the multiple application approach. Esau et al. 2007, given an approach for Real-Time Expression Recognition Using a Fuzzy based Model, that effectively recognizes emotions of the human face. Herrera and Magdalena 1997, published a tutorial on genetic Fuzzy Systems gives a detailed description on biological features of human faces using fuzzy logic. Jamshidnezhad 2011, designed an approach that learns Fuzzy Model for Emotion Recognition that recognizes the emotions of the human face. Khanum et al. 2009, contributed a research on Fuzzy case reasoning for expression recognition, that efficiently recognizes the human facial expressions based on Fuzzy rule. Klir and Yuan 2000, published a study material on Fuzzy logic and Fuzzy Set – Applications and Theory, that gives a detailed description on the applications and usage of the fuzzy rule. Kyoung-Man Lim et al., designed an algorithm for face recognition using A N N and Fuzzy Logic, that effectively recognizes the face in a given still image. Lee et al. 1996, introduced a method that recognizes the human faces using knowledge-based extraction of features and neuro-fuzzy technique, a multiple application approach, that uses geometrical facial features of the face along with Artificial neural networks plus fuzzy inference systems. Rasoulzadeh 2012, did a research on facial expression recognition using Fuzzy Inference System. Fuzzy inference tool box was developed by Math works in 1998, it is an organization consisting of qualified mathematicians and researchers. This toolbox consists of set of built in methods.

Mufti and Khanum 2006, proposed an approach on Fuzzy Based Recognition of Facial Appearance. This research is so effective that it recognizes the human facial expressions in a precise manner. Sawat and Hegadi 2018, described an approach on Lower Facial Curves Extraction for Unconstrained Face Detection in Video. Candemir et al. 2015, invented a method on RSILC: Rotation-and Scale-Invariant, Line-based Color-aware descriptor. Santosh and Wendling 2016, proposed an algorithm on Pattern Recognition Based on Hierarchical Description of Decision Rules Using Choquet Integral.

1.1   Fuzzy Inference System (FIS)

Fuzzy inference system is the process of formulating the mapping from a given input human face to an output facial expression using fuzzy logic and the membership function. In this paper, Mamdani's fuzzy inference method is implemented and it considers the output membership functions to be fuzzy sets. After the process of aggregation, there is a fuzzy set for each output facial expression that requires the defuzzification to make crisp decision on facial expression as shown in Fig. 1.

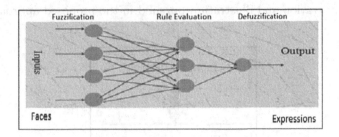

Fig. 1. Fuzzy Inference System for facial expressions.

1.2   Face Database

The projected algorithm is tested on the dataset of facial images of the individuals of various dimensions of happy expressions viz Normal, Bit smiley and Loud Laugh. This database is considered as the point of reference dataset to the performance estimations of prediction of human gender and age. Around 1000 human face images are included in the experimental dataset. Among them, 700 facial pictures are utilized as training set and 300 face images were employed for testing. Each face image dimension is normalized to 64×64 matrices, for the optimum computational cost. Sample facial expression images are shown in Fig. 2.

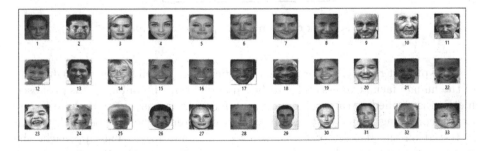

Fig. 2. Sample facial images in the image dataset

The upcoming sections of this paper are designed as follows. The Proposed Approach is explained in Sect. 2. The Proposed Algorithm is portrayed in Sect. 3. The Experimental Outcomes are presented in Sect. 4, and the Conclusions are given in Sect. 5.

2 Proposed Approach

This research paper proposes an effectual methodology for the quantification of human facial expression from facial image dataset. Figure 3 depicts the diagrammatic demonstration and illustration pursued in the proposed approach.

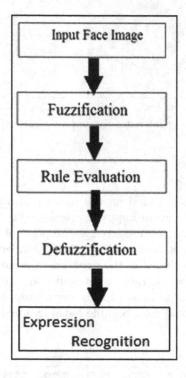

Fig. 3. Block diagram of the proposed approach.

The present research paper defines an efficient approach for the quantification of the human facial expression from the images of the human face. This proposed method is applied to categorize the input face pictures into one of three happy expression using Mamdani-type Fuzzy based system. The proposed algorithm is tested on the dataset of face images of persons. In current approach, the pictures containing the human face are scanned. To improvise the effectiveness, mean vector is obtained for each image is used in the fuzzy system as shown in Fig. 3.

2.1 Quantification of Facial Expressions

It is assumed that every human face is having the same geometrical configuration. An extremely crucial function in analysis of images is engaged by a term as **feature**. The proposed method of feature extraction is so simple that, instead of giving whole image in terms of matrix, the proposed algorithm gives the mean of the image as input to the fuzzy system. This improves the performance efficiency. Figure 4 illustrates the different forms of happy expressions which are used in the proposed method.

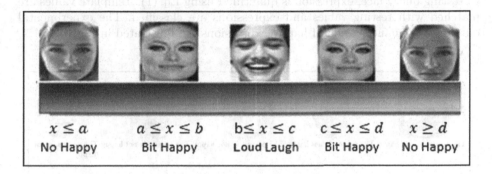

$x \leq a$	$a \leq x \leq b$	$b \leq x \leq c$	$c \leq x \leq d$	$x \geq d$
No Happy	Bit Happy	Loud Laugh	Bit Happy	No Happy

Fig. 4. Representation of quantification of happy expression.

In the current research paper, a new fuzzy methodology established on strategy of reasoning is invented for quantification of the happy expression on still images of the face. In this proposed method, geometrical features of the face are inputted to Mamdani type fuzzy rule system for quantification using trapezoidal membership functions. This approach is capable of quantifying 3 forms of happy expressions namely No Happy, Bit Smiley, Loud Laugh as shown in Figs. 4 and 5.

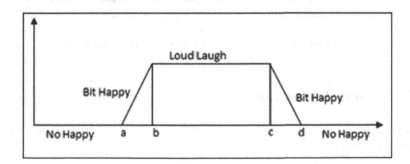

Fig. 5. Representation of trapezoidal membership function.

In the proposed methodology, the outcome of the F I S that is defuzzified value is either the given face is in not in a happy mood or it is in bit happy or it is in loud laugh. This can be represented by the below equation,

$$M(x, a, b, c, d) = \begin{cases} 0 & x \leq a & \text{No happy} \\ \frac{x-a}{x-b} & a \leq x \leq b & \text{Bit happy} \\ 1 & b \leq x \leq c & \text{Loud laugh} \\ \frac{x-d}{d-c} & c \leq x \leq d & \text{Bit happy} \\ 0 & x \geq d & \text{No happy} \end{cases} \quad (1)$$

In Fig. 5, the trapezoidal membership function, a, b, c, d values are determined empirically for quantification of expression as shown in the Eq. (1) and Fig. 5. In testing query face, expression is quantified using Eq. (1). Template values are matched with testing values and expressions are classified. The experimental outcomes of quantification of happy expressions are illustrated in Fig. 6.

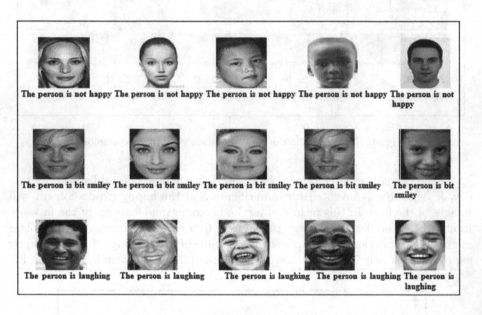

Fig. 6. Sample experimental results of happy expression.

3   Proposed Algorithm

The algorithm for the expression quantification from the inputted human face picture is as defined underneath:

Input: Query Human face
Output: Quantification of happy expression into No happy, Bit happy or Loud laugh

---

Algorithm 1. Expression Quantification

1: **Training**: Feed n pictures of human face to Fuzzy Inference System.
2: Define the Trapezoidal membership function for different expressions and compute membership values for a,b,c,d as templates as shown in Fig. 5.
3: **To Test**: Compute membership value x for the query face.
4: Evaluate facial expression using Eq. (1).

---

4 Experimental Results

In the present research problem, 1000 grayscale images of human faces are included for the trial. Each picture is normalized to size of 64 × 64 dimensions. Amongst 1000 facial pictures, 700 face pictures are operated on training dataset and 300 are employed on test dataset. This proposed algorithm has reflected a good robustness and reasonable accuracy for the test images. This proposed algorithm has a very fewer complexity and is appropriate for the real time accomplishment, namely animation, image surveillance, mood analysis applications. The processing time reduced by optimization (Fig. 7).

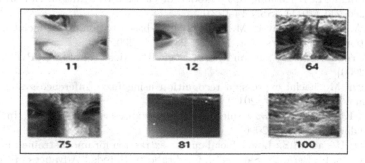

Fig. 7. Sample results for mis-detection.

In the testing part, the combination of all the three categories of happy expression images is considered. The rate of success for quantification of happy expressions is 94.00%, 95.00% and 96.00% for no happy, bit happy and loud laugh respectively. The overall success rate of experiment is 95%. The mean time of recognition for each face picture is less than 0.35 s on a Pentium @ Quad Core processor with RAM of 4 GB. However, this method not able to notice the faces with elevated-views, occluded face pictures and partial pictures of the faces. This condition is due to missing facial features in the process of recognition of facial expressions.

5 Conclusions

In this research paper, the fast and efficient quantification of expression algorithm is designed to classify the human face image into different modes of happy expression groups. Quantification of human expression of the face from the features of the face using Mamdani based fuzzy rule method is proposed. This algorithm recognizes three levels of same happy expression namely No Happy, Bit Happy and Loud Laugh. This method is comparatively superior in the form of accuracy and speed. Single front view faces with three modes of happy expression collections are classified productively with the success rate 95%.

In further research, misclassifications are reduced by supplementary development in proposed approach so that it turn out to be additional pertinent to the outline, draft and design of a real-time image and also video observation systems.

References

Esau, N., Wetzel, E., Kleinjohann, L., Kleinjohann, B.: Real-time facial expression recognition using a fuzzy emotion model. In: IEEE International Fuzzy Systems Conference, London, England, pp. 1–6 (2007)

Herrera, F., Magdalena, L.: Genetic fuzzy systems: a tutorial. Tatra Mt. Math. Publ. (Slovakia) **13**, 93–121 (1997)

Jamshidnezhad, A.: A learning fuzzy model for emotion recognition. Eur. J. Sci. Res. **57**(2), 206–211 (2011). ISSN 1450-216X

Khanum, A., Mufti, M., Javed, M.Y.: Fuzzy case-based reasoning for facial expression recognition. J. Fuzzy Sets Syst. **160**(2), 231–250 (2009)

Klir, G.J., Yuan, B.: Fuzzy Sets and Fuzzy Logic. Prentice Hall of India Pvt. Ltd., New Delhi (2000)

Rasoulzadeh, M.: Facial expression recognition using fuzzy inference system. Int. J. Eng. Innov. Technol. **1**(4) (2012)

Mufti, M., Khanum, A.: Fuzzy rule-based facial expression recognition. In: CIMCA 2006, Sydney, Australia (2006)

Sawat, D.D., Hegadi, R.S.: Lower facial curves extraction for unconstrained face detection in video. In: Bera, R., Sarkar, S., Chakraborty, S. (eds.) Advances in Communication, Devices and Networking. LNEE, vol. 462, pp. 689–700. Springer, Singapore (2018). https://doi.org/10.1007/978-981-10-7901-6_75

Candemir, S., Borovikov, E., Santosh, K.C., Antani, S., Thoma, G.: RSILC: rotation- and scale-invariant, line-based color-aware descriptor. Image Vis. Comput. **42**, 1–12 (2015)

Santosh, K.C., Wendling, L.: Pattern recognition based on hierarchical description of decision rules using Choquet integral. In: Santosh, K.C., Hangarge, M., Bevilacqua, V., Negi, A. (eds.) RTIP2R 2016. CCIS, vol. 709, pp. 146–158. Springer, Singapore (2017). https://doi.org/10.1007/978-981-10-4859-3_14

Experimental Study on Latent Fingerprint Matching Using Clustered Minutiae Patterns

Uttam U. Deshpande[1(✉)] and V. S. Malemath[2]

[1] Department of Electronics and Communication Engineering,
KLS Gogte Institute of Technology, Belagavi, Karnataka, India
uudeshpande@git.edu
[2] Department of Computer Science and Engineering,
KLE Dr. M. S. Sheshgiri College of Engineering and Technology,
Belagavi, Karnataka, India
veeru_sm@yahoo.com

Abstract. Use of Biometrics, especially fingerprints is increasing day by day due to its reliable verification and identification. This dominancy has resulted in establishment of Automatic Fingerprint Identification System (AFIS). Latent AFIS is in growing demand due to absence of manual human intervention in matching. But latent AFIS is still in its early stages and has to overcome many challenges. Thus, there is a need for developing a fast, reliable and robust latent AFIS. In this paper we have proposed a robust hash-based indexing technique for large latent fingerprint matching to speed up matching. In this system minutia local neighborhood information is extracted to create minutiae features. These obtained minutiae features are invariant to scale, translation, rotation and each minutia is represented as feature vector. This information further can be used to construct hash-table and matching can be performed. In the experiments, Correct Index Power (CIP) is used to measure the retrieval accuracy.

Keywords: Biometrics · AFIS · Latent fingerprint · Hash-based indexing · Feature vector · CIP · Retrieval accuracy

1 Introduction

Automated person identification of a person through automated systems is possible due to emerging biometric technologies. Research carried out in biometric identification includes face, IRIS, palm prints and fingerprints. Among all, finger print identification systems is considered as highly preferred biometric system due to its reliability and accuracy. Main aim of fingerprint identification system is to verify the identity of a person and the challenge arises in fingerprint matching. Fingerprints have achieved great success in past several decades and are widely used in personal verification and identification [1]. The use of fingerprint identification systems avoids the need of remembering passwords and they have been extensively used in applications like attendance systems, banks, building security, law enforcement, commercial, civilian and financial applications [2]. Due to good identification accuracy, fingerprint based

K. C. Santosh and R. S. Hegadi (Eds.): RTIP2R 2018, CCIS 1036, pp. 381–394, 2019.
https://doi.org/10.1007/978-981-13-9184-2_35

identification systems have put up themselves in a better market place than any other biometric systems.

In fingerprint recognition method [3] or identification method [4], following common steps are performed to build a database of templates:

1 Enrollment stage: Templates are formed by extracting features from a captured fingerprint and are stored in a database.
2 Verification stage: The template of query fingerprint is formed in the same manner and it is compared with the stored template. Matching decision is made based on the matching criteria.

Captured fingerprint images cannot be directly stored to perform matching. Rather, feature extraction is done before proceeding for matching stage. Feature extraction is a process of identifying important properties of the fingerprint which later can be used in fingerprint matching. Fingerprint features can be made up of three levels [5]. Level-1 features provide macro level information about the singular points, ridge flow, ridge orientation or ridge frequency. These features form the global feature set. Other global features containing the singular points include core point, delta point, and ridge counts. Level-2 features are made up of minutiae points. Minutiae points are the unique fingerprint features obtained from a person's fingerprint. Ridge bifurcations and endings form the local minutiae features. These minutiae points show significant variations from one fingerprint to another. Level-3 features comprises of low level minutiae features like sweat pores, ridge shape, ridge width, dots and curves.

These features are shown in Fig. 1. Left loop, right loop, whorl, arch and tented arch are the other ridge features shown in Fig. 2. Limitation of level-3 features is that it can only be observed in images captured from high resolution cameras.

Fig. 1. Fingerprint image features.

Most of fingerprint identification systems make use level-2 features. This is because fingerprint minutiae are the most unique and reliable features. After feature extraction step, next goal will be to identity a person's fingerprint from stored database. The task of matching step will be to provide match score between two fingerprints based on their likeliness. Most of the methods use minutiae based matching, but there are other methods of matching. These are based on image correlations of images, level-2 and level-3 features. Shape, edge and texture features [40] can also be used for matching

Fig. 2. Basic ridge features.

fingerprints. Pattern matching using minutiae features is being widely used to match the fingerprints from past decade. Results have shown that fingerprint storage is most efficiently used.

Also matching results have been proven in courts [6]. Recently, researchers have used local minutiae structures to align fingerprints and results at a global level are obtained from locally obtained results.

Latent fingerprint identification system is used in criminal investigation and forensic applications. Latent fingerprints are the impressions which are unintentionally left on the objects. The latent fingerprint matching techniques pose more difficult challenges compared to conventional fingerprint matching. This is mainly because minutiae in latent patterns may get missed out due to its nature or may get distorted by noise. Low image quality, poor texture, and non linear distortion in fingerprints create additional challenges in latent fingerprint identification system. Minutiae clustering based pattern matching technique for a latent fingerprint are unexplored. This paper proposes a study on minutiae clustering approach for pattern generation and recognition for latent fingerprint applications. Unique latent minutiae feature vector is defined from minutiae triangular features. A hash value is generated from different arrangements of feature vector. The obtained hash values are used to construct hash-table. Proposed hashing method shows significance improvement in the matching speed and accuracy. The paper is organized as following; Sect. 2 discusses about the related work carried out in conventional fingerprint matching. Section 3 introduces to the proposed pattern matching methods based on connected minutiae clusters. Section 4 describes the experiments carried out on latent database and Sect. 5 concludes the work.

2   Related Work on Fingerprint Matching

This section deals with work carried out by various researchers in the field of fingerprint matching.

2.1   Correlation Based Techniques

In most of the work carried out so far, fingerprint minutiae feature is most popularly used. This is because fingerprint minutiae contain unique and reliable features. In addition, the templates generated from minutiae information consumes less memory and are faster than the graphical based fingerprint matching. In correlation based technique, two fingerprint images are made to overlap on each other and the respective overlapping pixels at different alignment settings are used to compute the similarity score. This is simple operation but this did not yield acceptable results. This is mainly due to variations that occur in brightness and contrast of the image. Many researchers used different methods to improve this technique.

2.2 Minutiae-Based Features

Minutiae based matching algorithm are mainly designed to solve problems of connected minutiae and similarity score. Most of the indexing techniques make use of minutiae features [7–10] and are derived from its neighborhood relations. These features form geometrical patterns and are invariant to rotations and translations. Minutiae Cylinder Codes (MCC) [11] has been considered as state-of-art technique for minutiae neighborhood representation and is widely used in fingerprint matching or indexing.

2.3 Fingerprint Indexing and Clustering Technique

Fast access of fingerprint templates database for identification is possible using fingerprint indexing. Partially extracted minutiae information can be one indexing technique used to build local structures. This information can be used to calculate fingerprint similarity and to generate key indexes. Ordering of candidate templates done in this way increases the probability of match. Another indexing technique uses triangle characteristics like length, and angle [10]. This type of triangulation technique helped to improve the matching efficiency [12]. Delaunay triangulation technique connects the minutiae in its neighborhood using triangles to identify minutiae orientation and position. Triangular structure obtained using this technique doesn't much vary even if there are deformations in rotations or translations. Thus, minutiae matching using Delaunay triangle technique provide good performance with respect to match time and nonlinear distortions. Similarly, another Delaunay Triangulation [13] approach for fingerprint alignment and matching was proposed. Here global matching was done by comparing using minutiae points and singular points. Fingerprint matching using local features called as Voronoi neighbor structures (VNSs) [1] were used which are robust, rotation and translation invariant.

Several researchers worked on indexing techniques based on clustering and found out significant improvement in the matching accuracy. Fingerprint indexing using dynamic clustering technique [14] extracts translation and rotation invariant features from triangles constructed from triangle spiral from fingerprints. Automatic Fingerprint Identification using clustering technique [15] uses identification module to detect identical minutiae clusters from template images. Here the evaluation of minutiae cluster is done and based on the results the feedback module trains the core vector. Translation and rotation invariant cluster based template generation for fingerprint matching [17] stores fingerprint templates instead of minutia feature vectors. Latent fingerprint clustering using deformable minutiae [16] uses multiple merged and overlapping matching minutia pair clusters.

2.4 Fingerprint Based Pattern Matching Technique

This is a process of matching the individual persons fingerprint features against the features of database templates for person verification [21]. Pattern matching method is used in latent fingerprints for matching a set of primitive latent features with enrolled fingerprint database. Pattern matching is achieved by developing templates using global and local features [22] and matching against stored database template. These features are invariant to global transformation functions. Success of any fingerprint based pattern matching technique depends on absence of fingerprint distortion, noise, arch [23] and identification of core point and minutiae points. Core points and minutiae points are most commonly used as they are invariant to ridge patterns.

The combinations of Euclidean distance and ridge features [20] can be used for minutiae based fingerprint matching. Initially, minutia pairs were obtained using Euclidian distance for representation of transformation invariant features. Here, minutiae features were obtained from ridges and they were grouped together. Finally, histogram based modeling was used for similarity score computation. This method is suitable for AFIS with limited memory space. Fingerprint pattern matching in AFIS using minutiae neighborhood and singular points [18] have produced good matching accuracy. In this method Euclidian distance is used to calculate the distance between singular point and minutiae. This relationship between singular point and minutiae is utilized to calculate pattern matching scores. Pattern recognition and matching algorithm using core point was proposed in [24]. Such technique is suitable for real time AFIS.

A Partial fingerprint recognition system based on local features [25] was proposed to overcome miniaturization problem arising from fingerprint sensors capturing small sensing areas. This is because sensors capture only a part of the fingertip due to small area limitation or improper fingertip placement of the fingertip or scars on fingerprint due to injury [26, 27]. Partial fingerprint recognition system using global features [19] was proposed. Here pattern matching depends on core point, delta point and ridge flow. Fingerprint recognition using Dynamic Time Warping (DTW) [44, 45] was proposed to compress the fingerprint feature matrix into a single vector to save the information.

A detailed survey done on available computational algorithms [43] for fingerprint recognition helps in selecting suitable algorithm for a given fingerprint application.

Latent fingerprints pose more complex challenge compared to normal fingerprints due to presence of background noise and poor quality latent image. This affects the matching speed as the time taken to extract minutiae features from poor quality fingerprint is more. Robust minutiae algorithms [28–30] were proposed to overcome false minutiae detection and matching. Image segmentation technique [31] was proposed to reduce matching speed in AFIS. Segmentation is a process of partitioning fingerprint foreground information from irrelevant background noises. For classifying fingerprint match or no match, an active learning method [41] can be used. A supervised k-NN [42] can be used for classifying the fingerprint data. Here the data which is closest to the cluster center and the farthest from the cluster fingerprint is calculated and classified to reduce classification speed. Latent fingerprint based pattern matching is an unexplored field. This paper proposes a study on latent fingerprint based pattern matching, which is discussed in Sect. 3.

2.5   Recent Trends in Fingerprint Matching

Due to advanced hardware technology, the performance of fingerprint matching used in personal identification is improving. Different fingerprint features are being used in matching to achieve high matching accuracy. These methods are discussed in the next section:

Fingerprint Matching with Hardware Acceleration: Hardware acceleration can speed up the matching process and many researchers are working in this direction. Hardware acceleration using FPGA [32], distributed computing [33] and GPU [34] based architectures are successfully being used for this purpose.

Use of Embedded Systems: Different embedded systems using smart cards [36] and sensors [35] are being used to improve the performance of matching.

Palmprint Matching: This is achieved using minutiae and ridge features. Techniques followed in latent fingerprint matching [37, 38] can be utilized in palmprint matching.

Multi-feature Based Matching: Other human features like face recognition [39] and fingerprint features can be combined to improve matching accuracy.

3   Proposed System for Latent Fingerprint Matching Using Clustered Minutiae Patterns

Latent minutiae fingerprints suffer from low resolution, background noise and non-linear distortions. To address these problems, a minutiae feature based hash index method is adopted. First we begin with minutiae feature extraction. After obtaining minutiae features, a fixed length minutiae descriptor is created to perform matching. This descriptor is based on the neighborhood geometry pattern and is invariant to affine minutiae deformations. Hence, it can be directly applied on minutiae based templates. Another advantage of using this descriptor is that there is no need to detect singular points to align the templates.

3.1   Feature Extraction

Preprocessing of latent fingerprint is done before extracting features. Preprocessing contains steps like binarization and thinning. In binarization, a grey level image is converted into a binary image. This process involves the examination of each grey pixel value and thresholding it to binary '0' or '1' based on the grey level.

Thinning is carried out before proceeding to feature extraction. In this step, each pixels neighborhood from the binarized image may be deleted based on the pixel deletion criteria. It performs a morphological operation in which foreground pixels are removed until they become one pixel wide.

Finally, minutiae extraction is carried out after the preprocessing step. Crossing Number (CN) technique is used on the obtained skeleton image from the previous step. Ridge patterns are 8-connected. Hence, minutiae points are extracted by sliding a 3×3 window on the fingerprint image and ridge pixels are found from its local neighborhood. Based on the neighborhood connectivity, minutiae features such as ridge ending or bifurcations can be obtained.

Minutiae based indexing schemes provide better results than other indexing schemes. Hence indexing based on Minutia Cylinder Code (MCC) is most widely used and is considered to be state-of-art indexing technique. Major limitation of this method is that it is not invariant to affine transformations. A new fixed length local minutiae descriptor based indexing technique is proposed to overcome the above limitation. A latent minutiae feature vector is defined to capture the minutiae neighborhood information and is discussed next.

3.2 Latent Minutiae Feature Vector

A fixed length local latent minutiae descriptor is proposed to obtain the minutiae distinctiveness from its neighborhood. This representation is used to compare two minutiae points, which in turn helps to determine the fingerprint similarity. The procedure followed to create a latent minutiae feature vector is listed below:

- Consider 'n', which defines the total number of latent minutiae feature points (L) around a central latent minutia. Let n = 7, then L1, L2, L3, L4, L6, L7, and L8 are the 7 nearest latent minutiae neighbors obtained around L5. This is indicated in Fig. 3(a).

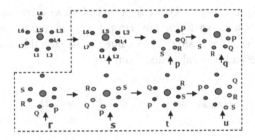

Fig. 3. (a) Nearest 'n' latent minutiae around L5 (n = 7) (b) 'm' invariants p, q, r, s, t, u obtained by rotating P, Q, R, S clockwise around L5.

- Consider 'm', which defines the possible local geometric arrangements around L5. These arrangements produce geometric invariant features which will help in the construction of feature vector. This is obtained by choosing four coplanar points P, Q, R, S on latent minutiae and rotating it in clockwise direction around L5. To decrease the probability of similar geometric arrangements, higher value of 'm' must be chosen. If the value of m = 4 (equal to number of coplanar points) is chosen to produce geometric arrangements, then it will become difficult to discriminate these similar arrangements from one another. Hence, to increase the discrimination power the value of m = 6 is chosen. As a result of this, six geometric arrangements namely p, q, r, s, t, u will be formed by rotating the points P, Q, R, S around the six nearest latent minutiae in clockwise cyclic order. Here, L3, L4, L2, L1, L7 and L6 are the six nearest latent minutiae obtained around L5. Since n = 7 and to obtain six invariant arrangements, the farthest latent minutia should be discarded. The 'm' invariants pqrstu indicates the feature vector shown in Fig. 3(b). This information will be used to generate feature vectors from combination of above arrangements. This is explained later.

3.3 Triangular Features

With P, Q, R, S points, different indexing features for construction of feature vector can be extracted. They are given below:

- The ratio of the triangle areas formed by latent minutiae triplets P, Q, R and P, Q, S.
- The ratio of the triangle lengths (larger-side) of latent minutiae triplets P, Q, R and P, R, S.
- The ratio of minimum and median triangle angles formed by latent minutiae triplets P, Q, R and P, R, S.

Features mentioned above are invariant to translation or rotation. Hence, the latent minutiae feature vector can be obtained by considering different combinations of above triangular features.

3.4 Fingerprint Hash-Table

Here we are creating hash table using feature vectors. The same was discussed in the previous section. By rotating P, Q, R, S points clockwise, possible permutations of these feature vectors are: pqrstu, qrstup, rstupq, stupqr, tupqrs, upqrst. Each feature vector 'fv' is considered for hashing. The values to be stored in the hash table depend upon the hash value and it is calculated from Eq. 1.

$$HT_{index} = \left(\sum_{i=1}^{m} f_u[i] \times k^i \right) \bmod HF_{size} \qquad (1)$$

Here, 'HTindex' is the hash index used in construction of hash-table, 'fv' is the feature vector having the length 'm', 'k' is the level of quantization to be chosen for invariant and 'HFsize' is the size of the hash function table. 'fv' is the discretized invariant feature vector obtained from weighted combination of triangular features and the quantization value chosen should be in the range of [0, k] to generate unique hash-value in the table.

Algorithm 1 Fingerprint Hash Table Construction Algorithm.

Input ← Latent Fingerprint image database ldb, n, m, k
Output ← Constructed Hash-Table
for each $lfp \in \{Latent\ Fingerprint\ image\ database\ ldb, n, m, k\}$ **do**
  p ← Extract minutiae feature points
  for each $p \in \{All\ latent\ minutiae\ feature\ points\ in\ lfp\}$ **do**
    P_n ← Set of n-nearest neighbors around minutia p
    L ← List of all possible combinations of m invariants obtained from P_n
    for each $P_m \in \{All\ possible\ combinations\ of\ m\ invariants\ from\ L\}$ **do**
      v ← feature vector
      CP ← list of all cyclic permutations of v
    end for
    for each v' $\in \{Feature\ vector\ in\ the\ list\ CP\}$ **do**
      HTindex ← Hash-value calculate from using eq.1 register fingerprint-item,
      (latent fingerprint-ID of lfp, minutia-ID of p, Feature vector v') using
      HTindex.
    end for
  end for
end for

These produce unique hash values which will be used in construction of hash-table. Construction of fingerprint hash-table is explained in Algorithm 1. Figure 4 shows the complete construction of fingerprint hash table. The latent fingerprint-ID, minutia-ID and feature vector are stored in the corresponding hash-table index. All these features are stored in the form of linked lists. Collisions can occur when multiple feature vectors map on to the same index of hash-table. To avoid this chaining technique is employed to resolve this problem. Size of the hash-table is calculated as below:

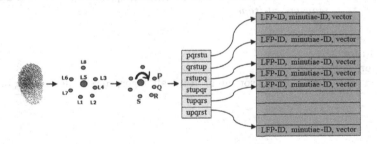

Fig. 4. Construction of fingerprint hash-table.

- n = 7, m = 6, and k = 15.
- Number of latent fingerprint images in the database is 80.
- Average number of feature-points from each latent fingerprint image is 6377.
- The latent fingerprint-ID, minutia-ID and feature vector are stored as 2-bytes, 2-bytes and 1-byte variables respectively in the corresponding hash-table index.

- A pointer to index this hash-table requires 8 bytes.

Size of hash-table, HFsize = $80 \times 6377 \times 7 \times (2 + 2 + 1 \times 15 + 8) = 96.42$ MB of memory.

3.5   Fingerprint Retrieval

Our next task is to perform the matching of query latent fingerprint with the fingerprints in the hash table. For query images, the fingerprint hash-table using feature vectors "v" is constructed in the similar manner as explained in the previous section. Using "v", latent fingerprint candidate list is retrieved from the hash-table database. Finally, every minutia of query fingerprint casts a vote in its candidate list and based on maximum votes the list of top latent fingerprints is obtained. The fingerprint retrieval procedure is given in Algorithm 2.

Algorithm 2 Fingerprint Retrieval Algorithm.

Input $\leftarrow$ Latent query image img, n, m, k, N
Output $\leftarrow$ List of sorted top latent fingerprints
p' $\leftarrow$ Extract minutiae feature points from img.
for each $p' \in \{All\ latent\ minutiae\ feature\ points\ in\ img\ \}$ **do**
    P'_n $\leftarrow$ Set of n-nearest neighbors around minutia p'
    L $\leftarrow$ List of all possible combinations of m invariants obtained from P'_n
end for
for each $P'_m \in \{All\ possible\ combinations\ of\ m\ invariants\ from\ L\}$ **do**
    v'' $\leftarrow$ feature vector for P'_m
    HTindex $\leftarrow$ Hash-value calculate from using eq.1, lookup in Hash Table with HTindex and retrieve the corresponding candidate list.
    for each $fingerprint\_item \in \{From\ retrieved\ candidate\ list\}$ **do**
        for all item in the retrieved list **do**
            if $v'' == fingerprint\_item.featurevector$ **then**
                Increase the vote count for latent fingerprint $-$ID corresponding to $fingerprint\_item$
            end if
        end for
    end for
end for
On the basis of vote counts sort and output all fingerprints in descending order

---

4   Result and Discussions

The results were obtained after conducting an experiment on latent fingerprint FVC_2002_DB4 database. This database contains 80 fingerprint impressions. These fingerprints were used to generate the latent fingerprint hash-table. For experimentation n, m and k values were set at 7, 6 and 15 respectively. The hash function table 'Hfsize' is set at 10,00,00,000. In the experiments, latent fingerprint retrieval accuracy is used for analysis. The accuracy is defined in terms of Correct-Index-Power (CIP). It is the percentage of correctly indexed query latent fingerprint images (Nciq) against the total number of latent fingerprint images stored in the database (Nid) as shown in Eq. 2.

$$\text{Hence, CIP} = (\text{Nciq}/\text{Nid}) \tag{2}$$

A plot in Fig. 5 compares the performance of proposed system on the normal latent fingerprint database FVC_2002_DB4_N against FVC_2002_M (25% Missing minutiae) and FVC_2002_S (25% spurious minutiae). It can be observed that the algorithm produces average 100% retrieval for normal query fingerprints,99% retrieval accuracy after adding 25% false (spurious minutiae) to the query fingerprints and 76% retrieval accuracy after removing 25% of minutiae (missing minutiae) from the query fingerprints.

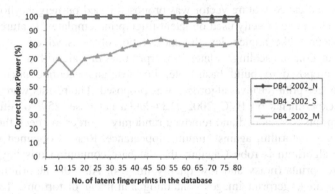

Fig. 5. Comparison of Normal (N) minutiae data vs. 25% Spurious (S) minutiae data vs. 25% Missing (M) minutiae data from FVC_2002_DB4 database.

A plot in Fig. 6 compares the performance of proposed system on the normal latent fingerprint database FVC_2002_DB4_N against FVC_2002_M (50% Missing minutiae) and FVC_2002_S (50% spurious minutiae). It can be observed that the algorithm

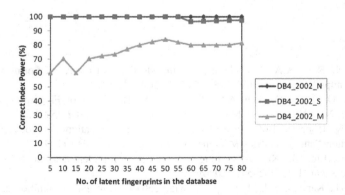

Fig. 6. Comparison of Normal (N) minutiae data vs. 50% Spurious (S) minutiae data vs. 50% Missing (M) minutiae data from FVC_2002_DB4 database.

produces average 100% retrieval for normal query fingerprints, 45% retrieval accuracy after adding 50% spurious minutiae to the query fingerprints and 22% retrieval accuracy after removing 50% of minutiae from the query fingerprints. It can be clearly observed from Figs. 5 and 6 that the proposed system produces better CIP even after adding and removing minutiae from query fingerprints.

5 Conclusion

Latent fingerprint matching using clustered minutiae patterns was proposed to tackle the fingerprint-indexing problem, where fingerprints are searched through large databases consuming lot of time. A literature survey was done on available minutiae based, index based and pattern recognition based methods. A new rotation and translation invariant method called feature vector was proposed based on neighborhood minutiae relation, which can be directly used on latent fingerprint templates. Feature vector uses triangular features like angles, length ratios, ratios of areas which are invariant to translation, rotation and scaling. A latent fingerprint hashing algorithm based on feature vector was proposed to build hash-table. For fast and accurate latent fingerprint retrieval, a finger-print retrieval algorithm was proposed. The results were obtained by conducting experiments on FVC_2002_DB4 latent database. 25% of minutiae from query fingerprints were added and removed randomly in order to check the robustness of the proposed algorithm against minutia-appearance. Results obtained showed that the retrieval algorithm is robust against the database containing missing or spurious minutiae fingerprints (noisy and low quality images). Also the algorithm effectively dealt with latent fingerprint images containing non-linear distortions. Thus the proposed algorithms are ideal in latent fingerprint matching. However, the performance of the algorithm got degraded when more than 50% noise was introduced. This was done by adding and removing 50% of minutiae from query fingerprint images randomly. More number of false minutiae appeared in the close vicinity of original neighbors in the process of generating spurious minutiae and many minutiae points were removed while creating fingerprint database containing missing minutiae. A new algorithm can overcome this problem.

References

1. Jain, A.K., Ross, A.A., Nandakumar, K.: Introduction to Biometrics. Springer, New York (2011). https://doi.org/10.1007/978-0-387-77326-1
2. Maltoni, D., Maio, D., Jain, A.K., Prabhakar, S.: Handbook of Fingerprint Recognition. Springer, New York (2009). https://doi.org/10.1007/978-1-84882-254-2
3. Jain, A.K., Hong, L., Bolle, R.M.: On-line fingerprint verification. In: Proceedings of IEEE Transaction Pattern Analysis Machine Intelligence, pp. 302–314 (1997)
4. Jiang, X., Liu, M., Kot, A.C.: Fingerprint retrieval for identification. IEEE Trans. Inf. Forensic Secur. 1(4), 532–542 (2006)
5. Hasan, H., Kareem, S.A.: Fingerprint image enhancement and recognition algorithms - a survey. Neural Comput. Appl. 23(6), 1605–1610 (2013)

6. Jain, A.K., Feng, J., Nandakumar, K.: Fingerprint matching, pp. 36–44. IEEE Computer Society (2010)
7. Germain, R., Califano, A., Colville, S.: Fingerprint matching using transformation parameter clustering. IEEE Comput. Sci. Eng. **4**(4), 42–49 (1997)
8. Iloanusi, O., Ross, A., Gyaourova, A.: Indexing fingerprints using minutiae quadruplets. In: Proceedings of IEEE Computer Society Workshop on Biometrics at the Computer Vision and Pattern Recognition (CVPR) Conference, pp. 127–133 (2011)
9. Boer, J., Bazen, A., Gerez, S.: Indexing fingerprint databases based on multiple features. In: Proceedings of ANN Workshop Circuits, Systems and Signal Proceedings, pp. 300–306 (2001)
10. Bhanu, B., Tan, X.: Fingerprint indexing based on novel features of minutiae triplets. IEEE Trans. Pattern Anal. Mach. Intell. **25**(5), 616–622 (2003)
11. Cappelli, R., Ferrara, M., Maltoni, D.: Fingerprint indexing based on minutia cylinder-code. IEEE Trans. Pattern Anal. Mach. Intell. **33**(5), 1051–1057 (2011)
12. Liang, X., Bishnu, A., Asano, T.: A robust fingerprint indexing scheme using minutia neighborhood structure and low-order delaunay triangles. IEEE Trans. Inf. Forensic Secur. **2**(4), 721–733 (2007)
13. Hawthorne, M.R.: Fingerprints analysis and understanding. CRC Press, Taylor and Francis Group, Boca Raton, London (2009)
14. Wang, C., Gavrilova, M.L.: Delaunay triangulation algorithm for fingerprint matching. In: 3rd International Symposium on Voronoi Diagrams in Science and Engineering, Banff, pp. 208–216 (2006)
15. Wencheng, Y., Hu, J., Song, W., Milos, S.: An alignment free finger print bio cryptosystem based on modified Voronoi neighbor structures. Pattern Recogn. **47**(3), 1309–1320 (2014)
16. Jain, A., Prasad, M.V.N.K.: A novel fingerprint indexing scheme using dynamic clustering. J. Reliable Intell. Environ. **2**(3), 159–171 (2016)
17. Qun, R., Jie, T., Yuliang, H., Jiangang, C.: Automatic fingerprint identification using cluster algorithm. In: ICS (2002)
18. Iwasokun, G., Charles, A.O., Officer, A.C.: Fingerprint matching using neighbourhood distinctiveness. Int. J. Comput. Appl. **66**, 1–8 (2013)
19. Jea, T.-Y., Venu, G.: A minutia-based partial fingerprint recognition system. Pattern Recogn. **38**(10), 1672–1684 (2005)
20. Iwasokun, G.B.: Fingerprint matching using minutiae-singular points network. Int. J. Signal Process. Image Process. Pattern Recogn. **8**(2), 375–388 (2015)
21. Iwasokun, G.B.: Development of a hybrid platform for pattern recognition and matching of thumbprints. J. Comput. Sci. Appl. **17**(2) (2010)
22. Zhang, Y., Jiao, Y., Li, J., Niu, X.: A fingerprint enhancement algorithm using a federated filter. In: Information Counter Measure Technique Institute. Harbin Institute of Technology (2008)
23. Lopez, A.C., Ricardo, R.L., Queeman, R.C.: Fingerprint pattern recognition. Ph.D. thesis, Electrical Engineering Department, Polytechnic University (2002)
24. Zhang, W., Wang, S., Wang, Y.: Pattern recognition and matching algorithm of fingerprint based on core point. National Laboratory of Pattern Recognition, Institute of Automation, Chinese Academy of Science, Beijing (2004)
25. Iwasokun, G.B., Akinyokun, O.C., Olabode, O.: A Block processing approach to fingerprint ridge orientation estimation. J. Comput. Technol. Appl. **3**, 401–407 (2012)
26. Akinyokun, O.C., Adegbeyeni E.O.: Scientific evaluation of the process of scanning and forensic analysis of thumb prints on ballot papers. In: Proceedings of Academy of Legal, Ethical and Regulatory Issue, vol. 13 (2009)

27. Jea, T.-Y., Govindaraju, V.: A minutia-based partial fingerprint recognition system. Pattern Recogn. **38**(10), 1672–1684 (2006)
28. Espinosa-Duro, V.: Minutiae detection algorithm for fingerprint recognition. In: Proceedings IEEE 35th Annual 2001 International Carnahan Conference on Security Technology (Cat. No. 01CH37186), London, England, UK, pp. 264–266 (2001)
29. Chikkerur, S., Wu, C., Govindaraju, V.: A systematic approach for feature extraction in fingerprint pattern recognition. Center for Unified Biometrics and Censors (CUBS), University at Buffalo, New York (2004)
30. Iwasokun, G.B., Akinyokun, O.C., Alese, B.K., Olabode, O.: Fingerprint image enhancement: segmentation to thinning. Int. J. Adv. Comput. Sci. Appl. (IJACSA) **3**(1) (2012)
31. Ali, S.M., Al-Zewary, M.S.: A new fast automatic technique for fingerprints recognition and identification. J. Islamic Acad. Sci. **10**, 55–60 (1997)
32. Jiang, R.M., Crookes, D.: FPGA-based minutia matching for biometric fingerprint image database retrieval. J. Real-Time Image Process. **3**(3), 177–182 (2008)
33. Peralta, D., Triguero, I., Sanchez-Reillo, R., Herrera, F., Benitez, J.M.: Fast fingerprint identification for large databases. Pattern Recogn. **47**, 588–602 (2014)
34. Gutierrez, P.D., Lastra, M., Herrera, F., Benitez, J.M.: A high performance fingerprint matching system for large databases based on GPU. IEEE Trans. Inform. Forensic Secur. **9**, 62–71 (2014)
35. Bayram, S., Sencar, H.T., Memon, N.: Efficient sensor fingerprint matching through finger print binarization. IEEE Trans. Inf. Forensic Secur. **7**, 1404–1413 (2012)
36. Benhammadi, F., Bey, K.B.: Embedded fingerprint matching on smart card. Int. J. Pattern Recogn. Artif. Intell. **27**(02), 1350006 (2013)
37. Jain, A.K., Feng, J.: Latent palmprint matching. IEEE Trans. Pattern Anal. Mach. Intell. **31**, 1032–1047 (2009)
38. Liu, E., Jain, A.K., Tian, J.: A coarse to fine minutiae-based latent palmprint matching. IEEE Trans. Pattern Anal. Mach. Intell. **35**, 2307–2322 (2013)
39. Hong, L., Jain, A.: Integrating faces and fingerprints for personal identification. IEEE Trans. Pattern Anal. Mach. Intell. **20**, 1295–1307 (1998)
40. Santosh, K.C., Antani, S.: Automated chest x-ray screening: can lung region symmetry help detect pulmonary abnormalities. IEEE Trans. Med. Imaging **37**(5), 1168–1177 (2018)
41. Bouguelia, M.R., Nowaczyk, S., Santosh, K.C., et al.: Agreeing to disagree: active learning with noisy labels without crowd sourcing. Int. J. Mach. Learn. Cybern. **9**(8), 1307–1319 (2018)
42. Szilárd, V., Santosh, K.C.: A fast k-nearest neighbor classifier using unsupervised clustering. In: Santosh, K.C., Mallikarjun, H., Vitoantonio, B., Atul, N. (eds.) RTIP2R 2016. CCIS, vol. 709, pp. 185–193. Springer, Singapore (2017). https://doi.org/10.1007/978-981-10-4859-3_17
43. Bhanu, B., Tan, X.: Computational Algorithms for Fingerprint Recognition. Springer, New York (2004). https://doi.org/10.1007/978-1-4615-0491-7
44. Santosh, K.C., Lamiroy, B., Wendling, L.: DTW-radon-based shape descriptor for pattern recognition. Trans. Pattern Anal. Mach. Intell. **27**, 1350008 (2013)
45. Santosh, K.C., Bart, L., L, Wendling: DTW for matching radon features: a pattern recognition and retrieval method. In: Jacques, Blanc-Talon, Richard, K., Wilfried, P., Dan, P., Paul, S. (eds.) ACIVS 2011. LNCS, vol. 6915, pp. 249–260. Springer, Heidelberg (2011). https://doi.org/10.1007/978-3-642-23687-7_23

Eye Like Landmarks Extraction and Patching for Face Detection Using Deep Neural Network

Dattatray D. Sawat[1]([✉]), Ravindra S. Hegadi[1], and Rajendra S. Hegadi[2]

[1] Department of Computer Science, Solapur University,
Solapur 413255, Maharastra, India
sawat.datta@gmail.com, rshegadi@gmail.com

[2] Department of Computer Science and Engineering, Indian Institute of Information
Technology, Dharwad 580029, Karnataka, India
rajendra.hegadi@gmail.com

Abstract. This paper presents a novel technique to extract facial Eye Like Landmarks (ELL) for face detection. The proposed technique pre-processes input images for skin segmentation using a minimal approach which segments all the skin regions in a fragmented manner. Morphological closing further enhances fragmented skin regions. Contrast Limited Adaptive Histogram Equalization (CLAHE) is used to enhance eye like regions. The ELL are extracted from skin regions By using divide and conquer method with a heuristically calculated threshold. While Extracting ELL the rotation and scale are taken into consideration. The pair of ELL which satisfy criteria to be a possible eye pair is then combined to form a patched image which is smaller in dimension, hence easier and faster to process. Patched images are used to train a Deep Neural Network. The classification accuracy on the face and non-face patches is 97%. In the result section, we discuss the training and validation loss. The Extracted Patches can be further processed to extract features for robust face detection.

Keywords: Landmark extraction · Face detection ·
Deep Neural Network (DNN)

1 Introduction

In computer vision applications, facial landmarks such as eyes, nose, mouth and ears are used for various tasks. Facial landmark localization is not a standalone problem, it has some consecutive use cases. Several researchers used it for the purpose of face tracking, face recognition, expression analysis, age estimation and 3D reconstruction. Apart from the uses mentioned in the previous sentence, very few attempts are made to use landmark localization for face detection, because it can be done without locating facial landmarks. In addition to this, the applications discussed earlier require the near perfect localization of these

© Springer Nature Singapore Pte Ltd. 2019
K. C. Santosh and R. S. Hegadi (Eds.): RTIP2R 2018, CCIS 1036, pp. 395–405, 2019.
https://doi.org/10.1007/978-981-13-9184-2_36

landmarks. For instance, if one wants to analyze the pose of the face, it is inherent to localize the landmarks perfectly. Similarly, the landmarks localization should be perfect in case of expression analysis where mouth endpoints change abruptly. While in face detection, the algorithm need not locate the facial landmarks perfectly, instead algorithm has to detect the presence of these landmarks in particular region of the image to conclude that particular region is a face. A face consists of several landmarks such as a pair of eyes, nose, mouth, eyebrows and ears. A feature vector can be formed by extracting features from an individual or a set of these landmarks.

The human face has some unique biological features called facial landmarks which includes a pair of eyes, nose, nose tip, nose trails, eyebrows, eye sockets, mouth, lips, chin and a pair of ears. So, instead of detecting face and then looking for landmarks, it could be feasible to locate a bunch of these landmarks and then by using individual or combination (patching) of these to detect the face. Since facial landmarks have a small footprint and unique structure they can be efficiently used to detect the face in images. In this work, we are going to extract landmarks pair that may constitute eyes in the face. Extracted landmarks can be used to train visual classifiers such as Deep Neural Network (DNN) or Networks that can be trained on numerical features. The only aim of this study is to extract Eye like Landmarks for face detection.

2   Related Works

Since landmark detection and localization is a subproblem in face based applications, we present a survey of the paper in which it is studied standalone as well as the papers in which it is used for face-based applications. After the impressive work on landmarks localizations by Cootes et al. [8], recently several researchers attempted to improve solution to this problem. We discuss these in the next paragraph.

Zhu et al. [26] proposed a tree-structured part model based on global mixtures of views, where the final model is the sum of all the trees for each view. Uřičář et al. [22] used local appearance model (combination of normalized intensities, intensity derivatives, Local Binary Pattern (LPB) and LBP pyramid) and deformation cost to classify the landmarks using the Structured Output Support Vector Machines (SOSVM) algorithm. Belhumer et al. [3] used a sliding window method over the face image, while sliding they used detectors with two different scales relative to the face size. Finally using concatenation of the SIFT features of two detectors, they classified the Face parts. Burgos et al. [4] proposed Robust Cascaded Pose Regression (RCPR) which helps to detect landmarks using regressors for various shapes. The regression is updated via voting, where voting is inversely proportional to occlusion. Zhang et al. [25] proposed Task Constrained Deep Convolution Network (TCDCN). Authors designed a loss function that was able to backpropagate errors related to each landmark. During the learning process, they designed criteria for landmark wise termination of the learning process. Deng et al. [9] proposed a Single Deep Neural Network defined with a group of three layers, each of which having Convolution followed by max pooling operations. At the end of the network, two fully connected layers were added for formation of the feature vector. By varying training parameters at every stage, the fast landmark location was achieved. Ranjan et al. [16,18] Designed a network that uses a combination of intermediate layers of Deep Convolutional Neural Network (DCNN) followed by a separate Convolutional Neural Network (CNN) designed to perform a specific task. The network was able to detect facial landmarks such as eyes, nose and mouth.

Facial landmarks extraction applies to a variety of computer vision tasks. Using landmarks location and their characteristics various applications has been designed. Cohn et al. proposed face tracking [7] using facial landmarks and feature points. Similarly, Dornaika et al. used facial landmarks [10] for face tracking. Face recognition also benefited a lot from landmarks based recognition methods [1,5,11,14]. Facial expression analysis is a primitive application which is based on detection and localization of facial landmarks [15,20], The extracted features from these landmarks are used to identify the expressions present of the face. Apart from these, there are few attempts made using landmarks detection and localization for face detection. The next paragraph discusses these works.

While the attempts discussed previously detected face first and then extracted landmarks, on the other hand detecting landmarks first and then making use of these landmarks to detect face is less explored. Yu et al. [24] proposed a mixture part model which marks landmarks as parts for each viewing angle and used a tree structure to represent the relation between parts. They assigned a score to each of these mixtures of trees and selected the best configuration to localize landmarks. Further using sparse learning method, they selected important landmarks for face detection. Candemir et al. [6] proposed rotation and scale invariant color-aware descriptor (RSILC) for line rich object matching, to describe the object using line segments the spatial relation needs to be described [17]. Since the face is a line rich object the same descriptor applies to face detection. Extraction of lower facial curves has been proposed by Sawat et al. [19], in which curves of lower facial landmarks are extracted, that includes chin curve, lip curves and ear curves. Yang et al. [23] calculated faceness of a region with the help of Deep Convolutional Network (DCN) using scores for facial part responses and their spatial arrangement. The scoring mechanism is carefully designed to tackle occlusion.

This work aims to make use of facial landmarks for face detection by extracting Eye like Landmarks (ELL) and Middle Facial landmarks (MFL) and further use of Deep Neural Network to classify the patched images of landmarks.

3   Methodology

To extract eye like landmarks the algorithm iterates through several steps. At every step, heuristics are used to remove unwanted elements which will reduce the processing overheads and leads to an optimum solution. The algorithm starts with pre-processing input images which include contrast enhancement, color space conversion and segmentation followed by a morphological closing. By using divide and conquer technique, only holes within segmented skin regions are processed for thresholding which results in Eye Like Regions (ELR). Then by using Algorithm 1, the pairs of ELR are formed. The coordinates for Eye Like Landmarks (ELL) are calculated using Algorithm 2. Once ELL are located, the Middle Facial Landmark (MFL) for each pair is located using the coordinates of ELL. To deal with rotation between ELL, Algorithm 2 uses a process defined in Eq. 2. The ELL are then concatenated (patched) together as discussed in Sect. 3.4. Finally, a fine-tuned Deep Neural Network is trained on the face and non-face patches obtained form IJB-A [12] dataset. The entire methodology is explained in the following sections.

3.1 Preprocessing

Unconstrained face images contain an improper distribution of contrast because of various factors such as outdoor illumination and the shadow effect. For the contrast enhancement, we used Contrast Limited Adaptive Histogram Equalization (CLAHE) with the clip limit set to three. The image is then converted to HSV color space which maintains hue, saturation and value information in separate channels. The hue and value information can be used to segment skin pixels effectively. For skin segmentation we use minimal approach, which uses a narrow range of threshold values defined in [2] to segment skin regions. Since in an unconstrained environment, the image may contain multiple faces and each with variable illumination, The minimal approach results into skin regions with several fragments. The resultant skin regions are then processed for morphological closing using 2×2 square structural element which fills isthmuses found in the skin regions. The resultant skin regions are then processed to threshold an Eye Like Regions (ELR). ELR is a nonskin region or hole in a segmented skin region. A nonskin region may contain an eye or eye socket which is one of the facial landmarks of the human face. To locate the ELR, the holes in the bounding box of the skin regions are thresholded using hue and value channel's threshold values for gray and darker intensities. Searching holes within skin regions save computation time since algorithms need not search entire image for an ELR. Some distant and unexpected skin regions are marked false for further processing using area-based threshold [19].

3.2 Formation of ELR Pairs

The eye-like regions which qualify the following criteria will form the pair. The formed pair will be processed to extract Eye Like Landmarks (ELL). Following heuristics will remove unwanted eye like regions and reduce processing overhead.

- Check whether ELR is enclosed within skin region only regions.
- As per anatomy of the eye, ELR should not be greater than ten percent of skin region being inspected.

- Center of the ELR should be greater than one.
- Scale proportion of eyes like regions should not be greater than five.
- Maximum of x, y coordinate difference should not be less than ten percent.

The preprocessing steps are given in the Algorithm 1.

Algorithm 1. Pair ELR

1: I: a input image
2: I_{he}: Histogram Equalization of input image I using CLAHE
3: I_{hsv}: conversion of I_{he} to HSV color space
4: I_s : skin segmentation using minimal approach
5: I_c: morphological closing of I_s by 2×2 square element
6: **for** each skin region i in I_c **do**
7:     **if** area(i) $\geq$ $athr$ **then**
8:         $skinpurned_i = i$
9:     **end if**
10: **end for**
11: **for** each skin region i in $skinpurned_i$ **do**
12:     $eyeregions_i$: non skin pixels which satisfy threshold criteria for eye like regions within the bounding of each skin regions.
13: **end for**
14: **for** each skin region i in $skinpurned_i$ **do**
15:     **for** each eye like region j in $eyeregions_i$ **do**
16:         **if** j is inside the bounding box of $skinpurned_i$ **then**
17:             $eyeregionspurned_i = j$
18:         **end if**
19:     **end for**
20: **end for**
21: **for** each skin region i in $skinpurned$ **do**
22:     **for** each eye like region j in $eyeregions_i$ **do**
23:         **for** each eye like region $j+1$ in $eyeregions_i$ **do**
24:             **if** center(j)$\leq$1 or center($j+1$)$\leq$1 **then**
25:                 continue
26:                 **if** scale($j, j+1$)$\leq$5 **then**
27:                     **if** dx($j, j+1$) $\geq$ 10 or dy($j, j+1$)$\geq$ 10) **then**
28:                         $pairinskin_i = j, j+1$
29:                     **end if**
30:                 **end if**
31:             **end if**
32:         **end for**
33:     **end for**
34: **end for**

To extract patches containing landmarks, a horizontal polygon mask is approximated over ELR bounding box coordinates of ELR pairs formed in Algorithm 1. Algorithm 2 explains the procedure to extract ELL and locate MFL coordinates.

Algorithm 2. Extract ELL

1: **for** each eye like region i, j in $pairinskin_i$ **do**
2:     **if** $i_{x1} \leq j_{x1}$ **then**
3:        $patch1=i$
4:        $patch2=j$
5:        $eyedistance$=maximun($distx$, $disty$)
6:        $patch3_{x2} = patch1x2\text{-}detpatch2_{y2}\text{-} patch1_{y2}$
7:        $patch3_{y2} = patch2_{y2}+ eyedistance\text{-}detpatch2_{y1}\text{-} patch1_{y2}$
8:        $patch3_{x1} = patch3_{x2}\text{-} \max(patch1_w, patch2_w)$
9:        $patch3_{y1} = patch3_{y2}\text{-}\max(patch1_h, patch2_h)$
10:    **end if**
11:    **if** $i_{x1} \geq j_{x1}$ **then**
12:       $patch1=i$
13:       $patch2=j$
14:       $eyedistance$=maximun($distx$, $disty$)
15:       $patch3_{x2} = patch2x2\text{-}detpatch2_{y2}\text{-} patch1_{y1}$
16:       $patch3_{y2} = patch2_{y2}+ eyedistance\text{-}detpatch2_{y1}\text{-} patch1_{y2}$
17:       $patch3_{x1} = patch3_{x2}\text{-} \max(patch1_w, patch2_w)$
18:       $patch3_{y1} = patch3_{y2}\text{-}\max(patch1_h, patch2_h)$
19:    **end if**
20: **end for**

Here patch1 and patch2 are extracted as ELL using coordinates of ELR and patch3 is extracted as MFL by calculating coordinates defined in Algorithm 2. For polygon mask definition 8 points of two ELR are used. The horizontal polygon mask is formed using point set calculated in Eq. 1.

$$
\begin{aligned}
HPOLY = \{ & Pt(ELL1_{x1}, ELL1_{y1}), Pt(ELL1_{x2}, ELL1_{y1}), \\
& Pt(ELL1_{x2}, ELL1_{y2}), Pt(ELL2_{x2}, ELL2_{y1}), \\
& Pt(ELL2_{x2}, ELL2_{y2}), Pt(ELL2_{x1}, ELL2_{y2}), \\
& Pt(ELL2_{x1}, ELL2_{y1}), Pt(ELL1_{x1}, ELL1_{y2}) \}
\end{aligned}
\tag{1}
$$

The coordinates for polygon will vary according to the positions of ELR. Figure 1 shows patches extracted using horizontal polygon mask over ELR.

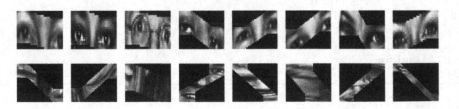

Fig. 1. Extracted horizontal polygons containing face and non-face ELR in first and second row respectively.

3.3 Extraction of ELL

Pairs of eye like regions are then processed to extract ELL using the procedure explained in Algorithm 2. The coordinates of ELL are used to find coordinates of Middle Facial Landmarks (MFL). In case, if there is a rotated or skewed face, Eq. 2 is used to calculate coordinates of MFL.

$$C'_x = C_x - (C'_y - C_y) \qquad (2)$$

The Fig. 2 shows Points A, B and C and the rotation of point B.

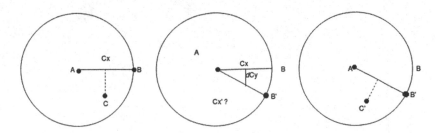

Fig. 2. (i) Initial positions of point A, B and C (ii) rotated point B and (iii) new position of point C respectively

Where points A, B and C represent coordinates of ELL and MFL. If A or B is rotated then C' needs to be localized, to do that y coordinate difference of C and C' is subtracted from C_x to calculate C'_x. Calculated C'_x is then used to determine the rest of the coordinates of the Middle Facial Landmark (MFL).

3.4 Patching of ELL

To discriminate between the face and nonface region, we propose to patch the extracted pair of ELL and MFL. Patching in this context involves the concatenation of a second ELL to the first ELL in a horizontal manner followed by the concatenation of resultant image by MFL vertically. The ultimate result of patching is a stitching of ELL to each other and MFL to ELLs. To keep a patched image in square shape, we resize the ELL and MFL appropriately. The ELL are resized to 16×16 and the MFL is resized to 16×32 before the patching. Hence the resultant image will be of the size of 32×32. Figure 3 shows patched images for both face and non-face.

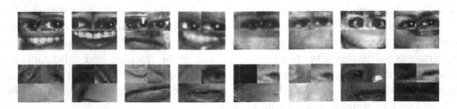

Fig. 3. Patched face and non-face images of extracted ELL in row 1 and 2 respectively.

4   Results

The Alexnet [13] is fine-tuned to discriminate between the face and nonface patched images. To fine-tune the network parameters we used the similar technique described in this [21] paper. The fine-tuned network uses a softmax classification layer with the loss to calculate the training accuracy and training loss. To reduce the number of parameters during training, the dropout layer is used after fully connected layer-6 and 7 with 0.5 dropout ratio. Figure 4 shows the accuracy along with training and validation loss.

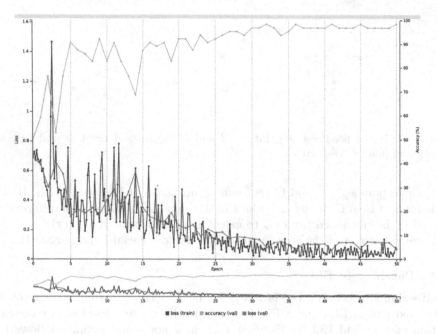

Fig. 4. Accuracy with training and validation loss using Deep Neural Network

Training is performed on 1400 face and 1400 non-face patches. While training, the learning rate is set to 0.01, and the epoch value is set to 50. The Deep Neural Network was able to discriminate the face and non-face patches with almost 97% accuracy along with 5.6% percent training loss and 5% validation loss. The ten-fold cross-validation is used while training the network. The training and validation accuracy can be increased with more number of training samples for face and-non face patch images. The graph in Fig. 4 shows the how accuracy is increasing after ten epoch, moreover the training and validation loss is started minimizing at the same time. The model gives its best performance approximately at 44 epochs. A subgraph in the same figure shows a smooth convergence of weights after initial epochs.

5 Platform Used

This work uses a fine-tuned Deep Neural Network which is inspired by Alexnet [13], implementation of this network is done using a high-performance Caffe library. The network is designed in NVIDIA-DIGITS Framework and algorithms proposed in this work are implemented using OpenCV C++. The algorithms are implemented with the support of parallel processing, to enable parallel processing NVIDIA Geforce 940 m GPU with CUDA support is used. The experiment has been carried out on 64-bit Linux operating system.

6 Conclusion and Future Work

The result shows how discrimination can be done between the face and nonface regions using patching of ELL together and training Deep Neural Network on these patched images. We used facial landmarks to discriminate between the face and nonface regions. The accuracy is approximately 97%. The patched images of facial landmarks can be further used for robust feature extraction. We look forward to extending this work for numerical and visual features extraction and combine them for robust face detection. To deal with side view faces the facial part response map can be used.

Acknowledgements. Authors thank the Ministry of Electronics and Information Technology (MeitY), New Delhi for granting Visvesvaraya Ph.D. fellowship through file no. PhD-MLA\4(34)\2014-15 Dated: 10/04/2015.

References

1. Ahonen, T., Hadid, A., Pietikainen, M.: Face description with local binary patterns: application to face recognition. IEEE Trans. Pattern Anal. Mach. Intell. **28**(12), 2037–2041 (2006)
2. Al-Tairi, Z.H., Rahmat, R.W., Saripan, M.I., Sulaiman, P.S.: Skin segmentation using yuv and rgb color spaces. JIPS **10**(2), 283–299 (2014)
3. Belhumeur, P.N., Jacobs, D.W., Kriegman, D.J., Kumar, N.: Localizing parts of faces using a consensus of exemplars. IEEE Trans. Pattern Anal. Mach. Intell. **35**(12), 2930–2940 (2013)
4. Burgos-Artizzu, X.P., Perona, P., Dollár, P.: Robust face landmark estimation under occlusion. In: Proceedings of IEEE International Conference on Computer Vision (ICCV), pp. 1513–1520. IEEE (2013)
5. Campadelli, P., Lanzarotti, R., Savazzi, C.: A feature-based face recognition system. In: International Conference on Image Analysis and Processing, pp. 68–73. IEEE (2003)
6. Candemir, S., Borovikov, E., Santosh, K., Antani, S., Thoma, G.: RSILC: rotation- and scale-invariant, line-based color-aware descriptor. Image Vis. Comput. **42**, 1–12 (2015)
7. Cohn, J.F., Zlochower, A.J., Lien, J.J., Kanade, T.: Feature-point tracking by optical flow discriminates subtle differences in facial expression. In: IEEE International Conference on Automatic Face and Gesture Recognition, pp. 396–401. IEEE (1998)

8. Cootes, T.F., Edwards, G.J., Taylor, C.J.: Active appearance models. IEEE Trans. Pattern Anal. Mach. Intell. **23**(6), 681–685 (2001)
9. Deng, Z., Li, K., Zhao, Q., Zhang, Y., Chen, H.: Effective face landmark localization via single deep network. arXiv preprint arXiv:1702.02719 (2017)
10. Dornaika, F., Davoine, F.: Online appearance-based face and facial feature tracking. In: International Conference on Pattern Recognition, vol. 3, pp. 814–817. IEEE (2004)
11. Juhong, A., Pintavirooj, C.: Face recognition based on facial landmark detection. In: Biomedical Engineering International Conference (BMEiCON), pp. 1–4. IEEE (2017)
12. Klare, B.F., et al.: Pushing the frontiers of unconstrained face detection and recognition: Iarpa janus benchmark a. In: Proceedings of the IEEE Conference on Computer Vision and Pattern Recognition, pp. 1931–1939 (2015)
13. Krizhevsky, A., Sutskever, I., Hinton, G.E.: ImageNet classification with deep convolutional neural networks. In: Advances in Neural Information Processing Systems, pp. 1097–1105 (2012)
14. Lanitis, A., Taylor, C.J., Cootes, T.F.: Automatic face identification system using flexible appearance models. Image Vis. Comput. **13**(5), 393–401 (1995)
15. Pantic, M., Rothkrantz, L.J.M.: Automatic analysis of facial expressions: the state of the art. IEEE Trans. Pattern Anal. Mach. Intell. **22**(12), 1424–1445 (2000)
16. Ranjan, R., Patel, V.M., Chellappa, R.: Hyperface: a deep multi-task learning framework for face detection, landmark localization, pose estimation, and gender recognition. IEEE Trans. Pattern Anal. Mach. Intell. **41**(1), 121–135 (2017)
17. Santosh, K., Lamiroy, B., Wendling, L.: Integrating vocabulary clustering with spatial relations for symbol recognition. Int. J. Doc. Anal. Recogn. (IJDAR) **17**(1), 61–78 (2014)
18. Sawat, D.D., Hegadi, R.S.: Unconstrained face detection: a deep learning and machine learning combined approach. CSI Trans. ICT **5**(2), 195–199 (2017)
19. Sawat, D.D., Hegadi, R.S.: Lower facial curves extraction for unconstrained face detection in video. In: Rabindranath, B., Subir Kumar, S., Swastika, C. (eds.) Advances in Communication, Devices and Networking. LNEE, vol. 462, pp. 689–700. Springer, Singapore (2018). https://doi.org/10.1007/978-981-10-7901-6_75
20. Tian, Y., Kanade, T., Cohn, J.F.: Recognizing action units for facial expression analysis. IEEE Trans. Pattern Anal. Mach. Intell. **23**(2), 97–115 (2001)
21. Ukil, S., Ghosh, S., Obaidullah, S.M., Santosh, K., Roy, K., Das, N.: Deep learning for word-level handwritten indic script identification. arXiv preprint arXiv:1801.01627 (2018)
22. Uřičář, M., Franc, V., Hlaváč, V.: Facial Landmarks Detector Learned by the Structured Output SVM. In: Csurka, G., Kraus, M., Laramee, R.S., Richard, P., Braz, J. (eds.) VISIGRAPP 2012. CCIS, vol. 359, pp. 383–398. Springer, Heidelberg (2013). https://doi.org/10.1007/978-3-642-38241-3_26
23. Yang, S., Luo, P., Loy, C.C., Tang, X.: Faceness-net: face detection through deep facial part responses. IEEE Trans. Pattern Anal. Mach. Intell. **40**(8), 1845–1859 (2017)
24. Yu, X., Huang, J., Zhang, S., Yan, W., Metaxas, D.N.: Pose-free facial landmark fitting via optimized part mixtures and cascaded deformable shape model. In: IEEE International Conference on Computer Vision (ICCV), pp. 1944–1951. IEEE (2013)

25. Zhang, Z., Ping, L., Chen Change, L., Xiaoou, T.: Facial landmark detection by deep multi-task learning. In: David, F., Tomas, P., Bernt, S., Tinne, T. (eds.) ECCV 2014. LNCS, vol. 8694, pp. 94–108. Springer, Cham (2014). https://doi.org/10.1007/978-3-319-10599-4_7
26. Zhu, X., Ramanan, D.: Face detection, pose estimation, and landmark localization in the wild. In: IEEE Conference on Computer Vision and Pattern Recognition (CVPR), pp. 2879–2886. IEEE (2012)

Development of Secure Multimodal Biometric System for Person Identification Using Feature Level Fusion: Fingerprint and Iris

Almas M. N. Siddiqui[1]($\boxtimes$), Rupali L. Telgad[2], Savita A. Lothe[3],
and Prapti D. Deshmukh[4]

[1] Department of Computer Science,
Vivekanand Art's Sardar Dalipsingh Commerce and Science College,
Aurangabad, India
siddiqui.almas29@gmail.com
[2] Shri Vyankatesh Arts, Commerce and Science College Deulgaon Raja,
District Buldhana, India
rupalitelgad@gmail.com
[3] Vasantrao Naik Mahavidyalaya, Aurangabad, India
savi_lo@rediffmail.com
[4] MGM's Dr. G. Y. Pathrikar College, Aurangabad, India
prapti.research@gmail.com

Abstract. Biometrics is a technology used to identify, analyze, and measure an individual's physical and behavioral characteristics. Multimodal biometric system can eliminate the limitation held by single biometric trait and provide improved precision so we have applied multimodal biometric system. Fingerprints and iris are two biometric features are used in this experimental work.

For both modalities, work is segmented into four divisions. Data acquisition is done in first segment, it helps for capturing images. For enhancement preprocessing is carried out in second segment. Features are extracted in third segment and matching is identified in last stage.

This paper describes the feature level fusion technique for fingerprint and iris. Multi instances are utilized for finger print and iris detection system. Then the equivalent gain is calculated and standardized.

We have developed the multimodal biometric recognition system for person identification by using fingerprint and iris modalities. For fingerprint detection can be through with the help of details by using minutiae matcher technique in which similarity and distance procedures are used. Iris features are extracted with the help of wavelet. Then the equivalent gain is standardized and attribute level combination is utilized to build up the system. The future system has been assessed on standard dataset and KVK dataset. On KVK dataset it gives an accuracy of 99.2% through FAR of 0.02% and FRR of 0.1%.

Keywords: Feature level fusion · False Acceptance Rate · False Rejection Rate

© Springer Nature Singapore Pte Ltd. 2019
K. C. Santosh and R. S. Hegadi (Eds.): RTIP2R 2018, CCIS 1036, pp. 406–432, 2019.
https://doi.org/10.1007/978-981-13-9184-2_37

1 Introduction

Biometrics is a tool which is utilized to recognize, analyze and assess an individual's bodily and behavioral description. Physical description based on t size of the human body, for example fingerprint, iris-scan, retina-scan, hand geometry and facial identification etc. Behavioral features such as voice recognition, keystroke- scan and signature-scan etc. [1, 44]. A Biometric configuration has two forms of procedures; it works like Verification form and Identification form. In verification form persons recognition can be completed by comparing the biometric data with the own biometric template [4, 42, 43]. In Identification form the system identifies the person by hunting the template of the entire user from the database [2]. The biometric system is categorized in following two methods.

A. Unimodal Biometric System.
B. Multimodal Biometric system.

A. Unimodal Biometric System

Unimodal Biometric system depends on solitary biometric attribute, which is utilized for individual's recognition or confirmation. This system is applied for various functions. It is as well utilized for the safety reasons. Despite the fact that the Unimodal biometric system has several applications still it has following drawbacks [3, 4].

Noisy Data: Due to Noisy information the matching is not promising and it provides fake rejection.

Intra class variation: Intra class variation raises the false refusal pace because it is arise due to the biometric statistics obtained is not identical as the information used to generate the pattern [4].

Inter class similarities: Inter class similarities occurs the repetition of characteristic due to several individuals.

Non universalities: Due to sickness or disability a few individuals cannot offer required Biometric.

Spoofing: The information can be fake because of spoofing.

B. Multimodal Biometric System

Multimodal Biometric system utilizes several physical or behavioural characteristics for individual's identification [4, 42, 43]. It applies different levels of fusion. Multimodal Biometric system has improved results and furnishes enhanced classification accurateness than the solitary Biometric classification [2].

Multimodal Biometric system has diverse design which move towards integrate information from numerous modalities, examples, sensors, trials, or any amalgamation of the five basis of the testimony like Multi replica, multi sensor, multi algorithm, multi illustration, multimodal [4, 5, 42, 43].

From the on top of diverse scheming approaches of multimodal biometric classification we have exercised multimodal scheme. Multiple trials are detained to augment the correctness and competence. This system confines impressions of all 10 fingers and both irises through dissimilar illustrations and merges them to figure precise biometric.

The fusion techniques are partitioned as Rule-based methods; Classification based methods, and Estimation-based methods [6].

Rule-Based Fusion Methods

The rule-based fusion method comprises a range of fundamental regulations of merging multimodal information. These comprises of statistical rule-based techniques such as linear weighted fusion (sum and product), MAX, MIN, AND, OR [4, 42, 43].

Classification-Based Fusion Methods

This class of methods consists of a variety of classification methods that have been utilized to categorize the multimodal surveillance into one of the pre-defined sets. The methods in this class are the support vector machine, Bayesian deduction, Dempster–Shafer hypothesis, dynamic Bayesian systems, neural networks and maximum entropy model [4, 41, 42].

Estimation-Based Fusion Methods

The assessment class comprises the Kalman filter; extended Kalman filter and particle filter fusion methods. These techniques have been mainly utilized to enhance estimate the position of a stirring object based on multimodal information [7, 8].

2 Survey of Multimodal Biometric System

Multimodal Biometric system make use of the fingerprint and iris characteristics by using amalgamation based on fingerprint minutiae Extraction and Iris pattern programming according to [9], this approach is based on two modalities. A multimodal biometric scheme is recommended by [10] with Scale Invariant feature Transformation [SIFT]. This paper explains both Face and Fingerprint technique at feature level synthesis which offers the enhanced outcome than the Unimodal Biometric system. Multimodal biometric scheme is built up using Fingerprint and Iris [11]. They use PCA (Principal Component Analysis) and FLD (Fisher Linear Discriminant) tactic for Biometric identification. The variance between board count method and logistic regression techniques. From the evaluation outcomes that rank-level union with the logistic regression approach presented the superior presentation in terms of inaccuracy rate and boost the detection rate of multi biometric systems, because in this approach, weights are allocated to different matchers as per their performance. Multi algorithm based multimodal biometric scheme by Face is developed by using the methodology PCA (Principal Component Analysis) RMPM (Reduced Multiple Polynomial Model) [12]. The stereo face identification formulation which merges form and intensity at feature level. A Reduced Multivariate Polynomial Model was accepted to combine the form and inequality images. RMPM is enlarged so that the difficulty of new user registration can be removed. The face detection approach is constructive for a few online functions such as visitors. Multimodal biometric system with Face and Finger layers use the LDA tactic for this scheme for low resolution face and finger layers acknowledgment scheme at score level fusion and multimodal identification scheme is extremely competent to decrease the FAR .000026 and augment GAR 97.4. The projected system is complicated due to the further processing needed for the characteristic spaces [13]. Multimodal biometric system is built up using fingerprint and voice. They use Leave-One-Out Cross Validation technique (LOOCV) and Gaussian mixture replica for score level fusion and Optimum reliability ratio based on incorporation of weight optimization system for fingerprint and voice modalities is applied. The working of the system is analyzed under different noise circumstances. One shortcoming of this technique is that under severe noise conditions it gives attenuating fusion [14]. Multimodal biometric system is expanded by Fingerprint and Finger layer by using the MHD (modified Hausdorff distance) algorithm as well as Minutia removal and matching founded on turnery vector at score level fusion based on finger print and finger vein [4, 15, 42, 43].

3 Methodology

1. Fingerprint biometrics

Fingerprint recognition is most popular biometrics, for authentication on computerized systems. The fingerprint images are classified into five categories: whorl, right loop, left loop, arch, and tented arch as shown in following figure (Fig. 1).

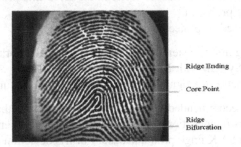

Ridge Ending

Core Point

Ridge Bifurcation

Fig. 1. A fingerprint image with the core and minutiae points.

Finger edge outlines do not modify all through the existence of a human being. This possession composes fingerprint an exceptional biometric classifier. The attribute significance characteristically communicates to the position and direction of definite significant points known as minutiae [16, 17].

In fingerprint biometric scheme the figures are confined by sensors as the taster points. The confined fingerprint images are then preprocessed, cleaned and then the minutiae spots are identified. Following figure shows the two features of fingerprint like, termination and bifurcation (Fig. 2).

2. Iris biometrics

Iris models also known as patterns are figured by merged layers of stained epithelial cells, muscles for managing the pupil, a stromal layer containing connective tissue, blood vessels and an anterior border coating [18, 19]. The physiological intricacy of the limb concludes results in the arbitrary patterns of the iris, which are statistically exclusive and appropriate for biometric dimensions [20]. Iris identification has been in the public interest for high- protection biometric functions.

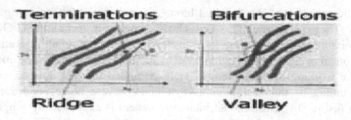

Fig. 2. A fingerprint termination and bifurcation.

The outcome obtained from the entity attributes is superior but the difficulty occurs due to measurements and place in support of iris images. Similarly, the difficulty faced by fingerprint recognition system is the occurrence of blemishes and scratches. The scratches add impurity to the fingerprint image which cannot be improved completely by means of enhancement module. Thus, the system takes noisy fingerprint as input which is not able to extract the minutiae points correctly and in turn, leads to false recognition of an individual.

Noisy fingerprint extracted by the system as an input unable to extract the minutiae points correctly which leads to false detection. To solve this problem combination of iris and fingerprint is proposed.

The phases occurred in projected scheme are:

1. Fingerprint Recognition System by means of Neural Network.
2. Iris Recognition System with Neural Network.
3. Multimodal Biometric Recognition System by means of Neural Network.

The above three recommended systems are estimated using three diverse databases of Fingerprint and Iris images. For fingerprint identification system we have used FVC 2002, FVC 2004 and KVK fingerprint database. For iris detection system we have used CASIA Iris database, MMU Iris Database and KVKR-Iris database. The research work is performed on Matlab R2012 with Core2Duo processor having 2.00 GHz speed (Tables 1 and 2).

Table 1. Database specification of fingerprint recognition system.

Characteristics	FVC 2002	FVC 2004	KVK fingerprint
Camera	Capacitive sensor	Optical sensor	L scan 500P
Illumination	569 dpi	500 dpi	700–900 NM
Detection	Automatically	Automatically	Automatically
File format	.JPEG	.JPEG	.BMP
Image size	300 × 300	640 × 480	480 * 480
	(88K pixels)	(307K pixels)	
No. of subjects	40	40	40
	1680	1440	2000
Image (Subject)	Images	Impressions	

Table 2. Database specification of iris recognition system.

Characteristics	CASIA	MMU	KVK Iris
Camera	-	LG Iris	ISCAN2 Cross Match
		Access 2200	
Illumination	-		700–900 NM
Detection	Automatically	Automatically	Automatically
File format	.BMP	.BMP	.JPEG
Size of image	320 * 280	320 * 280	480 * 480
No. of subjects	108 eyes	46	40
Image (Subject)	756 iris images	450 iris images	2000

Proposed Method 1: Fingerprint Recognition System Using Neural Network

Following figure shows that the proposed method for fingerprint recognition system using neural network (Fig. 3).

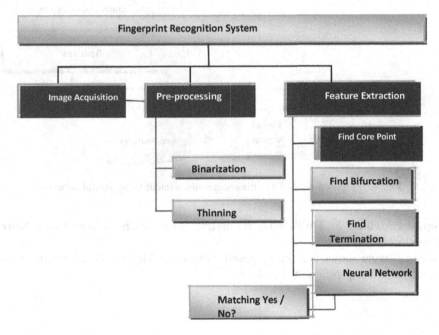

Fig. 3. Proposed method for fingerprint recognition system using neural network.

Proposed Method 2: Iris Recognition System Using Neural Network

Following figure shows that the proposed method for Iris recognition system using neural network (Fig. 4).

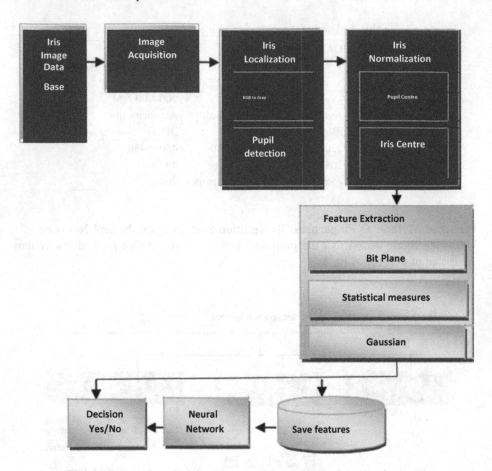

Fig. 4. Proposed method for Iris recognition system using neural network

Proposed Method 3: Multimodal Biometric Recognition System Using Neural Network

Following figure shows that the proposed method for Multimodal Biometric Recognition System using Neural Network (Fig. 5).

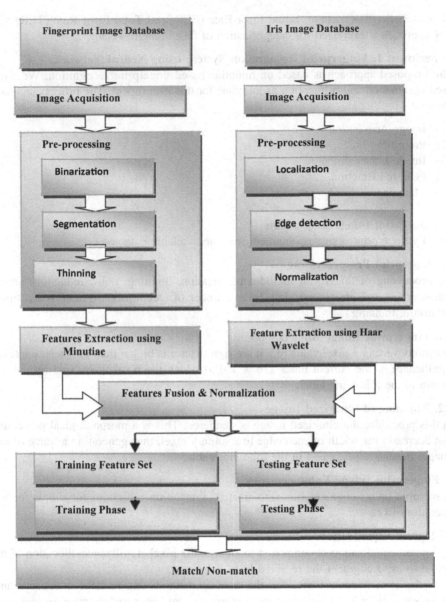

Fig. 5. Proposed method for multimodal biometric recognition system using neural network.

4 Experimental Results

We carried out these trials to estimate two dissimilar unimodal and multimodal biometric systems working. The working of the proposed systems is symbolized by the False Accept Rate (FAR), False Reject Rate (FRR) [44], Receiver Operation

Characteristic (ROC) curve, Equal Error Rate (EER) and Total Error Rate (TER) and NN approach is exercised for categorization of fused feature vector.

Experiment 1: Fingerprint Recognition System using Neural Network

The proposed approach is based on minutiae based fingerprint recognition. We have used termination and bifurcation as minutiae for this experiment. The process includes the following steps:

(1) Image Acquisition
(2) Pre-processing
(3) Image Enhancement
(4) Feature extraction
(5) Matching.

1. Fingerprint Image Acquisition

Initially fingerprint image is obtained from the available database.

2. Fingerprint Pre-processing

Pre-processing of fingerprint includes binarization, thinning, noise reduction, image enhancement and error correction. From number of techniques we have select binarization & thinning.

2.1. Fingerprint Image Binarization

In this converting a pixel value to 1 if the significance is bigger than the mean intensity significance of the current block (16 × 16) to which the pixel fit in. End results are shown in the following table.

2.2. Thinning of binarized Image

In this procedure the binarized image is slendered. This is a morphological procedure that decreases the width of each edge to a solitary pixel, thus generating a frame of the image. Results are revealed in the following table.

3. Fingerprint Image Enhancements

Fingerprint Image augmentation is to create the image unambiguous for simple further operations like,

Histogram Equalization

Histogram equalization is completed to enlarge the pixel significance allocation of an image so as to augment the received knowledge.

In this experiment we analyzed the existing methods and find the drawbacks and some issues. In order to overcome these issues and get better performance we proposed an efficient methodology for finger print pre-processing and evaluated by estimating the MSE and PSNR values. Following table shows that some statistical measurements used for fingerprint recognition (Table 3).

Table 3. Different filter techniques used for fingerprint recognition system

Methods	MSE	PSNR
Gabor filter	5.3	15.11
Wiener filter	205.26	28.59
Median	356.48	26.22
Gaussian	124.92	30.79
Proposed	1.1	42.62

Following figure shows that the performance evaluation for fingerprint recognition system (Fig. 6).

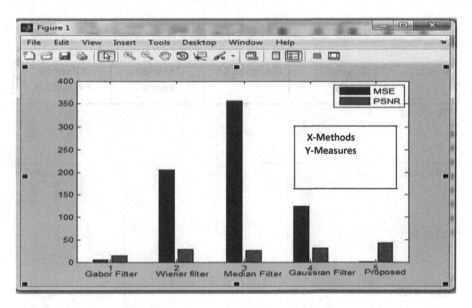

Fig. 6. Statistical performance evaluation for filter techniques on fingerprint system

4. Features Extraction

From above literature study we have utilized The Crossing Number (CN) technique to carry out minutiae withdrawal. This technique digs out the edges and bifurcations commencing the frame figure by investigating the local neighborhood of each edge pixel with a 3 × 3 window. The CN designed for a edge pixel P is represented by

$$CN = 0.5 \sum_{i-1}^{8} |P_i - P_{i+1}|$$

$$P9 = P1$$

Where Pi is the pixel significance in the region of P. Following the CN for a edge pixel has been calculated, the pixel be able to be classified as per concurrency to its CN assessment. A ridge pixel by means of a CN of one communicates to a ridge closing, and a CN of three communicates to a bifurcation for every extracted minutiae tip, the following information is obtained:

1. X and Y coordinate,
2. Direction of the combined ridge fragment, and
3. Type of minutiae (edge conclusion or bifurcation).

Following table shows the experimental results on FVC 2002 Fingerprint database, FVC 2004 Fingerprint database, KVK Fingerprint database for fingerprint recognition system (Table 4, 5 and 6).

Table 4. Fingerprint recognition system at feature extraction level at FVC2002 fingerprint database.

Sr.No.	Original Image	Binarization	Thinned Image	Minutiae Extraction
101_1.tif				
101_2.tif				
101_3.tif				
101_4.tif				

5. Matching: We have performed Neural Network to check whether they belong to the same person or not. By means of neural network minutiae supported fingerprint matching provides precise matching products. This recommended technique is utilized to feed forward back propagation network. Subsequent plots represent the outcome of fingerprint identification system at feature removal level at diverse fingerprint databases with neural network (Fig. 7).

Table 5. Fingerprint recognition system at feature extraction level at FVC2004 fingerprint database

Sr.No.	Original Image	Binariza tion	Thinned image	Minutiae Extraction
101_1.tif				
101_2.tif				
101_3.tif				
101_4.tif				

Table 6. Fingerprint recognition system at feature extraction level at KVK Fingerprint database

Sr. No.	Original Image	Binarization	Thinned image	Minutiae Extraction
Image_00				
Image_01				
Image_02				
Image_03				
Image_04				

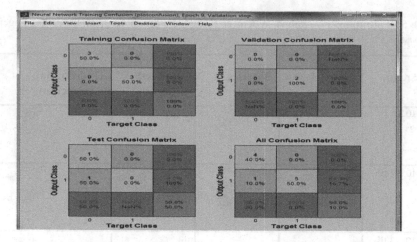

Fig. 7. Neural network training confusion for fingerprint recognition system. (Color figure online)

In the above figure the high numbers of accurate replies in the green squares and the low numbers of inaccurate replies in the red squares. The lower right blue squares demonstrate the overall accuracies. The blue bars correspond to training data, the green bars signify validation data, and red bars symbolize testing data.

The Receiver Operating Characteristic Curve

The ROC curve is a scheme of the factual positive pace (sensitivity) against the false positive pace (specificity) as the verge is diverse. The Receiver Operating Characteristic (ROC) curve plots FAR against FRR for a grouping of threshold values T. An ideal ROC curve merely has values that is present either on the x-axis (FAR) or the y-axis (FRR). For FRRs other than 0, the FAR is 1, and vice versa (Fig. 8).

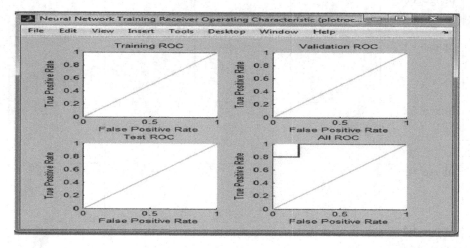

Fig. 8. Neural network training receiver operating characteristic for fingerprint recognition system

The performance of proposed technique is evaluated with an additional method, the corresponding and non-corresponding is based on Euclidean distance which comprises threshold value i.e., if Euclidean distance is less than the threshold significance then it is coordinated else not coordinated. If the Euclidean distance (Ed) among them is lesser than a given acceptance d0 and the direction difference (dd) between them is lesser than an angular tolerance θ0.

The total number of coordinated minutiae pair is calculated to generate match score. If match score is bigger than 90%, the current minutiae are fit in to the same person. The product provides the person matched with database image. Otherwise result is match not found.

$$Ed = \sqrt{(x_i - x_j)^2 + (y_i - y_j)^2} \leq d0$$

$$dd = \min\left(\left|\theta_i - \theta_j\right|, 360 - \left|\theta_i - \theta_j\right|\right) \leq 0$$

The projected minutiae feature-based fingerprint recognition system gives satisfactory precision. The precision of projected method is quantified in terms of the False Acceptance Rate (FAR) and False Rejection Rate (FRR). The identification pace of projected method of fingerprint recognition system with Euclidean distances is **90.02**. The recognition rate of this proposed method of fingerprint recognition system by means of neural network is **90.06**.

Subsequent table illustrates the accuracy rate of fingerprint recognition system using termination and bifurcation as a minutiae (Table 7).

Table 7. Accuracy rate of fingerprint recognition system using termination and bifurcation as a minutiae.

Trait	Algorithm	Accuracy	FAR (%)	FRR (%)
Finger print	Minutiae extraction (termination & bifurcation) & neural network	90.06	1.24	6.09

Experiment 2: Iris Recognition System using Neural Network
The proposed approach for iris recognition process includes the following steps.

(1) Image Acquisition
(2) Pre-processing
(3) Feature extraction
(4) Identification
(5) Matching.

1. Image Acquisition
In the first step image acquisition is taken from the available database. Like other techniques, noise sections were fragmented from novel iris images.

2. Preprocessing

The preprocessing component initially converts the factual color (RGB) into intensity picture. Preprocessing eliminates the result of spots/holes lying on the pupillary region. The revealing of pupil fails whenever there is a spot on the pupil region. The processing is made up of two steps:

1. Iris Localization.
2. Edge Detection.

1. Iris Localization: In this phase we should decide an iris fraction of the image by confining the location of the image resulting from within the limbos (outer border) and exterior the pupil (inner border), and finally change the iris part into a suitable representation. An integral-differential operator for detecting the iris boundary by searching the parameter space.

2. Edge Detection: It is used to discover composite entity limits by spotting prospective edge point's equivalent to places in an image where quick alter in brightness take places. After edge points have been noticed, they can be fused to form lines.

Following table shows that some experimental results for each database of iris.

Experiment 2.1: Results of Filter Techniques on CASIA Iris Database
See (Table 8).

Table 8. Iris recognition system using filtering techniques at CASIA Iris database

Sr. No.	Image Name	Original Image	Prewitt	Canny	Laplacian of Gaussian	Sobel
1	001_1_1.bmp					
2	001_1_2.bmp					
3	001_1_3.bmp					
4	001_2_1.bmp					
5	001_2_2.bmp					

Experiment 2.2: Results of Filter Techniques on MMU Iris Database
See (Table 9).

Table 9. Iris recognition system using filter techniques at MMU Iris database

Sr. No.	Image Name	Original Image	Prewitt	Canny	Laplacian of Gaussian	Sobel
1	aeval 1.bmp					
2	aeval 2.bmp					
3	aeval 3.bmp					
4	aeval 4.bmp					
5	aeval 5.bmp					

Experiment 2.3: Results of Filter Techniques on KVK Iris Database
See (Table 10).

Table 10. Iris recognition system using filter techniques at MMU Iris database

Sr. No.	Image Name	Original Image	Prewitt	Canny	Laplacian of Gaussian	Sobel
1	aeval 1.bmp					
2	aeval 2.bmp					
3	aeval 3.bmp					
4	aeval 4.bmp					
5	aeval 5.bmp					

422   A. M. N. Siddiqui et al.

Feature Extraction: In this method Gaussian filter are used to extract the features of given three databases as shown in the following table.

Experiment 2.4: Results of extracted features on CASIA Iris Database
See (Table 11).

Table 11. Iris recognition system for feature extraction at CASIA Iris database

Image Name	Original Image	Localize	Normalization	Enhancement	Feature Extraction
001_1_1 .bmp					
001_1_2 .bmp					
001_1_3 .bmp					
001_2_1 .bmp					
001_2_2 .bmp					

Experiment 2.5: Results of Extracted features on MMU Iris Database
See (Table 12).

Table 12. Iris Recognition System for feature extraction at MMU Iris database

Sr. No.	Original Image	Localize	Normalization	Enhancement	Feature Extraction
aeval 1.bmp					
aeval 2.bmp					
aeval 3.bmp					
aeval 4.bmp					
aeval 5.bmp					

Experiment 2.6: Results of extracted features on KVK Iris Database
See (Table 13).

Table 13. Iris recognition system for feature extraction at KVK Iris database

Sr. No.	Original Image	Localize	Normalization	Enhancement	Feature Extraction
L1.jpg					
L2.jpg					
L3.jpg					
L4.jpg					
L5.jpg					

4. Iris Matching

For classification purpose we used N.N classifier. A neural network is produced and the iris images are specified as the inputs. At this point the Neural Network has to evaluate the examined iris image with the accumulated image and recognize the person from the image. For doing this the nodes in the neural network has to discover and accumulate the features of the iris each time an image is being given as input (Fig. 9).

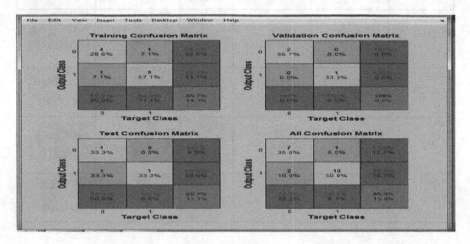

Fig. 9. Neural network training confusion matrix for iris recognition system (Color figure online)

In the above figure the high numbers of correct responses in the green squares and the low numbers of incorrect responses in the red squares. The lower right blue squares illustrate the overall accuracies (Fig. 10).

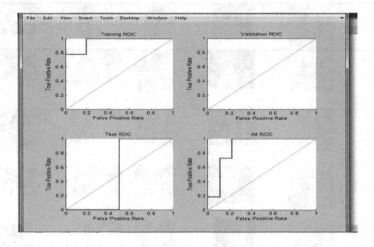

Fig. 10. Neural network training receiver operating characteristic for iris recognition

Experiment 3: Multimodal Biometric Recognition System Using Neural Network
The proposed approach is based on minutiae for fingerprint recognition system and Haar wavelet is used for iris recognition system. The first step is to collect the fingerprint and iris images. On these images various pre- processing steps are carried out. It includes conversion of color image to gray scale, histogram equalization and segmentation as discussed above two experiments. These templates are match with the stored one using neural network. Fingerprint recognition system minutiae method is used which discussed in experiment 2 and for iris recognition we have used Haar wavelet.

Feature Extraction (Haar Wavelet)
Features from the iris image are mined by Haar wavelet disintegration procedure. In the wavelet decomposition the image is decomposed into four coefficient i.e., horizontal, diagonal, vertical and approximation. The approximation coefficients are more decomposed into four coefficients. Proposed system believes Haar wavelet for removal of feature by means of four level decomposition procedures. Following table illustrates that mined Features using Haar wavelet.

Experiment 3.1: Results of Extracted Features on CASIA Iris Database Using Haar Wavelet
See (Table 14).

Table 14. Iris recognition system for feature extraction at CASIA Iris database

Image Name	Original Image	Approximate	Horizontal	Vertical	Diagonal
001_1_1. bmp					
001_1_2. bmp					
001_1_3. bmp					
001_2_1. bmp					
001_2_2. bmp					

Experiment 3.2: Results of Extracted Features on MMU Iris Database Using Haar Wavelet
See (Table 15).

Table 15. Iris recognition system for feature extraction at MMU Iris database

Image Name	Original Image	Approximate	Horizontal	Vertical	Diagonal
aeval1					
aeval2					
aeval3					
aeval4					
aeval5					

Experiment 3.3: Results of Extracted Features on KVK Iris Database Using Haar Wavelet
See (Table 16).

Table 16. Iris recognition system for feature extraction at KVK Iris database

Image Name	Original Image	Approximate	Horizontal	Vertical	Diagonal
L1.jpg					
L2.jpg					
L3.jpg					
L4.jpg					
L5.jpg					

Matching
Though in order to boost the exactness of the biometric system as a entire human being results are joint at matching score level with neural network (Fig. 11).

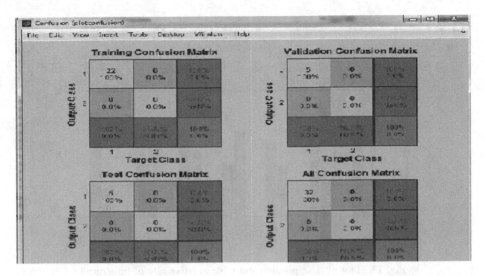

Fig. 11. Neural network training confusion using fingerprint & iris (Color figure online)

In the above figure the high numbers of correct responses in the green squares and the low numbers of incorrect responses in the red squares. The lower right blue squares illustrate the overall accuracies (Fig. 12).

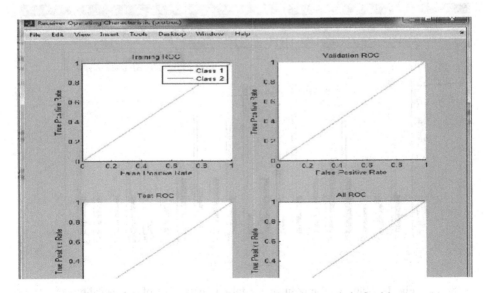

Fig. 12. Neural network training receiver operating characteristic for iris recognition

The subsequent table represents the False Negative and False positive percentage by means of the minutiae mining matching gain using threshold for fingerprint identification (Fig. 13).

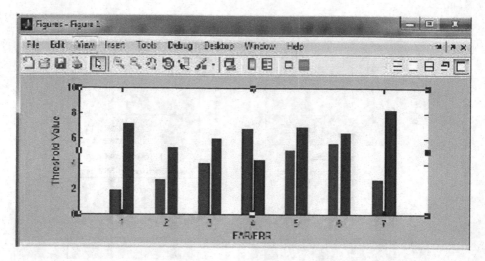

Fig. 13. Plot of false positive and false negative rate of fingerprint

Likewise the false positive rate and false Negative rate of the iris modality is as follows (Fig. 14).

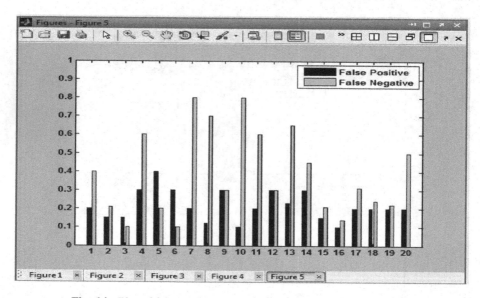

Fig. 14. Plot of false positive and false negative rate of iris modality.

At second phase of trial the matching scores from the entity traits are shared and the false negative rate and false positive rate is reduced while merging it as sum score combination is as follows (Fig. 15).

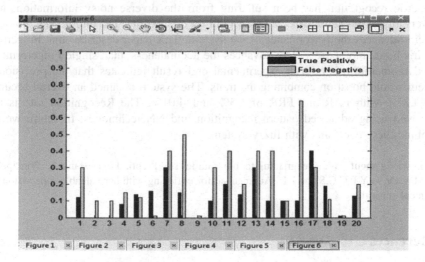

Fig. 15. Plot of false positive and false negative rate of fingerprint modality (minutiae) + iris modality (haar wavelet).

By using the feature level combination the False Acceptance rate and false refusal percentage is reduced. Final Table 17 shows the accuracy and error rates gained from the individual and combined system. The in general performance of the system has augmented presenting an accuracy of 99.23% with True positive of 1.30% and False Negative of 4.01% respectively. The allocation of genuine and imposter information illustrates that at threshold of 0.5 the system would give minimum FAR and FRR rates with maximum accuracy of 99.23%.

Table 17. Individual and combined accuracy of fingerprint and iris.

Trait	Algorithm	Accuracy (%)	FAR (%)	FRR (%)
Fingerprint	Minutiae matching	90.06	3.17	12.69
Iris	Gaussian feature	98.80	4.85	6.43
Fusion (Fingerprint + Iris)	**Minutiae + Haar + N. N.**	**99.23**	**0.9**	**0.23**

5 Conclusion

This paper developed a Multimodal biometric system for person identification using a combination of fingerprint and iris. The performance of solitary modality supported biometric recognition has been suffering from the diverse noisy information, non-universality of biometric data, and susceptibility of spoofing. The multimodal biometric system advances the performance of the system. This paper describes that Fingerprint and iris based biometric system advances the accurateness, than single Fingerprint/iris based biometric system. The experimental end result indicates that the precision of system would boost on combining the traits. The system obtained an overall accuracy of 99.23% with FAR and FRR of 1.3% and 4.01%. The Recognition rate is also improved using advanced pattern recognition and NN techniques. In future we can combine these two traits with fuzzy system.

Acknowledgement. We are grateful to our Guide Dr. Prapti. D. Deshmukh, Principal of MGM's, Dr. G.Y.P.C.C.S and I.T., Aurangabad for enriching with her valuable supervision and technical support.

References

1. Jain, A.K., Ross, A., Prabhakar, S.: An introduction to biometric recognition. IEEE Trans. Circuits Syst. Video Technol. **14**(1) (2004)
2. Mishra, A.: Multimodal biometrics it is: need for future system. Int. J. Comput. Appl. (0975-8887) **3**(4), 157–160 (2010)
3. Jain, A.K., Ross, A.: Commun. ACM **47**, 34–40 (2004)
4. Mitchell, H.B.: Multi-Sensor Data Fusion, p. 281. Springer, Heidelberg (2007). https://doi.org/10.1007/978-3-540-71559-7
5. Kaur, D., Kaur, G.: Level of fusion in multimodal biometrics: a review. Int. J. Adv. Res. Comput. Sci. Softw. Eng. **3**(2) (2013)
6. Kumar, A., Shekhar, S.: Personal identification using multibiometrics rank-level fusion. IEEE Trans. Syst. Man Cybern. — Part C: Appl. Rev.
7. Besbes, F., Trichili, H., Solaiman, B.: Multimodal biometric system based on fingerprint identification and iris recognition. In: Proceedings 3rd International IEEE Conference on Information and Communication Technologies: From Theory to Applications (ICTTA 2008), pp. 1–5 (2008). https://doi.org/10.1109/ICTTA.2008.4530129
8. Rattani, A., Kisku, D.R., Bicego, M., Tistarelli, M.: Feature level fusion of face and finger biometric
9. Radha, N., Kavitha, A.: Rank level fusion using fingerprint and iris biometric. Indian J. Comput. Sci. Eng. (IJCSE) **2**(6), 917–923 (2012). ISSN 0976-5166
10. Wang, J.-G., Toh, K.-A., Sung, E., Yau, W.-Y.: A feature-level fusion of appearance and passive depth information for face recognition. In: Delac, K., Grgic, M. (eds.) Source: Face Recognition, p. 558. I-Tech, Vienna, June 2007. ISBN 978-3-902613-03-5
11. Razzak, M.I., Khan, M.K., Alghathbar, K., Yusof, R.: Multimodal biometric recognition based on fusion of low resolution face and finger veins. Int. J. Innov. Comput. Inf. Control ICIC Int. **7**(8), 4679–4689 (2011). ISSN 1349-4198
12. Anzar, S.M., Sathidevi, P.S.: Optimal score level fusion using modalities reliability and separability measures. Int. J. Comput. Appl. (0975-8887) **51**(16), 1–8 (2012)

13. Snelick, R., Uludag, U., Mink, A., Indovina, M., Jain, A.: Large scale evaluation of multimodal biometric authentication using state-of-the-art systems. IEEE Trans. Pattern Anal. Mach. Intell. **27**(3), 450–455 (2005)

14. Kazi, M.M., et al.: Multimodal biometric system using face and signature: a score level fusion approach. Adv. Comput. Res. **4**(1) (2012). ISSN 0975-3273, E-ISSN 0975-9085

15. Zuev, Y.A., Ivanon, S.: The voting as a way to increase the decision reliability. In: Proceedings Foundations of Information/Decision Fusion with Applications to Engineering Problems, Washington, DC, pp. 206–210, August 1996

16. Ross, A., Jain, A.: Information fusion in biometrics. Department of Computer Science and Engineering, Michigan State University, 3115 Engineering Building, East Lansing, MI48824, USA

17. Ratha, N.K., Karu, K., Chen, S., Jain, A.K.: A real-time matching system for large fingerprint data bases. IEEE Trans. Pattern Anal. Mach. Intell. **18**(8), 779–813 (1996)

18. Jain, K., Hong, L.: Online fingerprint verification. IEEE Trans. Pattern Anal. Mach. Intell. **19** (4), 302–341 (1997)

19. Deshpande, A.S., Patil, S.M., Lathi, R.: A multimodel biometric recognition system. Int. J. Electron. Comput. Sci. Eng. (2012). ISSN 2277-1956

20. Wildes, R., et al.: A machine vision system for iris recognition. Mach. Vis. Appl. **9**, 1–8 (1996)

21. Adler, F.H.: Physiology of the Eye. Mosby, St. Louis (1965)

22. Daugman, J.G.: High confidence visual recognition of persons by a test of statistical indene dance. IEEE Trans. Pattern Anal. Mach. Intell. **15**(11), 1148–1161 (1993)

23. Flom, L., Safir, A.: Iris recognition system. U.S. Patent 4 641 349 (1987)

24. Ruili, J., Jing, F.: VC5509A based fingerprint identification preprocessing system. In: International Conference on Signal Processing, pp. 2859–2863 (2008)

25. Yun, E.K., Cho, S.B.: Adaptive fingerprint image enhancement with fingerprint image quality analysis. In: International conference of Image and Vision Computing, pp. 101–110 (2006)

26. Zhao, F., Tang, X.: Preprocessing and postprocessing for skeleton based fingerprint minutiae extraction. Pattern Recogn. **40**(4), 1270–1281 (2007)

27. Jain, A.K., Prabhakar, S., Hong, L.: A multichannel approach to fingerprint classification. IEEE Trans. Pattern Anal. Mach. Intell. **21**(4), 348–359 (1999)

29. Zebbiche, K., Khelifi, F.: Region-based watermarking of biometrics images: case study in fingerprint images. Proc. Int. J. Digit. Multimed. Broadcast. **2008**, 1–13 (2008)

30. Chen, Y., Jain, Anil K.: Beyond minutiae: a fingerprint individuality model with pattern, ridge and pore features. In: Tistarelli, M., Nixon, M.S. (eds.) ICB 2009. LNCS, vol. 5558, pp. 523–533. Springer, Heidelberg (2009). https://doi.org/10.1007/978-3-642-01793-3_54

31. Tachaphetpiboont, S., Amornraksa, T.: Applying FFT features for fingerprint matching. In: Proceedings of the IEEE Conference on Wireless Pervasive Computing, pp. 1–5 (2006)

32. You, J., Kong, W.K., Zhang, D., Cheung, K.H.: On hierarchical palmprint coding with multiple features for personal identification in large databases. IEEE Trans. Circuits Syst. Video Technol. **14**(2), 234–243 (2004)

33. Gayathri, R., Ramamoorthy, P.: Fingerprint and palmprint recognition approach based on multiple feature extraction. Eur. J. Sci. Res. **76**(4) (2012)

34. Ross, A., Jain, A., Reisman, J.: A hybrid fingerprint matcher. In: Proceedings of International Conference on Pattern Recognition, pp. 1661–1673 (2003)

35. Kin, J.K.U., Chae, S.-H., Lim, S.J., Pan, S.B.: A study on the performance analysis of hybrid fingerprint matching methods. Int. J. Future Gener. Commun. Netw. **1**, 23–28 (2008)

36. Mahadik, S., Narayanan, K., Bhoir, D.V., Shah, D.: Access control system using fingerprint recognition. In: International Conference on Advances in Computing, Communication and Control, pp. 306–311 (2009)
37. Jain, A.K., Prabhakar, S., Chen, A.: Combining multiple matchers for a high security fingerprint verification system. Pattern Recogn. Lett. **20**, 1371–1379 (1999)
38. Marana, A.N., Jain, A.K.: Ridge-based fingerprint matching using hough transform. In: Proceedings of the IEEE Brazilian Symposium on Computer Graphical and Image Processing, pp. 112–119 (2005)
39. Ma, L., Wang, Y., Zhang, D.: Efficient iris recognition by characterizing key local variations. IEEE Trans. Image Process. **13**(6), 739–750 (2004)
40. Daugman, J.: High confidence visual recognition of persons by a test of statistical independence. IEEE Trans. Pattern Anal. Mach. Intell. **15**, 1148–1161 (1993)
41. Daugman, J.: The importance of being random: statistical principles of iris recognition. Pattern Recogn. **36**(2), 279–291 (2003)
42. Wildes, R.: Iris recognition: an emerging biometric technology. Proc. IEEE **85**(9), 1348–1363 (1997)
43. Ross, A., Govindarajan, R.: SPIE Conference on Biometric Technology for Human Identification, vol. 5779, pp. 196–204 (2005)
44. Telgad, R.L., Deshmukh, P.D., Siddiqui, A.M.N.: Combination approach to score level fusion for multimodal biometric system by using face and fingerprint. In: Recent Advances and Innovations in Engineering (ICRAIE), pp. 1–8, 9–11 May 2014
45. https://www.tutorialspoint.com/biometrics/biometrics_quick_guide.htm

Optimal Band Selection for Improvement of Hyperspectral Palmprint Recognition System by Using SVM and KNN Classifier

Anita G. Khandizod$^{(\boxtimes)}$ and Ratnadeep R. Deshmukh

Department of Computer Science and Information Technology,
Dr. Babasaheb Ambedkar Marathwada University, Aurangabad, MH, India
anukhandizod@gmail.com, rrdeshmukh.csit@bamu.ac.in

Abstract. Palmprint recognition system is a promising biometric technology which received extremely large interest of researches. Many different algorithms and systems have been proposed and built. Although, great success has been achieved in palmprint research, however, the accuracy and spoofing mechanism are limited in some cases, as the palmprint feature may be similar for a given spectral illumination; hyperspectral Palmprint is a good recognition method to address this issue, it can provide more discriminate information under different illumination in a short time. Optimal band selection is a very important step to building hyperspectral palmprint recognition system. In this paper, for dimensionality reduction 2D2 LDA, is used on pre-processed palmprint images, for feature extraction 2D Gabor filter is being used, for pattern matching SVM, KNN classifiers are used. Experiment result showed that 2 spectral bands 700 nm and 960 nm could provide the highest recognition accuracy for hyperspectral palmprint recognition.

Keywords: Biometrics · 2D2 LDA · 2D Gabor filter · SVM · KNN

1 Introduction

In our society, person recognition is very important and in this field, biometrics is one of the most important and reliable methods. Biometric technologies are typically used for security purposes, these technologies depend on specific biometric characteristics that are used for recognition of individuals; they can be divided into two categories: Physiological Characteristics are related to the shape of the body (e.g. Finger-print, retina, iris, face, hand geometry, palmprint etc.), while Behavioral characteristics related to the behavior of the individuals, (e.g. voice, gait, signature, keystroke pattern) [1,17]. Palmprint recognition system is a promising biometric technology which received extremely large interest of researches. As compare to the other biometric identification technologies palmprint recognition system has been successful due to its simplicity, rich in

© Springer Nature Singapore Pte Ltd. 2019
K. C. Santosh and R. S. Hegadi (Eds.): RTIP2R 2018, CCIS 1036, pp. 433–442, 2019.
https://doi.org/10.1007/978-981-13-9184-2_38

features, low resolution, reliable, cost-effective, small size, real-time computation [2] Palmprint contains large numbers of feature as shown in Fig. 1, such as principal lines, secondary, wrinkles, minutiae, datum point, and ridges. In the case of identical twins, they also have different palmprint features [3]. The line structure of the palmprint remains unchanged throughout the life of individuals. Different individuals do not have the same palmprint features, therefore, it's very difficult to have their palmprint images taken for authentication. Whatever the system and algorithm proposed, they all use the white light as the source of illumination, and therefore accuracy is limited in some cases. This issue was addressed by 3-D imaging, but the device of the 3-D imaging bulky and expensive and very difficult to applied for real application [5]. One solution to these problems can be multispectral palmprint imaging. Multispectral palmprint collects spectral information in Blue (470 nm), Green (525), Red (660), and NIR (880) spectral bands, as compared to the multispectral palmprint, the hyperspectral palmprint gives more detail information because the sensor of hyperspectral palmprint collects spectral information in the range of 420 nm to 1100 nm. In palmprint research, the work related to hyperspectral palmprint recognition is quite limited in nature [6]. Hyperspectral imaging is rich in information, high dimension, because of this reason hyperspectral imaging may be useful for increasing the accuracy of the palmprint recognition system.

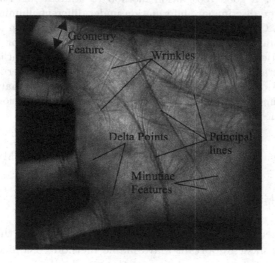

Fig. 1. Different features of palm [4]

2   Hyperspectral Palmprint Database

A huge HK-PolyU Hyperspectral Palmprint Database was collected from 190 persons from the same institute, the age of the persons vary from 20 to 60 years old, the database captured from 380 different palms at 69 spectral bands with a step-width of 10 nm over spectrum 420 nm–1100 nm is used for experiments. This database was collected in two different sessions first and the second session, the time interval between first and second session was about 1 month. In the first session, 2608 palmprint images were captured while in second session 2632 images [7].

3 Proposed Framework

Hyperspectral palmprint recognition system usually consists of four basic stages, preprocessing, dimensionality reduction, Feature Extraction, and matching. The following figure shows the proposed framework of palmprint recognition (Fig. 2).

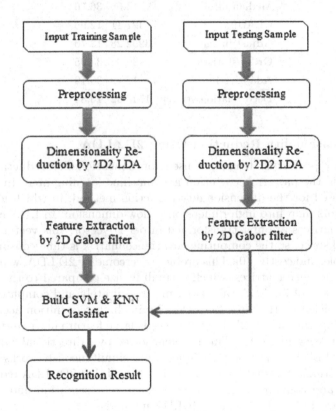

Fig. 2. The proposed framework for hyperspectral recognition

3.1 Preprocessing

Preprocessing plays very important role in palmprint recognition system, to removing noise and making images more visually appealing, in this paper five types of image enhancement techniques are used like Median Filter, CLAHE

(Contrast Limited Adaptive histogram Equalization), Adjust Image, Order-statistic, Adaptive filter, decorrelation stretch to enhance the quality of palm-print images. To decide the quality of image the objective image quality measure like MSE (Mean square error) and PSNR (Peak Signal to noise ratio) are used. Lower MSE Value provides Higher Quality and Higher PSNR value provides higher Quality [8]. The MSE value for median filter is 14.68, and PSNR = 36.69, therefore according to the Table 1 median filter is the best preprocessing technique for hyperspectral palmprint recognition system [9].

Table 1. Comparison of objective image quality metrics and image enhancement

Image enhancement	MSE	PSNR
Median filter	14.68	36.76
CLAHE	427.11	12.96
Adjust image	3292.28	12.76
Order-statistic	36.02	33.05
Adaptive filter	17.42	36.19
Decorrelation stretch	1747.32	15.84

3.2 Dimensionality Reduction Using 2D2 LDA

LDA is liner projection technique used for dimensionality reduction, feature extraction in the pattern recognition and machine learning area. In this paper 2D2 LDA used for the dimensionality reduction method, in which feature projection vectors map into high dimension to low-dimension. In LDA method the 2D image matrices are first transformed into 1 D vector; this vector gives high dimensional space, so the computing covariance matrix is time consuming, difficult, unstable, and costly [10]. This problem overcome by 2D LDA, which directly compute covariance matrices, which is small in size compared to the LDA. The scatter matrices of 2D LDA calculated more accurately and compute Eigenvectors more efficiently, therefore 2D LDA obtains high recognition accuracy than LDA, but the drawback of 2D LDA is it needs large feature matrices for the task of palmprint recognition [11]. This problem solved by 2 directional 2 dimensional LDA i.e. 2D2 LDA projection technique, which simultaneously works in row and column directions. Noushath and others [12] showed that 2D2 LDA requires fewer coefficients and computing time for face image representation and recognition than standard PCA, 2DPCA, and 2DLDA methods.

Steps to develop hyperspectral palmprint recognition system using 2D2 LDA: Let the training database of Ai j palmprint image, of size a * b, representing the ith sample in the jth class. The between-class scatter matrix Gb and within-class scatter matrix Gw are computed as.

$$Gb = \sum_{j=1}^{c} (uj - u)^T (uj - u) \tag{1}$$

$$Gw = \sum_{j=1}^{c} \sum_{Nj}^{i=1} (Ai^j - u)^T (Ai^j - uj) \tag{2}$$

Compute optimal projection axes denoted by X, so total scatter of the projected images are maximize.

$$\frac{X^T G_b X}{X^T G_w X} \tag{3}$$

Covariance matrix are calculated, then find largest eigenvalues p corresponds to eigenvectors, now the dimension of the optimal projection matrix is b * p. Project all training images onto the optimal projection axes i.e. X, which is result in feature matrix of training images, having a * p dimension, defined as fallows.

$$Z_i^j = A_i^j X \tag{4}$$

From Eqs. 1 and 2, scatter matrix of Gb and Gw are obtained by outer products of row vectors of images. Now we calculate scatter matrix of Gb and Gw column vectors of images. We propose that the between-class scatter matrix Hb and the within-class scatter matrix Hw be computed as.

$$H_b = \sum_{j=1}^{c} (u_j - u)(u_j - u)^T \tag{5}$$

$$H_b = \sum_{j=1}^{c} \sum_{i=1}^{Nj} (A_i^j - u)(A_i^j - u_j)^T \tag{6}$$

It can be observed that Hb and Hw in (5) and (6) are obtained in outer products of column directions, unlike Gb and Gw in (1) and (2) in the Row directions. Compute optimal projection matrix W of dimensions (w * p).

$$\frac{W^T H_b W}{W^T H_w W} \tag{7}$$

Calculate feature matrix of training images as fallows.

$$z_i^j = A_i^j W^T \tag{8}$$

Now, compute 2 directional 2 dimensional LDA for that, let X denote the a * p optimal projection matrix, from Eq. (4) and W denote the m * p matrix obtained by Eq. (11). Each training image is projected onto both X and W simultaneously to obtain the respective feature matrix which is of dimensions p × d as follows.

$$F_i^j = W^{T^{A_i^j}} X \tag{9}$$

Matrix F is called feature matrix, after projecting each training images in both row and column direction simultaneously, the feature matrix F_i^j can be obtained.

3.3   2D Gabor Filter

2D Gabor wavelet is used to extract the features of the hyperspectral palmprint database. This filter has multiplication convolution property i.e. convolution of the Fourier transform of the Gaussian function and Fourier transform of the harmonic function [13]. This filter has two components first real and second imaginary component; these two components formed complex number as follows [14]. Complex Component

$$g(x,y) = \exp(-\frac{x^2 + \lambda_2 y^2}{2\sigma^2})\exp(i(2\Pi\frac{x}{y} + \psi)) \tag{10}$$

Real component

$$g(x,y,\lambda,\Theta,\psi\sigma\gamma) = \exp(-\frac{x^2 + \gamma^2 y^2}{2\sigma^2})\cos(i(2\Pi\frac{x}{y} + \psi)) \tag{11}$$

Imaginary Component

$$g(x,y,\lambda,\Theta,\psi\sigma\gamma) = \exp(-\frac{x^2 + \gamma^2 y^2}{2\sigma^2})\sin((2\Pi\frac{x'}{y} + \psi)) \tag{12}$$

Gabor filter creates a column vector and these vectors are normalized to zero mean and unit variances. KNN and SVM these two classification techniques are used on feature matrix extracted by using 2D Gabor filter. KNN algorithm based on closest training images in the problem space [15]; if the value of k is 1 then the object is allocated to the class of its nearest neighbor. SVM is a linear or nonlinear classifier, mainly used to distinguished two or more different classes [16]. SVM builds a model; map the decision boundary for each and every class and with the help of hyperplane separates the different classes.

4   Experimental Result

We used HK-PolyU Hyperspectral Palmprint Database for experimentation work; the whole database is divided into two parts, training set and testing set. Firstly, the even number of samples is chosen as training set and the odd number of samples is used as testing set. Total 69 spectral bands (420 nm–1100 nm) are used for experimental work, database are collected from 380 persons with 2608 palmprint images, so total database used for the experimental work is 69×2608 $= 1,79,952$. The palmprint images first transfer to the preprocessing to enhance the quality of images, then 2D2 LDA was used for the dimensionality reduction in order to reduce the size and get most discriminative features of palmprint images. 2D Gabor filter is used in order to extract the features of palmprint images. The resulted vector is used as input to a KNN and SVM classifies. The recognition accuracy for SVM and KNN with different wavelength (420 nm to 1100 nm) curves is shown in Fig. 3. There are some observations that could be found from Table 2, the first 13 and last 3 bands have low recognition accuracy.

This is because of the poor image quality. The higher recognition accuracy is at 820 nm spectral band 99.70%, for SVM and spectral band 760 nm gives 99.98% higher recognition accuracy for KNN. As compare to the other 69 spectral bands, 820 nm and 760 nm this two bands are optimal band and give discriminate information of palmprint images. Experiment result showed that KNN perform well and achieve higher accuracy than SVM with sufficient database. We select the best five spectral bands out of total 69 spectral bands to enhance the recognition rate of hyperspectral palmprint system. For SVM the best five spectral bands are 820 nm, 780 nm, 760 nm, 730 nm, 740 nm and it could achieve the 99.70%, 99.62%, 99.54%, 99.47%, and 99.39%. In case of KNN the best five spectral bands 760 nm, 700 nm, 650 nm, 970 nm, 890 nm achieve recognition accuracy 99.98%, 99.92%, 99.85%, 99.77%, 99.70%, so these are optimal bands of hyperspectral palmprint recognition system used to enhance the technique for authentication of persons. Figure 4 shown the graphical user interface of palmprint recognition system.

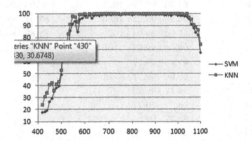

Fig. 3. Recognition rate vs wavelength

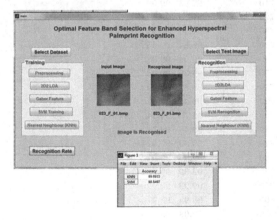

Fig. 4. Graphical user interface of hyperspectral palmprint recognition system

Table 2. Recognition rate corresponding to 2D2LDA

Spectral band	SVM	KNN	Spectral band	SVM	KNN
420	17.64	23.7	920	99.31	99.77
430	18.02	30.67	930	99.08	99.69
440	19.02	33.67	940	98.85	99.62
450	26.53	40.8	950	98.93	99.77
460	29.29	42.18	960	98.62	99.69
470	34.89	36.2	970	98.39	99.77
480	37.96	40.97	980	99	99.54
490	39.34	43.1	990	99.31	99.69
500	49.16	52.53	1000	98.24	99.62
510	71.01	74.46	1010	98.31	99.54
520	70.71	70.71	1020	98.31	99.62
530	73.01	82.98	1030	97.55	99.39
540	85.43	90.72	1040	97.32	99.62
550	91.33	97.55	1050	95.78	96.4
560	93.56	97.24	1060	90.72	94.86
570	84.36	93.17	1070	87.96	90.11
580	95.25	99.23	1080	83.97	88.5
590	95.94	99.08	1090	81.83	86.12
600	98.08	99.54	1100	67.18	74.08
610	97.69	99.77			
620	98.66	99.77			
630	96.17	99.39			
640	98.08	99.77			
650	98.77	99.85			
660	98.16	99.77			
670	98.93	99.85			
680	99	99.85			
690	98.85	99.85			
700	99	99.92			
710	98.93	99.92			
720	99.16	99.85			
730	99.46	99.92			
750	98.54	99.46			
760	99.54	99.98			
770	99.39	99.92			
780	99.62	99.92			
790	99.46	99.92			
800	98.7	99.54			
810	99.46	99.77			
820	99.69	99.92			
830	99.46	99.92			
840	99.39	99.92			
850	99.23	99.92			
860	98.93	99.77			
870	99.08	99.77			
880	99.23	99.77			
890	99	99.69			
900	99.31	9.69			
910	99.16	99.77			

5   Conclusion

In recent years there has been a growing demand for high accuracy and robust palmprint recognition systems. Large number of algorithm and system have been proposed and developed in palmprint recognition system, but a very little work carried out on how to determine the optimal number of spectral bands and how many bands are sufficient for palmprint recognition. For dimensionality reduction new projection technique 2D2 LDA, is used on pre-processed palmprint images, for feature extraction 2D Gabor filter is being used, for pattern matching SVM, KNN classifier are used. Based on our experiments, some findings can be concluded. Two optimal bands like 820 nm, 760 nm having highest recognition rate, 99.69% and 99.98%, for SVM and KNN Classifier as compare to other 69 spectral bands.

Acknowledgement. We would like to thank The Hong Kong Polytechnic University (PolyU) for sharing their database (PolyU Hyperspectral palmprint Database). This project is under UGC Maulana Azad National Fellowship, F1-17.1/2017-18/MANF-2017-18-MAH-82272/(SA-III/Website) sanction at Department of Computer Science & IT, Dr. Babasaheb Ambedkar Marathawada University.

References

1. Khandizod, A.G., Deshmukh, R.R., Borade, S.N.: Spectral biometric verification system for person identification. In: Mishra, D., Nayak, M., Joshi, A. (eds.) Information and Communication Technology for Sustainable Development. LNNS, vol. 10, pp. 103–111. Springer, Singapore (2018). https://doi.org/10.1007/978-981-10-3920-1_11
2. Santosh, K.C., Roy, P.P.: Arrow detection in biomedical images using sequential classifier. Int. J. Mach. Learn. Cybern., 993–1006 (2017)
3. Khandizod, A.G., Deshmukh, R.R.: Comparative analysis of image enhancement technique for hyperspectral palmprint images. Int. J. Comput. Appl. (0975–8887) **121**(23), 30–35 (2015). (Impact Factor = 0.752)
4. Elaydi, H., Alhanjouri, M., Abukmeil, M.: Palmprint recognition using 2-D wavelet, ridgelet, curvelet and contourlet. i-Manager's J. Electr. Eng. (JEE) **7**, 9–19 (2013)
5. Khandizod, A.G., Deshnukh, R.R., Manza, R.: Wavelet-based image fusion and quality assessment of multispectral palmprint recognition. In: 2nd National Conference on Computer Communication and Information Technology. Sinhgad Institute of Computer Science, Pandharpur, NC2IT (2013)
6. Morales, A., Ferrer, M., Kumar, A.: Improved palmprint authentication using contactless imaging. In: Fourth IEEE International Conference on Biometrics: Theory Applications and Systems (BTAS), pp. 1–6 (2010)
7. Ross, A.A., Nandakumar, K., Jain, A.K.: Handbook of Multibiometrics. Springer, New York (2006). https://doi.org/10.1007/0-387-33123-9
8. Department of Computing, the Hong Kong Polytechnic University (PolyU), Hyperspectral Palmprint database, PolyU. http://www4.comp.polyu.edu.hk/biometrics/HyperspectralPalmprint/HSP.htm. Accessed 22 Aug 2013
9. Saruchi, S.: Adaptive sigmoid function to enhance low contrast images. Int. J. Comput. Appl. (0975–8887) **55**(4), 45–49 (2012)

10. Khandizod, A.G., Deshmukh, R.R.: Comparative analysis of image enhancement technique for hyperspectral palmprint images. Int. J. Comput. Appl. (0975–8887) **121**(23), 30–35 (2015)
11. Santosh, K.C., Antani, S.K.: Automated chest x-ray screening: can lung region symmetry help detect pulmonary abnormalities? IEEE Trans. Med. Imaging **37**, 1168–1177 (2018)
12. Khandizod, A.G., Deshmukh, R.R.: Hyperspectral palmprint recognition: a review. In: International Conference on Recent Trends and Challenges in Science and Technology (RTCST 2014). Organized by Pravara Rural Education Society, Padmashri Vikhe Patil College of Arts, Science, and Commerce, Pravaranagar, 20–23 August 2014
13. Noushath, G., Hemantha, P.S.: (2D) 2LDA: an efficient approach for face recognition. Pattern Recogn. **39**(7), 1396–1400 (2015)
14. Somvanshi, P., Rane, M.: Survey of palmprint recognition. Int. J. Sci. Eng. Res. **3**, 1 (2012)
15. Wang, X., Lei, L., Wang, M.: Palmprint verification based on 2D-Gabor wavelet and pulse-coupled neural network. Knowl.-Based Syst. **27**, 451–455 (2012)
16. Koul, D., Alaria, S.K.: A new palm print recognition approach by using PCA & Gabor filter. Int. J. Future Revolution Comput. Sci. Commun. Eng. **4**(4), 38–45 (2018). ISSN 2454–4248
17. Jagtap, A.B., Hegadi, R.S.: Eigen value based features for offline handwritten signature verification using neural network approach. In: Santosh, K.C., Hangarge, M., Bevilacqua, V., Negi, A. (eds.) RTIP2R 2016. CCIS, vol. 709, pp. 39–48. Springer, Singapore (2017). https://doi.org/10.1007/978-981-10-4859-3_4

A Novel Study of Feature Extraction Techniques with Thermal Imaging for Face Recognition

Kiran P. Chaudhari[1], Ravindra B. Patil[1(✉)], and Sangeeta N. Kakarwal[2(✉)]

[1] MIT E, BAM University, Aurangabad 431005, India
kiran_chaudhari@rediffmail.com, ravindra.be2004@gmail.com
[2] PESCOE, BAM University, Aurangabad 431001, India
s_kakarwal@yahoo.com

Abstract. This paper presents a study of Thermal imaging and effective methods of feature extraction process for Face Recognition with thermal imaging. Thermal imaging method is chosen due to its robustness towards ambient light change. This system will be considered as most secure and reliable for accesses control/monitoring system. This system provides us three type images like thermal, visible and fused images of thermal plus visible. Among all biometric attributes, thermal faces trying to overcome accuracy issue of visible imaginary like background, illumination and luminance density, as well as Fake data like photographs or 3D object like statue. In this paper experiment is carried out on sensor level fused images i.e. thermal plus visible faces results are obtain by using PCA.

Keywords: Face recognition · Biometric · Thermal face · Skin

1 Introduction

A Human Face image, recognize as that are currently viewed Jain et al. [1]. The same face can be viewed differently in different images due to variation in view point. A strong recognizer must develop invariant recognition. By this mechanism no one can fool the person by changing outfits and unrecognizable facial visible view (Face with makeup/mask). The study demonstrates the ability of thermal infrared system can expand applications to face recognition system. In Security systems, which involving thermal face recognition is not possible to hack. As the identification process involves thermal signature identification method and thus only authorized users will be accepted. Normally face image is consisting of feature of lips, nose, ear, forehead, chin, cheek, teeth, beard, moustache, etc. There are some additional features like spectacles, ear ring, nose ring etc present. Some extraordinary features like expressions like happiness, sadness, surprise, etc. Again, few more features can be added with the help of

Supported by Organization x.

face makeup. The face is unique feature which used for distinguishing a person. Every person has special region or pattern on face, which is recognize by human eyes monitored by human brain. So, human have great face identification and recognition ability compared with computerized bio-metric identification system (Fig. 1).

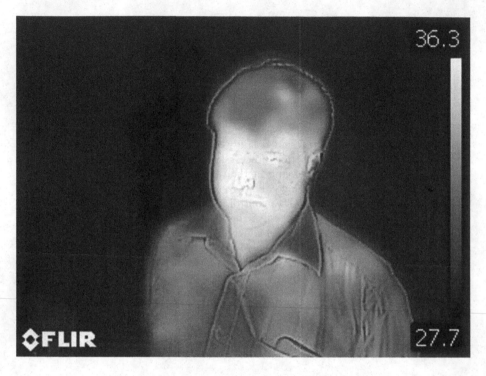

Fig. 1. Sample thermal image by FLIR [16] camera.

Thermal face recognition is capture heat emissivity of facial skin, which gives thermal image as input to system. Thermal image always consists of thermal heat pattern of human face or skin which is generated by blood vessels and tissues. So, the heat pattern generated by body is considered as unique pattern with temperature range 30 to 35.5 °C. Based on heat emissivity, features like ear ring, glasses, make up, etc can identified as noise and can be processed from image. Resulting output image, it will utilize for identification and recognition as input image. Hence IR images captured by the camera are also unique for the respective person.

2   Literature Survey

In prior work, Jain, Ross and Prabhakar [1] presented, According to this team to design unique and special system for individual, which can not be challenged by any other methods. Also unique individuals is able to access it. For an individuals identification the person biological characteristics definitely useful and these are categorized into physical and behavioral. Sheena, Mathu [2] presented a bio-metric system identify one and multiple biological characteristics which including fingerprint, face, etc. Also voice, to access & recognize said or unique person. A spacial computerized identification program for authentication, which is based on the use of different biological unique characteristics such as irises, facial patterns, palm prints and finger prints. Jain and Ross [3] says by using a combination of biometric features for human recognition or identification is more effective and challenging. They focused on the system known as multibiometric system. According to Deriche [4] many computer applications required a unique biological characteristics as key, for person identity. However, most monomodal biometrics [4] is proven to exhibit one and more weakness. With respect, evidence reconciliation from different bio-metric system has attracted much lately. Ives says [5] Bio-metrics is a upcoming future technology which uses users biological characteristics which serve for individuals identify. The bio-metrics required in security, e-commerce, etc. applications. They are using handwriting, fingerprint, hand geometry, face, iris and voice. System can compare with the traditional recognition methods, likely as photo identification or card identification, biometric reduces system operating cost and crime. In prior work Prabhakar, et al. [6], Presented that, the technique is widely acceptable, also acquired by public and industry. A security system requires low cost solutions and defined results from computer program. Many subject are characterized by good recognition rate. Many methods are listed for subject identification in his survey and on the basis of that a systematic approach & rules designed to measure sample quality. Karande, Talbar [7] showed a simple experimental procedure for facial feature recognition by implementing principal component analysis. For describing facial features Eigen faces uses sub sets of features for describing the difference between test and train image of face. This Eigen face/Eigen vector method is referred as principal component analysis procedure. The face recognition computer program depending on Eigen faces approach is defined. The motivation for Eigen images is to find a lower dimensional space. For explaining image cluster they identify cluster of direction in image space with maximum difference in the cluster. Zhao, Yuen [8] says face recognition is one of futuristic research domain in the pattern-recognition. Input image have a high dimension, and a procedure for dimensionality-reduction should be implemented before classification. Principal component analysis (PCA) is not only reduces the image dimensions, but also provides compact features for representing an image. In earlier work, Hermosilla, Ruiz-del-Solar, Verschae, Correa [9] presented article on comparatively - based face-recognition method in unconstrained environments. The comparative study was carried out using two data-bases: Equinox and UCH Thermal Face. The well-known Equinox databases have used as database line for comparison the result. Which compared with the experiment where carried out presented in previous work on images obtained under con-trolled conditions. The UCH Thermal- Face database includes aspects such as yaws and pitch rotations, environment illumination condition variations for indoor and outdoor images and facial expressions Sanjith Kumar and Saravanan [10] presented that thermal photographs as subjects by using MWIR sensors. These images are captured at various temperatures. The vasculature feature data recovered using matching process of image. In prior work, Turk et al. [11] presented Face recognition Using Eigen Faces. Thermal face recognition is relatively new concept. They discovered Eigen face can recognize face effectively. Bhowmik et al. [12] given that infra-red face recognition, for robust security system using fused face recognition technique. They have experimented different thermal spectrum. They used IRIS data base and PCA for thermal face recognition. Afterword by using visible images re-search reached at greater accuracy level. Proposed technique used to overcome face recognition using invisible wavelength i.e. thermal infrared band. Currently many researchers have experimented thermal infrared face features for person identification with change in illumination.

2.1 Infrared Thermography

'Infrared' means below red, this word is chosen from Latin language. 'Red' is longest wavelengths of visible illumination. Humans are not able to identify this infrared wavelength. 'Infrared' a wavelength between 0.7 and 300 µm. Hence these are longer than that of visible light. Thermo graphic cameras generally locate heat emitted by elements, which uses spectral rage 8–14 µm and derive images of that emissivity which is known as thermo grams.

The radiation emitted by a source, rises with temperature which results to allow difference in heat as an image. Infrared thermal imaging camera easily find low temperature back ground and hot blooded animals at foreground with their body part at sensor level display. Irrespective of surroundings light illumination condition. Therefore, thermograph is particularly useful surveillance and security. Thermal images are the result of infrared energy emitted, transmitted, and reflected by individuals. A thermal imaging camera can perform algorithms to understand that data and construct an image.

Even though the output shows observer an estimate present temperature of the output display, capturing camera is using sources of data, which are based on the surrounding areas of individuals to decide that value instead of detecting the definite heat. Sample image and Image capturing process shown in Fig. 2. This phenomenon may turn out to be clearer upon thought of the formula given in Eq. 1 [17]. For complete black body Assume that the received radiation power W from source of temperature Tsource output signal, Usource that is proportional to the power input. Hence equation is

$$Usource = CW(Tsource) \; Where \; C \; is \; a \; constant. \tag{1}$$

In thermal imaging camera, Incident Radiant Power is the sensor level values output on display. Emitted Radiant Power is usually what is intended to be calculated, Transmitted Radiant Power is the power transmitted through the individuals, Reflected Radiant Power is the reflected through individuals body. For gray body with emittance ϵ, and the received radiation is ϵ Wsource. For this we can now use following equation

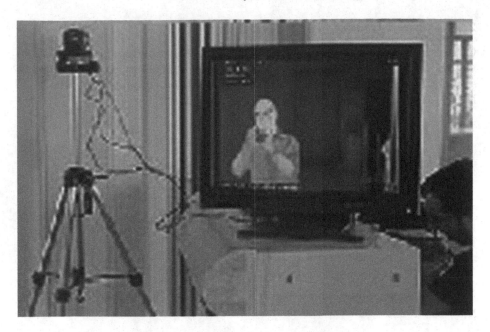

Fig. 2. Thermal Imaging Workstation [14].

$$Incident\ Radiant\ Power\ =\ A + B + C \tag{2}$$

Where A, B and C is A. Emitted Radiant Power $= \epsilon\tau Wobj$, where ϵ is the object emittance and τ is the atmosphere transmittance. The object temperature is Tobj. B. Transmitted Radiant Power $= (1 - \epsilon)\tau Wrefl$, where $(1 - \epsilon)$ is the object reflectance. The ambient sources have the temperature Trefl. C. Reflected Radiant Power $= (1 - \tau)\tau Watm$, where $(1 - \tau)$ is the atmosphere emittance. The atmosphere temperature is Tatm. The total received radiation power is given as:

$$Wtot = \epsilon\tau Wobj + (1 - \epsilon)\tau Wrefl + (1 - \tau)\tau Watm, \tag{3}$$

We multiply each term by the constant C of Eq. 3 and Hence U = CW, and we get (Eq. 4)

$$Utot = \epsilon\tau Uobj + (1 - \epsilon)\tau Urefl + (1 - \tau)\tau Uatm, \tag{4}$$

Solve Eq. 4 for Uobj

$$Uobj = Utot/\epsilon\tau - (1 - -\epsilon)Urefl/\epsilon - (1 - -\tau)\tau Uatm/\epsilon\tau \tag{5}$$

This is the general measurement formula used in all the FLIR Systems thermo-graphic equipment [17].

2.2 Infrared Bands

Objects/elements usually emit heat referred as infrared radiation. As per observations the infrared band separated in few divisions. The International Commission on Illumination stated the division of infrared radiation as IR-A, IR-B and IR-C. A namely used sub-division given in Table 1 as follows:

Table 1. Wavelength range for different spectrums

Spectrum name	Wavelength range	IR division
Near Infrared (NIR)	0.7–1 μm	IR-A
Short-Wave Infrared (SWIR)	1–3 μm	IR-B
Mid-Wave Infrared (MWIR)	3–5 μm	IR-C
Long-Wave Infrared (LWIR)	8–14 μm	IR-C

Developments and modification in infrared technology camera from last few years had given a new era and domain of imaging.

2.3 A Suggested Spectrum for Face Recognition

Human body has temperature of 37 °C normally. As per the fact skin with tissues has its thickness 4 cm. By considering this information human body skin temperature is as approximate 34 °C. As per the different weather condition and human body movements it varies from 30 to 35.5 °C. The heat emissivity for human skin is .91 to .98 as to absorb or radiate. Perfect Black body/object has emissivity as 1 [15].

Hence, facial features recognition in the thermal infrared range form 3–14 μm i.e. LWIR. Emission of LWIR is better than the MWIR. For studying human faces FLIR camera is also the mostly usable with the operating temperature 0 to 35 °C can be extended to −20 to 120 °C.

2.4 A Discussion over Thermal Face and Visual Face

In research work, Bhowmik et al. [12] presented Visual facial images are used for biometric identification process like extracting facial features. One more advantage of visual image is that the visual cameras are affordable. The visual imagery has considerable issues depend on light illumination conditions, change in facial expression of subject, and one more important problem of aging. By using multi-sensorial image capturing technique i.e. visible spectrum and IR spectrum can eliminate few issues. Considering the advantages of IR bands, researchers are taking interest in this multi-spectral modality. Face or skin detection and segmentation is possible by using thermal images, which gives less inter class variance, invariant behavior for changes in light illumination and facial expression and one more remarkable achievement that it works in total darkness.

3   Experimentation

This study created own set of databases for experimentation. In this research we used sensor level fused thermal images produced by Flir C2 camera available at lowest price in the series of Flir family. We have saved these images in JPEG file format for research.

3.1   Database

Most of the imagery used from local data set will used in this study, which collected in laboratory with ideal illumination.

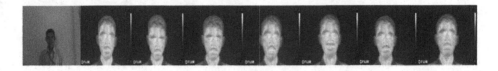

Fig. 3. Image with emotion for experimentation

For all sessions, select cooperative subjects, standing about in range from the cameras, and looking directly. The visible spatial data had resolution of 640×480 pixels. The thermal sensor is sensing between 8 to 12 μm with resolution of 80×60 pixels. Figure 3 show first experimental input, which have given initial success and lead to this experimentation. This input consists of few emotions. This gives us variation in facial image as input for research To avoid noise like self-one side shadowing of the face, due to illumination at source results in appearance variation. The pre-processing perform effective roll in PCA - based recognition. Subject images contain different mood like anger, sadness, astonishment, happy and change in facial expression, angle, and distance. By considering first sample images we have developed first sample data with four scenarios listed in Table 2. We used word CYCLE which defines iteration of input capturing. In which, we have captured five different images of subject. In first scenario and Cycle-I we captured images inside the lab. Input image captured as subjected arrived in lab. In second scenario and Cycle-II, Subject arrived outside lab. In third scenario and Cycle-III, after 20-min warm up input images are captured inside the lab. And finally, in forth scenario and Cycle-IV, again after warming up input images are captured outside the lab. FLIR sensor produces fused image which is give very good result in low cost setup. Figure 4 shows set of 20 images per person for experimentation. Following Table 2 is giving few database properties,

3.2   Principle Component Analysis (PCA) [13]

Principal component analysis is used for principal component known as linearly uncorrelated values, which is derived from correlated values. These values are

Scenario I	Scenario II	Scenario III	Scenario IV

Fig. 4. Subject images with scenarios

Fig. 5. Input set images per person

used to create orthogonal transformation. This is observed that these principal components resulting orthogonal vectors. By the dimensionality reduction, variation is possible in our training data. These help to find the low-dimensional space. Following is the PCA flow [7,13] used for experimentation This is the step by step execution of algorithm used for image matching.

1. Training images reshaped and formulated in single vector.
2. Test images reshaped and formulated in to a single vector.
3. The average of all image are calculated.
4. Then average image is subtracted from data of set of training and testing image, new derive image is recovered image.
5. Then using covariance matrix the eigenvector and eigen values are calculated.
6. Multiplexing of each eigenimages with the recovered images, we recover weight vector of the test image.
7. The weight of test image is subtracted from each weight vector of the recovered image.
6. And distance of each class of the images in the database is calculated.
9. If the class having less distance then input test image belongs to same class.

Table 2. Database properties

Parameter	Description
Number of persons	20
Subject acquisition cycle	Cycle I and II: as person arrived Cycle III and IV: after 20 min walk
Situations	i. Inside Lab ii. Outside Lab
Total images	No. of person *Cycle*Situation *Image
Total no of subjects	$= 20 * 2 * 2 * 5 = 400$

4   Results

This experimentation is performed for fused thermal image of face. We considered 20-person data with five types of expression. With this training and testing data set created for experimentation purpose. We used Flir C2, which is producing fused image. For the performance analysis of 20 persons, we are using five training images and two testing images per person per scenario. Following Table 4 is showing the results. Based on said procedure we have experimented input images and since success we observed it noted in Table 3.

Table 3. Performance analysis by PCA

Sr. no	Scenario	Accuracy in percent
1	Cycle I + Inside Lab	100
2	Cycle I + Outside	70
3	Cycle III + Inside Lab	90
4	Cycle IV+ Outside	55
	Average Results	79

Hence training purpose 400 images are used and for testing 40 images are used. On the basis small experimentation procedure, we have selected few threshold values like 0.5, 0.7, 1.0 and 1.5. These threshold values are giving comparatively positive observation results. Following Table 4 gives false acceptance and rejection values on data set images.

Table 4. FAR and FRR for Sample images

Sr. no	Threshold	A	R	Percent FAR = (A/T)*100	Percent FRR =(R/T)*100
1	0.5	40	45	10	10
2	0.7	83	58	21	15
3	1.0	120	20	30	05
4	1.5	42	43	10	11

Where A: Number of Forged Sample Accepted, T: Total Number of Verification Attempt Made is 400, R: Number of Genuine Sample Rejected. And following Fig. 5 gives the graph for observed threshold values.

We have observed that on threshold value of .5 and 1.5 our program gives good results up 70 to 90 (Fig. 6).

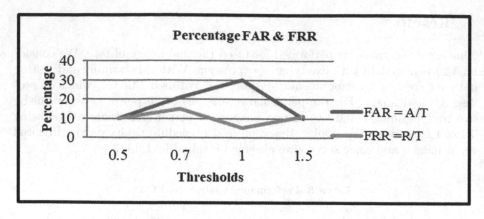

Fig. 6. SGraph showing FAR and FRR in percentage

5 Conclusion

This paper gives the discussions on Thermal Fused Face recognition using Principal Component Analysis (PCA). In this a tiny set of features thermal pictures of face data is describing the variation between input images i.e. test and training data. This gives good accuracy i.e. 79%. In future, we are going to collect more images by considering variation in angle, distance and outdoor shoot.

References

1. Jain, A.K., Ross, A., Prabhakar, S.: An introduction to biometric recognition. IEEE Trans. Circuits Syst. Video Technol. **14**, 4–20 (2004)
2. Sheena, S., Mathu, S.: A study of multimodal biometric system. IJRET **3**, 93–97 (2014)
3. Jain, A.K., Ross, A.: Multibiometric systems. IEEE Trans. **47**, 34–40 (2004)
4. Deriche, M.: Trends and challenges in mono and multi biometrics. In: IEEE Image Processing Theory, Tools and Applications, pp. 1–9 (2008)
5. Ives, R.W.: A Multidisciplinary approach to biometrics. IEEE Trans. Educ. **48**, 462–472 (2005)
6. Prabhakar, S., Kittler, J., Maltoni, D., O'Gorman, L., Tan, T.: Introduction to the special issue on biometrics: progress and directions. IEEE Trans. Pattern Anal. Mach. Intell. **29**, 513–516 (2007)
7. Karande, K.J., Talbar, S.N., Simplified and modified approach for face recognition using PCA. IET-UK ICTES 2007, Dr. M.G.R. University, Chennai, Tamil Nadu, India, pp. 523–526, 20–22 December 2007
8. Zhao, H., Yuen, P.C., Kwok, J.T.: A novel incremental principal component analysis and its application for face recognition. IEEE Trans. Syst. Man Cybern.-Part B: Cybern. **36**(4), 873–887 (2006)
9. Hermosilla, G., Ruiz-del-Solar, J., Verschae, R., Correa, M.: A comparative study of thermal face recognition methods in unconstrained environments. Pattern Recogn. **45**, 2445–2459 (2012). Elsevier

10. Sanjith Kumar, M.G., Saravanan, D.: A novel approach to face recognition based on thermal imaging. IJRET **03**(03), 141–145 (2014)
11. Turk, M.A., Pentland, A.P.: Face recognition using eigen faces, pp. 586–591. IEEE (1991)
12. Bhowmik, M.K., Saha, K., Majumder, S., Majumder, G., Saha, A., Sarma, A.N.: Thermal infrared face recognition -a biometric identification technique for robust security system, pp. 113–139. Intech (2011)
13. Wani, K.: Principal componant analysis to recognize face, iris and fingerprint image. IJSR **4**, 2343–2348
14. https://en.wikipedia.org/wiki/File:Airport_Thermographic_Camera.jpg
15. https://www.thermowork.com/emissivity_table
16. https://www.flir.com/product/c2
17. https://www.flir.com/globalassets/imported-assets/document/flir-ets320-user-manual.pdf, pp. 69–70

Optimal Search Space Strategy for Infrared Facial Image Recognition Using Capsule Networks

A. Vinay, Abhijay Gupta$^{(\boxtimes)}$, Aprameya Bharadwaj, Arvind Srinivasan,
K. N. Balasubramanya Murthy, and S. Natarajan

PES University, Bangalore, India
{a.vinay,knbmurthy,cprmi}@pes.edu, abhijayvgupta@gmail.com,
aprameya.bharadwaj@gmail.com, arvind.srini.8@gmail.com

Abstract. In this paper, we propose a highly accurate method for person identification in surveillance using infrared cameras. Our model performs well when faced with the challenges of variation in view, pose, expression, scale and lighting. It outperforms the Convolutional Neural Network in scenarios where there is a continuous change in the position and translation of the targeted individual. Our error rates were 1.5 times lower than the error rates of CNNs when tested on some standard infrared and thermal datasets. We have used Local Quantized Patterns to partition people based on their genders. The people in each gender group are identified by dynamic routing between capsules. Our contribution in this paper is a new approach to filter people based on their gender and classify them using the Capsule Network. The method was tested on two infrared datasets and four visible-light-based datasets and the average error rate converged between 1%–3%.

Keywords: Face recognition · Infrared images ·
Search space optimization · Local quantized pattern ·
Gender classification · Capsule networks · Person identification

1 Introduction

In the 21st century, face recognition has become a prominent area of research for the computer vision community. Its rapid progress is illustrated by its use in many real world applications today. Authentication systems like Apple's 'FaceID', Microsoft's 'Hello', OnePlus's 'Face Unlock' and Home Security systems like Pansonic's WV-ASF90 use Face Recognition technology to identify persons. Retail analytics is another field that relies on Face Recognition to solve problems like access permission and security. Applications that implement face recognition face many challenges like variation in illumination, scale, pose and background. For example, it is difficult for the same application to work in both indoor and outdoor situations where the lighting in the images will be drastically different. An application that has been designed for a particular purpose

© Springer Nature Singapore Pte Ltd. 2019
K. C. Santosh and R. S. Hegadi (Eds.): RTIP2R 2018, CCIS 1036, pp. 454–465, 2019.
https://doi.org/10.1007/978-981-13-9184-2_40

like home security might not continue to work if the position of the camera is changed. In any application, it is important to extract the important features from images. When the background in images differs, the system might extract these features and not capture important facial features. Other distractions like facial hair and spectacles might also mislead the feature extractor.

1.1 Contribution

Person Identification for surveillance is a novel application of the Capsule Network architecture. Security and surveillance cameras are very important in public areas and homes. Hence identifying the suspect is very important task that has to be done after the pictures are taken. Since these cameras operate in low-light conditions, the task of person identification becomes challenging. However, the method proposed in this paper delivers promising results for this problem. Local Quantized Patterns has been used for Face Recognition before, it has not been used for Gender Classification. Local Quantized Patterns are used here to reduce the subspace of our classification problem, and identify males and females separately. Once the males and females are segregated, these are passed separately into the Capsule Network for classification.

Some applications in surveillance require products to work in complete darkness. In these scenarios, images from normal cameras will not provide enough facial features. Infrared Cameras are used in these situations. They are sensitive to electromagnetic waves in the infrared region. This allows them to create a reasonable representation of the region of interest even in low light areas.

In our application, when we receive an input image, we first classify the person based on their gender using Local Quantized Patterns [1]. This is done to reduce the search space for our person identification problem. Then, we use a Capsule Network [2] that has been trained on infrared images and normal images of some standard datasets to recognize the person in the image. Traditional methods like the Convolutional Neural Network [3] have some drawbacks. First, they do not work well when there is variation in the pose and orientation of the object to be detected. Second, they look for certain features in images, like a face's nose and eyes, but do not consider the positions of these features in the image while classifying them. The dynamic routing algorithm used by Capsule Network promises to overcome these challenges.

2 Related Work

Given the huge amounts of visual data being generated from surveillance cameras, the task of person identification has become computationally expensive and exorbitantly time consuming. To solve this problem, sub sampling on gender must be accomplished to reduce the search space for person identification.

Local quantized pattern [1] computes a lookup table to apply visual word aggregation to discrete local patterns. Local quantized pattern finds its used in depth estimation [4], image based geographic localization [5] and face recognition

[6]. [4] tries to estimate single view depth and binocular stereo matching. For a given image, visual word representations are computed using LQP on which binocular stereo matching algorithm and single view depth algorithm are applied. In [5], a visual based look up table has been made for fast database search of mountain terrain. This automatic geo-tagging is important for historical and forensic sciences, intelligence applications and documentation purposes. Mountains in a give image are represented by LQP which contains visual words of the contours of the terrain. This vector is then used for recognition and parameter selection. [6] proposed a novel facial representation using local quantized pattern. For a given image, features are extracted using lqp with different filters. These features are then subjected to PCA projection and data sphering. Finally a cosine similarity score is computed to classify the person. This method outperforms LBP [7] and LTP [8] on LFW and FERET dataset.

Dynamic routing between capsules [2] proposed an iterative routing algorithm such that predictions from a lower level capsule are sent directly to a higher level capsule. In literature, this mechanism which consist of multi-layer capsules has proven to beat the state of the art performance on MNIST dataset. [2] find its use in monitoring of rice from UAV images [9], fault detection in a power system [10], object segmentation [11], brain tumor classification [12] and embedding model for search personalization [13]. [9] uses histogram equalization for converting rice images to gray scale, which are then subjected to CapsNet for reverse analysis. The paper also shows that CapsNet outperforms SVM and CNN for rice classification. In [10], capsule networks are used to detects faults in power system which can occur due to leakage, physical damage and high temperature. Visible, infrared and ultraviolet images are used to train the classifier. Dynamic routing has found it used in object segmentation as seen in [11], which uses a novel architecture of convolutional-deconvolutional capsule networks. The paper demonstrates the use of the above architecture for lung cancer segmentation from CT scans outperforming the state of the art U-Net by reducing the number of parameters by 95.4% and on an average providing 0.03% better accuracy. Another use-case of capsule networks in health care can be seen in [12] which classifies brain tumor into Meningioma, Pitiuitary and Glioma. CNN and its variants have been extensively used to solve this medical task in the literature but capsule network outperforms them as discussed in the above paper. A unique application of capsule networks can be seen in [13], where an embedding model is designed using capsule networks for search personalization. The paper proposes a 3 way relationship described by query, user and documents to train the Capsule Network, this returns a personalized result based on user's interest and preferences.

3   Method Proposed

The proposed method comprising of LQP which is used for gender-based filtering and capsule networks for face identification is described by Fig. 1

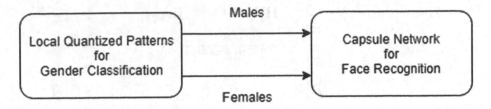

Fig. 1. Proposed pipeline

3.1   Local Quantized Patterns

The idea behind LQP is applying visual quantization for to discrete features using a pre-complied lookup table to enhance speed while the final coding occurs. LQP allows patterns much larger than existing methods such as LBP, LTP. These range from 8–24 pixels for binary coding and 5–16 pixels for ternary coding. Using these larger neighborhood of pixels, LQP has a very high discriminative power. Compared to other techniques, this also allows different kinds of neighborhood shapes for comparison and feature extraction. The LQP look-up table handles more patterns comparison metrics other than only L_2 comparison like, Earth Mover distances and coding under conditions of symmetry, rotation and reflection with no additional cost of run time. The lookup table also enhances learning of features.

During learning the training set is scanned once and the number of occurrences of each input feature is recorded in a has table. Then these values and the number of occurrences are passed to an algorithm that learns the codebook. This makes the task of training fast because most of the possible input features don't actually occur and for the ones which do occur, all occurrences are processed and calculated in a single operation.

As previously stated, LQP is very flexible and can use many different kinds of arrangements to extract features in a neighborhood of pixels. For the task of gender classification, LQP was tested in the following arrangements for feature extraction: vertical (V), horizontal (H), diagonal (D) and antidiagonal (A) strips of pixels and also the combinations of these like horizontal-vertical (HV), horizontal-vertical-diagonal-antidiagonal (HVDA) and diagonal-antidiagonal(DA). By default, the other pixels of the arrangement is compared with the central pixel.

The notation of the arrangement is a follows: Suppose the arrangement is HV_7^3, the HV represents the shape of the neighborhood, i.e horizontal-vertical the subscript indicates the diameter of the neighborhood which in this case is 7 pixels and the superscript indicates the quantization level. The codes from various feature sets or from various types of LQP neighborhoods can be combined at the cell level by concatenation of histograms. This allows more pixels to be incorporated in the local pattern neighborhood. The arrangement used in this pipeline was the $HVDA_7^3$ arrangement. Few arrangements of representing the neighborhood shape is represented in Fig. 2.

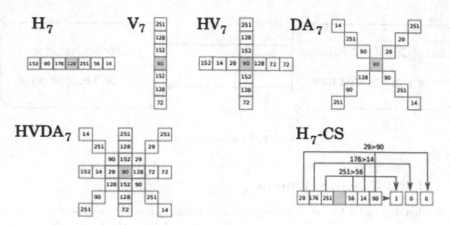

Fig. 2. Few arrangements of Local Quantized Patterns

3.2   Capsule Network

Capsules are a set of neurons such that the activity vectors of these represent different parameters of pose and whose length give the probability of existence of a specific entity. The Convolutional Neural Networks' shortcomings can be mainly attributed to the pooling layer. In this architecture however, there layers are replaced with a criteria known as "routing by agreement". Based on this, the outputs of the previous layer are sent to the parent capsules in the following layer but with different coupling coefficients. Each capsule predicts the output of the parent capsule. If this output is right, then the coupling coefficient between these two is increased. If u_i is the output from capsule i then its prediction for parent capsule j is computed as:

$$\hat{u}_{j|i} = W_{ij}u_i \tag{1}$$

Here $\hat{u}_{j|i}$ is the prediction vector. W_{ij} is the weighted matrix that needs to be learned through the backward pass. Based on the prediction, the coupling coefficient between the parent capsule and the layer below is calculated by the softmax function given by:

$$c_{ij} = \frac{exp(b_{ij})}{\Sigma_k exp(b_{ik})} \tag{2}$$

where b_{ij} indicates the log probability of whether capsules i and j should be coupled. This is initially set to 0 before the routing process starts. The input vector to the parent capsule j is calculated as:

$$s_j = \sum_i c_{ij}\hat{u}_{j|i} \tag{3}$$

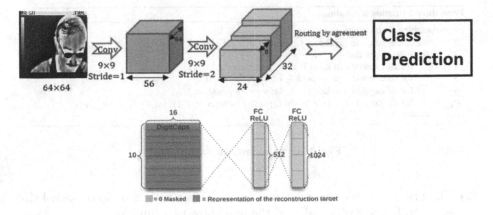

Fig. 3. Proposed capsule network architecture

Finally the "non linear Squashing" function is applied to prevent the lengths of the vectors from exceeding one and forms the final output.

$$v_j = \frac{||s_j||^2}{1 + ||s_j||^2} \frac{s_j}{||s_j||} \tag{4}$$

Here, s_j is the input vector to Capsule j and v_j is the output. The log probabilities are updated in the process of routing based on agreement between $\hat{u}_j$ given i and v_j and if this agreement happens, the inner product will be large. Hence this agreement $a + ij$ is given by:

$$a_{ij} = v_j.\hat{u}_{j|i} \tag{5}$$

Each capsule k in the previous layer is associated with a loss function l_k. This function puts a high loss value with long output parameters when the entity doesn't exist. This function is given by:

$$l_k = T_k max(0, m^+ - ||v_k||)^2 + \lambda(1 - T_k)max(0, ||v_k|| - m^-)^2 \tag{6}$$

Here, T_k is 1 when class k is present and 0 if its not present. m+, m−, and lambda are hyper parameters and are specified before the learning starts.

4   Designed Network

After experimenting with different architectures, the following was used on the datasets for the task of face recognition. Proposed architecture for capsule networks is depicted in Fig. 3. The proposed model is as follows (Fig. 4):

(1) The input images of the Infrared face Dataset and other face datasets were downsized to 64 × 64 from their original dimensions to reduce training time and number of features.

Procedure 1 Routing algorithm.

1: **procedure** ROUTING($\hat{u}_{j|i}, r, l$)
2:     for all capsule i in layer l and capsule j in layer $(l + 1)$: $b_{ij} \leftarrow 0$.
3:     **for** r iterations **do**
4:         for all capsule i in layer l: $c_i \leftarrow \text{softmax}(b_i)$
5:         for all capsule j in layer $(l + 1)$: $s_j \leftarrow \sum_i c_{ij}\hat{u}_{j|i}$
6:         for all capsule j in layer $(l + 1)$: $v_j \leftarrow \text{squash}(s_j)$
7:         for all capsule i in layer l and capsule j in layer $(l + 1)$: $b_{ij} \leftarrow b_{ij} + \hat{u}_{j|i}.v_j$
      return v_j

Fig. 4. Dynamic routing algorithm

(2) The first layer was a conventional conv2D layer with 256 filters, kernel size of 9, stride value of 1 and used the relu activation function.

(3) The second layer was the Primary Capsule layer. This layer was a conv2D layer which implemented the "squash" function and reshaped the input to none, num capsule, dim capsule where dim capsule $= 8$, n channels $= 32$, kernel size $= 9$, strides $= 2$.

(4) The third layer is the Capsule Layer. This is the layer in which the dynamic routing algorithm occurs. The number of capsules is determined by the number of classes and the dimension of each capsule is 16.

(5) The decoder comprises of two fully connected layers having 512 and 1024 neurons respectively.

Hence after gender classification is done by extracting the quantized features $HVDA_7^3$ and learning of the codebook generated, This dataset is now fed into the capsule network for the task of face recognition.

5   Datasets

The following datasets were used to test our method:

(1) IRIS Thermal Face Database - This infrared dataset consists of two parts. The first part is the Infrared images with complete motion of head and turning and the second part are the same images but not infrared. There is very large head scale variation and variation of illumination, expression and also the effects of shadow. The original dimensions of the images in this database are 320×240. This dataset contains pictures of both females and males (Fig. 5).

(2) Near Infrared Database - This infrared dataset has no female pictures in it. It consists of 102 males with 6 images each. Hence the total number of images in this dataset are 612. The number of pixels in each image is 768576. It poses the problem of large variation of head scale, distance between subject and camera and also variation of expression. Even though there are no female images in this dataset, it is used due to the fact that it is an open source IR dataset (Fig. 6).

Fig. 5. IRIS Thermal Face Database

Fig. 6. Near Infrared Database

(3) Georgia Tech Face Database - This dataset comprises of 50 different people with 15 pictures each. All the pictures in the database are 15 color JPEG images with cluttered background. The dimension of each image is 640 × 480. The pictures show frontal or tilted faces with variation of facial expressions, illumination and scale. This dataset contains pictures of both females and males (Fig. 7).

Fig. 7. Georgia Tech Face Database

(4) FACES95 Dataset - This dataset contains 20 images for 72 individuals each hence the total number of images being 1440. The dimensions of each image is 180 × 200. There is no background variation as all photos are taken against a red background. However, there is a large head-scale variation and illumination variance. There is also slight expression change between the images of the dataset. This dataset contains pictures of both females and males (Fig. 8).

(5) FACES96 Dataset - This dataset contains 20 images for 147 individuals each hence the total number of images being 2940. The dimensions of each image is 196 × 196. There is a large variation in illumination and head scale between the images. Adding to these challenges, the images of this dataset

Fig. 8. Faces95 database

Fig. 9. Faces96 database

Fig. 10. Grimace database

are also taken against complex backgrounds. This dataset contains pictures of both females and males (Figs. 9 and 10).

(6) Grimace Dataset - This dataset contains of 18 different people with 20 pictures each. The total number of images is 360 and the size of each image is 180×200. The pictures are taken against a plain background. There is little head scale variation and change in lighting between images. There is a large variation of expression and motion of face in the images. This dataset contains pictures of both females and males.

Table 1. Gender classification using Local Quantized Pattern

Dataset	Male	Female
IRIS	0.98	0.99
NIR	1.00	None
GaTech	0.97	0.97
Face95	0.98	0.99
Face96	0.96	0.98
Grimace	0.99	0.99

Table 2. Facial classification - male and female

Dataset	Male	Female	CNN
IRIS-thermal	0.9779	0.9845	0.9664
IRIS-nonthermal	0.991	0.9954	0.9743
NIR	0.9688	None	0.9531
GaTech	0.9867	0.9934	0.9764
Face95	0.9889	0.9912	0.9634
Face96	0.971	0.9823	0.9623
Grimace	0.9921	0.9945	0.9889

Table 3. Classification based on different Capsule Network Architectures on IRIS dataset

Architecture	Accuracy
Original architecture	0.9779
Two convolutional layers with 256 and 512 feature maps and 16 primary capsules	0.9710
One convolutional layers with 256 feature maps and 32 primary capsules	0.9823
Two convolutional layers with 256 and 512 feature maps and 32 primary capsules	0.9742

6 Results and Conclusion

The proposed model was tested on two infrared datasets and four non-infrared datasets. The first part of our model categories the data into male and female. The percentage of correctly classified genders can be seen in Table 1. Since one of the genders (female) is very scarce in all the datasets we get a better gender classification rate.

The results obtained using our model on the six datasets are tabulated in Table 2. Even though the IRIS dataset contains thermal image with significant head rotation, scale variation, illumination variation, variation in expression and shadow effects, our proposed model is able to correctly identify the person with an accuracy of 97.79%. The IRIS dataset containing non-thermal images when tested gives an accuracy of 99.1%. The results obtained on near-infrared images of the NIR dataset are also impressive, the dataset shows wide variations in pose and expression. The model was also tested on four non-infrared benchmark datasets namely FACES95, FACES96, GRIMACE and GaTech Face database. They perform extremely well with our application of capsule networks, achieving an average accuracy of 98% with no background. The four datasets in combination provides significant head movement, series of various expression, changing background lights, variance in scale and pose.

Table 3 shows the accuracy of our model on IRIS thermal dataset for different capsule network architecture. The original architecture comprised of one convolutional neural network have 256 features maps, 16 primary capsules and a decoder model with two hidden layer. Changing the number of convolutional layers, number of features maps and number of primary capsules displayed interesting results. Changing the number of convolutional layers and feature map does not effect the accuracy of the system by huge margin because one convolutional layer produces enough feature maps to incorporate all the facial regions. On increasing the number of primary capsule the accuracy of the model increases because more features can be accommodated in the capsule and more relations between them can be defined.

A direct comparison with a CNN model containing two convolutional layers, eight hidden layers and 128 neurons on all the datasets can be seen in Table 2. This shows that our proposed model outperforms the vanilla-CNN in conditions having significant variations in pose and viewpoint changes. This is due to the dynamic routing algorithm as proposed in [2], where capsules are created for each individual features and each capsule is related to another capsule by a specified rule which are calculated during the training phase of the network. A direct comparison of our proposed method with the state of the art models cannot be made because implementation and testing of the modules are performed on different hardware which might result in different efficiency results and Datasets used and pre-processing steps followed in the state of the art models and our proposed methods are different which leads to variation in results. [14] developed a method for deep hypersphere embedding for face recognition to achieve a remarkable accuracy of 95% on the YTF dataset. Chain code based local descriptors are proposed in [15] for the task of face recognition. [15] was tested on CAS-PEAL, ColorFERET and FG-NET resulting in an average accuracy of 98%. A deep learning approach for face recognition was developed in [16] where a trunk branch ensemble convolutional neural network was designed to solve the problem of pose variation and occlusions resulting in an average accuracy of 95% on PaSC, COX Face and YouTube faces datasets. A different approach to face recognition was proposed in [17] using multi-resolution wavelet combining discrete cosine transform and Walsh transform which resulted in an accuracy of 99.24% on FACES94 dataset. The results obtained above show that our method is PESI invariant i.e Pose, Expression, Scale, and Illumination invariant.

References

1. Hussain, S., Triggs, B.: Visual recognition using local quantized patterns. In: Fitzgibbon, A., Lazebnik, S., Perona, P., Sato, Y., Schmid, C. (eds.) ECCV 2012. LNCS, pp. 716–729. Springer, Heidelberg (2012). https://doi.org/10.1007/978-3-642-33709-3_51
2. Sabour, S., Frosst, N., Hinton, N.: Dynamic routing between capsules. In: Advances in Neural Information Processing Systems, pp. 3859–3869 (2017)
3. Krizhevsky, A., Sutskever, I., Hinton, G.E.: ImageNet classification with deep convolutional neural networks. In: Advances in Neural Information Processing Systems, pp. 1097–1105 (2012)
4. Hane, C., Ladicky, L., Pollefeys, M.: Direction matters: depth estimation with a surface normal classifier. In: Proceedings of the IEEE Conference on Computer Vision and Pattern Recognition, pp. 381–389 (2015)
5. Saurer, O., Baatz, G., Köser, K., Pollefeys, M.: Image based geo-localization in the alps. Int. J. Comput. Vis. **116**(3), 213–225 (2016)
6. Hussain, S.U., Napoléon, T., Jurie, F.: Face recognition using local quantized patterns. In: British Machive Vision Conference, p. 11 (2012)
7. Ahonen, T., Hadid, A., Pietikainen, M.: Face description with local binary patterns: application to face recognition. IEEE Trans. Pattern Anal. Mach. Intell. **28**(12), 2037–2041 (2006)
8. Tan, X., Triggs, B.: Enhanced local texture feature sets for face recognition under difficult lighting conditions. IEEE Trans. Image Process. **19**(6), 1635–1650 (2010)
9. Li, Y., et al.: The recognition of rice images by UAV based on capsule network. Cluster Comput. 1–10 (2018)
10. Li, Y., et al.: Image fusion of fault detection in power system based on deep learning. Cluster Comput. 1–9
11. LaLonde, R., Bagci, U.: Capsules for object segmentation. arXiv preprint arXiv:1804.04241 (2018)
12. Afshar, P., Mohammadi, A., Plataniotis, K.N.: Brain tumor type classification via capsule networks. arXiv preprint arXiv:1802.10200 (2018)
13. Nguyen, D.Q., Vu, T., Nguyen, T.D., Phung, D.: A capsule network-based embedding model for search personalization. arXiv preprint arXiv:1804.04266 (2018)
14. Liu, W., Wen, Y., Yu, Z., Li, M., Raj, B., Song, L.: Sphereface: deep hypersphere embedding for face recognition. In: The IEEE Conference on Computer Vision and Pattern Recognition (CVPR), vol. 1, p. 1 (2017)
15. Karczmarek, P., Kiersztyn, A., Pedrycz, W., Dolecki, M.: An application of chain code-based local descriptor and its extension to face recognition. Pattern Recogn. **65**, 26–34 (2017)
16. Kaur, R., Sharma, D., Verma, A.: An advance 2D face recognition by feature extraction (ICA) and optimize multilayer architecture. In: 2017 4th International Conference on Signal Processing, Computing and Control (ISPCC), pp. 122–129. IEEE (2017)
17. Choudhary, A., Vig, R.: Face recognition using multiresolution wavelet combining discrete cosine transform and Walsh transform. In: Proceedings of the 2017 International Conference on Biometrics Engineering and Application, pp. 33–38. ACM (2017)

Incept-N: A Convolutional Neural Network Based Classification Approach for Predicting Nationality from Facial Features

Masum Shah Junayed$^{(\boxtimes)}$, Afsana Ahsan Jeny, Nafis Neehal, Eshtiak Ahmed, and Syed Akhter Hossain

Daffodil International University, Dhaka 1207, Bangladesh
{junayed15-5008,ahsan15-5278,nafis.cse,eshtiak.cse}@diu.edu.bd,
aktarhossain@daffodilvarsity.edu.bd

Abstract. Nationality of a human being is a well-known identifying characteristic used for every major authentication purpose in every country. Albeit advances in application of Artificial Intelligence and Computer Vision in different aspects, its' contribution to this specific security procedure is yet to be cultivated. With a goal to successfully applying computer vision techniques to predict a human's nationality based on his facial features, we have proposed this novel method and have achieved an average of 93.6% accuracy with very low mis-classification rate.

Keywords: Nationality · Artificial intelligence · Computer vision

1 Introduction

Facial recognition is a complex process in which information and experience are used to set the average face for measuring other faces. The ability to detect face is very important in many aspects of life. It does not help us identify the people near us, but allow us to recognize those people so that we do not know that we can become more aware of possible dangers.

The human face is a very rich inspiration that gives amazing information for social communication adapted to people. For the last several decades, many attempts have been made in the biological, psychological, and cognitive science fields to understand, remember, and remember how humans are brainwashed [11].

The human face is a complex visual pattern that identifies the primary, specific, along with the general in-structure and the unrealistic. With this unique information, we can say that some aspects of the face are not employed in the person's face, but are shared by facial symptoms. These aspects can be used to make both known and unidentified people in general paranormal groups such as district or nationality.

© Springer Nature Singapore Pte Ltd. 2019
K. C. Santosh and R. S. Hegadi (Eds.): RTIP2R 2018, CCIS 1036, pp. 466–475, 2019.
https://doi.org/10.1007/978-981-13-9184-2_41

With the development of computer technology and digital image processing technology, people start exploring the automatic national identification system through computers, this process is mainly compared to distance, angle, and other characteristics between people, and then people determine the picture and then the people [5]. The oldest method of identifying nationality is to look at living habits, metaphor formations and other characteristics of the people. This classification method is fully artificial, massive, and professionally staffed by professionals who have lots of resources for professional knowledge and experience.

In this paper, we use the technique of learning to review the Inception-V3 [8] model of tensor flow [1] in the datasets (China, Germany, India, Jamaica and Zimbabwe) in 5 countries. We meet an efficient national identity model using a short training time and achieve a high accuracy. The remaining paper is sorted in the following manner: Details of Convolutional Neural Network (CNN) and Inception-v3 [8] model are discussed in Sect. 2. The comparison with other papers is discussed in Sect. 3. Data collection and training are discussed in Sect. 4. Performance analysis is done in Sect. 5. Finally conclusion with some future work scopes is described in Sects. 6 and 7.

2 Background Study

Our work is based on the Inception-v3 [8] model of TensorFlow [1] platform and also used CNN [2].

TensorFlow [1] Like Google's artificial intelligence to learn artificial intelligence, it has received many interesting and presentable introductions for learning across the globe. TensorFlow [1] So far all the deep learning and machine learning programs got the first place. Tensorflow [1] has the advantage of superior facilities and higher facilities, and with the help of TensorFlow [1] researchers, TensorFlow [1] floatation capacity has been improved. To facilitate the use of researchers from various sectors today, TensorFlow [1] has opened many trained models on the official website.

Inception- V3 [8] is one of the TensorFlow [1] training models. This is a reconsideration of the initial framework for computer vision after Inception-V1 [9], Inception-V2 [9] in 2015. Intention-V3 [8] model is trained in ImageNet datasets, contains information that can detect 1000 classes in ImageNet. Inception-V3 [8] consists of two parts: a fully-connected and softmax layer [10] with classification properties and feature extraction part with a convolutional neural network (CNN) [2].

Convolutional Neural Networks (CNN) [2] is a class of neural networks that improve effectively in areas such as image classification and recognition. ConvNets have been effective to detect objects, face, and traffic. Generally, three main types of layers are used to create ConvNet architectures:

- Convolutional Layer,
- Pooling Layer and
- Fully-Connected Layer (Fig. 1).

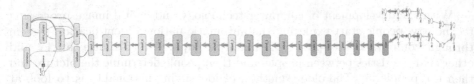

Fig. 1. Main graph of Inception v3 model.

The first layers of a CNN [2] strain (big) features can be accepted and illustrated relatively simply. As a result of the convolution of the neuronal network, the image is divided into emotion, made in local receptor file and compressed synthesis of the size m2 × m3 features map is briefly compressed. In this way, where this map stores the feature in the image and stores how much it is compatible. Therefore, in each file, it is applied to the location of the volume of trained spatial [2].

At each layer, there is a bank of m1 filters. The output characteristics are similar to the depth of the volume on the number of filters applied at one stage. Identifies a particular feature in each location of the input in each file. The output $Y_i^{(l)}$ of layer 1 consists of $m_1^{(l)}$ feature maps of size $m_2^{(l)} * m_2^{(l)}$. The ith feature map, denoted $Y_i^{(l)}$, is computed as

$$Y_i^{(l)} = B_i^{(l)} + \sum_{(j=1)}^{m_1^{(l-1)}} K_{i,j}^{(l)} * Y_i^{(l-1)}$$

where $B_i^{(l)}$ is a bias matrix and $K_i^{(l)}$, j is the filter of size $2h_1^{(l)} + 1 * 2h_2^{(l)} + 1$ connecting the jth feature map in layer $l-1$ with i^{th} feature map in layer [2].

Later the layers detect increasingly (smaller) features that are more abstract (and are usually present in many of the larger features detected by earlier layers). The pooling layer 1 has two hyper parameters, the spatial extent of the filter $F^{(l)}$ and the stride $S^{(l)}$. It takes an input volume of size $m_1^{(l-1)} * m_2^{(l-1)} * m_3^{(l-1)} *$ and provides an output volume of size $m_1^{(l)} * m_2^{(l)} * m_3^{(l)}$ where;

$$m_1^{(l)} = m_1^{(l-1)}$$
$$m_2^{(l)} = (m_2^{(l-1)} - F^{(l)})/S^{(l)}$$
$$m_3^{(l)} = (m_3^{(l-1)} - F^{(l)})/S^{(l)} + 1$$

CNN's last level [2] input data allows a highly-specific classification by combining all the specific features detected by the previous level. It has an image of a specific degree translation, rotation and distortion image. It has made great progress in the field of painting [2].

If $l-1$ is a fully connected layer;
$y_i^{(l)} = f(z_i^{(l)}) \, with \, z_i^{(l)} = \sum_{(j=1)}^{m_1^{(l-1)}} w_{i,j}^{(l)} * y_i^{(l-1)}$
Otherwise (Fig. 2);

$$y_i^{(l)} = f(z_i^{(l)}) \; with \; z_i^{(l)} = \sum_{(j=1)}^{m_1^{(l-1)}} \sum_{(r=1)}^{m_2^{(l-1)}} \sum_{(s=1)}^{m_3^{(l-1)}} W_{i,j,r,s}^{(l)}(Y_i^{(l-1)}) r_l^s$$

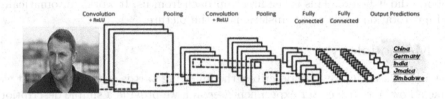

Fig. 2. Structure of Convolutional Neural Network (CNN).

TensorFlow [1] is available in the detailed tutorials. Repeat the final level of inspection for new sections using transfer learning. Learning movements is a new ma-chine learning process that can use existing knowledge from an environment and can answer another new problem, but it is related to the old problem. Measuring with the traditional Neural Network, it requires only fewer data for model training and high accuracy with minimal training time [5,7].

3   Literature Review

Inception v3 [8] model used in many experiments. Among them:

In 2017, Xiaoling Xia and Cui Xu used the transfer learning technique to retrain the Inception-v3 [8] model of TensorFlow [1] on the datasets of flower [11, 13] of Oxford-I7 and Oxford-102 for Flower Classification. The accuracy of the classification of the model was 95% on the dataset of Oxford-I7 flower and 94% on the dataset of Oxford-I7 flower [5].

In 2017, Alwyn Mathew, Jimson Mathewa, Mahesh Govindb, Asif Mooppanb from bVuelogix Technologies Pvt Ltd. used Google's TensorFlow [1] deep learning a framework to train, validate and test the network for Intrusion Detection and the accuracy was 95.3%. But the proposed network is found to be harder to train due to vanishing gradient [3] and degradation problems [3].

In 2017, Brady Kieffer, Morteza Babaie Shivam Kalra, and H. R. Tizhoosh used CNN and Inception v3 [8] model for Histopathology Image Classification [6]. All experiments are done on Kimia Path24 dataset and the accuracy was 56.98% [6]. In 2017, Xiao-Ling Xia, Cui Xu, Bing Nan worked for Facial Expression Recognition based on the Inception-v3 [8] model of TensorFlow [1] platform. They used CK+ dataset [15] and selected 1004 images of facial expression. Their accuracy was 97% but it wasn't based on dynamic sequences [7].

In 2016, Bat-Erdene and Ganbat worked on Effective Computer Model for Recognizing Nationality from Frontal Image [4]. They used SVM, AAM, ASM and the accuracy was 86.4%. Their experiment was worked manually and images must be the frontal face image that has smooth lighting and does not have any rotation angle.

Our experiment is based on the Inception-v3 [8] model of TensorFlow [1] platform for Nationality Recognition based on facial features with Deep Learning. Nobody did it before. This is the first approach from us. It works automatically and has rotation angle and translations in the picture.

4   Methodology

Now in this section, we have described the following part is as follows: at first we make a flowchart [12] of our experiment; second, we provide a simple description on the dataset; third, we give about the data pre-processing; then, we describe the model installation; eventually, we discuss about the train model.

A flowchart [12] is a diagram that represents a workflow or process. Flowchart [12] connects boxes with the steps and arrows of the box and shows their order. Flowcharts [12] is used in a process analysis, design or management. The following diagrammatic representations illustrate the solution model of our system (Fig. 3).

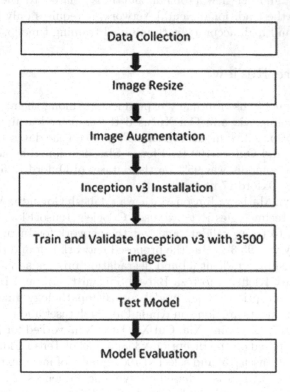

Fig. 3. Flowchart of the system model.

4.1 Dataset

There are many countries in this world with many people. There is a similarity in the outlook of human faces. For recognizing nationality, we have used 600 images of five countries for our experiment. They are China, Germany, India, Jamaica, and Zimbabwe.

4.2 Data Preprocessiong

Image promotion is a very important step to promote the effect of image classification. The Convention Neural Network's [2] learning process follows our activity monitoring and guidance on machine learning, so we need to label the data in the image pre-processing step. Then we have resized the data and also augmented (Rotate +30, Rotate −30, Translation, Lighting and Flip). Finally, we have found 3600 images for training (Fig. 4).

Fig. 4. The example of our dataset.

4.3   Model Installation

This test is based on the Tensorflow [1] platform's Inception-V3 [3] model. Processor 2 GHz Intel i3, Memory 4 GB 1600 MHz DDR 3, System Type: 64-bit OS, X-64 based processor.

After all, we've downloaded Tensorflow [1]. Then we've downloaded the Inception V3 [8] model. We also used the transfer learning process, which had previously set up the level parameter and removed the final level of the inception V3 [8] model, then re-marked a final layer.

4.4   Train Model

At this step, we have to keep the parameters of the previous level, then the last layer will be removed and we need to input dataset to restore the next level.

By the backing promotion algorithm, the cross-entropy spend function is used to synthesize weight parameters by calculating the last level of the model and counting the output error of the texture level, and the label vector used for the given test category is used [5,7].

For final accuracy, we have also created Confusion Matrix. From Confusion Matrix, we have calculated Precision, Recall, Accuracy, and F1-Score. And finally, we have determined Macro Average Accuracy of our experiment.

Here is the Confusion Matrix of our model. From the following Confusion matrix of Table 1, we can say that our model has given a very high number of True Positive values.

Table 1. Confusion matrix

	China	Germany	India	Jamaica	Zimbabwe
China	18	1	0	1	0
Germany	0	19	1	0	0
India	1	0	16	1	2
Jamaica	0	1	0	16	3
Zimbabwe	1	1	0	3	15

5   Result Analysis

Figures 5 and 6 show the variation of accuracy and the cross-entropy which are based on our training dataset. The orange line represents the training set while the blue line represents the validation set.

Table 2 shows two statistical descriptions. For our dataset, training acupressure can reach 95% and validity accuracy can be maintained at 89%–90%.

Figure 7 shows the precision, withdrawal, accuracy and F1 scores graph of China, Gar-On, India, Jamaica, and Zimbabwe and shows macro-average accuracy, withdrawal, accuracy, and F1 scores [14,15].

accuracy_1

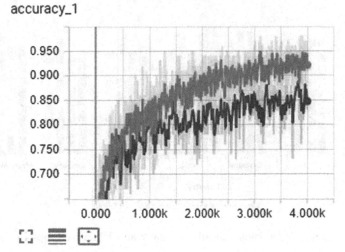

Fig. 5. The accuracy graph of Inception-V3 model.

cross_entropy_1

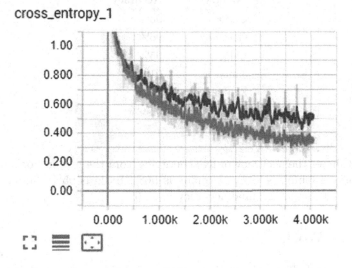

Fig. 6. The cross-entropy graph of Inception-V3 model.

Table 2. Description of the two figures.

	Index	Performance
Dataset	The accuracy of the training set	95%
	The accuracy of the validation set	89%-90%
	The cross-entropy of the training set	0.24
	The cross-entropy of the validation set	0.41

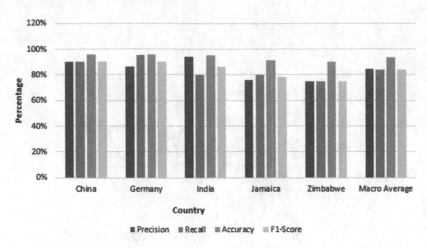

Fig. 7. Precision, Recall, Accuracy and F1-Score graph.

Table 3. The accuracy of five countries and final accuracy.

Country	Accuracy
China	96%
Germany	96%
India	95%
Jamaica	91%
Zimbabwe	90%
Macro average	93.6%

Table 3 shows the accuracy of five countries from the graph. For our dataset, the accuracy of China is 96%, Germany is 96%, India is 95%, Jamaica is 91%, Zimbabwe is 90% and the final accuracy is 93.6%.

6 Future Work

Google creates the TensorFlow [1] platform of the Inception-V3 [3] model and we used it. So our future work is to read and develop a more effective model so that we can use that model and enhance our accuracy.

7 Conclusion

Based on the Tensor Flow [1] platform Inception-V3 [3] model, we use transfer learning technology to identify the nationalities of five countries based on our datasets. And we get 93.6% accuracy of the model. We hope that our work will be useful for the future.

References

1. Abadi, M., Agarwal, A., et al.: TensorFlow: large-scale machine learning on heterogeneous distributed systems. CoRR abs/1603.04467 (2016)
2. Krizhevsky, A., Sutskever, I., Hinton, G.E.: ImageNet classification with deep convolutional neural networks, 03–06 December 2012
3. Mathew, A., Mathew, J., Govind, M., Mooppanb, A.: An improved transfer learning approach for intrusion detection. In: 7th International Conference on Advances in Computing & Communications, ICACC 2017, Cochin, India, 22–24 August 2017 (2017)
4. Bat-Erdene, B., Ganbat, T.: Effective computer model for recognizing nationality from frontal image (2016)
5. Xia, X., Xu, C.: Inception-v3 for flower classification. In: 2017 2nd International Conference on Image, Vision and Computing (2017)
6. Kieffer, B., Babaie, M., Kalra, S., Tizhoosh, H.R.: Convolutional neural networks for histopathology image classification: training vs. using pre-trained networks. arXiv:1710.05726v1 [cs.CV] 11 October 2017
7. Xia, X.-L., Xu, C., Nan, B.: Facial expression recognition based on TensorFlow platform. In: ITM Web of Conferences, vol. 12, p. 01005 (2017)
8. Szegedy, C., Vanhoucke, V., Ioffe, S., Shlens, J., Wojna, Z.: Rethinking the inception architecture for computer vision. arXiv: 1512.00567v1 [cs.CV] 2 December 2015
9. https://datascience.stackexchange.com/questions/15328/what-is-thedifference-between-inception-v2-and-inception-v3
10. https://codelabs.developers.google.com/codelabs/tensorflow-forpoets/#0
11. Lu, X., Jain, A.K.: Ethnicity identification from face images. In: Proceedings of SPIE, vol. 5404. Biometric Technology for Human Identification (2004). https://doi.org/10.1117/12.542847
12. Charntaweekhun, K., Wangsiripitak, S.: Visual programming using flowchart. In: 2006 International Symposium on Communications and Information Technologies (2006)
13. https://datascience.stackexchange.com/questions/15989/microaverage-vs-macro-average-performance-in-a-multiclass-classification
14. Salmon, B.P., Kleynhans, W., Schwegmann, C.P., Olivier, J.C.: Proper comparison among methods using a confusion matrix. In: 2015 IEEE International Geoscience and Remote Sensing Symposium (IGARSS) (2015)
15. Sawat, D.D., Hegadi, R.S.: Unconstrained face detection: a deep learning and machine learning combined approach. CSI Trans. ICT **5**(2), 195–199 (2017)

Biometric Recognition System Based on Minutiae on the Dorsal Hand Vein

B. M. Sontakke[1]($\boxtimes$), V. T. Humbe[2]($\boxtimes$), and P. L. Yannawar[3]($\boxtimes$)

[1] Computer Science, COCSIT, Latur, MH, India
bmsontakke@gmail.com
[2] School of Technology, SRTMU, Sub Campus, Latur, MH, India
vikashumbe@gmail.com
[3] Department of Computer Science and IT, Dr. BAMU, Aurangabad, MH, India
pravinyannawar@gmail.com

Abstract. A innovative algorithm has been devloped for dorsal hand vein that forms template and matching in automatic dorsal hand vein recognition. Now a days, dorsal hand vein having uniqueness and liveness traits because of these properties biometrics is a promising component of biometric study. Our goal is to obtain less equal error rate, false acceptance and false rejection rate for this we require such a method that matches and generates template. The identical terminology used here is based on minutiae features and distance between them. We used hand vein dataset from Bosphorus university for the evaluation purpose. Our system obtained 1.26% FAR and 1.60% FRR. accuracy after result is 97.00%.

Keywords: Dorsal · Vein · Minutiae · KNN · Euclidean distance

1 Introduction

Now Biometrics are used for the purpose of providing verification, authentication, security, trust, uniqueness and increases the performance criteria. So many biometric systems are used now a days like fingerprint, face recognition, voice recognition, DNA, hand pattern for the criminal identification, cloud computing solutions, e-commerce. Some older technique are also present that are designed to recognise persons [4,5]. Some systems we can purchase at very low cost, ease to use for identification like fingerprint identification system. Iris and Retina scan are also used now a days because they have better results than finger print, face detection, DNA and speech. Persons dorsal hand vein is the most advanced identification terminology. Now a days Dorsal vein of the hand has a key point in biometric [7]. The vein network shown on the dorsal hand, referred to as the hand vein. Cephalic and basilic are the two types. We can ease to obtain the network of dorsal veins of the hand because veins engross full light as compared to tissues in this way we can obtain efficient image [5,7]. For vein imaging some environmental conditions are affected like temperature around us, human

© Springer Nature Singapore Pte Ltd. 2019
K. C. Santosh and R. S. Hegadi (Eds.): RTIP2R 2018, CCIS 1036, pp. 476–484, 2019.
https://doi.org/10.1007/978-981-13-9184-2_42

skin, age and other surface conditions like hairs, permanent mark on the body, wound, burn [8]. As the name suggest (superficial vein) which helps drain parts of the hand and forearm. The basilic are lies on the upper limb of the back of hand, whereas Cephalic are attached with the elbow of the hand [22]. The two technologies for vein imaging are FIR and NIR. FIR having $14\,\mu m$ range and it is very effective in capturing veins on the dorsal hand but FIR technology is very sensitive to the environmental conditions thats why it does not gives better and complete info of image [26]. The NIR has a continuum from 700–1200 nm. It is very beneficial for capturing img. Later in the paper having some sections. Section 2 highlights existing techniques for hand vein recognition. Section 3 shows description of our work 4. Finally summarise the work (Fig. 1).

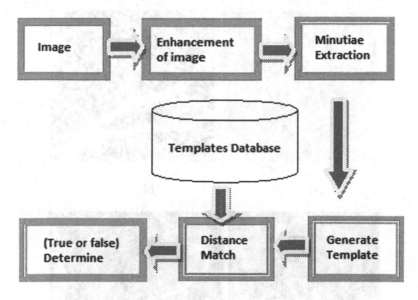

Fig. 1. Flow of our system

2   Correlated Works

Maximum recognition systems are devided into fusion of multiple features, statistical, geometry based. Global features consist in most of the appearance based approaches [9–11] that are derived from all pixels in an image. Lin and Fan [12] did multi-resolution thermal imaging analysis using thermal cameras. Lakshmi and Kandaswamy [13] got a low error rate (EER) because they combined morphological and statistical feature sets. Hsu et al. [14] extracts global and local feature sets to create more efficient recognition system. Radzi et al. [16] tested and evaluated CNN approach for biometric identification using finger print. The shown good accuracy result in experiment using CNN approach. local features of principal veins [17] and minutiae points like position, location and local statistics consisted in geometry based approach [7,18–21]. Wang et al. [18] used Hausdorff distance to calculate matching scores to classify vein pattern. They obtained excellent result by using thermal images. In [18], they represented the shapes observed in vein patterns by geometric based approach. They used bifurcaion and end points as a feature set and match using Euclidean distance. Lee et al. [20] LBP patterns used for finger vein recognition as well as bifurcation and end points alignment Yuksel et al. Hsu et al. [10] used PCA to obtain eigainveins for low dimensional reduction. Lee [15] took features and make fusion that increased recognition accuracy which is time- consuming, he also proposed two dimensional gabor filtering used. Kumar and Prathyusha [9] applied knuckle shape and related triangulation for recognition of person (Fig. 2).

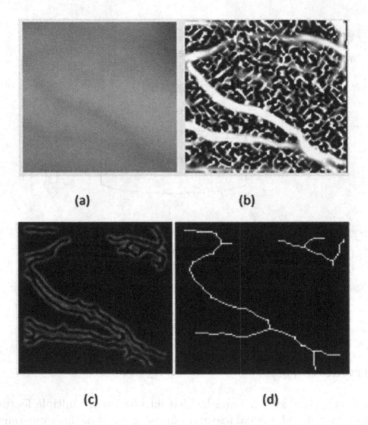

(a)   (b)

(c)   (d)

Fig. 2. Results feature enhancement: (a) Source image, (b) filtered image, (c) Gaussian low-pass filter, (d) After canny edging and removing spur

3   Vein Image Preprocessing

Vivid function of the ratio of eigenvalues is stronger then eigenvalues of less value [23]. We can achieve the homogenous image without any change of the object contrast by the characteristics of vivid function [20]. Volume ratio is nothing but ratio of eigenvalues. Following is volume ratio.

$$VR = \lambda 1 \lambda 2 \lambda 3 \left[\frac{3}{\lambda 1 + \lambda 2 + \lambda 3} \right]^3 \qquad (1)$$

To obtain the cylindrical model of vein we can use modified volume ratio. The range of volume ratio is in between zero and one $\lambda 1 \rightarrow (\lambda 2 - \lambda 1)$, vivid function change to following ratio:

$$v = (\lambda 2 - \lambda 1) \lambda 2 \lambda 3 \left[\frac{3}{2\lambda 2 - \lambda 1 + \lambda 3} \right]^3 \qquad (2)$$

To obtain VR in between zero and one we can equate magnitudes of eigenvalues, since $|\lambda 1| = |\lambda 2|$. $\lambda 2$ and $\lambda 3$ should be of maximum value to produce clear and non- luminous image without any noise. We can choose value of $\lambda 1$ as follows to get clear image inspite of getting maximum value of $\lambda 2$ and $\lambda 3$.

$$\lambda \rho = \begin{cases} \lambda 3 & if \ \lambda 3 < \tau min_x \lambda 3 \ (x, \ s), \\ \tau \min_x \lambda 3 \ (x, \ s) & otherwise, \end{cases} \qquad (3)$$

So, τ specifies value within zero and one [20]. To get the image with maximum intensity on black background all the eigenvalues must be negative i.e. $(\lambda 2 < \lambda 3)$, therefore $\lambda 3$ having maximum value which is minx $\lambda 3(x, s)$. now this function becomes:

$$v = \lambda_2^2 \lambda \rho \left[\frac{3}{2\lambda 2 + \lambda \rho} \right]^3 \qquad (4)$$

To account vascular structure sufficient condition is $(\lambda 2 < \lambda 3/2), \lambda 3$ is changed to $\lambda \rho$. For $\lambda 2 \geq \lambda \rho/2 >$ following condition is computed to get 1. To obtain this by specifying next formula $\lambda \rho \rightarrow (\lambda \rho - \lambda 2)$ into (4) and fixing the response to 1 for $\lambda 2 \geq \lambda \rho/2$. Lastly, the vivid method is calculated as:

$$vp = \begin{cases} 0 & if \ \lambda_2 > 0 \wedge \lambda_3 > 0, \\ \lambda_2^2 \ (\lambda \rho - \lambda 2) \ (\frac{3}{\lambda 2 \ + \lambda \rho})^3 & if \ \lambda_2 \leq \lambda \rho/ \ 2, \\ 1 & otherwise, \end{cases} \qquad (5)$$

In the above equation $\tau < 1$ to get the bright structure.
After getting the bright vein structure on the black background, We need to obtain edges for further operation on image we have used following algorithm.

- Smooth the image f(t, z) by gaussian method $f \wedge (t, z)$.

$$f \wedge (t, z) = f(t, z) * G(t, z, 6) \qquad (6)$$

- Secondly, accomplish value, robustness and direction for edges using operator (gradient).
- Thirdly, acute destruction to the gradient value.
- Lastly, compute threshold of suppressed image.

The output obtained by above algorithm having short spurs that affect on feature extraction process, so we need to remove those spurs by using some pruning methods [27] Fig. 3 shows structuring element for this operation. First obtain a binary image by setting gray level value to 60 by thresholding. Next revolve this algorithm, that resulted single pixel width. Remaining spurs can be removed using pruning. The image (Fig. 3(c)) pruning (using thinning) for five iterations. Almost no spurs are there.

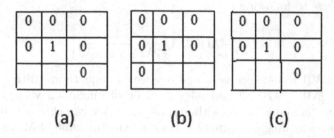

(a)   (b)   (c)

Fig. 3. How to thin vein structure. (a) detect object in the form of binary (b) and (c) illustrate technique of pruning.

4   Proposed Method Used for Recognition of Dorsal Hand Veins

After preprocesing and segmenting the dorsal hand vein next step is to extract the features. In our research minutiae features are used. Normally the features of minutiae having veins bifurcation points, veins ending points, veins continuing points, loops and so on.

4.1   Feature Extraction

Generally dorsal hand vein image minutiae features having vein bifurcation, vein ending and vein continuing. here we have used those feature for dorsal hand vein recognition. To determine branching or termination (ending line) crossing number algorithm has been used.

$$CN = 0.5 \sum_{n=1}^{8} |(Pi - Pi + 1)| \tag{7}$$

The above formula checks each and every pixel in the image. If crossing number is three then that point is branching and crossing number is one then that point is termination. After calculating those features next step is to detect and remove false minutiae. For this refer following equation [16].

$$\sqrt{(x1 - x2)^2 + (y1 - y2)^2} < Dist1 \tag{8}$$

Here, value of dist1 is stable, we can put any value rather then one and three for dist1. It resulted type, location, angle etc.

4.2   Feature Matching

Geometric based approach is effcient for matching minutiae features on dorsal hand vein as compared with statistic approach. Features of distinct biometric is not same [18]. Thats why no one can use same matching approach like distinct biometric. Euclidean distance categorised into distance matching. What happen in matching process simply comparing two feature vector set, where the overall feature from template set are compared by feature of testing set, Then compute nearest distance between them for further matching process.

4.3   Classification

Similarity metrics is usually represented by a distance function based on the attribute values. A variety of distance functions are in use in various communities. Many distance function have been proposed for the task of KNN classification such as Euclidean distance, Squared Euclidean distance, Minkowski distance, Manhattan distance [28].

In classification simply find out the true identity of the person who wants to retrieve the system by comparing image with existing image [30]. In our research Euclidean distance is used to classify the object on the basis of distance between minutiae features. Formula of Euclidean Distance is given below:

$$dij = \sqrt{\sum_{k=1}^{x}(xik - xjk)^2}$$

5   Experiment Results and Discussion

We used standard database for experiment purpose [10]. Firstly we devided bosphorous hand vein dataset into testing and training, 75% of images are the part of training and 25% of images are part of testing (Figs. 4 and 5). With those we calculate FAR, FRR and EER at a specific threshold. Our system is tested on a recent computer configuration and recent version of matlab (Table 1).

Table 1. Differentiating our system with correlated systems

Author	Identity rate	EER	FAR	FRR
Yuksel et al. [8]	96.67	3.17	–	–
Usman [21]	96.97	3.03	1.30	1.75
Proposed system	97.00	3.00	1.26	1.60

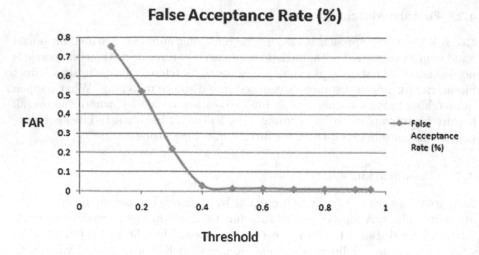

Fig. 4. FAR curve

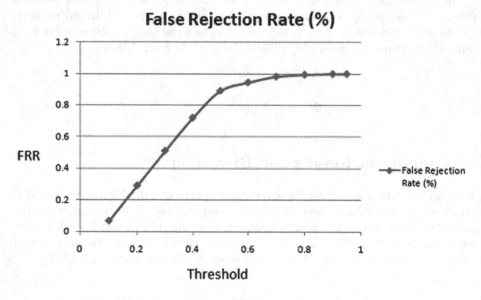

Fig. 5. FRR curve

6   Conclusion

The german filter is a good option for image enhancement and crossing Number with euclidean distance is the better choice for similarity method. When all techniques are get together, or we trust that proposed system would be a robust system for verification and identification of dorsal hand vein. Security is a major issue now a days that get attracted to every personnel working in private or public sector so biometric based recognition systems are used. Persons are theirthe image in existing template. Simply we extracted minutiae features from persons dorsal hand vein. In pre-processing vivid function used for getting brighter image. Finally, we obtained 97.00% accuracy and 3.0% of error.

References

1. Jain, A.K., Bolle, R., Pankanti, S.: Biometrics Personal Identification in Networked Society. Kluwer (1999)
2. Cross, J.M., Smith, C.L.: Thermographic imaging of subcutaneous vascular network of the back of the hand for biometric identification. In: IEEE Proceeding of 29th International Carnahan Conference on Security Technology, pp. 20–35 (1995)
3. Wilson, C.: Vein Pattern Recognition: A Privacy-Enhancing Biometric. CRC Press, Boca Raton (2010)
4. Im, S.-K., Park, H.-M., Kim, Y.-W., Han, S.-C., Kim, S.-W., Kang, C.-H.: Improved vein pattern extracting algorithm and its implementation. J. Korean Phys. Soc. **38**(3), 268–272 (2001)
5. Im, S.-K., Park, H.-M., Kim, S.-W., Chung, C.-K., Choi, H.-S.: Improved vein pattern extracting algorithm and its implementation. In: 2000 Digest of Technical Papers, International Conference on Consumer Electronics. Nineteenth in the Series Cat. No. 00CH37102, pp. 2–3 (2000)
6. Tanaka, T., Kubo, N.: Biometric authentication by hand vein pattern
7. Jain, A.K., Flynn, P., Ross, A.A.: Handbook of Biometrics. Springer, Boston (2008). https://doi.org/10.1007/978-0-387-71041-9
8. Yuksel, A., Akarun, L., Sankur, B.: Biometric identification through hand vein pattern
9. Hsu, C.B., Hao, S.S., Lee, J.C.: Personal authentication through dorsal hand vein patterns. Opt. Eng. **47**, 067205-1–067205-10 (2011)
10. Lee, Y.P.: Palm vein recognition based on a modified $(2D)^2LDA$. Sig. Image Video Process. **9**, 229–242 (2015)
11. Lin, C.L., et al.: Biometric verification using thermal images of palm-dorsa vein patterns. IEEE Trans. Circ. Syst. Video Technol. **14**, 199–213 (2004)
12. Lakshmi, D.C.: An algorithm for improved accuracy in unimodal biometric systems through fusion of multiple feature sets. ICGST-GVIP J. **9**, 33–40 (2009)
13. Hsu, C.B.: Dorsal hand vein recognition using Gabor feature-based 2-directional 2-dimensional PCA. Adv. Sci. Lett. **8**, 813–817 (2012)
14. Lee, J.-C.: A novel biometric system based on palm vein image. Pattern Recogn. Lett. **33**, 1520–1528 (2012)
15. Radzi, S.A., Hani, M.K., Bakhteri, R.: Finger-vein biometric identification using convolutional neural network. Turk. J. Electr. Eng. Comput. Sci. **24**(3), 1863–1878 (2016)
16. Wang, L.: Infrared imaging of hand vein patterns for biometric purposes. IET Computer. Vis. **1**, 113–122 (2007)
17. Wang, L.: Minutiae feature analysis for infrared hand vein pattern biometrics. Pattern Recogn. **41**, 920–929 (2008)
18. Kumar, A.: Personal authentication using hand vein triangulation and knuckle shape. IEEE Trans. Image Process. **18**, 2127–2136 (2009)
19. Lee, J.C.: Dorsal hand vein recognition based on 2D Gabor filters. Imaging Sci. J. **62**, 127–138 (2014)

20. Jerman, T.: Beyond Frangi: an improved multiscale vesselness filter. In: Ourselin, S., Styner, M.A. (eds.) Medical Imaging 2015: Image Processing. Proceedings of SPIE, vol. 9413, pp. 94132A-1–94132A-11 (2015)

21. Usman, A.U.: Dorsal hand veins based person identification. In: Image Processing Theory, Tools and Applications. IEEE (2014)

22. Kumar, A., et al.: Personal authentication using hand vein triangulation and knuckle shape. IEEE Trans. Image Process. 18, 2127–2136 (2009)

23. Peeters, T.H.J.M., Rodrigues, P.R., Vilanova, A., ter Haar Romeny, B.M.: Analysis of distance/similarity measures for diffusion tensor imaging. In: Laidlaw, D., Weickert, J. (eds.) Visualization and Processing of Tensor Fields. MATHVISUAL, pp. 113–136. Springer, Heidelberg (2009). https://doi.org/10.1007/978-3-540-88378-4_6

24. Pierpaoli, C., Basser, P.J.: Toward a quantitative assessment of diffusion anisotropy. Magn. Reson. Med. 36, 893–906 (1996)

25. Wiemker, R., Klinder, T., Bergtholdt, M.: A radial structure tensor and its use for shape-encoding medical visualization of tubular and nodular structures. IEEE Trans. Vis. Comput. Graph. 19, 353–366 (2013)

26. Muthukrishnan, R., Radha, M.: Edge detection techniques for image segmentation. Int. J. Comput. Sci. Inf. Technol. (IJCSIT) 3 (2011)

27. Gonzalez, R., et al.: Digital Image Processing, pp. 518–518. Addison-Wesley Publishing Company, Reading (1992)

28. Davies, E., et al.: Machine Vision: Theory, Algorithms and Practicalities, pp. 149–161. Academic Press, Cambridge (1990)

29. Haralick, R., Shapiro, L.: Computer and Robot Vision, vol. 1. Addison-Wesley Publishing Company. Chap. 5, pp. 168–173 (1992)

30. Fix, E., Hodges, J.: Discriminatory analysis. Nonparametric discrimination: consistency properties. Technical report 4, USAF School of Aviation Medicine, Randolph Field, Texas (1951)

A Biometric Recognition Model Based on Palm Vein Feature Characteristics

Shriram D. Raut[1]($\boxtimes$) and Vikas T. Humbe[2]

[1] Department of Computer Science, Solapur University, Solapur, India
sdraut@sus.ac.in
[2] School of Technology, S.R.T. M. U. N., Sub-centre Latur, Latur, India
vikashumbe@gmail.com

Abstract. The biometric is an automated approach to recognize a person based on unique and distinct physiological or behavioral feature characteristics. The era of biometric characteristics starts from inked palm print to iris based person recognition system. The era has been evolved to use blood vascular structure lies at palm region of hand which is found unique and distinct from person to person. The paper discuss about the proposed biometric recognition model based on the palm vein feature characteristics. The various techniques were used for enhancement, feature extraction and detection being experimented to come to a certain conclusion. The paper also discuss about the visualization and localization of blood vessel structure using Harris Stephen's corner point detection. The recognition model is proposed to extract, detect and classify a subject based on the unique palm vein feature characteristics. This paper also includes about an interpretation of work and results comparison of proposed palm vein based biometric recognition model.

Keywords: Blood vein · Gabor · Canny · Euclidean distance · ROC · Confusion matrix

1 Introduction

The biometric system is designed, while considering physiological as external characteristic of an individual such as structure of lines on finger [3], unique characteristic set of palm vein [17], dorsal vein [13] and palm [14] region of hand, face [7], iris [12] and ear [5], finger knuckle [15] etc. The palm vein biometrics is practiced as new advancement in the family of physiological biometrics. The palm vein characteristic is superior to other traditional biometrics; in view that it is time invariant and nontransferable means of medical identity [1,14]. The multispectral palm print database developed by Hong Kong Polytechnic University [14] is used for the purpose of experimentation. The palm vein image preprocessed using Gabor filter and canny edge detector [8] to extract edges of the blood vascular structure. An output of this phase is given to the Harris-Stephens corner point detector [9] to localize the features in an image. The feature detected

© Springer Nature Singapore Pte Ltd. 2019
K. C. Santosh and R. S. Hegadi (Eds.): RTIP2R 2018, CCIS 1036, pp. 485–495, 2019.
https://doi.org/10.1007/978-981-13-9184-2_43

image is processed through Euclidean distance metric computation [10]. Then key generation algorithm is applied on image to gain unique and distinct thirty bit binary feature vector, later used for the purpose of identification. The experimental result with proposed methodology gives prominent results in terms of high accuracy rate. An analysis and interpretation of results is prepared with an increase in train sample from total 3000 images of 200 known subjects (2400 images) and 50 unknown subjects (600 images). An existing image database was enrolled 250 individual subjects out of them 195 male and 55 female were participants. The calculation of FAR, FRR, GAR are also discussed with respect to the proposed work experimentation.

1.1 Palm Vein

Palm vein images are acquired from the palm region of the hand. It's a simply flat, has a region of interest with no hair and also gets expose well to the camera setup. Blood veins carry deoxygenated hemoglobin (a pigment containing iron) from body parts to heart. The near infrared source is used as a light to trace and scan hemoglobin in the blood [17]. Deoxygenated hemoglobin in the blood absorbs a near-infrared light illumination and reflects a black shade of light [1]. The image acquisition tool can be used to capture or scan vein pattern lies at palm region of the hand. It has been empirically found that light in the range of 700 to 1000 nm (infrared light) range get pass through an epidermis of the skin, whereas an illumination range between 880 to 930 nm gives a view of the vein at palm region of hand [4]. An optical penetration depth for near-infrared (NIR) imaging at 850 nm has shown to offer higher contrast for subcutaneous veins while imaging. A palm and back side of hand has much multipart vein structure than fingers that gives more unlike pattern for its validation.

2 Proposed System

The system consists of two phases as enrollment and identification phase. In enrollment phase, the result of feature extraction using Gabor filter, canny edge is processed. The Euclidean distance metric of the processed image matrix is computed and stored in the system database. The result of distance computation metric is further processed to generate biometric trait key. The step by step procedure for the recognition system is given below-

1. Read an image from the database.
2. Now apply Gabor filter with passed and accepted parameter followed by the canny edge detector to extract edges of the blood vessel.
3. Perform Harris-Stephens corner point detection on the resultant image of the Step (2). These operations are useful to extract edges of the blood vein.
4. Compute Euclidean distance metric of feature extracted and detected image matrix.

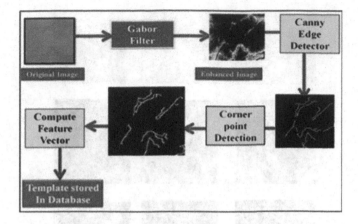

Fig. 1. Workflow for generation of template for matching

5. Compute the binary vector from image resulted of Euclidean distance metric. The process is described clearly in Fig. 1 Using these steps a template would be generated to do matching.

The use of Gabor filter and canny edge detector prominently lead to extract the blood vein structure from original image. The step is useful to extract the palm vein feature characteristic in an efficient manner as shown in Fig. 2. The procession of experimentation is followed as once an edge of the blood vessel is extracted using Gabor filter and canny edge detector; then Harris-Stephens corner point detection algorithm is applied to map the feature vector set. The points are nothing but the feature points detected in an image sample as shown in Fig. 3. The recognition technique would be robust using biometric trait key derived from Local Binary Pattern generate unique binary pattern mapping of palm vein characteristics. An image matrix processed under feature extraction algorithm would be applied with Euclidean distance metric computation.

This would result in a generation of the distance matrix and consist of distance between pixels in fraction values. The Euclidean distance computation result in a distance matrix of the same size as of palm vein image. The key is encoded by selecting a row of length thirty bit from image distance matrix. The experiments were performed on 250 Right palms (3000 images) having 200 train set (2400 images) as known sample and 50 un-train set (600 images) as an unknown sample of the different subject. The varying number of gallery image sample of a different subject with respect to a different number of known and unknown samples from an image database was tested for the purpose of identification. Based on this testing, the false rate of acceptance and rejection and genuine acceptance rate were computed. The generated biometric trait keys are unique and distinct among different kinds of subject. The proposed approach is the most promising as compared to results of existing algorithms. The experimentation has resulted in higher rate of genuine acceptance and a lower rate of false rejection and false acceptance.

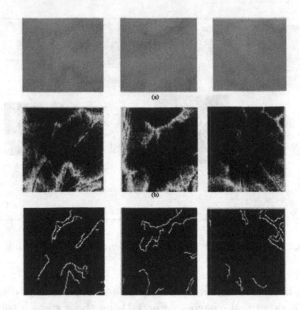

Fig. 2. Extraction of blood vessel

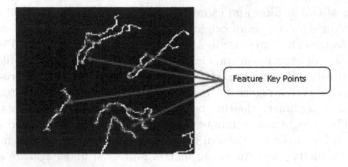

Fig. 3. Detection of feature key points

3   Analysis and Interpretation

The concept of information security is used to assure consistency of using key for encoding of the palm vein information to do image registration. The same would be recalled from the database at the time of matching to verify the template key with claimant generated key to identify an individual with high accuracy rate. The result and work is further applied with computation formula for false acceptance, rejection and computation of genuine acceptance rate. The analysis of recognition accuracy is computed through formula given below for FAR [16], FRR [2] and GAR [4]. Total 200 subjects about $200 \times 12 = 2400$ sample images are processed as train or known set (Authorized user) and $50 \times 12 = 600$ sample images as an unknown set (Unauthorized user). The total sample size of 3000 images was sampled altogether.

The false acceptance rate could be calculated as-

$$FAR = \frac{Incorrect\ Accepted\ Individual}{Total\ Number\ of\ Incorrect\ Matching} \tag{1}$$

The false rejection rate could be calculated as-

$$FRR = \frac{Incorrectly\ Rejected\ individual}{Total\ Number\ of\ Correct\ Matching} \tag{2}$$

The genuine acceptance rate could be calculated as-

$$GAR = 1 - FRR \times 100 \tag{3}$$

An analysis and interpretation of results is prepared with an increase in train sample from total 3000 images of 200 known subjects (2400 images) and 50 unknown subjects (600 images). An existing image database was enrolled 250 individual subjects out of them 195 male and 55 female were participants. The following Table 1 shows the calculation of FAR, FRR, GAR and RR with respect to the proposed work experimentation.

For example as shown above in the Table 1 and row 4, from an individual subject out of 12 images, 9 images (from First Session: 05 and from Second Session: 04) are acquired for processing. Total 200 subjects about $200 \times 9 = 1800$ sample images are processed as train or known set (Authorized user) and $50 \times 9 = 450$ sample images as an unknown set (Unauthorized user). The total sample size of 2250 images was sampled altogether.

As per Eq. 1, there are total 08 image samples are incorrectly accepted individual, those are unauthorized user (from unknown set 50×9) but are accepted as authorized user. And there are total 20 incorrect matching from the sample set.

$$FAR = \frac{8}{20} = 0.4$$

To consider the count of the incorrectly rejected individual, the value 08 (Unauthorized user) subjects are incorrectly accepted individual. This is subtracted from 20 incorrect matching including samples from image sample set. Hence, $20 - 8$ would result as 12 are incorrectly rejected individuals from image sample set. Now to consider the total number of Correct matching, the count 20 (total number of incorrect matching from sample set) are subtracted from 2250 (Total number of the sample) would result into 2230 as the total number of correct matching as per Eq. (2), so

$$(FRR = \frac{12}{2230} = 0.0053)$$

The GAR is computed as per Eq. (3) would be

$$GAR = (1 - 0.0053) \times 100 = 99.47\%$$

Table 1. Computation of FAR, FRR and GAR

No. of gallery sample per class	Set of known and unknown subjects	Subject and per session image	Total no. of image	FAR	FRR	GAR (%)
1	200 train 50 unknown	200 × 12 = 2400 50 × 12 = 600	3000	0.0345	0.009	**99.06**
2	200 train 50 unknown	200 × 11 = 2200 50 × 11 = 550	2750	0.3043	0.006	**99.42**
3	200 train 50 unknown	200 × 10 = 2000 50 × 10 = 500	2500	0.0476	0.008	**99.2**
4	200 train 50 unknown	200 × 9 = 1800 50 × 9 = 450	2250	0.4	0.005	99.47
5	200 train 50 unknown	200 × 8 = 1600 50 × 8 = 400	2000	0.0769	0.006	**99.4**

3.1   Goodness of the Fit

The ROC plot of FAR versus FRR is shown in Fig. 4 using curve fitting technique. The FRR on y-axis has values 0.0094, 0.008, 0.006, 0.0069, 0.0066, 0.0058, 0.0053, 0.004, 0.004 and FAR on x-axis having sorted values 0.0345, 0.0476, 0.0769, 0.0769, 0.0909, 0.3043, 0.4, 0.4444, 0.5. The different curve fitting techniques such as Exponential, Gaussian, Power, and Smoothing Spline, Rational were applied to fit curve into a plot. The Goodness of fit by considering Root Mean Square Error (RMSE) of each technique is compared. These values are analyzed and found for Exponential - RMSE: 0.0009123, Gaussian - 0.001009, Power - 0.000681 and Smoothing Spline - 0.0005065, Rational - 0.0008833. The Goodness of the fit of the Smoothing Spline based on Root Mean Square Error is minimum compared to other curve fitting techniques. So the plot of FRR versus FAR is generated using Gaussian curve.

Table 1 shows values of FAR and FRR and the plot in Fig. 4 shows the relationship between FRR is inversely proportion to the FAR. As per Table 1, the False Rejection Rate is much less than the False Acceptance Rate of subject identification. As there is an increase in the False Acceptance Rate, then the False Rejection Rate is at a minimum. As shown in Table 1, if FAR is highest has 0.5 value then counter wise FRR is lowest, has 0.004 value.

The ROC plot of GAR versus FAR is shown in Fig. 5 using curve fitting technique. The FAR on x-axis having sorted values 0.0345, 0.0476, 0.0769, 0.0769, 0.0909, 0.3043, 0.4, 0.4444, 0.5 and GAR on y-axis having corresponding to these values 0.9906, 0.9920, 0.9940, 0.9931, 0.9934, 0.9942, 0.9947, 0.9960, and 0.9960. The different curve fitting technique such as an Exponential, Gaussian, Power, Smoothing Spline, rational etc. was applied to fit curve into a plot. The Goodness of fit by considering the Root Mean Squared Error (RMSE) of each technique is compared. These values are analyzed and found for Exponential- RMSE: 0.00093, Gaussian- 0.000987, Power- 0.000742 and Smoothing Spline- 0.000507, Rational-

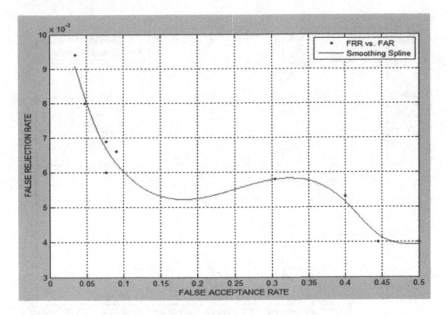

Fig. 4. FAR versus FRR plot of ROC

0.002333. The Goodness of fit of Smoothing Spline curve based on Root Mean Square Error is minimum compared to other curve fitting techniques. So the plot of GAR versus FAR is generated using Smoothing Spline.

Table 1 shows values of GAR and FAR and the plot in Fig. 5 shows the relationship between GAR is directly proportion to FAR. As there is an increase in Genuine Acceptance Rate, then False Acceptance Rate is at maximum. As shown in Table 1, if GAR is highest has 0.9906 value then counter wise FAR is highest has 0.5 value. The dataset of 200, 500, 700 and 1000 image samples has processed using Neural Pattern Recognition approach. The sample of data in the form of biometric trait key 0and related binary representation as "0" and "1" is being mapped with the claim of authentication. The Neural Pattern Recognition approach is useful for analytical validation and test data utilities. The sample of data in the form of biometric trait key and related binary representation as "0" and "1" is being mapped with the claim of authentication. The MATLAB has Neural Network Pattern Recognition Tool useful for analytical validation and test data utilities. Using this approach, the data set has randomly divided up "500 sample" of images into a training set of "70%" means "350 samples" and considering validation and testing rate are at "15%" of sample size means "75 sample". With these inputs, the Neural Network Pattern Recognition approach gives a 99.40% recognition rate. The confusion matrix as shown below Fig. 6 is generated using Neural Network Pattern Recognition approach.

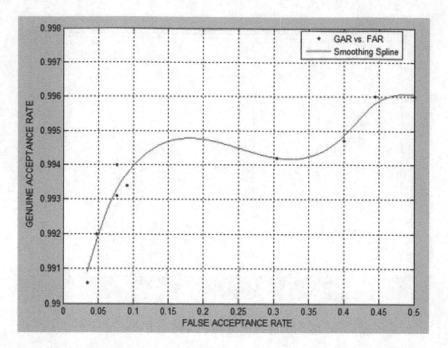

Fig. 5. FAR versus GAR plot of ROC

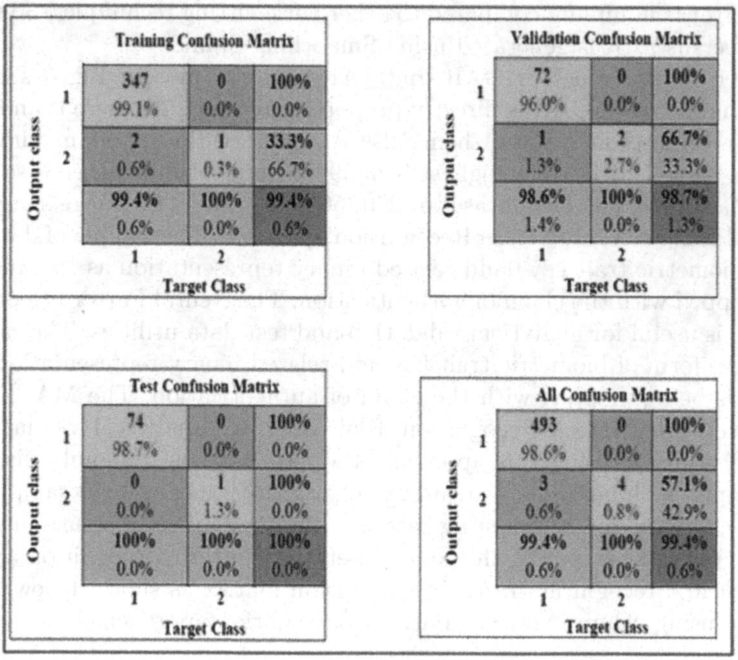

Fig. 6. Plot of confusion matrix

3.2   Results and Comparison

The proposed work and existing algorithms are compared to observe the success rate of experimentation. The research work is compared with approaches such as Laplacian palm [11], SIFT [6] and Comp Code [18] based experimentations. In the first instance, the identification rate based on 200 Right palms acquired through PolyU image database having 200 train set (3000 images) as known sample and 50 test set (650 images) as unknown sample of different number of subject images were processed. The table below shows that the proposed approach is found most promising compared to results of all other approaches. Table 2: Comparison of Results.

Table 2. Comparison of results

No. of gallery samples per class	Proposed approach	Laplacian palm [11]	SIFT [6]	Comp code [18]
1	99.42%	77.42%	85.58%	92.67%
2	99.20%	84.21%	91.08%	96.00%
3	99.47%	86.83%	92.50%	97.67%
4	99.40%	88.50%	93.92%	98.83%
5	99.31%	89.17%	94.92%	98.92%
6	99.34%	89.00%	95.42%	99.00%

The proposed novel approach of deriving a biometric trait key using a sequence of operation is found most promising in comparison with existing evolved biometric identification technique. The result of proposed research work claim higher rate of accuracy for the identification of subject from among a population. The rate of accuracy of proposed research work range from lowest as 99.20% and highest as 99.47%; whereas accuracy rate based on Laplacian palm algorithm, Comp code and Scale Invariant Feature Transform (SIFT) ranges from lowest as 77.42%, 92.67% and 85.58% and highest as 89.17%, 99.00% and 95.42% respectively. The best so far result is found prominent using proposed research in comparison with existing methodologies.

4   Conclusion

This research work is carried in view to develop palm vein characteristic based recognition technique. As human palm vein are easier to present for imaging and recognize, this could be beneficial for biometric authentication in civilian security application. The work is contribution to be a multimodal biometric system; in which research is proposed for recognition of an individual on the basis of fusion of physiological and behavioral characteristics. For experimentation purpose, the steps of standard research process are followed along with pattern recognition.

The research work mainly focuses on palm vein feature analysis and extraction followed by the development of the recognition technique. In this proposed research work, a novel algorithm is defined for recognition of person based on the structural uniqueness of palm vein characteristics. An existing image database developed by the Hong Kong Polytechnic University (PolyU) was obtained to test the hypothesis and to perform experimentation based on proposed algorithms. The multispectral palm vein image was processed to gain the feature set that would be useful for the purpose of recognition. The digital image processing techniques such as image enhancement, spatial filtering was performed to enhance the image and highlight the key characteristics of palm vein image. The Gabor filter and canny edge detector were applied in sequence to extract edge [19] of the palm vein vascular structure lies at palm region of the hand. Further feature key points were localized by sampling Harris-Stephens corner detection method. And finally, representation of detected feature key points of palm vein using Euclidean distance and Local Binary Pattern derivative, which is useful in formation of a Biometric trait key is put into operation for recognition of the subject. The database provided by the Hong Kong Polytechnic University (PolyU) is only for 250 subjects having 3000 set of images, thus more data are required to be collected for better evaluation of proposed work.

References

1. Fujitsu: Palm vein pattern biometric authentication technology. Whitepaper-Fujitsu Computer Products of America, p. 45 (2007)
2. Hao, Y., Sun, Z., Tan, T., Ren, C.: Multispectral palm image fusion for accurate contact-free palmprint recognition. In: 2008 15th IEEE International Conference on Image Processing, ICIP 2008, pp. 281–284. IEEE (2008)
3. Humbe, V., Gornale, S.S., Manza, R., Kale, K.V.: Mathematical morphology approach for genuine fingerprint feature extraction. Int. J. Comput. Sci. Secur. (IJCSS) **1**, 53–59
4. Kabaciński, R., Kowalski, M.: Human vein pattern segmentation from low quality images - a comparison of methods. In: Choraś, R.S. (ed.) Image Processing and Communications Challenges 2. AINSC, vol. 84, pp. 105–112. Springer, Heidelberg (2010). https://doi.org/10.1007/978-3-642-16295-4_12
5. Kandgaonkar, T.V., Mente, R.S., Shinde, A.R., Raut, S.D.: Ear biometrics: a survey on ear image databases and techniques for ear detection and recognition. IBMRD's J. Manag. Res. **4**(1), 88–103 (2015)
6. Ladoux, P.-O., Rosenberger, C., Dorizzi, B.: Palm vein verification system based on SIFT matching. In: Tistarelli, M., Nixon, M.S. (eds.) ICB 2009. LNCS, vol. 5558, pp. 1290–1298. Springer, Heidelberg (2009). https://doi.org/10.1007/978-3-642-01793-3_130
7. Nooreyezdan, S.S.: Plastic surgery: a new dimension to face recognition. IEEE Trans. Inf. Forensics Secur. **5**, 441–448 (2010)
8. Raut, S.D., Humbe, V.T.: A novel approach for palm vein feature extraction using Gabor and canny edge detector. In: 2015 IEEE International Conference on Computational Intelligence and Computing Research (ICCIC), pp. 1–4. IEEE (2015)

9. Raut, S.D., Humbe, V.T.: Palm vein recognition system based on corner point detection. In: 2015 IEEE International WIE Conference on Electrical and Computer Engineering (WIECON-ECE), pp. 499–502. IEEE (2015)

10. Raut, S.D., Humbe, V.T.: A distance metric computation model for the design and development of palm vein recognition system. Int. J. Bio-Sci. Bio-Technol. 7(6), 201–206 (2016)

11. Wang, J.-G., Yau, W.-Y., Suwandy, A., Sung, E.: Person recognition by fusing palmprint and palm vein images based on "Laplacianpalm" representation. Pattern Recogn. 41(5), 1514–1527 (2008)

12. Wei, Z., Qiu, X., Sun, Z., Tan, T.: Counterfeit IRIS detection based on texture analysis. In: ICPR, no. 1, pp. 1340–1343 (2008)

13. Yüksel, A., Akarun, L., Sankur, B.: Biometric identification through hand vein patterns. In: 2010 IEEE 18th Signal Processing and Communications Applications Conference (SIU), pp. 708–711. IEEE (2010)

14. Zhang, D., Guo, Z., Guangming, L., Zhang, L., Zuo, W.: An online system of multispectral palmprint verification. IEEE Trans. Instrum. Measur. 59(2), 480–490 (2010)

15. Zhang, L., Zhang, L., Zhang, D., Guo, Z.: Phase congruency induced local features for finger-knuckle-print recognition. Pattern Recogn. 45(7), 2522–2531 (2012)

16. Zhao, Q., Zhang, D., Zhang, L., Luo, N.: Adaptive fingerprint pore modeling and extraction. Pattern Recogn. 43(8), 2833–2844 (2010)

17. Zhou, Y., Kumar, A.: Human identification using palm-vein images. IEEE Trans. Inf. Forensics Secur. 6(4), 1259–1274 (2011)

18. Zhou, Y., Liu, Y., Feng, Q., Yang, F., Huang, J., Nie, Y.: Palm-vein classification based on principal orientation features. PLoS One 9(11), e112429 (2014)

Design and Development of New Algorithm for Person Identification Based on Iris Statistical Features and Retinal Blood Vessels Bifurcation Points

Yogesh Rajput[1]($\boxtimes$), Shaikh Abdul Hannan[2]($\boxtimes$), Mohammad Eid Alzahrani[2]($\boxtimes$), Dnyaneshwari Patil[1]($\boxtimes$), and Ramesh Manza[3]($\boxtimes$)

[1] MGM's Dr. G.Y. Pathrikar College of CS and IT, Aurangabad, India
yogeshrajput128@gmail.com, dnyaneshwari03patil@gmail.com
[2] Department of Computer Science, Al Baha University, Albaha, Saudi Arabia
abdulhannan05@gmail.com, meid@bu.edu.sa
[3] Department of CS and IT, Dr. B.A.M.U., Aurangabad, India
manzaramesh@gmail.com

Abstract. Biometric are the trait, measured features used to tag and describe persons. Physical distinctiveness is interconnected to the structure of the body. Biometric recognition are, face recognition, fingerprint, DNA, palm print, iris, and retina. We describe a novel algorithm for the identification and measurement of iris statistical features and identify the bifurcation points of retinal blood vessels for person recognition, by applying DIP techniques. Iris algorithm is performed on CASIA database and local database collected from KVKR (Department of CS and IT, Dr. B.A.M.U, Aurangabad) research lab, overall 100 iris image databases. For localization and extraction of inner iris is done by using different image processing techniques. After extraction of inner iris, statistical features are calculated such as, area, diameter, length, thickness, and mean. Performance analysis of the algorithm is done by using ROC curve. This algorithm achieves sensitivity of 94.92% and specificity of 100%. Afterwards retinal blood vessels bifurcations points are extracted. Retinal image database is collected by Dr. Manoj Saswade (Ophthalmologist, Saswade Netra Rugnalaya, Aurangabad (MH)), overall 500 retinal image databases is collected. Then-after apply DIP techniques, like image enhancement and otsu's method, minutia techniques etc. Retinal blood vessels bifurcations points achieves a TP rate of 98%, FP rate of 20%, and overall accuracy score of 0.9702.

Keywords: Person identification · Biometric · Iris · Retina

1 Introduction

The Iris recognition has proposed into an important permitting technology in our society. While an iris pattern is genuinely an absolute identifier, the development of a high-performance iris recognition algorithm and conveying it from research

© Springer Nature Singapore Pte Ltd. 2019
K. C. Santosh and R. S. Hegadi (Eds.): RTIP2R 2018, CCIS 1036, pp. 496–504, 2019.
https://doi.org/10.1007/978-981-13-9184-2_44

lab to practical petitions is still a demanding task. Iris is a meaningful biometric trait. It contains idiosyncratic texture and is complex generous to be used as a biometric signature. Connected with other biometric features such as fingerprint, face, iris patterns are more constant. It is inimitable to people and constant with age. Also, iris recognition systems can be non-invasive. For localization of inner iris we have collected the 40 Iris images from CASIA image dataset [1]. And KVKR iris database is having 100 iris images. This database is collected in department of CS and IT, Dr. B.A.M. University, Aurangabad. Nowadays e-security are in vital need of finding correct, secure substitution to passwords and personal identification numbers as financial reimbursement increase intensely year over year from computer-based scam such as computer hacking and identity stealing [2]. Biometric account these fundamental problems, because a person's biometric data is typical and cannot be pass. Biometrics which mentions to discern a person by his or her physiological or behavioral appearances has potential to discriminate between recognized user and an impersonator. A advantage of using biometric authentication is that it cannot be unwise or pass, as the person has to be physically present during at the point of identification activity [3]. Biometrics is usually more reliable and conclude than conventional information based and recognition based techniques [4].

2   Methodology

The present algorithm is design for localization of inner iris, shown in below Fig. 1. Prepossessing is done by renovating the image into gray. Afterwards apply histogram equalization for image enhancement. After image enhancement, image complement operation is done for highlighting the iris. Accordingly image adjustment is done by using contrast stretching algorithm. Afterwards some noise is get added, to remove the unwanted noise, median filter is used. After removing the salt and pepper noise, Otsu's operation is done for extraction of inner iris. In the Fig. 2, high resolution fundus image is taken then perform preprocessing operation on fundus image. Afterward perform image processing operation for enhancement of blood vessels. Afterward applying threshold operation for separating foreground to background. Then perform morphological skeletonozation for getting the centerline of vessels. Then-after performs minutia technique for getting the bifurcation points of the vessels. For executing this techniques, database is taken from Dr. M. Saswade and Dr. N. Deshpande, this database have the 500 high resolution retinal fundus images.

Following are the mathematical formulations are use for extraction and localizing of inner iris.

Histogram equalization function for enhancing the iris image:

$$h(v) = round \frac{cdf(v) - cdf_{min}}{(M \times N) - cdf_{min}} X(L - 1) \tag{1}$$

Here cdf_{min} is the minimum value of the cumulative distribution function, M × N gives the image's number of pixels and L is the number of grey levels. 2D median filter is use for removing the salt and pepper noise.

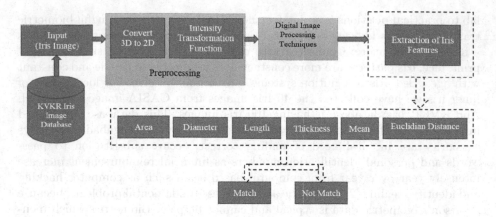

Fig. 1. Workflow for inner iris extraction and localization

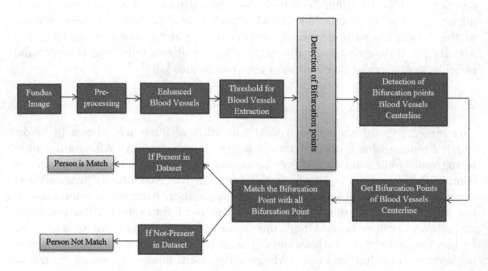

Fig. 2. Workflow for retinal blood vessels extraction and calculating the bifurcation points

$$y[m,n] = median\ x[i,j], (i,j) \in w \qquad (2)$$

Here w indicates a neighborhood centered around location (m, n) in the image.

Threshold function is used for segmenting foreground from background for getting the retinal blood vessels.

$$T = \frac{1}{2}(m1 + m2) \qquad (3)$$

Here m1 and m2 are the Intensity Values of the retinal image.

3   Result

3.1   Inner Iris Extraction and Localization

By applying digital image processing techniques, we have extracted the inner iris following Fig. 3 shows the output of inner iris localization. After extraction of inner iris we have calculated the statistical features like area, diameter, length, thickness and mean (Figs. 4 and 5).

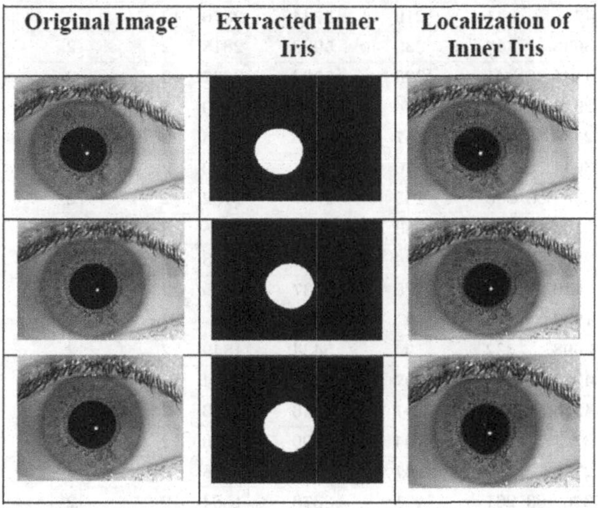

Original Image	Extracted Inner Iris	Localization of Inner Iris

Fig. 3. Statistical features of CASIA

Following Table 1 show the statistical features of CASIA and KVKR Iris image database.

3.2   Retinal Blood Vessels Extraction and Calculating the Bifurcation Points

Below is the complement function for converting image into negative.

$$A^c = \omega | \omega \notin A \qquad (4)$$

Here A^c is a complement, w is the element of A, $\notin$ stands for not an element of A and A is set.

Below formula represents the histogram equalization function.

$$h(v) = round \frac{cdf(v) - cdf_{min}}{(M \times N) - cdf_{min}} X(L-1) \qquad (5)$$

Here cdf_{min} is the minimum value of the cumulative distribution function, $M \times N$ gives the image's number of pixels and L is the number of grey levels.

Table 1. CASIA (C) and KVKR (K) feature database

Sr. No.	Area (C)	Area (K)	Diameter (C)	Diameter (K)	Length (C)	Length (K)	Thickness (C)	Thickness (K)	Mean (C)	Mean (K)
1	5999.88	9836.63	247	316	3000	4918	2	2	77.46	99.18
2	6518.13	7222.88	257	271	3259	3611	2	2	80.73	84.99
3	6746.38	9180	261	305	3373	4590	2	2	82.14	95.81
4	6969.38	3152.75	266	179	3485	1576	2	2	83.48	56.15
5	6963	1060.25	266	104	3482	530	2	2	83.44	32.56
6	10853.88	1915.38	332	139	5427	958	2	2	104.18	43.76
7	11487.38	4751	341	219	5744	2376	2	2	107.18	68.93
8	11876.88	5636.5	347	239	5938	2818	2	2	108.98	75.08
9	11448.63	4194.38	341	206	5724	2097	2	2	107	64.76
10	9874.25	2980.88	316	174	4937	1490	2	2	99.37	54.6
11	3877.38	2430.25	198	157	1939	1215	2	2	62.27	49.3
12	3414.75	3044.75	186	176	1707	1522	2	2	58.44	55.18
13	4174.13	2363.38	206	155	2087	1182	2	2	64.61	48.61
14	4235.63	3309	207	183	2118	1655	2	2	65.08	57.52
15	4418	2975.5	212	174	2209	1488	2	2	66.47	54.55
16	10474.13	3110.88	326	178	5237	1555	2	2	102.34	55.78
17	10492.5	1952.13	326	141	5246	976	2	2	102.43	44.18
18	10867.88	3698	332	194	5434	1849	2	2	104.25	60.81
19	10971.75	4228.13	333	207	5486	2114	2	2	104.75	65.02
20	10892.5	4225.5	332	207	5446	2113	2	2	104.37	65
21	5999.88	3711.75	247	194	3000	1856	2	2	77.46	60.92
22	6518.13	2756	257	167	3259	1378	2	2	80.73	52.5
23	6746.38	3767.63	261	195	3373	1884	2	2	82.14	61.38
24	6969.38	3823.38	266	197	3485	1912	2	2	83.48	61.83
25	6963	3825	266	197	3482	1913	2	2	83.44	61.85
26	10853.88	6241.88	332	251	5427	3121	2	2	104.18	79.01
27	11487.38	9836.63	341	316	5744	4918	2	2	107.18	99.18
28	11876.88	7222.88	347	271	5938	3611	2	2	108.98	84.99
29	11448.63	9180	341	305	5724	4590	2	2	107	95.81
30	9874.25	3152.75	316	179	4937	1576	2	2	99.37	56.15
31	5999.88	1060.25	247	104	3000	530	2	2	77.46	32.56
32	9874.25	1915.38	316	139	4937	958	2	2	99.37	43.76
33	3877.38	4751	198	219	1939	2376	2	2	62.27	68.93
34	3414.75	5636.5	186	239	1707	2818	2	2	58.44	75.08
35	4174.13	4194.38	206	206	2087	2097	2	2	64.61	64.76
36	4235.63	2980.88	207	174	2118	1490	2	2	65.08	54.6
37	9874.25	2430.25	316	157	4937	1215	2	2	99.37	49.3
38	11876.88	3044.75	347	176	5938	1522	2	2	108.98	55.18
39	11448.63	2363.38	341	155	5724	1182	2	2	107	48.61
40	9874.25	3309	316	183	4937	1655	2	2	99.37	57.52

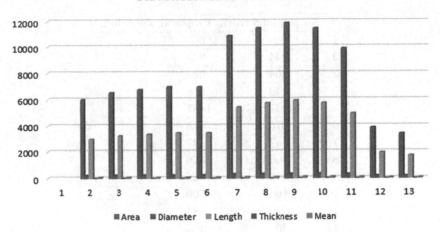

Fig. 4. Statistical features of CASIA

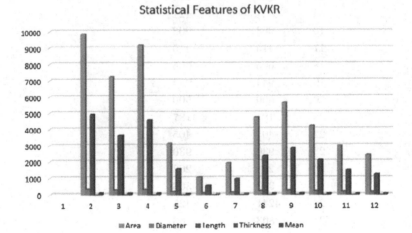

Fig. 5. Statistical features of KVKR

Below are morphological operations.

$$I_{dilated}(i,j) = max_{f(n,m)=true} I(i+n, j+m) \qquad (6)$$

$$I_{eroded}(i,j) = min_{f(n,m)=true} I(i+n, j+m) \qquad (7)$$

Perform erosion and dilation for combining the corrupted blood vessels. After applying these operations, the result is shown in Fig. 6.

Following Table 2 shows the comparison between ground truth and proposed algorithm.

Table 2. CASIA (C) and KVKR (K) feature database

Sr. No.	Ground truth	Proposed algorithm
1	605	605
2	944	944
3	304	304
4	649	649
5	1431	1431
6	675	675
7	147	147
8	306	306
9	616	616
10	500	500
11	187	187
12	405	405
13	930	930
14	203	203
15	421	421
16	903	903
17	415	415
18	605	605
19	944	944
20	500	500
21	187	187
22	405	405
23	930	930
24	203	203
25	303	303
26	4241	425
27	803	803
28	425	427
29	655	655
30	994	995
31	507	507
32	182	183
33	475	476
34	939	940
35	223	224
36	344	344
37	345	346
38	182	184
39	105	105
40	730	731

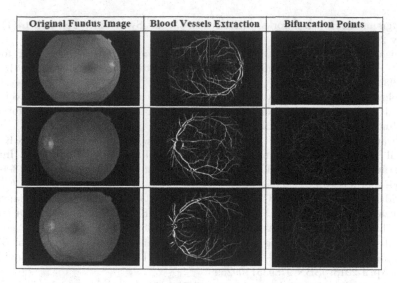

Fig. 6. Bifurcation points on extracted blood vessels

4   Conclusion

For localization and extraction of inner iris we have use digital image processing techniques, for analysis of the algorithm, online CASIA database and local database collected from KVKR research lab (Department of Computer Science and IT, Dr. Babasaheb Ambedkar Marathwada University, Aurangabad). Subsequent to extraction of inner iris, statistical feature extraction is done with the assist of area, diameter, length, thickness, and mean. For performance analysis of the proposed algorithm, receiver operating characteristic curve is used. The algorithm achieves sensitivity of 94.92% and specificity of 100%. And for retinal blood vessels bifurcations points, achieves a TP rate of 98%, FP rate of 20%, and accuracy score of 0.9702.

Acknowledgements. We are thankful to Multimodal System Development laboratory entrenched under UGC's SAP scheme, Department of CS IT, Dr. B.A.M. University, Aurangabad for giving KVKR iris image database. We are also grateful to Dr. Manoj Saswade, Director "Saswade Eye Clinic" Aurangabad and Dr. Neha Deshpande, Director "Guruprasad Netra Rungnalaya Pvt. Ltd.", Samarth Nagar, Aurangabad for providing the fundus image database and verify the result.

References

1. CASIA Iris Image Database. http://biometric.idealtest.org/db. Accessed 23 Feb 2018
2. Kevin, R.: E-Security for E-Government, A Kyberpass Technical White Paper, April 2001. www.kyberpass.com. Accessed 25 Jan 2018

3. Daugman, J.: How Iris works. IEEE Trans. Circuit Syst. Video Technol. **14**(1) (January 2004)
4. Manza, R., Patwari, M., Rajput, Y.: Understanding GUI Using MATLAB for Students. Shroff Publisher and Distributer Pvt. Ltd., Navi Mumbai (2013). ISBN 9789351109259
5. Khobragade, K., Kale, K.V.: Iris edge detection with bit plane slicing technique. In: IJCA Proceedings on National Conference on Recent Advances in Information Technology (0975-8887) (2014)
6. Patwari, M.B., Manza, R.R., Rajput, Y.M., Saswade, M., Deshpande, N.K.: Personal identification algorithm based on retinal blood vessels bifurcation. In: 2014 International Conference on Intelligent Computing Applications (2014). 978-1-4799-3966-4/14 © 2014
7. "Understanding MATLAB" by Karbhari Kale, Ramesh R. Manza, Ganesh R. Manza, Vikas T. Humbe, Pravin L. Yannawar, Shroff Publisher and Distributer Pvt. Ltd., Navi Mumbai, April 2013. ISBN 9789350237199
8. "Understanding GUI Using MATLAB for Students" by Ramesh Manza, Manjiri Patwari and Yogesh Rajput, Shroff Publisher and Distributer Pvt. Ltd., Navi Mumbai, April 2013. ISBN 9789351109259
9. "Understanding Programming Aspects of Diabetic Retinopathy Lesion Using MATLAB" by Dr. Yogesh Rajput, Dr. Ramesh R. Manza, Dr. Pravin L. Yannawar, Dr. Dnyaneshwari D. Patil, Shroff Publisher and Distributer Pvt. Ltd., June 2017. ISBN 9789352135202

Multispectral Palmprint Biometric Verification System Using Deep CNN

H. D. Supreetha Gowda[1]([✉]), Mohammad Imran[2]([✉]),
and G. Hemantha Kumar[1]([✉])

[1] Department of Computer Science, University of Mysore, Mysore 560 007, India
supreethaghd@gmail.com, ghk2007@yahoo.com
[2] Ejyle Technologies, #125 & 126, 1st Floor, Brigade Arcade, Brigade Metropolis,
Mahadevapura Post, Garudacharpalya 560048, Bengaluru, India
emraangi@gmail.com

Abstract. In this paper, we are proposing deep biometric verification system based on multispectral palmprint images. We have built our own deep learning architecture, deployed on multispectral palmprint unimodal system, deep features and traditional texture namely log-gabor features are compared in this work. Multispectral palmprint biometric are captured under different illumination. Images from multispectral database provides images obtained from varied wavelength in four different instance colours Red, Green, Blue and Near InfraRed. Deep learning is gaining extraordinary leaps by consolidating most of the machine learning algorithms essence and proved tremendous in computer vision and image recognition. We have evaluated our system by performing feature level fusion on traditional log gabor method and from the novel deep learning CNN (Convolution Neural Network) architecture proposed in our work. We have compared the performance of the system under different spectral wavelengths. Experimental results demonstrate the effectiveness of our proposed system by outer performing the traditional method results and shown reliable.

Keywords: Deep learning · CNN · Multispectral · Log-gabor · Palmprint

1 Introduction

In modern society, proposing an automated biometric verification system to identify individuals in real time unconstrained environment is still a challenging issue and a fundamental requirement in many applications such as border security, financial transactions, surveillance, computer security, when covert recognition is required or in cases where an individual may try to conceal their true identity. To ensure complete reliable system, it is better to practice multi-factor authentication mechanism which combines passwords with biometric traits (physiological

© Springer Nature Singapore Pte Ltd. 2019
K. C. Santosh and R. S. Hegadi (Eds.): RTIP2R 2018, CCIS 1036, pp. 505–513, 2019.
https://doi.org/10.1007/978-981-13-9184-2_45

or/and behavioural traits) [2]. Choosing biometric traits for a particular application is truly a critical issue, however the trait that fulfils all the necessary fundamental properties such as uniqueness, permanence, universality, collectability, resistance to spoof, performance, user acceptance, throughput, and integration to a certain threshold matters. The primary concerns is developing a biometric recognition system is to design a suitable acquisition device, efficient feature extraction techniques and classifiers that classifies between the users accurately.

Many drawbacks faced by unimodal biometric systems such as resiliency to noise, spoofing, fault tolerance are being successfully addressed by multibiometrics, which fuses multiple evidence of identities from individuals. Multi-sensor approach, captures biometric traits from different sensors (for example, fuse images obtained from visible and IR light). Multi-sample approach, fuses multiple samples of same trait (for example, facial images from a surveillance system). Multi-instance approach, fuses more than one instance of same biometric modality (for example, image obtained from left and right iris scan). Multi-modal approach, different modalities captured from same individual provides stronger level of evidence and this is one of the most practiced approach in literature (for example, combining face and iris modality). Multi-algorithmic approach, more than one algorithm is adopted for feature extraction of same modality in judging its performance on the dataset (for example, texture based and subspace based algorithms applied on facial modality). Fusion at multimodal biometric systems takes place majorly at four levels namely sensor level, feature level, matching score level and decision level. At sensor level fusion, raw data is acquired from multiple sensors and then fused to form a single image, finally the further processing of feature extraction and classification is done on the fused image. At feature level fusion, features extracted from multiple traits are concatenated, as it contains the rich set of features and forms a higher dimension, obviously this level of fusion is promising but 'curse of dimensionality' and compatibility of feature set to combine has to be diagnosed properly. At score level fusion, multiple classifiers output match scores which are combined to obtain a single scalar score [13]. At Decision level fusion, each matcher outputs its own decision like accept or reject, finally the ultimate decision is obtained by techniques such as majority voting, behaviour knowledge space, etc.

In the recent literature, deep learning has gained interest of researchers due to its superior performance in computer vision, natural language processing and translation, sentiment analysis, visual tracking etc. Deep learning exhibits multiple level of representation, learns from a series of nonlinear feature mapping functions and transforms the representation from lower level to higher level layers and thus generating graph of hierarchies [12]. Traditional machine learning approaches had their own limitations and required careful engineering in selecting suitable feature extractors and classifiers in processing raw data and data obtained from unconstrained environment. Though deep learning has scarce theoretical backing [14], it owes its resurgence to efficient optimization techniques and proved highly successful for both supervised and unsupervised learning. Deep learning requires large amount of training data to disentangle all the variations and suppress irrelevant variations which leads in handling large intraclass variations and noisy biometric data. Deep learning has the capability of outperforming Previous state-of-the-art methods and generalization ability [15].

Offering Palmprint trait for biometric security is widely accepted in society, as it provides accurate and robust recognition. Workers and elderly people may have poor fingerprint, in such a case palmprint would be a choice as it provides more information than fingerprint. Red (660 nm), blue (470 nm) and green (525 nm) are the three primary colours that generates various composite lights in visible spectrum. Infrared light has more penetrability [1]. Multispectral imaging is widely used in Healthcare, aerospace, military and machine vision. In this article, we investigate the impact of deep learning in the field of biometrics adopting multispectral palmprint database of huge training samples.

2 Literature Review

Hu et al. [3] conducted extensive experimentation on three different CNN based architectures applied on facial trait by adopting LFW (Labeled Faces in the Wild) public database. Due to lack of theoretical background in generalizing the CNN architecture for all kinds of constrained and unconstrained environmental datasets, authors have tried different architectures including number of filters and layers and made work easily reproducible. Experiments showed that fusion of multiple CNNs and metric learning is performing well.

CNN architectures have given remarkable results in the field of biometric identification/verification systems irrespective of traits. Facebook AI group people have trained an 8- layer CNN named DeepFace which contains four million facial images of 4,000 subjects [4], the DeepFace architecture is designed as follows; first three layers are conventional convolution-pooling-convolution layers, followed by three locally connected and two fully connected layers. Pooling layers make learned features insensitive to local transformations and makes the architecture robust by preserving maximum texture details. Authors have also contributed to 3D face alignment.

Yi et al. [5] proposed biometric face recognition system using a CNN by collecting which 10,000 subjects - 500,000 images and named it has WebFace database (made publicly available). WebFace trains deep CNN of 17-layers which includes 10 convolutional layers, 5 pooling layers and 2 fully connected layers, having a smaller convolution filter size preserve all the necessary texture information, which is a crucial thing in powerful feature extraction. Grey-level v/s RGB colour images were used in training CNN, the recognition accuracies obtained from both the kind of images yielded very close results. Though the RGB images contain more information, results did not prove greater marginal impact.

Ranjan et al. [6] developed CNN framework for face detection, localization of facial marks, head pose variation estimation and gender recognition. As we know that in CNNs features are hierarchically organized throughout the network, initial layers extract edges and corner points, subsequent higher layers learn complex information that are class specific. During this process, every layer gathers hundreds of feature maps and it is too high to encode. Hence the authors have handled hyper features using feature fusion techniques and have trained them in parallel using multiple loss functions which leads the system in greater performance. Raghavendra et al. [7] proposed a multispectral palmprint biometric recognition system. Multispectral palmprint system is usually developed at the image level (not sensitive to motion blur and illumination) or score level fusion approaches. Firstly they proposed novel ROI extraction method to locate the valley regions, then the fusion is done on dependency measure and finally feature extraction and classification is done by the novel proposal of the authors.

Multispectral palmprint images of desired spectral band can be obtained from different cameras where each camera represent one spectral band or the same camera could be used with different illuminators [8], probably the second idea is worth doing for the following reasons; each time the image does not require registration and movement of the modality could be eliminated. Multispectral images at 4 spectral band is captured - red, green, blue and near infrared. The images are subjected to competitive code analysis and score level fusion is performing superior than the results of analysing single spectral bands alone. Wang et al. [9] proposed multispectral palmprint adopting palmprint and palm vein traits. The traits were acquired from two different cameras (image registration increases excess computation) under two spectral bands namely visible and infrared spectrum, then the image is fused using Wavelet transform. Laplacian palm representation is done before performing classification. Xu [10] performed experimentation of multispectral palmprint at four different spectral bands. Score level fusion using weighted sum rule performed better when compared with individual bands.

3   Methods and Materials: Gabor and CNN

Log-Gabor(LG): Gabor filter was firstly proposed by Dennis Gabor in 1946 [11], log gabor is a STFT which aims in processing the higher frequency band to narrowly localized oriented filters, so the highest and lowest frequencies would uniformly cover the Fourier domain. 2D Gabor filtering is a Gaussian envelope multiplied by complex sinusoidal carrier wave generating signal at a specific frequency and orientation. The Gabor representation of an image is obtained by convolution with a filter that in turn generates the impulse response like Gaussian radius $= 2\pi$, expressed in terms of real and imaginary parts. 2D Gabor filter is given by,

$$f(x, y, \omega, \theta, \sigma_x, \sigma_y) = \frac{1}{2\pi\sigma_x\sigma_y} exp^{-\frac{1}{2}(\frac{x}{\sigma_x})^2 + (\frac{y}{\sigma_y})^2} + j\omega(cos\theta + ysin\theta) \quad (1)$$

σ is the spatial spread, ω is the frequency, θ is the orientation. By using Gaussian as the spread function,2D log Gabor function has the transfer function and is given by,

ω_0 is the filter's centre frequency and σ_r controls the filter's bandwidth. θ_j is the orientation angle, j is the no of orientations, σ_θ is the angular bandwidth of a filter.

Convolutional Neural Networks (CNN): CNN architecture is inspired by mammalian visual cortex. Convolution layers perform convolution with the input image and kernels are usually determined during learning and the kernel weights are shared across the entire image, predefined kernels are not used in CNN.

$$c[n] = (i * k)[n] = \sum_{r=-\infty}^{\infty} i[r]k[n-r] \qquad (2)$$

Output of one convolution layer is input to the next convolution layer. The convolved matrix is processed with a nonlinear activation function ReLu (Rectified linear unit). Pooling layer also called subsampling layers are in interspersed among the convolution layers, these layers performs statistical operation (e.g., max pooling, min pooling that replaces a mXm neighbourhood with the maximum, minimum values respectively) and ensures invariance to scaling, distortion. Finally fully connected layers are appended to last pooling layer which gives the representation of output. Fig. 1, illustrate comparison of both log-gabor and deep features for given multispectral palmprint image.

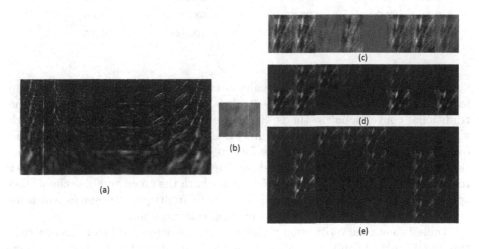

Fig. 1. (a) Log gabor feature with 4 number of wavelet scales and 8 filter orientations (b) Original multispectral palmprint image (c) First convolution layer with 8 filters (d) Second convolution layer with 16 filters (e) Third convolution layer with 32 filters

4   Experimental Results

We have considered PolyU multispectral palm dataset of 250 users of 500 palmprint images are collected, each user has 12 samples and out of which we have split the number of samples for training (8 samples) and testing (4 samples) [16]. The input image size is 128×128, gray scale image. Fixing number of convolution layers to the CNN framework often depends on the nature of data, usually we fix up the layers by trail method looking into its performance. In our framework we have fixed the number of convolution layers to three, Initial Learn Rate is 0.001, maximum epochs is 30, Mini Batch Size is 32. In the first CNN layer, we have considered 8 output filters with 3×3 kernel size and padding is set to 1 because $(k - 1) = 2$ is padding, where k is kernel size. Then the $ReLu$ activation function is executed that retains the positive values and ignores the negative values setting them to zero. In pooling we have chosen Max pooling as our choice with 2×2 size and stride 2. Output from the first layer is input to the second layer and hence this layer has 16 kernels with 3×3 kernel size and the last convolution layer performs convolution with 32 filters of 3×3 kernel followed by one fully-connected layer. $Softmax$ function performs the categorical probability distribution and classifies the sample to its corresponding classes.

Table 1. Results of deep features on multiple-spectral palmprint biometric verification system

FAR%	RED (GAR%)	GREEN (GAR%)	BLUE (GAR%)	NIR (GAR%)
0.01	89.75	96.50	95.25	81.25
0.1	98.00	98.75	98.50	91.00
1	99.75	100.00	100.00	97.75

Table 1 shows the verification results obtained from our novel proposed deep learning framework for the multispectral palm images. When compared to the traditional conventional methods of feature extraction and classification work which we find in the literature, definitely deep learning has the strong essence in overcoming all the hurdles which we face in selection of suitable feature extraction algorithms, kernel selection, addressing noisy data etc., From the Table 1 results we have got 100% GAR at 1% FAR for both the green and blue channelled spectral images and we can claim almost all the multispectral images which we have employed in our work have yielded appreciating results.

Table 2 shows the verification performance of multispectral palm images captured at 4 spectral band - red, green, blue and near infrared, log gabor is adopted as feature extraction algorithm. Images of blue spectral band is performing better and gradually the verification rate is increasing from 0.01% to 1% FAR rate when compared to other spectral images. At 1% FAR threshold almost all the four spectral band images are performing satisfactorily (Figs. 2, 3 and 4).

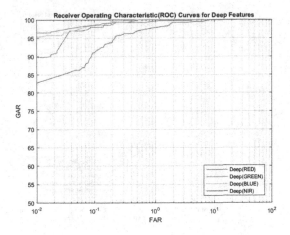

Fig. 2. ROC Curve of deep features on multiple-spectral palmprint biometric

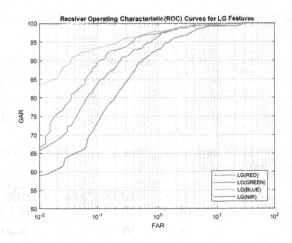

Fig. 3. ROC Curve of log-gabor features on multiple-spectral palmprint biometric

Table 3 depicts the feature level fusion (Log Gabor features) performed by the combination of two (RED+GREEN), (GREEN+BLUE), (RED+BLUE), three (RED+GREEN+BLUE) and four (RED+GREEN+BLUE+NIR) spectral images. NIR images have underperformed in unimodal verification experiments and hence we have not combined NIR features with the other spectral images. In bimodal feature level fusion of Red and Blue, we have obtained good results and this shows combination of these two spectral image features is a feasible choice. We further explored our experimentation from the combination of two spectral images to four spectral images and we can see clearly that as the new features add up, verification rate is also going high to 99.50% GAR at 1% FAR for the combination of all the four spectral images.

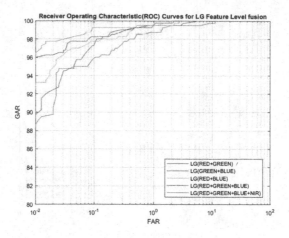

Fig. 4. ROC Curve of log-gabor features on multiple-spectral palmprint biometric

Table 2. Results of log-gabor features on multiple-spectral palmprint biometric verification system

FAR%	RED (GAR%)	GREEN (GAR%)	BLUE (GAR%)	NIR (GAR%)
0.01	65.75	66.50	83.50	59.00
0.1	84.25	89.50	93.25	70.75
1	96.25	97.25	97.75	93.00

Table 3. Results on feature level fusion of log-gabor feature on multiple-spectral of palmprint biometric verification system

FAR%	RED+ GREEN	GREEN+ BLUE	RED+ BLUE	RED+GREEN+ BLUE	RED+GREEN+ BLUE+NIR
0.01	89.75	88.75	93.25	96.00	96.50
0.1	97.25	96.00	97.75	98.25	99.25
1	99.50	98.75	99.50	99.50	99.50

5 Conclusion

Our proposed biometric verification system has shown the performance of both conventional and deep learning method and our results shows comparison between them. Traditional ways of developing a biometric model requires proper selection of feature extraction techniques, classifiers and hence the complexity is more while performing pre and post classification fusion process. Whereas in deep learning we have to provide huge data with all kinds of variations, so that the architecture itself decides selection of kernels for feature extraction implicitly and learns from edges, gradients and network itself adjusts the weights until the desirable results are attained, so in our experimentation we have observed the above said nature in the system. Though the time complexity in deep learning is high but gives the best performance results. The performance of proposed deep learning architecture take more time, we would try to optimize this in our future work.

References

1. Raghavendra, R., Dorizzi, B., Rao, A., Kumar, G.H.: Particle swarm optimization based fusion of near infrared and visible images for improved face verification. Pattern Recogn. **44**(2), 401–411 (2011)
2. Gowda, H.D.S., Kumar, G.H., Imran, M.: Robust multimodal biometric verification system based on face and fingerprint. In: 2017 International Conference on Advances in Computing, Communications and Informatics (ICACCI), Udupi, pp. 243–247 (2017)
3. Hu, G., et al.: When face recognition meets with deep learning: an evaluation of convolutional neural networks for face recognition. In: 2015 IEEE International Conference on Computer Vision Workshop (ICCVW), Santiago, pp. 384–392 (2015)
4. Taigman, Y., Yang, M., Ranzato, M., Wolf, L.: DeepFace: closing the gap to human-level performance in face verification. In: 2014 IEEE Conference on Computer Vision and Pattern Recognition (CVPR), pp. 1701–1708. IEEE (2014)
5. Yi, D., Lei, Z., Liao, S., Li, S.Z.: Learning face representation from scratch. arXiv preprint arXiv:1411.7923 (2014)
6. Ranjan, R., Patel, V.M., Chellappa, R.: HyperFace: a deep multi-task learning framework for face detection, landmark localization, pose estimation, and gender recognition. IEEE Trans. Pattern Anal. Mach. Intell. **41**, 121–135 (2017)
7. Raghavendra, R., Busch, C.: Novel image fusion scheme based on dependency measure for robust multispectral palmprint recognition. Pattern Recogn. **47**(6), 2205–2221 (2014)
8. Zhang, D., Guo, Z., Lu, G., Zhang, L., Zuo, W.: An online system of multispectral palmprint verification. IEEE Trans. Instrum. Meas. **59**(2), 480–490 (2010)
9. Wang, J.-G., Yau, W.-Y., Suwandy, A., Sung, E.: Person recognition by fusing palmprint and palm vein images based on "Laplacian Palm" representation. Pattern Recogn. **41**(5), 1531–1544 (2008)
10. Xu, X., Guo, Z., Song, C., Li, Y.: Multispectral palmprint recognition using a quaternion matrix. Sensors **12**(4), 4633–4647 (2012)
11. Gabor, D.: Theory of communication. J. Inst. Electr. Eng. **93**, 429–457 (1946)
12. Ukil, S., Ghosh, S., Obaidullah, S.M., Santosh, K.C., Roy, K., Das, N.: Deep learning for word-level handwritten Indic script identification. CoRR abs/1801.01627 (2018)
13. Vajda, S., Santosh, K.C.: A fast K-nearest neighbor classifier using unsupervised clustering. In: Santosh, K.C., Hangarge, M., Bevilacqua, V., Negi, A. (eds.) RTIP2R 2016. CCIS, vol. 709, pp. 185–193. Springer, Singapore (2017). https://doi.org/10.1007/978-981-10-4859-3_17
14. Sawat, D.D., Hegadi, R.S.: Unconstrained face detection: a deep learning and machine learning combined approach. CSI Trans. ICT **5**(2), 195–199 (2017)
15. Sawat, D.D., Hegadi, R.S.: Lower facial curves extraction for unconstrained face detection in video. In: Bera, R., Sarkar, S.K., Chakraborty, S. (eds.) Advances in Communication, Devices and Networking. LNEE, vol. 462, pp. 689–700. Springer, Singapore (2018). https://doi.org/10.1007/978-981-10-7901-6_75
16. Zhang, D., Guo, Z., Lu, G., Zhang, L., Zuo, W.: An online system of multi-spectral palmprint verification. IEEE Trans. Instrum. Meas. **59**(2), 480–490 (2010)

SOM-VLAD Based Feature Aggregation for Face Recognition Using Keypoint Fusion

A. Vinay, Ajaykumar S. Cholin, Aditya D. Bhat, Arnav Ajay Deshpande[(✉)],
K. N. Balasubramanya Murthy, and S. Natarajan

Centre for Pattern Recognition and Machine Intelligence, PES University,
100 Feet Ring Road, BSK Stage III, Bengaluru 560085, Karnataka, India
`arnav.deshpande6@gmail.com,cprmi@pes.edu`

Abstract. Over the years, quite a lot of research has been done on using different machine learning techniques for Face Recognition (FR) for identifying the faces of different people. Current FR techniques are still not accurate enough for real world scenarios and pose a lot of problems against varying illumination levels, pose variations, noise and occlusion in the image. Thus, a single keypoint extraction technique may not be suitable for all cases. Hence, in this paper a novel technique is proposed for Keypoint Fusion (KF) obtained by fusing SIFT, SURF and ORB keypoints which is more accurate and suitable for real time application. The paper is also focused on proposing a novel technique of using a Self-Organizing Map (SOM) and Vector of Locally Aggregated Descriptors (VLAD) for image clustering. VLAD is used to extend the SOM's ability to cluster keypoint descriptors. Image classification is carried out using a SGD (Stochastic Gradient Descent) based SVM (Support Vector Machine) classifier. The performance of classification of the proposed framework on benchmark datasets (Grimace, Faces95 and Faces96) has been tabulated and compared with other standard techniques. It is seen that the proposed framework performs better than the BOW (Bag Of Words) model and the KF technique was accurate and quick enough to beat the traditional keypoint extraction techniques.

Keywords: Face recognition · SOM · VLAD · Keypoint Fusion

1 Introduction and Related Work

Face recognition (FR) has become a trending area of interest for research in the fields of computer vision and image processing. It has become a crucial part of biometric systems, where it is used for authenticating a person for privileged access, bank transactions, etc. During the recent years, FR has gained immense attention with lot of research being done to improve the ability of computers to recognize faces in images with varying pose, scale, rotation and illumination levels of the face. Present day research has also focused on developing new machine learning (ML) based models for FR and improving their accuracy metrics [18].

© Springer Nature Singapore Pte Ltd. 2019
K. C. Santosh and R. S. Hegadi (Eds.): RTIP2R 2018, CCIS 1036, pp. 514–523, 2019.
https://doi.org/10.1007/978-981-13-9184-2_46

Feature descriptors used in FR are ranked based on their robustness towards variations in scale, pose, viewpoint, brightness, noise, shearing etc. The various keypoint detection techniques have their own advantages and disadvantages [3,5,7,17]. In FR applications (both in streaming or batch processing), the background and foreground variations of the images are typically unpredictable. So it is important to extract maximum information from the images for better image processing. Hence, we present a novel Keypoint Fusion (KF) technique to achieve this.

In this paper, SOM [9,10] is used as a novel unsupervised clustering technique for FR along with VLAD [11,12]. It is used to form a vocabulary of features that simplifies the images and eases the classification task. It is proven to be quicker than BOW [4] technique and can be of practical usage as shown in Sect. 5. VLAD further extends the vocabulary created by SOM by accumulating the residuals of descriptors with respect to the centroid of its closest cluster which improves the classification ability of the classifier. It is proven in Sect. 5 that SOM-VLAD technique works more efficiently than BOW technique in terms of accuracy and speed.

The classification is then carried out by training a SVM classifier using SGD (Stochastic Gradient Descent) approach. It is shown in the results section of the paper that using such a classifier is quicker and more accurate than other classification techniques.

2 Proposed Framework

Keypoints are extracted from the input image using various techniques such as SIFT, SURF and ORB. These keypoints are then combined using keypoint fusion. The fused set of keypoints is now used to generate a descriptor (Fig. 1). The SOM can now be used to create a vocabulary from the descriptors generated. VLAD is used to extend the SOM. The features extracted after clustering and aggregation using the proposed SOM-VLAD is now used to train a SGD based SVM Classifier. Finally, the match is classified as TP (true-positive), TN (true-negative), FP (false-positive) or FN (false-negative).

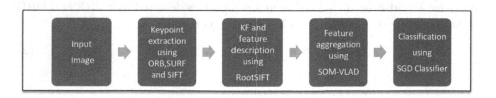

Fig. 1. Block diagram of the proposed framework

2.1 Keypoint Fusion (KF)

The concept of feature fusion has been often used in recent frameworks of FR [1,2] to enhance the reliability and improvise recognition rate. In FR systems, feature fusion involves fusing two or more different feature vectors obtained from different detection techniques on the same image to get a single feature vector which is more informative. The limitation of feature fusion is that, both the feature vectors to be fused should be of the same number of rows. This drawback is generally handled by using techniques like zero padding, dimensionality reduction techniques, concatenating the feature vectors, etc. which may lead to loss of data and complexity. The novel Keypoint Fusion (KF) is similar to feature fusion but here the keypoints (interest points) detected by different detector techniques are combined together (rather than combining the feature vectors) and a single feature description technique is used to calculate descriptors of all keypoints. This novel technique is found to work efficiently both in terms of accuracy and speed overcoming the complexity of feature fusion as we are computing a single feature descriptor. Some of the well known keypoint detector techniques used for FR are SIFT (Scale-Invariant Feature Transform) [3,4], SURF (Speeded Up Robust Features) [5,6] and ORB (Oriented-Fast and Rotated BRIEF) [7].

The SIFT keypoint detection technique, although more than a decade old, is used till today for it's tremendous success in various vision based applications [8]. SIFT keypoint detectors are scale invariant and detect the keypoints with same location but in different directions accounting to stability of matching. SURF keypoint detectors handle the images which are blurred and subjected to rotation but performs poorly in handling changes in brightness and viewpoint of the images. ORB keypoint detection technique is the quickest when compared to SIFT and SURF, is computationally inexpensive and is proven to be rotation invariant and robust towards noise [7]. Hence, we combine all three keypoint detection techniques as explained below. Thus, we obtain a much more informative set of keypoints which are highly robust towards almost all of the variations mentioned above.

The Keypoint Fusion (KF) can be achieved as follows. Let X be the set of keypoint co-ordinates detected by SIFT detector. It is represented as,

$$X = \{x_1, \ldots, x_n\} \tag{1}$$

Where, $x_1, \ldots, x_n$ represents the n keypoint co-ordinates detected by SIFT. Similarly let Y and Z be the sets of keypoint locations detected by SURF and ORB of length m and k respectively. They can be represented as,

$$Y = \{y_1, \ldots, y_n\} \tag{2}$$

$$Z = \{z_1, \ldots, z_n\} \tag{3}$$

Now, our aim is to combine these three sets of keypoint co-ordinates (X, Y and Z) to yield the resultant keypoint co-ordinates set denoted by R, which is more informative. It is achieved by simply taking the union of X, Y and Z, thereby avoiding the duplication of keypoints. The resultant set R is represented as,

$$R = \{X \cup Y \cup Z\} \tag{4}$$

Though the proposed technique adds little overhead of time, it increases the overall efficiency of the system and can be efficiently used in both batch processing and Real-Time FR applications like face authentication at ATMs, fraud detection, etc, where accuracy is given at most importance. From Fig. 2, it is evident that the order of the number of keypoints detected is SURF > SIFT > ORB and these keypoints are fused to get more informative features (Fig. 2.(d)).

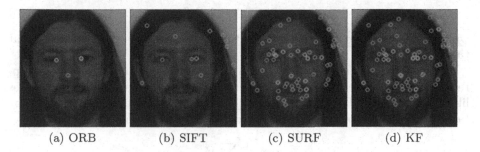

(a) ORB   (b) SIFT   (c) SURF   (d) KF

Fig. 2. Fusion of ORB, SIFT and SURF keypoints

2.2 Self-organizing Maps (SOM)

Self-Organizing Maps (SOM) [9,10] belong to a class of unsupervised competitive learning networks. They help in visualizing high dimensional data by reducing the data to a map while also clustering the data. The input layer is fully connected to the network which is formed by a 2D-lattice of nodes. Figure 3 shows the architecture of the SOM. Each node has an n -dimensional weight vector $[w_1, \ldots, w_n]$ associated with it, where n is the dimension of the input vector i.e $[v_1, \ldots, v_n]$. The network is initialized with random weights. A random data point is selected and the Best Matching Unit (BMU) is determined using Euclidean distance, given as,

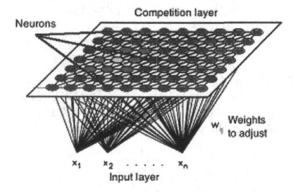

Fig. 3. The SOM architecture

$$Dist = \sqrt{\sum_{i=0}^{n}(V_i - W_i)^2} \qquad (5)$$

where, V and W are the input and weight vectors respectively.

Weights are updated in the neighbourhood of the BMU, where the neighbourhood is an exponential decay function over time, given as,

$$\sigma(t) = \sigma_0 e^{\frac{-t}{\lambda}}, t = 1, 2, 3, \ldots \qquad (6)$$

where λ denotes time constant, t is the time step, and σ_0 is the lattice width at time t_0.

The SOM can thus be used to create a vocabulary to simplify images by representing it as a collection of features that best describe it as done in techniques like Bag of Words (BOW) [12]. Here, the codebook of k clusters is formed by taking the centroid of the clusters formed by SOM, given as,

$$c_i = \frac{1}{N}\sum_{j \in i}V_j \qquad (7)$$

Where, c_i is the centroid of the ith cluster, V_j is the jth vector of the cluster i, and N is the total number of vectors in the cluster.

2.3   VLAD (Vector of Locally Aggregated Descriptors)

The SOM is used to create a codebook of k clusters given as,

$$C = \{c_1, \ldots, c_k\}, where \; k \; is \; the \; number \; of \; centroids. \qquad (8)$$

VLAD (Vector of Locally Aggregated Descriptors) [11,12] which is a first-order extension of the BOW model can now be used to extend the codebook created by the SOM. The local descriptor of the image x is assigned to its nearest neighbor. Feature quantization is carried out by accumulating the residuals $x - c_i$, for each x, where $c_i = NN(x)$.

The VLAD vector v of the image is now given by:

$$v_{i,j} = \sum_{NN(x)=c_i}(x_j - c_{i,j}) \qquad (9)$$

where $i = 1, 2, 3, \ldots, k$ and $j = 1, 2, 3, \ldots, d$

Here, d is the dimension of the local descriptor and k is the number of clusters. The dimension of our representation is given by $D = k \times d$.

L_2 norm is applied on the vector v,

$$v = \frac{v}{\|v\|_2} \qquad (10)$$

Aggregating the residuals adds more information to the descriptors which can help increase the discrimination power of the classifier.

2.4 Stochastic Gradient Descent (SGD)

Stochastic Gradient Descent (SGD) [16] is a simple and effective approach to learning linear classifiers such as Support Vector Machines. This approach seeks to minimize an error or a loss function that models the error in the estimation. It uses one data point at a time in order to minimize a loss function which indicates how good the prediction of the model is. Thus, the model parameters are updated after each training instance. This is in contrast to true Gradient Descent (GD) where the model parameters also known as weights, are updated after having gone through the entire training dataset instance by instance i.e., after every epoch. This would rather take a long time in the case of very large datasets for the learning process to converge. Instead, the SGD method of updating the weights after every instance is adopted. SGD has been shown to converge to the minimum of the error function faster than GD.

Let w be the vector of the weights of the model (the model parameters), η be the learning rate, Q_i be the error function associated with the ith observation in the data.

To start with, random initial weights are assigned. Then the update rule for the weights is given by:

$$w = w - \eta \nabla Q_i(w) \tag{11}$$

where $\nabla Q_i(w)$ is the gradient of the error function associated with the ith observation with respect to w. The update in Eq. 11 has to be repeated for all values of i, i.e. for all instances in the training data until an approximate minimum of the error function is reached. The training samples need to be randomly shuffled after every epoch.

3 Algorithm

This section shows the algorithm used in the proposed framework in both the training and testing phase.

4 Databases

Three Benchmark databases namely Faces95 [13], Faces96 [14] and Grimace [15] each consisting of images of equal size were used to test and evaluate the framework. Figure 4 shows a sample of images from the above mentioned databases. The Faces95 database has images from 72 classes having 20 images per class, each having variation in background, lighting, illumination, head-scale and expression. The Faces96 database has images from 152 classes having 20 images per class, each having variation in illumination, inclination and size of face area. The Grimace database is similar to the Faces96 database and has images from 18 classes having 20 images per class, containing sequences of images where the subject makes grimaces and adjusts his/her head.

1: **procedure** TRAINING PHASE
2:     1. For each image in the dataset:
3:       i) Extract keypoints using various techniques such as SIFT, SURF, ORB
4:       ii) Fuse the keypoints
5:       iii) Generate and aggregate descriptors on the resulting keypoints of the image after KF.
6:     2. Construct a vocabulary of k-clusters from the aggregated descriptors by using SOM
7:     3. For each image:
8:       i) Initialize a VLAD Matrix ($K \times D$)
9:       ii) For each feature in the image:
10:         a) Find the closest centroid to the feature vector by using L_2 norm.
11:         b) Calculate the difference between feature vector and the centroid.
12:         c) Add the difference to the row of the VLAD matrix.
13:         d) Use L2 norm to normalize the VLAD matrix.
14:         e) Reshape the matrix to a 1× KD row matrix.
15:     4. The VLAD vectors and the labels corresponding to them are used to train a SGD Classifier.
16: **procedure** TESTING PHASE
17:     1. For each image :
18:       i) Generate descriptors as done in training phase.
19:       ii) Calculate the VLAD vector as given above.
20:       iii) Use the classifier to predict the label corresponding the obtained vector.

(a) Faces96 Database   (b) Grimace database   (c) Faces95 Database

Fig. 4. Sample of images from the databases used.

5   Results and Inferences

In this section, we report the experimental results obtained on the benchmark databases making use of the proposed technique. Comparison of accuracy obtained using Keypoint Fusion (KF) with different description techniques such as SIFT, SURF and RootSIFT have been tabulated in Table 1. It can be seen that KF with RootSIFT descriptor (KF-RootSIFT) technique gives the highest accuracy.

Table 2 shows the comparison of the average accuracy obtained on benchmark datasets by using various techniques. It can be seen from Table 2 that using KF improves accuracy over methods using a single keypoint detection by about 3%. Techniques using SOM to create the vocabulary are seen to perform better than

Table 1. Comparison of results obtained on Benchmark Datasets.

Method	Accuracy (%)		
	Grimace	Faces95	Faces96
KF-SIFT	95.08	94.34	89.8
KF-SURF	95.44	95.39	90.86
KF-RootSIFT	**98.2**	**97.04**	**93.3**

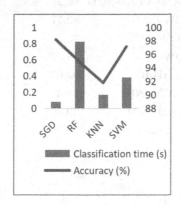

Fig. 5. Accuracy and time for some well known classifiers in our pipeline.

those using BOW/VLAD. Also, the SGD based classifier is seen to give better classification accuracy when compared with other classifiers.

Figure 5 Shows the average classification time taken (in seconds) by various well-known classification techniques, as seen when tested on Grimace dataset with the proposed technique. It is evident from the graph that SGD classifier is the most efficient both in terms of speed and accuracy than SVM. Although the RF classifier is having considerable accuracy, it is much slower than other classifiers. The KNN classifier has the least accuracy. SGD performs better than other techniques due to the fact that the model parameters are updated after each training instance in contrast to other techniques (GD) where model parameters (weights) are updated after every epoch. Figure 6 shows the average time taken (in seconds) by various techniques to create the vocabulary for the Grimace dataset. It is observed that the SOM is quicker than BOW in creating a vocabulary for the whole dataset. In addition to being quicker, SOM technique can also be used with VLAD to perform better and ease the classification technique. SOM-VLAD takes less time than the BOW model for the same. From the results shown, SOM is seen to be a better feature aggregation technique to ease classification when compared with other techniques. Using VLAD with SOM helps further improve the accuracy of classification of the SGD classifier. Performing KF helps improve accuracy as it is able to give a better set of keypoints for the description technique to work on. Incorporating these, we see that the accuracy and efficiency of classification is improved.

Table 2. Comparison of accuracy of proposed method with some standard techniques.

Method	Average accuracy (%)
SIFT-SOM-VLAD-SGD	94.56
SURF-SOM-VLAD-SGD	95.90
SURF-SOM-VLAD-RF	93.16
SIFT-SOM-VLAD-RF	92.8
DSIFT-SOM-VLAD-SGD	96.29
SIFT-BOW-GNB	82.77
SURF-BOW-GNB	92.22
SIFT-BOW-SVM	74.37
SURF-VLAD-RF	93.05
SIFT-VLAD-GNB	85.3
SIFT-VLAD-SVM	90.56
Proposed framework	**98.2**

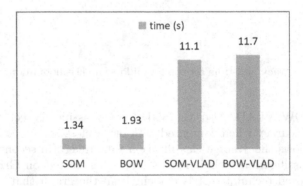

Fig. 6. Time taken to form the vocabulary using various clustering techniques

6   Conclusion and Future Work

While training a system for image classification, it is crucial to consider varying background, lighting, illumination, rotations and translations. A single keypoint extraction technique may not be suitable in all cases. Therefore, the proposed framework addresses the issue and was tested on benchmark databases having the above mentioned variations. Incorporating SOM-VLAD into the technique improved accuracy of classification when compared to other standard techniques by about 5% while also being efficient in terms of time taken (as observed in the results). Thus, it could be a suitable solution in a real-time classification system where we would need an accurate and real-time response.

Future work is focused on improving the accuracy of the framework and also making the framework more efficient in terms of speed by making use of various optimization techniques. Different keypoint extraction and description techniques could be considered to improve the same. Also, different descriptors like RSILC [17] which take into account information about local factors like intensity and color and global factors like line inter-positioning can be included in keypoint fusion.

References

1. Lei, Y., Bennamoun, M., El-Sallam, A.A.: An efficient 3D face recognition approach based on the fusion of novel local low-level features. Pattern Recogn. **46**(1), 24–37 (2013)
2. Tan, X., Triggs, B.: Fusing gabor and LBP feature sets for Kernel-based face recognition. In: Zhou, S.K., Zhao, W., Tang, X., Gong, S. (eds.) AMFG 2007. LNCS, vol. 4778, pp. 235–249. Springer, Heidelberg (2007). https://doi.org/10.1007/978-3-540-75690-3_18
3. Geng, C., Jiang, X.: Face recognition using sift features. In: 2009 16th IEEE International Conference on Image Processing (ICIP), pp. 3313–3316. IEEE (2009)
4. Jurie, F., Schmid, C.: Scale-invariant shape features for recognition of object categories. In: CVPR, vol. II, pp. 90–96 (2004)
5. Bay, H., Tuytelaars, T., Van Gool, L.: SURF: speeded up robust features. In: Leonardis, A., Bischof, H., Pinz, A. (eds.) ECCV 2006. LNCS, vol. 3951, pp. 404–417. Springer, Heidelberg (2006). https://doi.org/10.1007/11744023_32
6. Bay, H., Ess, A., Tuytelaars, T., Van Gool, L.: Speeded-up robust features (SURF). Comput. Vis. Image Underst. **110**(3), 346–359 (2008)
7. Rublee, E., Rabaud, V., Konolige, K., Bradski, G.: ORB: an efficient alternative to SIFT or SURF. In: 2011 IEEE International Conference on Computer Vision (ICCV), pp. 2564–2571. IEEE (2011)
8. Lowe, D.G.: Object recognition from local scale-invariant features. In: The Proceedings of the Seventh IEEE International Conference on Computer Vision, vol. 2, pp. 1150–1157. IEEE (1999)
9. Kohonen, T.: Self-Organizing Maps. SSINF. Springer, Heidelberg (1995). https://doi.org/10.1007/978-3-642-97610-0
10. Kohonen, T.: The self-organizing map. Proc. IEEE **78**(9), 1464–1480 (1990)
11. Arandjelovic, R., Zisserman, A.: All about VLAD. In: Proceedings of the IEEE Computer Society Conference on Computer Vision and Pattern Recognition, pp. 1578–1585 (2013). https://doi.org/10.1109/CVPR.2013.207
12. Jegou, H., Perronnin, F., Douze, M., Sanchez, J., Perez, P., Schmid, C.: Aggregating local image descriptors into compact codes. IEEE Trans. Pattern Anal. Mach. Intell. **34**, 1704 (2012)
13. The Faces95 database. http://cswww.essex.ac.uk/mv/allfaces/faces95.html
14. The Faces96 database. http://cswww.essex.ac.uk/mv/allfaces/faces96.html
15. The Grimace database. http://cswww.essex.ac.uk/mv/allfaces/grimace.html
16. Bottou, L.: Large-scale machine learning with stochastic gradient descent. In: Lechevallier, Y., Saporta, G. (eds.) Proceedings of COMPSTAT 2010, pp. 177–186. Physica, Heidelberg (2010). https://doi.org/10.1007/978-3-7908-2604-3_16
17. Candemir, S., Borovikov, E., Santosh, K., Antani, S., Thoma, G.: RSILC: rotation- and scale-invariant, line-based color-aware descriptor. Image Vis. Comput. **42**, 1–12 (2015)
18. Sawat, D., Hegadi, R.: Unconstrained face detection: a deep learning and machine learning combined approach. CSI Trans. ICT. **5**, 1–5 (2016). https://doi.org/10.1007/s40012-016-0149-1

A Class Specific Representation Learning for Illumination Tolerant Face Recognition

Tiash Ghosh[(✉)] and Pradipta K. Banerjee[(✉)]

Department of Electrical Engineering,
Future Institute of Engineering and Management, Kolkata, India
tiazghosh@gmail.com, pradiptak.banerjee@ieee.org

Abstract. An approach of class specific representation based learning for illumination tolerant face recognition is reported in this paper. Autoencoder based representation and class specific reconstruction along with phase correlation in frequency domain for classification is proposed. Autoencoder based representation is evaluated as very few number of training images are sufficient to handle the entire variation of test face subspace. Phase correlation is used at the classification stage to handle the illumination problem as intensity is the primary concern. This judicial combination of representation and classification shows improved recognition accuracy on benchmark databases. The performance of the proposed approach compared to another state-of-the-art technique on other representation based learning is established with extensive experimental. Advantage of the proposed approach is also shown by the performance analysis with single training image, which is necessary for some real time applications.

Keywords: Representation learning · Class autoencoder ·
Phase correlation · Face recognition

1 Introduction

Representation based classification shows significant performance in face recognition. Several approaches have been made in recent years regarding face recognition using representation learning. Sparse representation [6] is widely used technique in face recognition. In sparse representation, a test face/image is considered to be the linear combination of training images. This type of representation shows prominent success in face recognition with large number of available training images. However, in real face recognition system, there is a limitation of having sufficient number of training images. Face classification based on pseudo-full face representation is reported in [3]. Apart from sprase representation, some different approaches can be found in local binary pattern representation of face [8], metaface learning [9], vocabulary based representation [4] and collaborative

K. C. Santosh and R. S. Hegadi (Eds.): RTIP2R 2018, CCIS 1036, pp. 524–537, 2019.
https://doi.org/10.1007/978-981-13-9184-2_47

representation [7], Unconstrained face detection [1] and Rotation- and Scale-Invariant, Line-based Color-aware descriptor [2].

Another widely used approach of representation based learning is based on principal component analysis (PCA) [5,10]. From 1D to 2D, distinguished PCA based approaches are reported in face recognition. Some approaches are gabor kernel based PCA [12], modular eigenfaces [11]. PCA is good for representation and not for classification. PCA is generally used for data representation through dimensionality reduction. One approach of PCA based representation and face recognition is successfully reported in Core-face [13]. In this paper a frequency domain approach is exploited. The principal component analysis is carried out in frequency domain. Phase spectrum of the images are taken to form the subspace. Further the phase of test faces are projected and reconstructed through subspace. Input phase and corresponding reconstructed phase is then correlated for decision purpose. Several frequency domain approaches are there for face recognition [16,17]. Among these, Coreface provides promising results in illumination tolerant face recognition. Motivated by Coreface approach, in this paper a simple representation/reconstruction based class specific approach is proposed for illumination tolerant face recognition. The representation of faces evaluated through autoencoder learning. Most of the recent researches on representation learning are based on encoders, auto encoders and sparse encoders. Vincent et al. [14] in his paper introduced a new training principle for unsupervised learning of a representation to make it robust to partial corruption of the input pattern. Many representation based learning based on sprase autoencoder can be found in [15] Simple autoencoders are trained for specific classes. Hence a class based reconstruction is available from the trained autoencoders. This class based reconstruction develops class specific subspaces which are responsible for the representation of respective classes. This further helps to classify the test face, by means of reconstruction, when projected onto the class specific subspaces. Here projection stands for the encoding and decoding of the test face onto the trained weights of the autoencoder. During the training phase the training images are encoded to a very low dimensional space. These low dimensional features are further decoded to represent the input face image. Nature of the reconstructed face image depends on the reduced dimension through encoding. To exploit the entire information of both input image and reconstructed image, phase spectra are evaluated for both. Phase correlation approach is then performed at the decision stage. As phase contains more information than magnitude, the advantage of phase correlation is exploited for further classification. This proposed is evaluated on two benchmark databases (1) PIE [20] and (2) Extended YaleB [19] which contains frontal faces with different illumination. Extensive results show the robust performance of the proposed approach in face recognition accuracy. Comparative result is also given to establish the out-performance of the proposed strategy over state-of-the-art Coreface approach. In Sect. 2 the methodology of the proposed approach is described in detail. Detail extensive result is given in Sect. 3.

2   Methodology

2.1   Class Specific Representation Learning

In representation based studies the relevant information are extracted for classifiers or predictors through learning the data representation. Recent years, in
representation based learning, two parallel approaches are studied. One from
probabilistic graphical approach and one from neural network approach. Perhaps
more common characteristics are present in both the approaches than separating them. Restricted Boltzman Machine (RBM) is used in probabilistic model
where as auto-encoder and its variants are exploited in neural network based representation learning. Principal component analysis (PCA), probably the oldest
feature extraction and dimensionality reduction algorithm, illustrates all probabilistic, auto-encoder and manifold views in representation learning. PCA learns
the linear transformation

$$h = f(x) = W^T x + b$$

for input $x \in \Re^{n_x}$, where, n_h is the number of orthogonal directions of largest
variance in the training data and W matrix contains the orthogonal basis for n_h
orthogonal directions along its columns. Reconstruction error of the test images
is minimized by PCA for representation learning as:

$$L = \sum_j \|x^{(j)} - \hat{x}^{(j)}\|^2 \tag{1}$$

where, L represents the loss which is to be minimized for perfect reconstruction.
$x^{(j)}$ represents the jth original sample and its reconstructed version is denoted
as $\hat{x}^{(j)}$, which is further expressed as:

$$\hat{x}^{(j)} = \sum_{i=1}^{n_p} z_i^{(j)} u_i + \sum_{i=n_p+1}^{n_x} b_i u_i \tag{2}$$

The objective given in Eq. (1) is minimized when

$$z_i^{(j)} = (x^{(j)})^T u_i \tag{3}$$

and

$$b_i = \bar{x}^T u_i \tag{4}$$

where, $\bar{x}$ is the mean of $x^{(j)}$, $j = 1, 2, 3 \ldots, m$

Figure 1 shows the special architecture of PCA, known as auto-encoder. The
architecture of auto-encoder is very much similar to PCA architecture as shown
in Fig. 1(a). The only difference is that, PCA is used as a linear mapping and
autoencoders, a neural network technique, is used as a non-linear mapping. In the
auto-encoder framework feature extraction is done with a parametrized function
in closed form. This function, f_θ is called the encoder. This encoder efficiently
compute the feature vector from the input x, as

$$h = f_\theta(x)$$

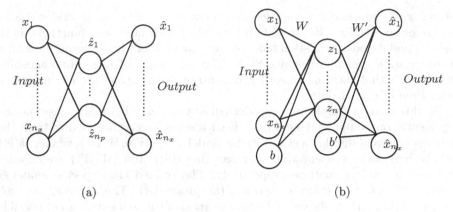

Fig. 1. (a) Graphical representation of PCA in form of neural network architecture (b) Auto-encoder with bias and weights.

Each input vector $x^{(j)}$ from a data set $\{x^{(1)}, x^{(2)} \dots, x^{(m)}\}$ is encoded nonlinearly in lower dimensional space or latent space as

$$h^{(j)} = f_\theta(x^{(j)})$$

where $h^{(j)}$ is the feature vector representation of $x^{(j)}$. For reconstruction, it is needed to map the feature space back to input space. This is done by decoder g_θ, parametrized by weights as

$$\hat{x} = g_\theta(h) \tag{5}$$

Different strategies exist in auto-encoder training. A set of weights, Θ, known as parameters of encoder and decoder are learned simultaneously. The objective is to minimize the reconstruction loss $L(x, \hat{x})$ – a measure of the difference between the original target x and its reconstruction $\hat{x}$ – over all training examples $\{x^{(1)}, x^{(2)} \dots, x^{(m)}\}$. A regularized auto-encoder, one of the variants of auto-encoder family, can be used for preventing the zero reconstruction loss responsible for reconstructing the identity function everywhere in all respects. When this regularization is made by constraining the data dimensionality reduction in latent space, it resembles to PCA.

In general, auto-encoder training consists of finding the parameters θ minimizing the reconstruction error

$$L_{AE}(\theta) = \sum_j L(x^{(j)}, g_\theta(f_\theta(x^{(j)}))) \tag{6}$$

This minimization of reconstruction error is carried out by adaptive momentum gradient descent approach. The advantage of using encoder-decoder structure is that a nonlinear mapping from input space to lower dimensional latent space can be achieved as:

$$f_\theta(x) = s_f(Wx + b)$$
$$g_\theta(h) = s_g(Wh + c) \tag{7}$$

where, s_f, s_g are encoder and decoder activation functions, generally chosen as sigmoid or rectified linear unit. If instead of such nonlinear functions, both encoder and decoder activation functions are linear, then the above auto-encoder representation system resembles PCA. The advantage of using auto-encoders over PCA is that these nonlinear activation functions extract the nonlinear features from the input data.

In this study a simple auto-encoder (shown in Fig. 1(b)) based class specific representation system is done. Inputs from a specific class are used to train the auto-encoder and develop a class specific model $\Theta = \{W, b, W', b'\}$, where, $\{b, b'\}$ are the bias vectors at encoding and decoding stage and $\{W, W'\}$ are encoder and decoder weight matrices respectively. The trained class specific model Θ is used at the testing stage to represent the probe data. The advantage of such representation is that, the faithful representation of the probe data is not possible from the specific class model Θ if there is a class difference between Θ (from which it is trained) and probe data (from which it is taken for testing). This strategy helps the classifier at the decision stage for classification.

Figure 2 shows the class specific reconstruction for two different class of images (class-1, class-2) while the class specific model Θ is developed by training of class-1 images. The advantage of class specific representation is clearly visible in Fig. 2 and the purpose of such representation is prominent.

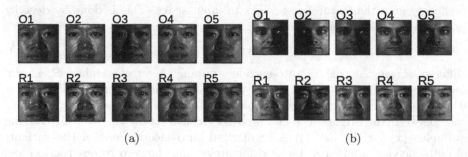

Fig. 2. (a) Probe image and model Θ from same class. Original probe images are shown in top row. Bottom row represents the reconstruction version of respective probe images. (b) Probe image and model Θ from different class. Ri is the reconstruction form of Oi

2.2 Phase Correlation Classifier

Authentication or classification in face recognition system in frequency domain is carried out by cross-correlation technique as pictorially given in Fig. 3. Fourier transform of the test face image is cross-correlated with synthesized correlation filter (CF) designed from a set of Fourier transformed training images. The response of the cross-correlation is inverse Fourier transformed and is searched for distinct peak. If any such distinct peak is found in the response surface, the authentication is made.

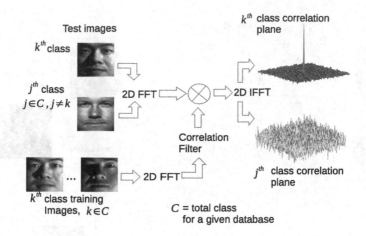

Fig. 3. Frequency domain cross-correlation technique for face classification

The process given in Fig. 3 can be mathematically summarized as: $\mathbf{F}_x$ and $\mathbf{F}_{h_f}$ denote the frequency domain counterpart of spatial domain image x and CF h_f. $\mathbf{F}_y$ is the Fourier transformed test image. The cross-correlation surface s in spatial domain can be expressed as

$$s = FFT^{-1}[\mathbf{F}_y \circ \mathbf{F}_{h_f}^*] \tag{8}$$

where, $\circ$ represents the element wise array multiplication, $*$ stands for complex conjugate operation, and FFT is an efficient algorithm to perform DFT. A distinct peak with high value of peak to sidelobe ratio (PSR) [18] is obtained, when any kth class image is correlated with kth CF as shown in Fig. 3. This indicates authentication of test face. Whereas no such peak will be found in case of imposter as shown in Fig. 3 when test face is from jth class.

Instead of designing any correlation filter with several training images, this approach exploits simple matched filtering technique in frequency domain where only single image is used for training and testing. As phase contains more information than magnitude, the phase spectrum of both input and output images from auto-encoders are extracted and then correlated:

$$s = FFT^{-1}[e^{j\phi_y} \circ e^{-j\phi_{\hat{y}}}] \tag{9}$$

where, $j = \sqrt{-1}$, ϕ_y and $\phi_{\hat{y}}$ represents the phase angle of input image to the trained auto-encoder model (Θ) and phase angle of output obtained from the model. It is obvious from the Fig. 4 i.e. when probe image from class k is encoded and decoded with model Θ_j (model Θ is trained with class-j training images), Eq. (9) gives the surface s with no such distinct peak, whereas probe images from jth class after encoded-decoded with Θ_j gives a surface s having distinct peaks.

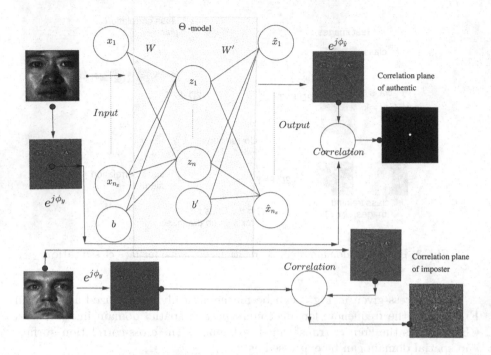

Fig. 4. Detail block diagram of the proposed approach

2.3   Detail Description of the Proposed Approach

A model Θ is developed from a trained encoder for a specific class of images (say jth class). This model Θ is originated from encoding weights W of dimension 32×10000 and decoding weights W' of dimension 10000×32 where 32 stands for the encoding dimension and 100×100 sized image is lexicographically ordered. When the test image from jth-class is encoded and decoded with the model Θ, a corresponding reconstructed image is obtained as shown in Fig. 4. The phase spectrum of both input and projected image are extracted and correlated. The resulting correlation plane is depicted in Fig. 4 with bright white spot (corresponding to the peak value) at the center position. Similar approach is made with the test image from class k as shown in Fig. 4. The decoded image looks like jth-class image. Phase correlation of both kth image and its decoded image, results in a correlation plane, shown in Fig. 4, provides no such peak in the correlation plane.

3   Experimental Results

3.1   Database and Preparation

Two benchmark databases are used in this paper. (1) PIE database which contains 65 classes each having 21 frontal face images under poor lighting conditions.

Each images are grey-scaled as the illumination information is the primary concern. Further each images are resized to 100×100 for experimental studies. (2) The Extended Yale Face Database B [19] contains 38 individuals under 64 different illumination conditions with 9 poses. Only frontal face images are taken for experiments. Frontal cropped face images are readily available from website[1]. All grey-scale images are down-sampled to a size of 64×64 for experimental purpose.

3.2 Representation Performance

In the first phase of experiment, the correlation peak values for authentic and imposter images are observed. On the correlation surfaces, PSR values are computed and evaluated for all class of images. Figure 5 shows the reconstruction effect of autoencoder. In each case PSR values from the correlation planes are evaluated. A hard threshold value of $PSR = 10$ has been chosen for classification purpose; $PSR < 10$ termed as imposters and $PSR \geq 10$ termed as authentic. It has been observed that, with the increase of encoding dimension at the latent space, recognition accuracy increases. Two different encoding dimensions have been chosen as 8 and 32. As 100% accuracy is obtained with encoding dimension $= 32$, the latent space dimension is maintained at this dimension for further studies.

3.3 PSR Distribution

In this phase of experiment performance of the proposed approach is measured using PSR distribution over the entire PIE database. Different training set randomly chosen for 20 times, such as 10%, 15% and 20% training images are taken for different classes and Θ model is evaluated through auto-encoder fit. 90%, 85% and 80% test images from all classes are taken and reconstructed through class specific Θ model. Phase spectrum of both probe image and its reconstructed version are evaluated and correlated. From the correlation surface PSR value is calculated and plotted in Fig. 6. It is observed from Fig. 6 that a clear demarcation between authentic and imposter classes exists even if a very few number of training images are taken for training such as 10%.

3.4 ROC Plots

Performance of classifiers can be efficiently measured by receiver operating characteristic curve (ROC) analysis. ROCs are made by plotting true positive rate (along Y-axis) vs false positive rate (along X-axis). For an ideal classifier the ROC provides a step function. The following experiments show the ROC response of different classes from PIE data-set. Taking all of these curves, the mean area under curve is calculated, and variance of the curve is critically observed when the training and testing set is split into different subsets and different encoding dimensions are chosen (Fig. 8). Figure 8 shows how the proposed

---

[1] http://vision.ucsd.edu/~leekc/ExtYaleDatabase/ExtYaleB.html.

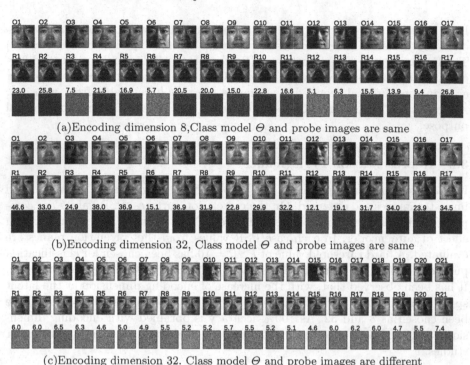

Fig. 5. Reconstruction and corresponding PSR values. Reconstruction of probe images from class-1 are shown for test size 0.8, i.e. 20% of total images are taken as training purpose. It is observed from the above figures, (a) five (5) images having PSR values less than 10 and hence misclassified. (b) No images are misclassified if hard threshold is selected as PSR = 10. Interesting result is obtained in case (c) where the probe images are taken from class-19 whereas the Θ model is developed with class-1. After phase correlation between O_i and R_i, all the correlation surfaces having $PSR < 10$ indicate imposters

classifier output is affected by changes in the training data as well as encoding dimensions. In each case mean ROCs of all 65 classes for PIE and 39 classes for YaleB are plotted to ensure the overall performance of the whole database under the proposed scheme. In addition to ROC plots, mean AUCs are also calculated and given in the figures. It is observed from Fig. 8 that, $AUC = 0.99$ with $std = \pm 0.2$ obtained from the proposed approach, while encoding dimension is set to 16 or 32, 20% training images are taken and auto-encoder is trained with 100 epochs. Obtained AUC value close to 1 indicates an appreciable performance of the proposed method over the entire PIE and YaleB database under poor illumination conditions.

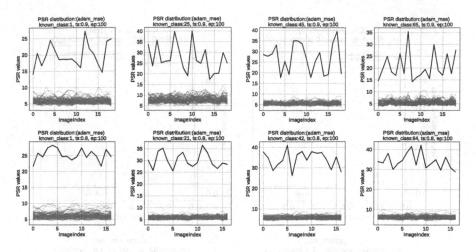

Fig. 6. Average PSR distribution of different classes of PIE database while 10% (top row), and 20% (bottom row) of training images are randomly taken 20 times. With the authentic class all 64 imposter class PSR values are plotted (as red). It can be easily observed that a clear demarcation line is present anywhere above the $PSR = 10$ line to distinguish authentic and imposters in all the cases. (Color figure online)

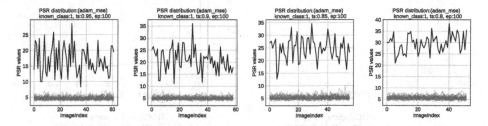

Fig. 7. Average PSR distribution of different classes of YaleB database. It is observed from the PSR distribution that with lesser number of training images, better recognition accuracy can be achieved with $PSR = 10$ taken as hard threshold.

3.5   Comparative Study

The performance of proposed approach is compared with another frequency domain representation based learning: Corface. Figure 9 shows the PSR distribution of a specific class (authentic) of images against rest of the classes (imposter). Three different training sets are taken for both the cases as 10%, 15% and 20%. Same training images are used to train both the classifiers. From Fig. 9, it is clear that, much overlap of PSRs for authentic and imposters is found in case of Coreface comparing to proposed method. This is due to the fact, the non-linearity in subspace analysis is considered in case of proposed method, whereas Coreface exploits only the linear subspace based representation. Hence better modeling of the unseen images is obtained in case of proposed method. Another advantage of the proposed approach is single image training which is not possible

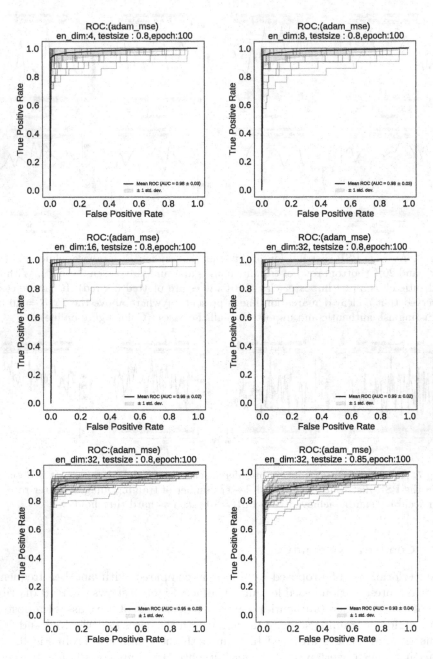

Fig. 8. ROC plots for all 65 classes of PIE database at different training sizes and for different encoding dimensions. It is observed that almost 99% recognition accuracy is obtained with the proposed method when encoding dimension is chosen 32 and 20% training images are taken to fit class specific auto-encoder. Bottom row shows ROC plots for all 39 classes of Yale database using proposed method.

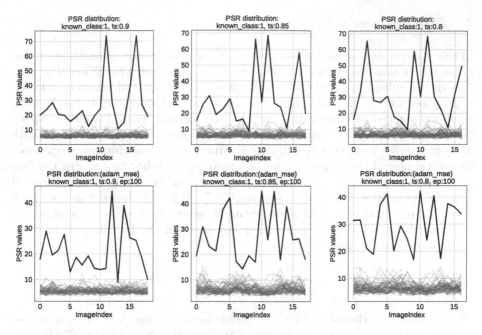

Fig. 9. Comparative results on PSR distribution of Coreface (top row) and proposed method (bottom row). It is evident from the figure that improved PSR distribution is obtained in the proposed method.

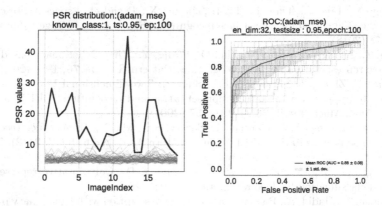

Fig. 10. Advantage of the proposed system is shown in terms of PSR distribution and ROC plots where single training image is taken to train the proposed system

for linear PCA based analysis in Coreface. Figure 10 shows the PSR distribution of an authentic class alongwith the imposter images while a single training image is considered to design the class specific model Θ. Figure 10 also shows the ROC plot of 65 classes of PIE dataset and mean AUC for single training image training.

4   Conclusions

In this paper a representation learning based on autoencoder reconstruction and phase correlation is proposed for illumination tolerant face recognition. The nonlinearity of subspace based representation is carried out and phase correlation is proposed at the decision stage. The proposed method shows outperforming results compared to state-of-the-art Coreface method. From extensive experimental study, the performance of the proposed approach on both benchmark data-sets shows improved recognition accuracy as well as imposter rejection. The advantage of the proposed approach is also established by showing the face recognition performance on a single training system. From Fig. 10 it is observed that the mean ROC plot of 65 class has $AUC = 0.88$, which is not desired for a good face recognition system. However the proposed system can be improved by incorporating stacked autoencoder. Further research can be extended for pose and noise variation face recognition.

References

1. Dattatray, S., Hegadi, R.: Unconstrained face detection: a deep learning and machine learning combined approach. CSI Trans. ICT **5**(2), 195–199 (2017)
2. Candemir, S., Borovikov, E., Santosh, K.C., Antani, S.K., Thoma, G.R.: RSILC: rotation-and scale-invariant, line-based color-aware descriptor. Image Vis. Comput. **42**, 1–12 (2015)
3. Yang, X., Liu, F., Tian, L., Li, H., Jiang, X.: Pseudo-full-space representation based classification for robust face recognition. Sig. Process.: Image Commun. **60**, 64–78 (2018)
4. Fan, Z., Zhang, D., Wang, X., Zhu, Q., Wang, Y.: Virtual dictionary based Kernel sparse representation for face recognition. Pattern Recogn. **76**, 1–13 (2018)
5. Yang, J., Zhang, D., Frangi, A.F., Yang, J.: Two-dimensional PCA: a new approach to appearance-based face representation and recognition. IEEE Trans. Pattern Anal. Mach. Intell. **26**(1), 131–137 (2004)
6. Wright, J., Yang, A.Y., Ganesh, A., Sastry, S.S., Ma, Y.: Robust face recognition via sparse representation. IEEE Trans. Pattern Anal. Mach. Intell. **31**(2), 210–227 (2009)
7. Zhang, L., Yang, M., Feng, X., Ma, Y., Zhang, D.: Collaborative representation based classification for face recognition. CoRR (2012)
8. Ahonen, T., Hadid, A., Pietikainen, M.: Face description with local binary patterns: application to face recognition. IEEE Trans. Pattern Anal. Mach. Intell. **28**(12), 2037–2041 (2007)
9. Yang, M., Zhang, L., Yang, J., Zhang, D.: Metaface learning for sparse representation based face recognition. In: IEEE International Conference on Image Processing, Hong Kong, pp. 1601–1604 (2010)
10. Turk, M.A., Pentland, A.P.: Face recognition using eigenfaces. In: Proceedings of IEEE Computer Society Conference on Computer Vision and Pattern Recognition, CVPR 1991, pp. 586–591. IEEE (1991)
11. Pentland, A., Moghaddam, B., Starner, T.: View-based and modular eigenspaces for face recognition. In: Computer Vision and Pattern Recognition (1994)

12. Liu, C.: Gabor-based Kernel PCA with fractional power polynomial models for face recognition. IEEE Trans. Pattern Anal. Mach. Intell. **26**(5), 572–581 (2004)
13. Sawides, M., Kumar, B.V.K.V., Khosla, P.K.: Corefaces - robust shift invariant PCA based correlation filter for illumination tolerant face recognition. In: Proceedings of the 2004 IEEE Computer Society Conference on Computer Vision and Pattern Recognition, CVPR 2004, vol. 2, pp. II-834–II-841 (2004)
14. Vincent, P., Larochelle, H., Bengio, Y., Manzagol, P.A: Extracting and composing robust features with denoising autoencoders. In: ICML, pp. 1096–1103 (2008)
15. Zhang, Z., Li, J., Zhu, R.: Deep neural network for face recognition based on sparse autoencoder. In: 2015 8th International Congress on Image and Signal Processing (CISP), Shenyang, pp. 594–598 (2015)
16. Banerjee, P.K., Datta, A.K.: Generalized regression neural network trained preprocessing of frequency domain correlation filter for improved face recognition and its optical implementation. Opt. Laser Technol. **45**, 217–227 (2013)
17. Banerjee, P.K., Datta, A.K.: Class specific subspace dependent nonlinear correlation filtering for illumination tolerant face recognition. Pattern Recogn. Lett. **36**, 177–185 (2014)
18. Kumar, B., Savvides, M., Xie, C., Venkataramani, K., Thornton, J., Mahalanobis, A.: Biometric verification with correlation filters. Appl. Opt. **43**(2), 391–402 (2004)
19. Lee, K., Ho, J., Kriegman, D.: Acquiring linear subspaces for face recognition under variable lighting. IEEE Trans. Pattern Anal. Mach. Intell. **27**(5), 684–698 (2005)
20. Sim, T., Baker, S., Bsat, M.: The CMU Pose, Illumination, and Expression (PIE) database. In: Proceedings of the Fifth IEEE International Conference on Automatic Face and Gesture Recognition (2002)

Author Index

Printed in the United States
By Bookmasters